CRYSTAL GROWTH: AN INTRODUCTION

North-Holland Series in Crystal Growth 1

Edited by

W. Bardsley, D. T. J. Hurle and J. B. Mullin

NORTH-HOLLAND PUBLISHING COMPANY – AMSTERDAM • LONDON
AMERICAN ELSEVIER PUBLISHING COMPANY, INC. – NEW YORK

Crystal Growth:
An Introduction

Edited by

P. HARTMAN

Geological and Mineralogical Institute,
State University of Leiden, The Netherlands

1973

NORTH-HOLLAND PUBLISHING COMPANY – AMSTERDAM • LONDON
AMERICAN ELSEVIER PUBLISHING COMPANY, INC. – NEW YORK

Library of Congress Catalog Card Number: 72-93091
ISBN North-Holland, for this series: 0 7204 1820 8
for this volume: 0 7204 1821 6
ISBN American Elsevier 0 444 10463 1

PUBLISHERS:

NORTH-HOLLAND PUBLISHING COMPANY – AMSTERDAM
NORTH-HOLLAND PUBLISHING COMPANY, LTD. – LONDON

SOLE DISTRIBUTORS FOR THE U.S.A. AND CANADA:

AMERICAN ELSEVIER PUBLISHING COMPANY, INC.
52 VANDERBILT AVENUE
NEW YORK, N.Y. 10017

PRINTED IN THE NETHERLANDS

PREFACE 1

This volume is the first of a series of monographs to be known as the North-Holland Series in Crystal Growth.

Although crystal growth has been a subject of absorbing interest for very many years, much of the recent development in both the understanding and the technology has been stimulated by the increasing commercial importance of the subject. The spectacular growth of solid state electronics, so critically dependent on crystal growth and perfection, is a case in point. The increase in activity throughout the world is reflected in the International Conferences on Crystal Growth (ICCG) and the establishment, through the Comité International de Croissance Cristalline, of the International Organisation of Crystal Growth (IOCG).

The rapid rise in the number of research publications in crystal growth has led to the appearance of several new journals including the International Journal of Crystal Growth in 1967. The field has already grown to the point where no one person can reasonably hope to keep abreast of all developments and the newcomer to the field faces a formidable task in his attempt to familiarise himself with all the literature on a particular topic. Such a task would be greatly eased by the existence of authorative reviews. We believe that a series devoted exclusively to crystal growth could make a valuable contribution at this time. It is appropriate that the publication of the series should be undertaken by North-Holland since, as publishers of J. Crystal Growth and of two volumes of ICCG Proceedings, they have particularly associated themselves with this field.

Our general aim is to satisfy two criteria for each monograph: – That a review of the subject treated is timely and that its author is a recognised authority in the field capable of giving a balanced and comprehensive review. We envisage that some of the monographs will be suitable for use in undergraduate teaching whilst others, by the specialised nature of their subject matter, will be of interest primarily to the research worker.

This first volume differs from the others planned for the series in that, instead of being a review of a particular topic by a single author, or group of authors working together, it derives from the lectures given at the International Summer School on Crystal Growth at Noordwijkerhout, The Netherlands in June 1971.

We are delighted that Dr. Hartman has given us the opportunity to publish the work of the Summer School as our introductory volume since we believe that it represents a valuable introduction to a broad spectrum of the topics which make up the inter-disciplinary subject of crystal growth. We feel that it is particularly comprehensive in its treatment of the theoretical aspects of the subject and that Dr. Hartman, who organised the Summer School and has edited this volume, has made an important contribution to the dissemination of crystal growth knowledge.

W. Bardsley
D.T.J. Hurle
J.B. Mullin

Malvern UK. 1972.

PREFACE 2

The contents of this book derive from a series of lectures given at the International Summer School on Crystal Growth held at Noordwijkerhout, The Netherlands, from 21st June to 2nd July, 1971. The idea that this School should be held was put forward by Professor R. Kern, chairman of the Commission on Crystal Growth of the International Union of Crystallography. A programme was made up by the members of the IUCr Commission and advice was sought and obtained from several well-known scientists. It is a pleasure to gratefully acknowledge assistance from the following colleagues: Dr. A.A. Chernov, Dr. B. Honigmann, Prof. Dr. R. Kaischev, Prof. Dr. R. Kern, Prof Dr. W. Kleber (now deceased), Dr. R.A. Laudise, Dr. A.N. Lobachev, Dr. J.B. Mullin, Prof. Dr. A. Neuhaus, Prof. Dr. W.G. Perdok, Prof. Dr. N.N. Sheftal', Dr. R.F. Strickland-Constable, Prof. Dr. E.A.D. White and Dr. G.A. Wolff. The Summer School was finally realized under the sponsorship of the International Union of Crystallography and of the Comité International de Croissance Cristalline with financial and organizational assistance from several industries and institutions. I feel sure that I speak not only for myself, but also on behalf of the participants, when I express our deep gratitude towards all those who contributed in this way. I also wish to express my most sincere thanks to Mr. C.F. Woensdregt and Mr. R.O. Felius for their work in organising the Summer School.

The scope of the programme was chosen to be rather broad. For publication in the North-Holland Series on Crystal Growth several lectures were completely reworked.

The chapters in Part I are concerned with nucleation. Dr. S. Toschev considers homogeneous nucleation; both the thermodynamic and the atomistic approach. Next, Prof. B.K. Chakraverty deals with several aspects of heterogeneous nucleation. Success and failure of classical nucleation theory are considered and new experimental results and recent developments in the theory are presented. Dr. M. Gebhardt deals with epitaxy, which can be considered as a special case of heterogeneous nucleation. As an application, the formation of epitactic layers in the corrosion of metals was chosen, a subject not often considered from this point of view, but certainly of value when considered in its broader aspects.

Part II deals with techniques of crystal growth. Chemical transport reactions from the vapour are treated by Prof. Dr. H. Schäfer, while Prof. Dr. A. Rabenau stresses the possibilities of multi-phase equilibria with this method. The field of hydrothermal growth is covered by Dr. R.A. Laudise for the usual methods in more or less alkaline solutions. The technique to be used with acid solutions is treated by Prof. Dr. A. Rabenau. Dr. D.T.J. Hurle gives a review on the various methods which can be used in melt growth and treats the problems connected with segregation and convection, such as the faceting effect and the occurrence of striations. This part is concluded with a short review on problems in industrial mass crystallization by Dr. M.A. van Damme–van Weele. This last subject was included because it was felt that somehow in the future the link between single-crystal-growth theories and the problems encountered in mass crystallization has to be made.

The third part deals with various theoretical approaches to crystal growth phenomena. The first chapter, by Dr. P. Bennema and Dr. G.H. Gilmer, is concerned with current theories of the kinetics of crystal growth. First, the structure of an interface on the atomic scale is considered by means of the Temkin approach, then the almost classical theory of Burton, Cabrera and Frank is explored in all its implications and finally, on a larger scale, the kinematic wave theory is reviewed. The next three chapters are devoted to a theoretical concept that played a great role in various theories, namely the thermodynamic equilibrium form. The classical thermodynamic theory is presented by Dr. S. Toschev. Then Dr. P. Bennema gives a review of the work of Herring and of Wolff, while in the third chapter Dr. P. Hartman records the work of Kern and Bienfait who formulated the conditions under which a stable equilibrium form can exist and can be observed. In the next chapter Dr. P. Hartman shows how the external morphology, and therefore the growth rates, can be related to the crystal structure and to environmental influences. The penultimate chapter, by Prof. Dr. R.F. Sekerka, deals with the morphological stability of a surface in contact with a non-crystalline phase.

The assessment of single crystals produced by any of the many methods described, is usually performed by means of crystallographic techniques. Therefore, in the last chapter, Dr. A.R. Lang deals with the properties and observation of dislocations. After a survey on dislocations and their interactions a brief introduction to the dynamical theory of diffraction of X-rays and electrons is presented and various methods of observing dislocations are reviewed.

In this book, where the contributions have more or less the character of reviews, many of the diagrams and illustrations have been copied from various sources. The permission for reproduction, obtained from the authors, editors and publishers concerned, is gratefully acknowledged herewith. In all cases reference is made to the original publication.

The process of crystal nucleation and growth is encountered in various disciplines, not only in solid state physics or solid state chemistry, but also in industrial mass crystallization, in metallurgy, in ceramics, in polymer science, in mineralogy and petrology, even in the study of viruses and in some aspects of medical science such as the formation of bone material. From applications as homely as the regulation of the crystallization of table salt or the prevention of the crystallization of lactose in condensed milk, to applications as sophisticated as those used in the manufacture of solid state electronic devices, a whole spectrum of crystal growth problems is at the base of much of everyday life.

It is hoped that the present volume will give the reader a clear insight into the various problems connected with so seemingly a simple process as the nucleation and growth of a crystal.

P. Hartman
Leiden, Netherlands
1972

CONTRIBUTORS

Dr. P. Bennema, Laboratory of Physical Chemistry, Technical University of Delft, The Netherlands

Prof. Dr. B.K. Chakraverty, Laboratoire d'Electrostatique et de Physique du Métal, Grenoble, France

Dr. Ir. M.A. van Damme–van Weele, N.V. Centrale Research KNZ, Hengelo (O), The Netherlands

Dr. M. Gebhardt, Mineralogisches-Petrologisches Institut of the Rheinische Friedrich Wilhelms-Universität, Bonn, Germany

Prof. Dr. G.H. Gilmer, Physics Dept., Washington and Lee University, Lexington, Virginia, U.S.A.

Dr. P. Hartman, Geologisch en Mineralogisch Instituut der Rijksuniversiteit, Leiden, The Netherlands

Dr. D.T.J. Hurle, Royal Radar Establishment, Great Malvern, Worcs., England

Dr. A.R. Lang, H.H. Wills Physics Laboratory, Bristol, England

Dr. R.A. Laudise, Bell Telephone Laboratories, Murray Hill, New Jersey, U.S.A.

Prof. Dr. A. Rabenau, Philips Research Laboratories, Aachen and Max-Planck-Institut für Festkörperforschung, Stuttgart, Germany

Prof. Dr. H. Schäfer, Anorganisch-Chemisches Institut der Westfälische Wilhelms-Universität, Münster, Germany

Prof. Dr. R.F. Sekerka, Department of Metallurgy and Materials Science of the Carnegie-Mellon University, Pittsburgh, Penna., U.S.A.

Dr. S. Toschev, Institute of Physical Chemistry of the Bulgarian Academy of Sciences, Sofia, Bulgaria

CONTENTS

P. Hartman, ed., Crystal growth: an introduction. © (1973) North-Holland Publ. Co.

HOMOGENEOUS NUCLEATION

† S. TOSCHEV *

1.1. Introduction

Nucleation is a very important branch of physical science nowadays. Despite the fact that homogeneous nucleation, i.e. the spontaneous formation of nuclei in the bulk of supersaturated systems is a comparatively rare occurrence, its basic principles form the necessary background for the understanding of numerous processes in science and technology as well as in nature when phase transitions are involved. We shall mention the fields of thin film formation and epitaxy, metal electrodeposition, atmospheric condensation, industrial crystallization, the preparation of metal alloys and partly crystallized glassy materials etc. Nucleation is the precursor of crystal growth and of the overall crystallization processes.

This chapter offers an outline of the present state of homogeneous nucleation theory and practice. It is hoped that the brevity and incompleteness of this treatment imposed by the limited space will be partly compensated by the attached list of references which should stimulate a more thorough examination of the subject.

*) Dr. Svetoslav D. Toschev died in a mountain accident on the 12th December, 1971, at the age of 34 years. He was Senior Research Assistant at the Institute of Physical Chemistry of the Bulgarian Academy of Sciences. Dr. Toschev was one of the most talented young scientists at that Institute. He made many valuable contributions in the field of phase transitions and latterly worked especially in electro-crystallization. His untimely death will be deeply regretted by all those who knew him. His modest and friendly personality and his great abilities as a scientist will be long remembered.

1.2. Thermodynamic Theory

1.2.1. Preliminary Remarks

Gibbs was the first to realize that the formation of small embryonic *clusters* with some critical size is a prerequisite for a macroscopic phase transformation to take place. Due to their increased surface/volume ratio such clusters, termed *critical nuclei,* have more energy than bulk phases of the same mass. Hence, they have a chance to survive and to produce macroscopic entities of the new phase only within supersaturated or undercooled systems. As a matter of fact, the equilibrium between the critical nucleus and its environment is unstable provided that the volume of the whole system is not

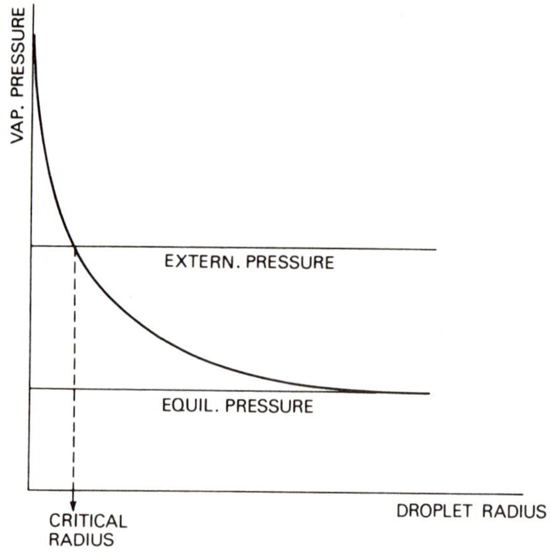

Fig. 1.1. Vapour pressure of small droplets in relation to their radii.

very small. Fig. 1.1 represents the well known relationship between the vapour pressure of liquid droplets and their size. If by any chance the radius of the critical droplet r^* increases slightly, the droplet will continue to grow until a macroscopic two phase equilibrium is attained. In the reverse case the droplet will disappear. In other words, minor fluctuations of the critical radius release an irreversible process destroying the initial state. This shows that the energetics of phase formation is governed by the energetics of the critical nucleus. Nucleation is a very rare example of a process whose activation energy can be calculated solely on the basis of phenomenological thermodynamics. Thermodynamic treatments of nucleation are given by Volmer [1], Defay et al. [2], Russanov [3], Kaischew [4, 5], Turnbull [6].

1.2.2. The Critical Nucleus/Ambient Phase Equilibrium

Let us consider an arbitrary cluster within a supersaturated/undercooled medium. A simple thermodynamic analysis of the system depicted in fig. 1.2 is possible if the real cluster/ambient phase interface is replaced by an imaginary geometrical surface, the so called *dividing surface* [2, 3, 7]. If n is the total number of moles in the whole system

$$n = n' + n'' + n''' . \tag{1.1}$$

where n' and n'' are the numbers of moles of the initial and the new phases respectively, assuming that these phases extend with uniform density on both sides of the geometrical surface, while n''' is the remaining excess*. Since the position of the dividing surface can be arbitrary n''' can be positive, negative

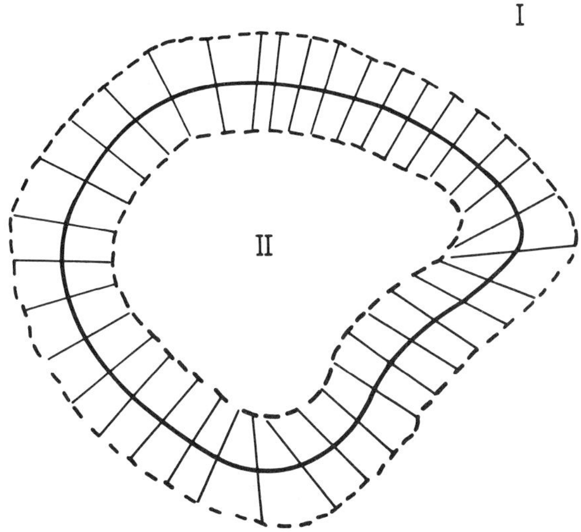

Fig. 1.2. Schematic representation of the cluster/ambient phase system
I – region with bulk properties of phase I
II – region with bulk properties of phase II
full line contour – geometrical (non-physical) dividing surface; dashed area – physical interface layer.

* This treatment is confined to one component systems.

or zero. If the dividing surface is chosen in such a way that $n'''=0$, it is called the *equimolecular surface* (ES). This treatment originating from Gibbs makes it possible to consider formally the cluster and the ambient phase as a three phase system. The total Helmholtz free energy is then ($'$-initial phase, $''$-second phase, $'''$-"surface" phase)

$$F = F' + F'' + F''' . \tag{1.2}$$

where

$$F''' = n'''\mu''' + \phi , \tag{1.3}$$

μ''' being the chemical potential of the "surface" phase and ϕ – the total surface energy. To arrive at the equilibrium conditions at constant temperature T and external pressure p we impose minimum variation of the total Gibbs free energy:

$$\mathrm{d}G = \mathrm{d}F + p\mathrm{d}V = 0 .$$

Since

$$\mathrm{d}F' = \mu'(p')\mathrm{d}n' - p'\mathrm{d}V' ,$$

$$\mathrm{d}F'' = \mu''(p'')\mathrm{d}n'' - p''\mathrm{d}V'' ,$$

$$\mathrm{d}F''' = \mu'''\mathrm{d}n''' + \mathrm{d}\phi ,$$

and $\mathrm{d}V=\mathrm{d}V'+\mathrm{d}V''$, V being the total volume, it follows that $\mathrm{d}G=\mu'(p')\mathrm{d}n'+$ $+\mu''(p'')\mathrm{d}n''+(p-p')\mathrm{d}V'+(p-p'')\mathrm{d}V''+\mu'''\mathrm{d}n'''+\mathrm{d}\phi$. Under the assumption that $n=$const, we have additionally:

$$\mathrm{d}n' = -\,\mathrm{d}n'' - \mathrm{d}n''' ,$$

$$\mathrm{d}V' = n'\mathrm{d}v' + v'\mathrm{d}n' = n'\mathrm{d}v' - v'\mathrm{d}n'' - v'\mathrm{d}n''' ,$$

$$\mathrm{d}V'' = n''\mathrm{d}v'' + v''\mathrm{d}n'' ,$$

where v' and v'' are the corresponding molar volumes. In this way we find

$$(p-p')n'\mathrm{d}v' + \{[\mu''(p'') - \mu'(p')] + (p-p'')v'' - (p-p')v' + \frac{\mathrm{d}\phi_{v''}}{\mathrm{d}V''}\}\,\mathrm{d}n'' +$$

$$+ [(p-p'') + \frac{\mathrm{d}\phi}{\mathrm{d}V''}]n''\mathrm{d}v'' + \{[\mu''' - \mu'(p')] - (p-p')v'\}\,\mathrm{d}n'' = 0 .$$

The equilibrium conditions are

$$p' \equiv p \,. \tag{1.4}$$

$$\mu'(p') = \mu''(p'') = \mu''' \,. \tag{1.5}$$

$$p'' - p = \frac{\mathrm{d}\phi}{\mathrm{d}V''} \,. \tag{1.6}$$

Conditions (1.4) and (1.5) are trivial. Expression (1.6) shows that the pressure within a cluster exceeds the pressure in the bulk of the ambient phase. Let us introduce the *surface tension* σ. For an isotropic sphere

$$\phi = 4\pi r^2 \sigma \qquad \text{and} \qquad V'' = \tfrac{4}{3}\pi r^3 \,. \tag{1.7}$$

The total surface energy ϕ can be regarded as a function of two independent variables, namely the physical radius of the droplet and the surface tension which for curved surfaces depends on the location of the dividing surface. Hence

$$\mathrm{d}\phi = \left(\frac{\partial \phi}{\partial r}\right)_\sigma \mathrm{d}r + \left(\frac{\partial \phi}{\partial \sigma}\right)_r \left[\frac{\mathrm{d}\sigma}{\mathrm{d}r}\right] \mathrm{d}r \,.$$

The unusual notation with square brackets indicates that it is the non-physical variation of σ with the radius corresponding with the non-physical choice of the dividing surface which is taken into account. It follows then that

$$p'' - p = \frac{2\sigma}{r} + \left[\frac{\mathrm{d}\sigma}{\mathrm{d}r}\right] .$$

This is a generalization of the well known *Kelvin equation,* a rigorous derivation being given in [7]. There is a particular surface named *surface of tension* (ST) subjected to the condition $(\mathrm{d}\sigma/\mathrm{d}r)_s = 0$. The surface of tension is the equivalent substitute of the real interface with respect to surface tension and mechanical properties, since in this particular case

$$p'' - p = \frac{2\sigma_s}{r_s} \,. \tag{1.6$'$}$$

Ono and Kondo [7] have shown that for the ES

$$\left[\frac{\mathrm{d}\sigma}{\mathrm{d}r}\right]_{\mathrm{e}} = \frac{\mathrm{d}\sigma_{\mathrm{e}}}{\mathrm{d}r_{\mathrm{e}}}$$

and

$$p'' - p = \frac{2\sigma_{\mathrm{e}}}{r_{\mathrm{e}}} + \frac{\mathrm{d}\sigma_{\mathrm{e}}}{\mathrm{d}r_{\mathrm{e}}} . \tag{1.8}$$

These considerations become important when the physical dependence of σ on the radius of curvature is introduced in nucleation theory. Normally this effect is neglected and σ is identified with the surface tension at the plane interface, for which the location of the dividing surface is immaterial for the value of σ, i.e.

$$p'' - p = \frac{2\sigma}{r} . \tag{1.9}$$

1.2.3. The Work of Formation of Critical Nuclei

We calculate first the Gibbs free energy change of a system upon the formation of an arbitrary cluster under isobaric and isothermal conditions:

$$\Delta G(n'') = G_2 - G_1 = F_2 - F_1 + p(V_2 - V_1) .$$

In the initial state 1

$$F_1 = n\mu'(p) - pV_1 .$$

In the final state 2

$$F_2 = F_2' + F_2'' + F''' ,$$

$$F_2' = n'\mu'(p) - pV_2' ,$$

$$F_2'' = n''\mu''(p'') - p''V_2'' .$$

Considering relations (1.1) and (1.3), as well as

$$n = \text{const.},$$

$$V_2 = V_2' + V_2'' \qquad (V_2'' \equiv V''),$$

$$\Delta G(n'') = [\mu''(p'') - \mu'(p)]\,n'' + [\mu''' - \mu'(p')]\,n''' + \\ + (p - p'')V'' + \phi .$$

For the critical nucleus conditions (1.4), (1.5) and (1.6) are fulfilled so that

$$\Delta G^* = \phi^* - (p'' - p)\,V^{*''} \tag{1.10}$$

the asterisk denoting that the relevant quantities refer to the critical size. Eq. (1.10) is the general expression for the energy barrier of nucleation, irrespective of the choice of the dividing surface. It can be further rewritten in the following form [4]

$$\Delta G^* = \phi^* - \left(\frac{\mathrm{d}\phi}{\mathrm{d}V''}\right)^* V^{*''} . \tag{1.11}$$

If σ = const. (plane interface approximation),

$$\phi = \mathrm{b}V''^{2/3} \quad \text{and} \quad \frac{\mathrm{d}\phi}{\mathrm{d}V''} = \frac{2\phi}{3V''},$$

where b is a geometrical factor. The substitution of these relations in (1.11) gives

$$\Delta G^* = \tfrac{1}{3}\,\phi^* , \tag{1.12}$$

i.e. the activation energy of the nucleation process is one third of the total surface energy of the critical nucleus. This very basic result was obtained many years ago by Gibbs. For a spherical nucleus

$$\Delta G^* = \tfrac{4}{3}\,\pi r^{*2}\sigma , \tag{1.13}$$

The complete analysis should take into account the $\sigma(r)$ relationship. At this stage it is very important to specify the dividing surface in order to arrive at correct results. Since expressions (1.6′) and (1.9) are formally identical, eq. (1.13) retains the same form for the surface of tension, i.e.

$$\Delta G^* = \tfrac{4}{3}\,\pi r_s^{*2}\sigma_s .$$

To account for the physical dependence of σ_s on the radius of curvature, the σ_s (r_s) function (if known) can be directly inserted in the above written expression.

For the ES the combination of (1.11), (1.7) and (1.8) yields (see [7])

$$\Delta G^* = \tfrac{4}{3}\pi r_e^{*2}\sigma_e\left(1 - \frac{\mathrm{d}\ln\sigma_e}{\mathrm{d}\ln r_e}\right), \tag{1.14}$$

where the derivative in (1.14) should be evaluated for $r_e = r^*_e$. In the theory of nucleation it is more appropriate to operate with σ_e defined for the ES. The expression for $\Delta G(n'')$ reduces then to

$$\Delta G(n'') = [\mu''(p'') - \mu'(p)]\, n'' + (p - p'')V'' + \phi\,. \tag{1.15}$$

1.2.4. Some Typical Cases

1.2.4.1. Nucleation of Droplets from Supersaturated Vapours. The phase diagram of a single component system is depicted in fig. 1.3. The droplets appear in the stable domain of the liquid, say at point A. The vapours with pressure p_g can be regarded as supersaturated or undercooled, depending on the chosen reference equilibrium state at M or M_1 respectively.

To obtain an explicit formula for $\Delta G(n_l)$ we start from the general equation (1.15) which we rewrite in the following manner (g – vapour, l – liquid)

$$\begin{aligned}\Delta G(n_l) = {} & [\mu_l(p_l) - \mu_l(p_{eq})]\, n_l - [\mu_g(p_g) - \mu_g(p_{eq})]\, n_l + \\ & + (p_g - p_l)V_l + \phi\,.\end{aligned} \tag{1.16}$$

Here p_{eq} is the equilibrium pressure at the same temperature and accordingly $\mu_l(p_{eq}) = \mu_g(p_{eq})$. Taking into account that the liquid is practically non-compressible and considering the vapours as ideal gas, we have

$$\mu_l(p_l) - \mu_l(p_{eq}) = \int_{p_{eq}}^{p_l} v_l\,\mathrm{d}p = v_l(p_l - p_{eq})\,. \tag{1.17}$$

$$\mu_g(p_g) - \mu_g(p_{eq}) = \int_{p_{eq}}^{p_g} v_g\,\mathrm{d}p = RT\ln(p_g/p_{eq})\,. \tag{1.18}$$

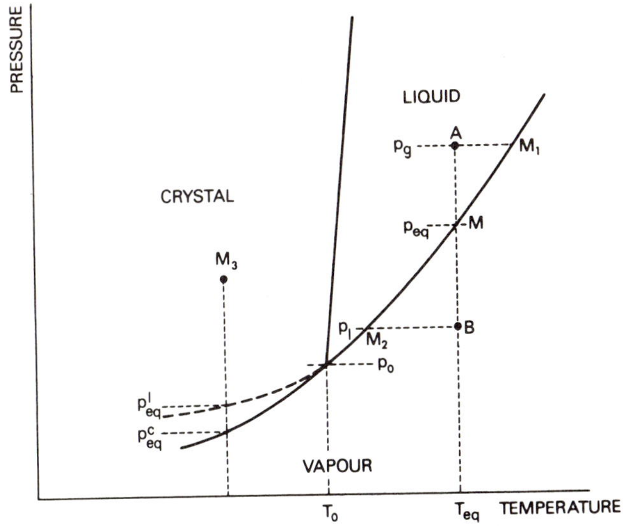

Fig. 1.3. Phase diagram of a single component system.

Expression (1.18) defines the thermodynamic driving force of nucleation from supersaturated vapours at constant temperature; p_g/p_{eq} is the supersaturation ratio. We obtain then from (1.16), (1.17) and (1.18)

$$\Delta G(n_l) = [v_l(p_g - p_{eq}) - \Delta\mu]\, n_l + \phi \,.$$

A closer inspection of the expression in the square brackets reveals that

$$v_l(p_g - p_{eq}) - \Delta\mu = p_{eq}(v_{eq}\ln p_{eq} - v_l) - p_g(v_g \ln p_g - v_l)\,,$$

where $v_{eq} = v_g p_g/p_{eq} = RT/p_{eq}$ is the molar volume of the saturated vapours at at the given temperature. As

$$v_l \ll v_{eq}\ln p_{eq} \qquad \text{and} \qquad v_l \ll v_g \ln p_g\,,$$

$$v_l(p_g - p_{eq}) - \Delta\mu \approx -\Delta\mu\,.$$

It should be noted that this approximation is very good. The final result for an arbitrary droplet takes the form

$$\Delta G(n_l) = \phi - \Delta\mu n_l\,. \qquad (1.19)$$

Introducing in (1.19) the radius of the droplet and recalling (1.13), invoking the plane interface approximation

$$\Delta G(r) = 4\pi r^2\sigma - \tfrac{4}{3}\pi r^3 \frac{\Delta\mu}{v''} = \Delta G^* \left[3\left(\frac{r}{r*}\right)^2 - 2\left(\frac{r}{r*}\right)^3 \right] . \tag{1.20}$$

This function (see fig. 1.4) passes through a maximum at

$$r^* = \frac{2\sigma v_1}{RT\ln(p_g/p_{eq})} , \tag{1.21}$$

which is the critical radius at a given supersaturation (*Thomson-Gibbs equation*). The work of formation of critical nuclei

$$\Delta G^* = \frac{16\pi v_1^2\sigma^3}{3(RT\ln(p_g/p_{eq}))^2} , \tag{1.22}$$

corresponds to the maximal value of $\Delta G(r)$. In undersaturated vapours $p_g < p_{eq}$, $\Delta\mu < 0$ and $\Delta G(r)$ increases monotonically with the radius (fig. 1.4), i.e. the formation of clusters is impeded over all dimensions.

To account for the influence of the curvature on σ and on ΔG^* we can use a relation derived by Tolman [8]

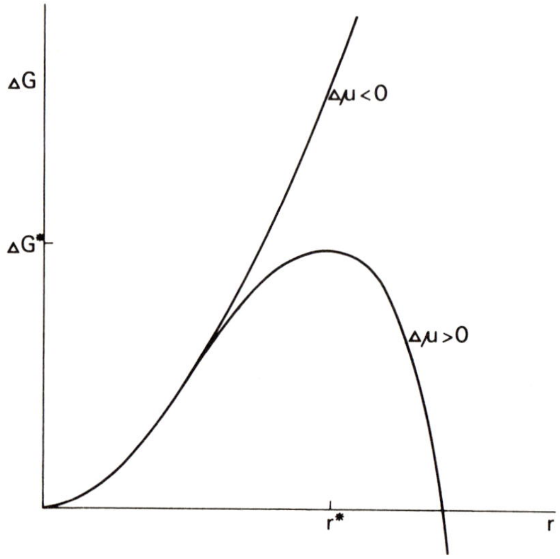

Fig. 1.4. Gibbs free energy change $\Delta G(r)$ as a function of droplet radius.

$$\sigma_e = \sigma\left(1 + \frac{2\delta}{r_e}\right)^{-1} , \qquad (1.23)$$

δ being the difference in the radii of ES and ST. Hence, it can be shown on the basis of (1.14) that

$$\frac{\Delta G^*}{\Delta G^{*,0}} = \tfrac{1}{4}(\sigma_e/\sigma)^4(3 - \sigma_e/\sigma)^2 , \qquad (1.24)$$

where $\Delta G^{*,0}$ is given by (1.22). An equivalent result can be found elsewhere [9, 10]. Expressions (1.23) and (1.24) are presented graphically in fig. 1.5.

1.2.4.2. Nucleation of Bubbles from Superheated Liquids. Bubbles can form only in the domain of the vapour phase. (Fig. 1.5). If we refer the state B to the two-phase equilibrium at the same temperature (state M), the liquid can be described as underpressured. With reference to the equilibrium at the same pressure (state M_2) the liquid is superheated.

The conditions defining the equilibrium of a bubble in a superheated liquid read

$$\begin{aligned} \mu_g(p_g) &= \mu_l(p) , \\ p_g - p_l &= 2\sigma/r , \end{aligned} \qquad (1.25)$$

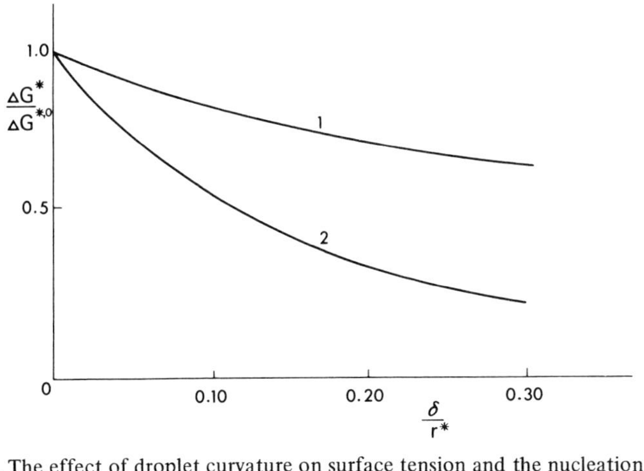

Fig. 1.5. The effect of droplet curvature on surface tension and the nucleation barrier 1 – Plot of $\frac{\sigma_e}{\sigma}$ vs $\frac{\delta}{r^*}$ after Tolman's equation 2 – Plot of $\frac{\Delta G^*}{\Delta G^{*,0}}$ vs $\frac{\delta}{r^*}$ (eq. (1.24)).

As in the previous treatment it can be shown that

$$v_l(p_l - p_{eq}) = RT\ln(p_g/p_{eq}) \approx v_l(p_l - p_g)\,. \tag{1.26}$$

Combining (1.25) and (1.26) we get

$$p_g = p_{eq}\exp(-2\sigma v_l/RTr)\,. \tag{1.27}$$

It is seen that the pressure within a bubble p_g is lower than the equilibrium pressure at the plane liquid/vapour interface. The radius of the critical bubble at constant external pressure p_l and temperature is defined by the non-explicit equation

$$p_l + \frac{2\sigma}{r^*} = p_{eq}\exp(-2\sigma v_l/RTr^*)\,.$$

All relations pertaining to the case under consideration are illustrated in fig. 1.6 (see [11]). The calculation of ΔG^* yields

$$\Delta G^* = \frac{16\pi\sigma^3}{3(p_g - p_l)^2} \approx \frac{16\pi\sigma^3}{3(p_{eq} - p_l)^2}\,. \tag{1.28}$$

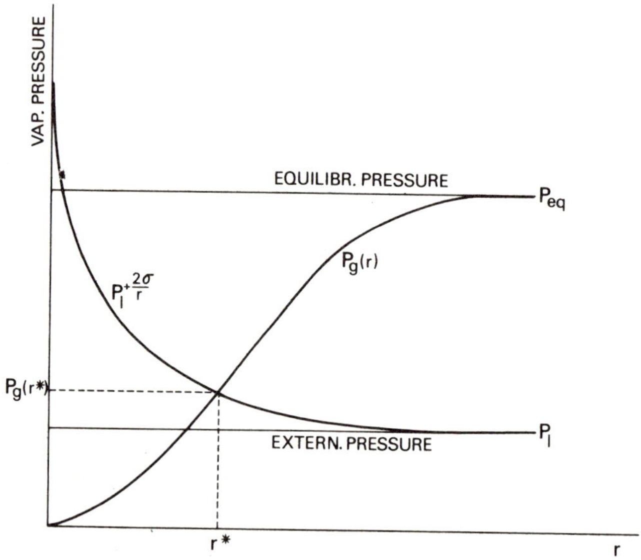

Fig. 1.6. Schematic diagram illustrating the thermodynamic relations in the nucleation of bubbles (after Kaischew and Stranski [11]).

The thermodynamic driving force of bubble formation is

$$\Delta\mu = (p_g - p_l)v_l \approx (p_{eq} - p_l)v_l \,.$$

1.2.4.3. Nucleation of Crystals from Supercooled Liquids. Crystal nuclei can be produced within liquid phases below the point of triple equilibrium (fig. 1.3). For the thermodynamic treatment it is immaterial whether the crystal nuclei originate from the undercooled melt or from vapours in equilibrium with the melt below T_0 (fig. 1.3, dotted line). The thermodynamic driving force is then

$$\Delta\mu = \mu(p_{eq}^{l}) - \mu(p_{eq}^{c}) = [\mu(p_0) - \mu(p_{eq}^{c})] - [\mu(p_0) - \mu(p_{eq}^{l})] \,,$$

where $\mu(p_0)$ denotes the chemical potential of each of the three phases in mutual equilibrium at temperature T_0 and pressure p_0. In accordance with the Clausius-Clapeyron equation

$$d\mu_{l,c} = -S_g dT + v_g dp_{eq}^{l,c} = -S_g dT + v_g \frac{S_g - S_{l,c}}{v_g - v_{l,c}} dT \,,$$

S_g, S_l and S_c being the molar entropies of the vapour, of the liquid and of the crystal respectively. It follows that

$$\Delta\mu = \int_T^{T_0} v_g \left[\frac{S_g - S_c}{v_g - v_c} - \frac{S_g - S_l}{v_g - v_l}\right] dT \,.$$

As $v_c \ll v_g$, $v_l \ll v_g$ and noting $S_l - S_c = \Delta S$,

$$\Delta\mu = \int_T^{T_0} \Delta S dT. \tag{1.29}$$

The integrand in (1.29) is a function of the temperature since

$$\Delta S = \int_0^T \frac{\Delta C_p}{T} dT = \Delta S_{T_0} - \int_T^{T_0} \frac{\Delta C_p}{T} dT \,, \tag{1.30}$$

where $\Delta C_p = C_p^l - C_p^c$ is the difference in the corresponding molar heat capacities and ΔS_{T_0} refers to the melting point. If Q_m is the molar heat of fusion

$$\Delta S_{T_0} = Q_m/T_0 = \text{const.}$$

The insertion of (1.30) in (1.29) gives

$$\Delta\mu = \int_T^{T_0} \Delta S_{T_0} \mathrm{d}T - \int_T^{T_0} \mathrm{d}T \int_T^{T_0} \frac{\Delta C_p}{T} \mathrm{d}T .$$

Usually the second term is omitted, so that

$$\Delta\mu \approx \frac{Q_m}{T_0} \Delta T , \tag{1.31}$$

$\Delta T = T_0 - T$ being the undercooling .

At higher undercoolings eq. (1.31) does not give correct results. The detailed analysis of the temperature dependence of the thermodynamic driving force of nucleation in undercooled liquids and particularly in glassforming melts reveals some essential features of the crystallization mechanism in such systems. It was shown for instance that the shape of the $\Delta S(T)$ function in condensed systems predetermines an upper limit of $\Delta\mu$ irrespective of the undercooling, i.e. the critical size cannot be less than a given value [12].

The barrier of nucleation at moderate undercoolings taking the critical nucleus as an isotropic sphere is

$$\Delta G^* = \frac{16\pi v_c^2 \sigma^3 T_0^2}{Q_m^2 (\Delta T)^2} . \tag{1.32}$$

It is interesting to compare the nucleation barriers of liquid droplets and of crystals from the vapour phase below T_0 (fig. 1.3, state M_3). We have

$$\frac{\Delta G_l^*}{\Delta G_c^*} = \left(\frac{v_l}{v_c}\right)^2 \left(\frac{\sigma_l}{\sigma_c}\right)^3 (RT \ln (p^l/p^c))^2 , \tag{1.33}$$

where $p^l > p^c$. Usually the specific surface energy of the liquid is lower than that of the crystal. Hence not very far below T_0, $\Delta G_1^* < \Delta G_c^*$ while at higher undercoolings the reverse situation occurs.

The transition liquid solution → crystal can be treated in the same way if the new phase is single component. In this case ΔG^* is given by the same expression as, for instance (1.22) writing in the denominator $RT\ln(a/a_{eq})$, a and a_{eq} being the actual and the equilibrium concentrations (activities) respectively at the temperature T.

1.3. Kinetic Theory

1.3.1. General Formulation

The ultimate task of nucleation theory is to calculate the *rate of nucleation*, that is to say, the number of the particles which are formed per unit volume and time within a supersaturated system. This requires a detailed analysis on a microscopic scale of the elementary processes leading to the formation of new phase entities. The reader is again referred to the classical monograph of Max Volmer [1], as well as to a series of more recent works [13–20] where theory and experiment are reviewed with preference for a given type of phase transitions.

Nucleation is treated as a chain reaction advancing through a multitude of *clusters* with the result that some of them reach macroscopic dimensions. According to general kinetic principles [21], the time dependence of the concentration of clusters containing i atoms each can be expressed by the following set of equations:

$$\frac{dc_i}{dt} = \sum_{k=1}^{M} D_{k\to i}\, c_k - \sum_{k=1}^{M} D_{i\to k}\, c_i \qquad i = 1,2\ldots M\,. \tag{1.34}$$

The first sum in (1.34) accounts for all reactions giving rise to an i-cluster (birth processes), while the second sum considers all possible ways in which such clusters are being decomposed (death processes), $D_{k\to i}$ and $D_{i\to k}$ denoting the corresponding transition probabilities. The chain is cut off at some upper limit M at which the clusters leave the system irreversibly and accordingly their concentration is extremely small.

To simplify the problem, only transitions between nearest states are considered, i.e.

$$i-1 \underset{D_i^-}{\overset{D_{i-1}^+}{\rightleftharpoons}} i \underset{D_{i+1}^-}{\overset{D_i^+}{\rightleftharpoons}} i+1\,.$$

If the concentration of single atoms n_1 is held constant with time, the equation system (1.34) transforms into:

$$\frac{\mathrm{d}c_i}{\mathrm{d}t} = (D_{i-1}^+ c_{i-1} + D_{i+1}^- c_{i+1}) - (D_i^+ c_i + D_i^- c_i) , \tag{1.35}$$

$$i = 2,3 \ldots M-1, \qquad c_M = 0 .$$

The rate constants D_i^+, D_i^- are defined in the following way

$$D_i^+ = A_i \beta . \tag{1.36}$$

$$D_i^- = A_i \alpha_i . \tag{1.36'}$$

Here A_i is the surface area of the i-cluster and β and α_i are respectively the numbers of the atoms joining and leaving the cluster per unit time and surface area. The ratio D_i^+/D_i^- can be expressed by the difference in the chemical potentials of the critical nucleus and the chosen cluster [1]

$$\frac{D_i^+}{D_i^-} = \exp\left(\frac{\mu^* - \mu_i''}{RT}\right) .$$

This can be further modified to give

$$\frac{D_i^+}{D_i^-} = \exp\left\{\frac{B}{i^{*1/3}}\left[1 - \left(\frac{i^*}{i}\right)^{1/3}\right]\right\}, \tag{1.37}$$

as

$$\mu^* - \mu_i'' = (\mu^* - \mu_{\mathrm{eq}}) - (\mu^* - \mu_{\mathrm{eq}}) = v''\left[\left(\frac{\mathrm{d}\phi}{\mathrm{d}V''}\right)_i - \left(\frac{\mathrm{d}\phi}{\mathrm{d}V''}\right)^*\right] .$$

For spherically shaped nuclei

$$B = \frac{2\sigma v}{RT} N_{\mathrm{A}}^{1/3} \left(\frac{4\pi}{3v}\right)^{1/3} .$$

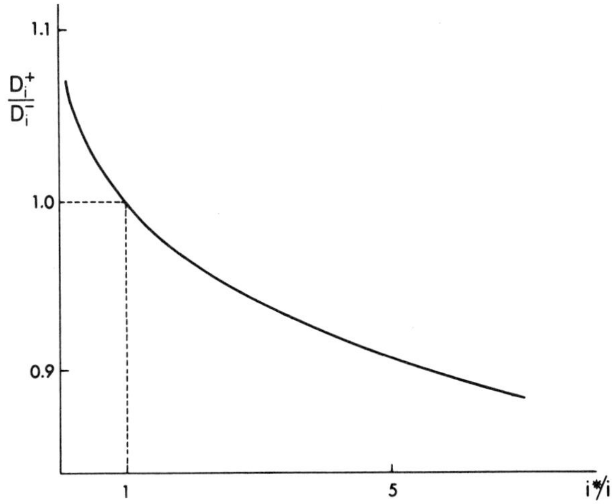

Fig. 1.7. Plot of eq. (1.37) calculated for water at $T = 300°$K and $i^* = 64$.

For the critical nucleus $D^{+*} = D^{-*}$. If $i \gtrless i^*$, $D_i^+ \lessgtr D_i^-$ respectively (see fig. 1.7). Hence, it is again demonstrated that the growth of subcritical clusters being at variance with the overall thermodynamic tendency in the system can proceed only in a fluctuating manner (random walk).

Introducing the net flux of the clusters through size i,

$$J_i = D_{i-1}^+ c_{i-1} - D_i^- c_i , \tag{1.38}$$

into equations (1.35) we get

$$\frac{\mathrm{d}c_i}{\mathrm{d}t} = J_i - J_{i+1} . \tag{1.39}$$

The rate of nucleation is defined by the flux J^* through the critical size. The quantitative description of nucleation kinetics is completed if the $c(i,t)$ functions are known. The special case $\mathrm{d}c_i/\mathrm{d}t = 0$, i.e. the steady state approximation has been considered in 1935 by Becker and Döring [22], their analysis being further refined by other authors [1, 23]. A general, time-dependent solution to the system (1.35) has been attempted by Turnbull [24] under the additional assumption that only single atoms are present at the initial moment $t = 0$. However, the mathematics of the problem requires tedious numerical calculations for each separate case under consideration when the theoretical results are to be compared with experimental data. There is also a significant uncertainty in the choice of the rate constants D_i^- for very small clusters in terms of the essentially thermodynamic relation (1.37). The continuity approach to nucleation kinetics which will be discussed in the following is therefore more advantageous for practical purposes.

1.3.2. The Zeldovich-Frenkel Approach

Considering i as a continuous variable, Zeldovich [25] and Frenkel [13] replaced the equations system (1.35) and (1.39) by the partial differential equations

$$\frac{\partial c(i,t)}{\partial t} = \frac{\partial}{\partial i}\left\{D^+(i)\,n(i)\frac{\partial}{\partial i}\left[\frac{c(i,t)}{n(i)}\right]\right\}, \tag{1.40}$$

$$J(i,t) = -D^+(i)\,n(i)\frac{\partial}{\partial i}\left[\frac{c(i,t)}{n(i)}\right], \tag{1.40'}$$

satisfying the continuity condition

$$\frac{\partial c(i,t)}{\partial t} = -\frac{\partial J(i,t)}{\partial i}.$$

Here $n(i)$ denotes the equilibrium distribution function of the clusters by size introduced from the principle of detailed balance

$$D^+(i)\,n(i) = D^-(i+1)\,n(i+1) \tag{1.41}$$

eliminating in this way the decay constants $D^-(i)$. Formally, equations (1.40) and (1.40′) describe the "diffusion" of the size i in a thermodynamic field of force. Imposing the initial and boundary conditions [26]

$$c(i,t) = 0 \qquad t = 0,\ i > 1\ , \tag{1.42}$$

$$\frac{c(i,t)}{n(i)} = 1 \qquad t \geqslant 0,\ i = 1\ , \tag{1.42'}$$

$$\frac{c(i,t)}{n(i)} = 0 \qquad t \geqslant 0,\ i = M \gg i^*\ , \tag{1.42''}$$

the problem under study is unambiguously stated physically and mathematically. Condition (1.42) requires only the presence of single atoms at the initial moment, i.e. pre-existing clusters are ignored. This assumption is not critical in principle in the sense that condition (1.42) can be appropriately modified to account for more complicated cases [44]. Also, it is stated that for very small clusters the steady state and equilibrium concentrations coincide, as well as that the actual concentration of the clusters approaches zero when i increases irrespective of the time of observation. The Zeldovich-Frenkel formulation makes it possible to derive all the basic features of nucleation kinetics.

1.3.3. The Equilibrium State

In the absence of an irreversible molecular flux within a supersaturated phase $J = 0$. Combining equation (1.40′) with (1.42′), we obtain the evident result

$$c(i) \equiv n(i) \, ,$$

To arrive at an explicit expression for $n(i)$ starting from eq. (1.41) we compose the product

$$\prod_{x=1}^{i-1>i^*} \frac{D_x^+}{D_{x+1}^-} = \frac{n_i}{n_1} \, . \tag{1.43}$$

Writing $D_x^+/D_{x+1}^- \approx D_x^+/D_x^-$ and substituting the resulting sum in the exponent by an integral, it can be readily shown that (see eq. (1.20))

$$n(i) = n_1 \exp\left(-\frac{\Delta G(i)}{kT}\right), \tag{1.44}$$

$\Delta G(i)$ being the Gibbs free energy change associated with the formation of an arbitrary cluster*. The ratio n_i/n_1 can be also viewed as the equilibrium constant of the polymolecular reaction

$$i\mathrm{A}_1 \rightleftarrows \mathrm{A}_i^{**}$$

* If we stick to eqs. (1.37) and (1.43) the final expression for $n(i)$ will be slightly modified [20].

** Here A is used as a chemical symbol.

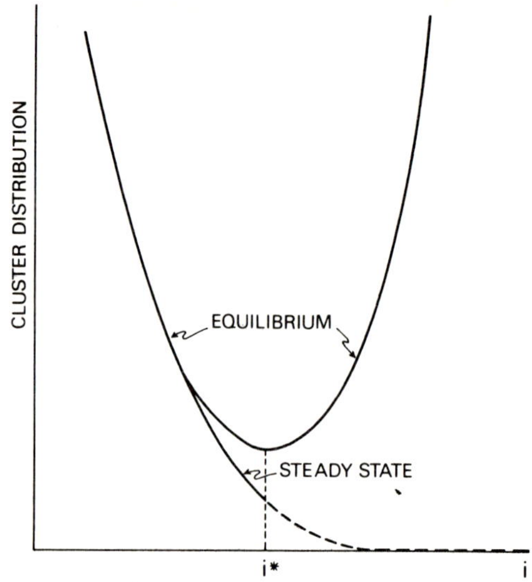

Fig. 1.8. Schematic representation of the equilibrium and steady state distribution functions of clusters by size.

assuming that $n_i \gg n_1$ [17]. A rigorous thermodynamic derivation of the equilibrium cluster concentration is given by Frenkel [13]. The $n(i)$ function is presented in fig. 1.8.

1.3.4. The Steady State

In this case $J = J_0$, a constant, i.e. a time-independent flux of clusters is established through all dimensions. The integration of eq. (1.40′) within limits i and M gives

$$\frac{c(i)}{n(i)} = J_0 \int_i^M \frac{\mathrm{d}i}{D^+(i)\,n(i)} \,. \tag{1.45}$$

Considering (1.42′)

$$J_0 \int_1^M \frac{\mathrm{d}i}{D^+(i)\,n(i)} = 1 \,. \tag{1.46}$$

The approximations made to evaluate the definite integral in (1.46) deserve some attention as they give an insight into the difficulties in the mathematical treatment of nucleation kinetics. These approximations are inherent in one or another form in all derivations of the stationary rate of nucleation. First, the exponent of the $n(i)$ function is expanded in Taylor series in the vicinity of its maximum:

$$\Delta G(i) \approx \Delta G^* - \tfrac{1}{2}\lambda(i - i^*)^2 . \qquad (1.47)$$

$$\lambda = -\left(\frac{d^2 \Delta G(i)}{di^2}\right)_{i=i^*} > 0 . \qquad (1.48)$$

Introducing the auxiliary variable $x = i - i^*$ the limits of the integral become $1 - i^*$ and $M - i^*$ respectively. However, the maximum of the $1/n(i)$ function is very sharp, so that the limits of integration can be extended to $-\infty$ and $+\infty$ respectively, i.e. only slight variations of i around i^* count. On the other hand, $D^+(i)$ as a not sensitive function of i can be replaced by $D^+(i) \equiv D^{+*} =$ const. In this way

$$\int_1^M \frac{di}{D^+(i)\, n(i)} \approx \frac{1}{D^{+*} n_1} \exp(\Delta G^*/kT) \int_{-\infty}^{+\infty} \exp(-\lambda x^2/2kT)\, dx .$$

Finally

$$J_0 = n_1 D^{+*} Z \exp(-\Delta G^*/kT) . \qquad (1.49)$$

The value of Z frequently termed as *Zeldovich factor* is given by

$$Z = \left(\frac{\lambda}{2\pi kT}\right)^{1/2} .$$

The discontinuity approach of Becker and Döring yields the same result, Z being expressed in a somewhat different form:

$$Z = \frac{1}{i^*}\left(\frac{\Delta G^*}{3\pi kT}\right)^{1/2} .$$

Since $\lambda = 2\Delta G^*/3i^{*2}$, the two formulations of Z are identical. It is worth noting that the first equation for the steady state rate of nucleation proposed by Volmer and Weber in 1926 [27] is very close to (1.49), since the value of Z which has been omitted by these authors is only of the order of 10^{-2} (see table 1.1). Nevertheless, the exact evaluation of the pre-exponent contributed very much to the deeper understanding of nucleation kinetics, the more so that the Zeldovich factor, accounting for the deviation of the actual cluster concentration from the equilibrium concentration within the critical region becomes extremely important in the transient case.

The actual concentration of the clusters $c(i)$ is given by eq. (1.45). The steady state distribution function deviates perceptibly from the equilibrium distribution only in the vicinity of the critical size (fig. 1.8). When $i = i^*$, then it can be shown that c^* is equal to $\frac{1}{2}n^*$ (see also eq. (1.58) and fig. 1.8).

Eq. (1.49) predicts a very steep dependence of the steady state nucleation rate on the supersaturation, the pre-exponent being nearly constant

$$J_0 = K_1 \exp\left(-\frac{K_2}{RT(\Delta\mu)^2}\right) . \tag{1.50}$$

This makes it possible to define a critical supersaturation under which the nucleation process is practically arrested. In the case of vapour → liquid condensation, the pre-exponential factor in (1.50) takes the form

$$K_1 = \underbrace{\frac{pN_\mathrm{A}}{RT}}_{n_1} \underbrace{4\pi\left(\frac{2\sigma v}{2.3\,RT\log(p/p_\mathrm{eq})}\right)^2}_{A^*} \underbrace{\frac{pN_\mathrm{A}}{(2\pi MRT)^{1/2}}}_{\beta} \underbrace{\frac{(RT)^{3/2}(2.3\log(p/p_\mathrm{eq}))^2}{8\pi v N_\mathrm{A}^{1/2}\ \sigma^{3/2}}}_{Z} .$$

Table 1.1 summarizes calculated numerical values pertaining to *water* condensation at 300°K, the critical supersaturation being 3.1.

In condensed systems transport hindrances of the building units within the mother phase play a significant role in nucleation kinetics. Treating nucleation in terms of the absolute reaction rate theory Turnbull and Fisher [28] derived*

$$J_0 = n_1 \frac{kT}{h} \exp(-\Delta U/RT) \exp(-\Delta G^*/RT) ,$$

* A more precise equation is given by Buckle [41].

Table 1.1
Calculated values for water at 300°K

p/p_{eq}	i^*	Z	n^* cm^{-3}	$\Delta G^*/kT$	K_1 $cm^{-3}\,sec^{-1}$	J_0 $cm^{-3}\,sec^{-1}$
2	469	4.1×10^{-3}	2.8×10^{-53}	162.8	5.0×10^{25}	1.7×10^{-71}
3	115	2.1×10^{-2}	3.0×10^{-10}	64.3	2.3×10^{26}	1.2×10^{-28}
3.1	106	5.1×10^{-2}	1.5×10^{-8}	60.4	3.1×10^{26}	1.7
4	58	3.4×10^{-2}	12.2	40.3	4.0×10^{26}	1.4×10^{9}
5	37	4.6×10^{-2}	5.1×10^{5}	29.7	6.5×10^{26}	7.8×10^{13}
6	30	5.4×10^{-2}	4.7×10^{7}	25.4	9.1×10^{26}	8.5×10^{15}

where ΔU is the activation energy (per mole) of self diffusion and h – Planck's constant. However, ΔU is frequently unknown and it is besides temperature dependent. It is more appropriate therefore to introduce the bulk viscosity η of the melt via the Stokes-Einstein relationship, retaining at the same time the very instructive formalism of equation (1.49). The collision frequency is then approximately (see [29])

$$\beta = \frac{RTN_A^{2/3}}{v^{5/3}\eta}. \tag{1.51}$$

Assuming spherical symmetry and isotropic interfacial tension the pre-exponent in the rate eq. (1.50) for condensed systems can be written as

$$K_1 = \underbrace{\frac{v}{N_A}}_{n_1} \underbrace{4\pi\left(\frac{2\sigma v T_0}{Q_m \Delta T}\right)^2}_{A^*} \underbrace{\frac{RTN_A^{2/3}}{v^{5/3}}}_{\beta} \underbrace{\frac{(Q_m \Delta T)^2}{8\pi v (RT)^{1/2} N_A^{1/2}\, T_0^2\, \sigma^{3/2}}}_{Z}.$$

Inserting numerical values for *ice* nucleation from undercooled water at 233°K (ΔT = 40°) in the above given formula, $K_1 = 5.8 \times 10^{32}$ cm^{-3} sec^{-1} is obtained. Unlike the preceding case the factor K_1 is now very temperature sensitive. It must be also noted that the $J_0(T)$ function goes through a maximum at $T > T_0/3$.

It has been implicitly assumed so far that each molecule reaching the surface of the critical nucleus becomes a part of it. This cannot be true when crystalline clusters are involved as building units join crystal surfaces only at energetically favoured sites. Formally we can write

$$D^{+*} = A^*\beta\beta', \tag{1.52}$$

where $\beta' \leqslant 1$ is the sticking probability of the impinging molecules. The value of β' depends strongly on the growth mechanism of the crystalline clusters. An approximate estimate of β' is possible on the basis of crystal growth data for the system under consideration [30]. It is worth noting that when two-dimensional nucleation is the rate controlling factor β' can take very low values. Hence, the above given numerical value of the pre-exponent in ice nucleation calculated with $\beta' = 1$ should be regarded as an approximate estimate of its upper limit.

1.3.5. The Transient State

The kinetics of nucleation is not fully described unless the time needed to attain a steady state distribution of clusters by size, i.e. a steady state rate of nucleation is evaluated. In terms of the Zeldovich-Frenkel formulation of nucleation kinetics the rate of generation of stable nuclei is defined by

$$J \equiv J(i^*,t) = -D^{+*}n^*\left[\frac{\partial}{\partial i}\frac{c(i,t)}{n(i)}\right]_{i=i^*}. \tag{1.53}$$

Consequently, the calculation of the transient rate of nucleation requires an explicit solution of (1.40) with respect to the variables i and t. This is an extremely complicated mathematical problem which can be only approximately solved. Surveying some of the contributions in this field [31–36] Lyubov and Roitburd [26] point out that most of them are based on the tacit assumption that the processes in the vicinity of the critical size are rate determining in the transient regime. This would imply that outside a very narrow region around i^* the derivative in (1.53) is practically zero (fig. 1.9). A very detailed mathematical examination of the problem carried out by Kashchiev [38] shows that this approximation is sufficiently good, although for the sake of simplicity we shall follow the analysis in [26]. Performing the differentiation in the right side of eq. (1.40) and recalling (1.44) we have

$$\frac{1}{n(i)}\frac{\partial c(i,t)}{\partial t} = D^+(i)\frac{\partial^2}{\partial i^2}\left[\frac{c(i,t)}{n(i)}\right] + \left[\frac{\mathrm{d}D^+(i)}{\mathrm{d}i} - \frac{D^+(i)}{kT}\frac{\mathrm{d}\Delta G(i)}{\mathrm{d}i}\right]\frac{\partial}{\partial i}\left[\frac{c(i,t)}{n(i)}\right]. \tag{1.54}$$

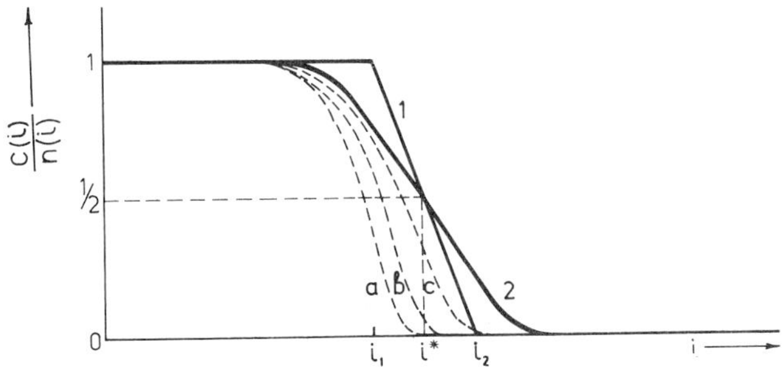

Fig. 1.9. Plots illustrating the analysis of the transient problem in nucleation kinetics 1,2 – the steady state after eq. (1.45) and the approximate solution (1.58) respectively. a,b,c – transient states attained after different times in increasing order.

This equation is valid within the whole interval of variation of i. In the vicinity of the critical size $D^+(i) \approx D^{+*}$ = const and $\mathrm{d}\Delta G(i)/\mathrm{d}i = 0$ so that

$$\frac{1}{n(i)} \frac{\partial c(i,t)}{\partial t} = D^{+*} \frac{\partial^2}{\partial i^2} \left[\frac{c(i,t)}{n(i)}\right] . \tag{1.55}$$

To proceed further it is necessary to specify the width Δi^* of the region around i^* in which equation (1.55) is valid. After Zeldovich [25] within this region $\Delta G(i)$ varies by kT in both directions from the peak of the energy profile (fig. 1.10), i.e.

$$\Delta G(i^*) - \Delta G(i_{1,2}) = kT .$$

Using the development (1.47) we find

$$\Delta i^* = \left(\frac{2kT}{\lambda}\right)^{1/2} . \tag{1.56}$$

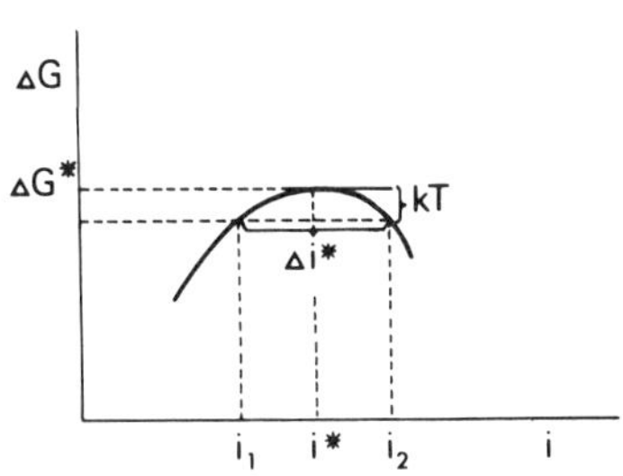

Fig. 1.10. The critical region concept after Zeldovich.

The major approximation involved consists in applying the initial and boundary conditions (1.42), (1.42′) and (1.42″) defined for the whole range of variation of i to eq. (1.55) which as already mentioned holds only for the critical region, i.e. it is assumed that outside this region the steady state is instantaneously attained. With this approximation the solution to (1.55) reads

$$\frac{c(i,t)}{n(i)} = 1 - \frac{i - i_1}{\Delta i^*} - \frac{2}{\pi} \sum_{s=1}^{\infty} \frac{1}{s} \sin\left(s\pi \frac{i - i_1}{\Delta i^*}\right) \exp\left(- \frac{s^2 \pi D^{+*} t}{(\Delta i^*)^2}\right). \tag{1.57}$$

It follows that for $t \to \infty$ i.e. under steady state conditions

$$\frac{c(i)}{n(i)} = \frac{1}{2} - \frac{i - i^*}{\Delta i^*}, \quad i_1 \leqslant i \leqslant i_2\,, \tag{1.58}$$

which is an approximate solution of (1.45). The combination of (1.53) and (1.57) yields

$$J = J_0 \left[1 + 2 \sum_{s=1}^{\infty} (-1)^s \exp(-s^2 t/\tau)\right]. \tag{1.59}$$

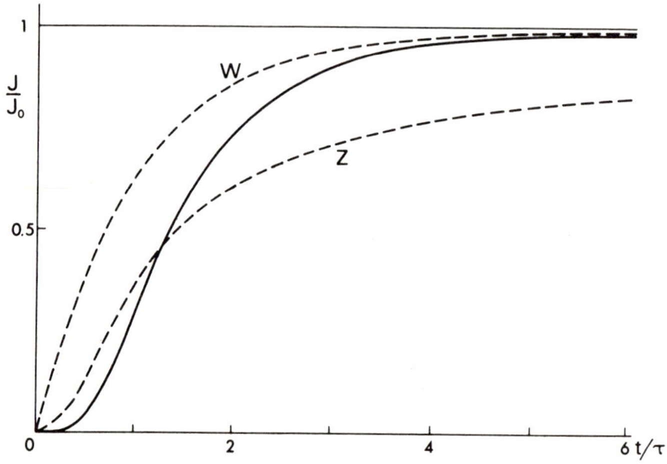

Fig. 1.11. The nuclation rate as a function of time
Full line – plot of eq. (1.59)
Dashed curve Z – solution of Zeldovich [25]
Dashed Curve W – solution of Wakeshima [33].
(From Ber. Bunsenges. phys. Chem. 73 (1969) 184).

Here J_0 differs from (1.49) only by the factor $\pi^{1/2}/2$. In the more precise treatment this negligible "correction" disappears [38]. The parameter τ called the *induction time* or *non-stationary time lag* characterizes the capacity of the system to reorganize itself until a steady flow of nuclei is produced. As can be directly seen, the theory predicts that

$$\tau = \frac{8kT}{\pi\lambda D^{+*}}. \tag{1.60}$$

An approximate estimate of τ can be found by formally considering nucleation as uni-dimensional diffusion of the size i with a generalized diffusion coefficient D^{+*}. Taking τ to be the time to cover the "distance" Δi^* the application of Einstein's relation gives

$$\tau = \frac{4kT}{\lambda D^{+*}}.$$

Fig. 1.11 shows the plot of eq. (1.59) in dimensionless units. For the sake of comparison two other approximate solutions of the problem are also presented on the same figure.

Normally it is the total number of the nuclei N formed within a finite time interval and not the rate of nucleation which is the experimentally available quantity. As per definition

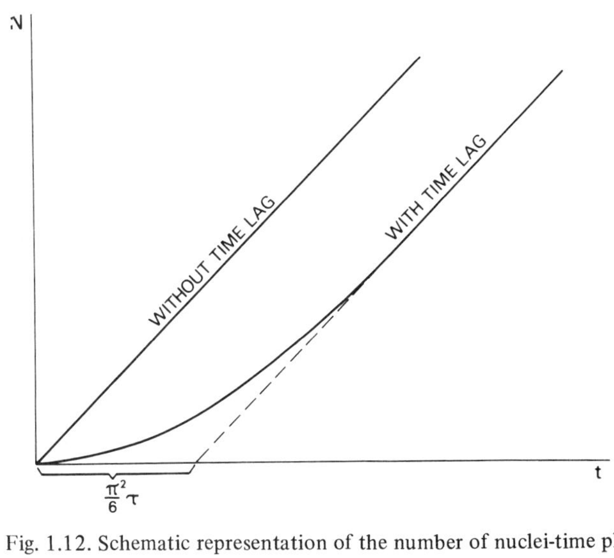

Fig. 1.12. Schematic representation of the number of nuclei-time plot.

$$N = \int_0^t J\mathrm{d}t',$$

the integration of (1.59) gives

$$N = J_0\left[t - \frac{\pi^2}{6}\tau - 2\tau \sum_{s=1}^{\infty} \frac{(-1)^s}{s^2} \exp(-s^2 t/\tau)\right]. \tag{1.61}$$

If $t \geqslant 5\,\tau$ the sum in (1.61) can be left out and N then increases further linearly with time [38].

$$N = J_0(t - \tfrac{1}{6}\pi^2\tau). \tag{1.62}$$

When the induction time is small as compared to the observation time

$$N = J_0 t. \tag{1.63}$$

Fig. 1.12 illustrates relationships (1.61), (1.62) and (1.63).

It must be pointed out again that eqs. (1.60) and (1.61) are derived on the assumption that only single molecules are present at the initial moment, i.e. pre-existing clusters are completely destroyed prior to the establishment of the supersaturation. This would imply that a melt for instance is subjected to prolonged annealing high above the melting point before crystallization. It can be expected that pre-existing clusters sometimes called *athermal nuclei*, will bring about an increase in the transient nucleation rate and a shortening of the time required to attain a stationary distribution of clusters if the critical supersaturation is once surpassed (fig. 1.13). Although this idea is not new [39–43] a lucid treatment based on the Zeldovich-Frenkel equation has been only recently proposed [44, 45]. In contrast to Frisch's analysis predicting unrealistically high induction times (up to 10^{10} sec) for vapour → liquid condensation in the absence of pre-existing clusters, appreciable effects should be expected only in the nucleation kinetics of condensed systems.

1.3.6. Discussion of the Induction Time

Inserting the value of λ evaluated from (1.48) and (1.20) in (1.60) we obtain the general expression

$$\tau = 0.7\mathrm{b}^3 \frac{RT\sigma^3 v^2 N_\mathrm{A}}{\Delta\mu^4 D^{+*}}. \tag{1.64}$$

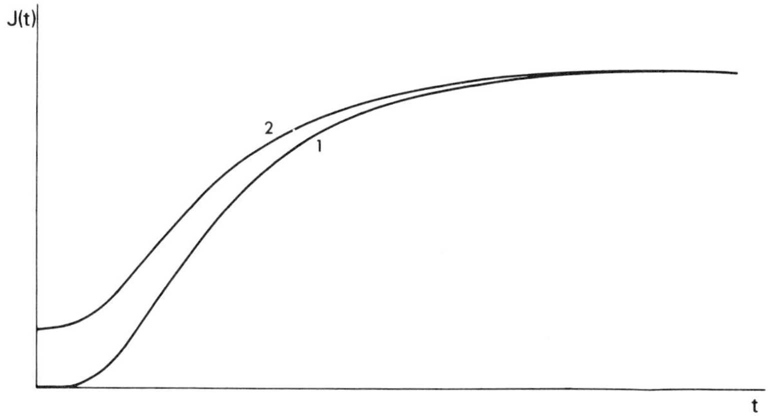

Fig. 1.13. The effect of pre-existing clusters on nucleation kinetics after Kashchiev [44]
Curve 1 – Nucleation rate in the absence of pre-existing clusters
Curve 2 – Nucleation rate in the presence of pre-existing clusters.

For spherical nuclei $b=6^{2/3}\pi^{1/3}$. Expressing accordingly $\Delta\mu$ and D^{+*}, the following formulae can be written for the induction time in vapours, melts and solutions respectively:

$$\tau = 0.75\left(\frac{M}{RT}\right)^{1/2} \frac{\sigma}{p(\log(p/p_{\mathrm{eq}}))^2\beta'} . \tag{1.65}$$

$$\tau = 0.07 \frac{v^{5/3} N_{\mathrm{A}}^{1/3} T_0^2 \eta}{Q_{\mathrm{m}}^2 (\Delta T)^2 \beta'} . \tag{1.66}$$

$$\tau = 0.3 \frac{v^{5/3} \sigma N_{\mathrm{A}}^{1/3} \eta}{(RT\log(a/a_{\mathrm{eq}}))^2 \beta'} . \tag{1.67}$$

The inspection of expressions (1.65), (1.66) and (1.67) leads to the following conclusions. If $\beta' = 1$, i.e. if no additional impediments across the nucleus/ambient phase interface exist the induction time for nucleation from supersaturated vapours and from ordinary liquids is extremely small. In the case of water condensation at $T = 300°\mathrm{K}$, and $p/p_{\mathrm{eq}} = 4$, $\tau = 3 \times 10^{-8}$ sec. Observable induction times should be typical only for highly viscous media ($\eta \approx 10^{10}$ poise), i.e. in glass-forming systems. In such instances the process of phase formation can go sometimes to completion before the establishment of a steady state rate of nucleation. However, as in the majority of occurrences the new phase is anisotropic, i.e. $\beta' < 1$ an additional enhancement of τ by several orders of magnitude should be expected apart from the effect of transport limitations in the interior of the mother phase. From this point of view the precipitation of a phase with ordered structure will be kinetically impeded despite the fact that it may represent the thermodynamically stable modification [30]. This argument is complementary to that raised by Stranski and Totomanow [37] to explain the Ostwald Rule of Stages by means of steady state nucleation theory (see eqs. (1.33) and (1.49)). Theoretical estimations of the magnitude of the induction time in various types of phase transitions are also given by Gindt and Kern [46] on the basis of a similar approach. Concluding this paragraph we note that the role of non-steady state effects in nucleation kinetics can be higher than it is usually believed*.

1.3.7. Probabilistic Aspects

Nucleation is a random phenomenon. The number of nuclei formed within a fixed time interval, as well as the time needed for the first nuclei to appear, are random quantities. Their statistical distribution is governed by the formal probability laws while the average values are predicted by the kinetic theory of nucleation. The combination of these two approaches reveals some interesting probabilistic features of the nucleation process offering a possible way for the quantitative experimental study of nucleation kinetics [48–50].

Nucleation can be regarded as a Poisson sequence of independent random events along the time axis (fig. 1.14a). The probability of finding m events within the time interval 0, t can be calculated from the Poisson formula

$$W_m = \frac{N^m \exp(-N)}{m!}, \tag{1.68}$$

where N is the average (time-dependent) number of events or nuclei within the chosen period of time. Let us determine the probability of finding the m-th nucleus within the infinitesimal interval $t, t + \mathrm{d}t$. This is equivalent to deriving the distribution of the times separating the initial moment from the first event in a new sequence of random events which originates from that depicted in fig. 1.14a, all points but those with serial numbers $m, 2m, 3m$ etc.

* Some recent developments in the field of transient nucleation are reviewed in a forthcoming paper by Toschev and Gutzow [47].

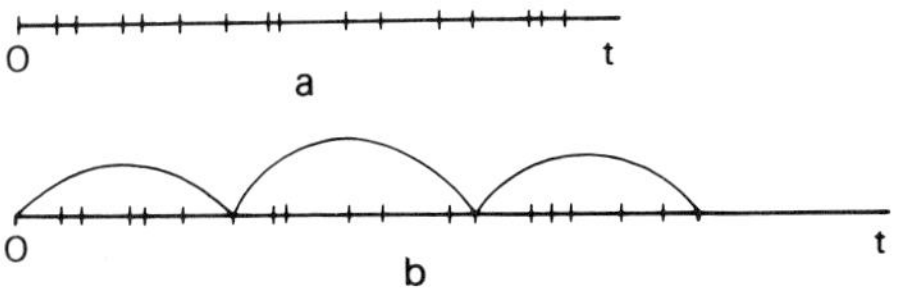

Fig. 1.14. Nucleation as a series of random events along the time axis.

being eliminated (fig. 1.14b). We then have*

$$\mathrm{d}W_{\geqslant m} = \frac{\mathrm{d}N}{\mathrm{d}t}\frac{N^{m-1}}{(m-1)!}\exp(-N)\,\mathrm{d}t\,. \tag{1.69}$$

Eq. (1.69) integrates to

$$W_{\geqslant m} = 1 - \exp(-N)\sum_{s=0}^{m-1}\frac{N^s}{s!}\,, \tag{1.70}$$

giving the probability of formation of a minimum of m nuclei before the expiration of the time t. We introduce now the theory of nucleation. Assuming as limiting case steady state nucleation we have to insert eq. (1.63) in (1.69) and (1.70).

Setting $m = 1$ we obtain

$$\mathrm{d}W_{\geqslant 1} = J_0 \exp(-J_0 t)\mathrm{d}t \tag{1.71}$$

$$W_{\geqslant 1} = 1 - \exp(-J_0 t) \tag{1.72}$$

Expression (1.71) defines the probability of the occurrence of the first nucleation event within the time interval $t, t + \mathrm{d}t$ whereas $W_{\geqslant 1}$ is the probability that at least one nucleus has been formed before the time t is elapsed. The mean time of expectation is under these premises

$$\bar{t} = \int_0^\infty t\mathrm{d}W_{\geqslant 1} = \frac{1}{J_0},$$

* It is assumed that the supersaturation remains constant and that N and J refer to the whole volume under consideration.

i.e. this time is exactly equal to the reciprocal of the steady state nucleation rate. Referring to eq. (1.63) it is seen that in this case $\bar{t}$ is also equal to the time t_1 needed to form one nucleus on an average. The transient case is more complex since the combination of (1.71) and (1.72) with (1.61) yields very complicated expressions. However, there is a distinct difference between steady state and transient nucleation as far as the probabilistic aspects considered are concerned (fig. 1.15). It must be also added that when non-stationary effects cannot be ignored, the mean time of expectation is no meaningful characteristic feature of the process, neither is $\bar{t}$ equal to t_1

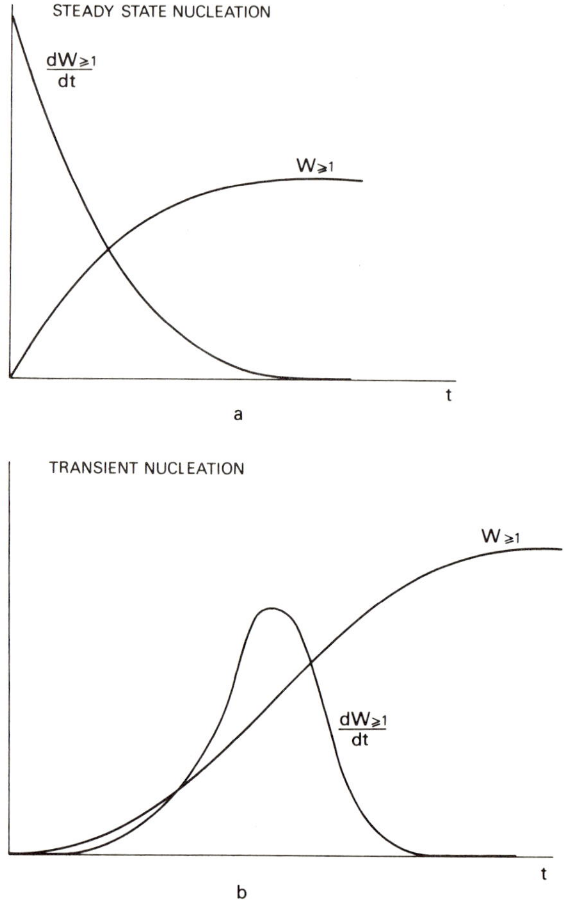

Fig. 1.15. Schematic plots of $dW_{\geqslant 1}/dt$ and $W_{\geqslant 1}$ as a function of time for steady state and transient nucleation. (After [50]).

though under certain circumstances their numerical values can be very close together [49].

1.3.8. Some Further Developments Derived from Statistical Thermodynamics

The theoretical description of the nucleation process presented so far is not complete as it ignores some additional terms in the evaluation of the entropy decrement associated with the transformation of part of the single molecules into dimers, trimers etc. The basic idea to be briefly discussed is that the clusters translate and rotate freely throughout the volume of the initial phase contributing thus to a higher degree of disorder than if the clusters were frozen in space. As a result, a reduction of the free energy change required for clustering should be expected. This problem was posed long ago but the earlier contributions [51–53] have remained unnoticed for some time. In 1962 Lothe and Pound [54] analyzed again the statistical mechanical corrections and reformulated accordingly the theory of nucleation. Their derivation was based on the following thought process*:

1. Isothermal expansion of 1 mole of vapour containing only single molecules to the equilibrium pressure p_{eq} and conversion of the gas into bulk liquid.

2. Mechanical comminution of the liquid into equal droplets of i molecules each. A new interface is created but the droplet loses internal degrees of freedom of motion in the bulk liquid as a whole.

3. Energization of the mist of droplets produced each cluster acquiring thus 3 translational and 3 rotational degrees of freedom.

The corresponding free energy changes referred to i molecules are:

$$\text{Step 1}: \Delta G_{vol} = -ikT\ln(p/p_{eq}).$$

$$\text{Step 2}: \Delta G_{surf} + \Delta G_{rep} = 4\pi r_i^2\sigma - TS_m ,$$

where S_m is the molecular entropy of the liquid.

$$\text{Step 3}: \Delta G_{tr} + \Delta G_{rot} = -kT\ln\frac{i^{2/3}}{\kappa^3} - kT\ln\frac{bv_0 i^{5/2}}{\kappa^3},$$

where $\kappa = \dfrac{hN_A}{\sqrt{2\pi MRT}}$ and b is a geometrical coefficient. The algebraic sum

* A more refined analysis along the same lines of thought is given in [55].

$$\Delta G(i) = \Delta G_{vol} + \Delta G_{surf} + \Delta G_{rep} + \Delta G_{tr} + \Delta G_{rot} , \qquad (1.73)$$

gives the total free energy change upon the formation of a cluster containing i atoms. As is seen, the classical approach considers only the first two terms in (1.73). It must be noted that since the additional corrections are only slightly dependent on the size i, the Thomson-Gibbs equation satisfying the condition $d\Delta G(i)/di = 0$ is practically unaffected by these considerations. The same is true for the Zeldovich factor Z and for the induction time [56]. Hence, the rate equation (1.49) will be multiplied by a fairly constant factor Γ (*Lothe-Pound factor*). In the case of water condensation under typical experimental conditions the calculations based on this model yield $\Gamma = 10^{17}$. A similar result was obtained by Hirth [57]. The reliability of this estimate depends primarily on the correct evaluation of the so called replacement term ΔG_{rep} which is the subtlest stage in this analysis. The discussion of this particular point which aroused a lot of controversial arguments in the literature [58–61] is still in progress.

1.4. Comparison with Experiment

1.4.1. Basic Principles

A straightforward test of nucleation theory would require an experimental plot of the number of the nuclei formed within a given volume of the mother phase against the time at constant supersaturation. As is seen from fig. 1.12, the steady state rate of nucleation and the induction time are readily obtained from the slope and the intercept respectively of the linear part of the $N(t)$ curve. The values of J_0 and τ obtained can be further interpreted by means of the expression given in the preceding. However it is not an easy task to carry out the necessary measurements because : 1. The determination of the new phase particles within the interior of a parent phase especially in opaque bodies is a difficult problem. 2. The rapid growth of the nuclei once formed changes the initial conditions in a very complicated manner. In vapours and solutions the concentration is depleted in the vicinity of the growing nuclei thus causing a reduction in the supersaturation. In melts the volume available for progressive nucleation diminishes, bringing about a steady decrease in the effective nucleation rate. The local liberation of heat as well as the possibility of coalescence and ripening phenomena should also be taken into account. At the present stage reliable direct measurements of nucleation kinetics can be performed in systems in which the early stages of phase formation can be frozen and subsequently studied by a suitable technique.

This is possible for instance in glass-forming melts distinguished by a low rate of growth of the new phase and by a steep temperature dependence of the viscosity. Measurable induction times have been detected in such systems [29, 62]. Another serious problem arises from the fact that extraneous solid particles, the walls of the container and other surfaces inevitably present in the investigated systems strongly promote nucleation, obscuring in this way the observation of spontaneous phase formation. This is the reason why earlier studies on the freezing of metals, for instance, revealed a marked dependence of crystallization behaviour on sample purity and thermal pre-history, the critical supercoolings rarely surpassing 10–20 degrees [63, 64]. Apart from the progress in purification techniques nowadays there are ways to suppress or even to completely eliminate uncontrolled heterogeneous nucleation. Some of the methods used will be mentioned further on.

In view of the difficulties enumerated, indirect methods for the study of nucleation kinetics can prove useful. A very promising and simple method is based on the statistical analysis of the times needed for the first nucleus to appear. This approach was successfully applied by Skripov et al. [65, 66] for the quantitative study of bubble formation and for the nucleation of *tin* in undercooled droplets. In both cases equation (1.71) was confirmed which among the other things indicated that within the limits of the experimental time scale non-steady state induction times were unimportant (fig. 1.16). It is

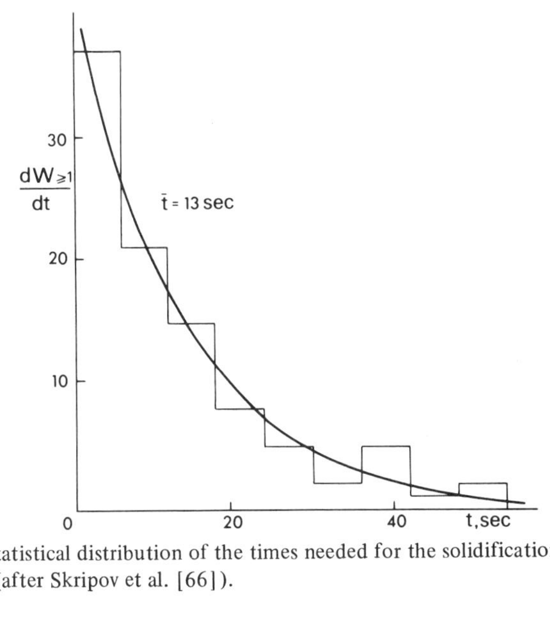

Fig. 1.16. Statistical distribution of the times needed for the solidification of tin droplets at 109.2°C (after Skripov et al. [66]).

obvious that correct results will be obtained only if the nucleation time is much longer than the time needed for the nucleus to reach detectable size.

The stationary rate of nucleation and the induction time can in principle be evaluated from measurements of the overall crystallization rate [67, 68], the size distribution of the particles produced [69], the time required for the visible appearance of the new phase [70], or from any other experiments in which nucleation plays a major role, provided that all factors contributing to the effect under study are accurately accounted for. The mathematical analysis of the indirect experiments is sometimes very complicated which is more or less reflected in the reliability of the numerical values obtained. As was already pointed out, the theory predicts a drastic increase in the nucleation rate beyond some critical value of the supersaturation. If the induction time is neglected, the critical threshold is defined by

$$(\Delta\mu)_{\text{crit}} = \left(\frac{\mathrm{b}\nu^2\sigma^3}{2.3RT\log(K_1t/N)}\right)^{1/2}, \tag{1.74}$$

wherein $\mathrm{b} = 16\ \pi/3$ and $\mathrm{b} = 32$ for spherical and cubic particles respectively. A common procedure is to measure $(\Delta\mu)_{\text{crit}}$ and to calculate σ from (1.74), t and N being experimentally available, while K_1 must be estimated theoretically. This brings about some uncertainty, since neither K_1 nor N and sometimes t can be accurately determined. However, an error by one order of magnitude in the value of the term K_1t/N affects usually the value of σ by one per cent only. This is a very favourable situation when the interfacial tension is unknown. In the reverse case only a very close coincidence with the value of the surface tension can be regarded as a meaningful test of the nucleation theory. It is seen that the critical supersaturation is not a characteristic constant for a given substance as its value depends on the sensitivity of the experimental technique employed to detect the onset of nucleation.

Before proceeding further with the experimental evidence on homogeneous nucleation we shall illustrate some of the above outlined basic principles by experimental results obtained in the study of the initial stages of metal *electrodeposition.* In fact, metal electrodeposition on inert electrodes is a heterogeneous process but, since the classical treatment of heterogeneous nucleation is essentially the same and considering that in this particular case a quantitative and reproducible control of the factors governing the kinetics of nucleation is possible, we believe that electrolytic nucleation is a field where some basic predictions of the theory can be conveniently tested. The so-called double pulse potentiostatic technique serves very well to this purpose [71, 72]. Electric pulses possessing the form shown in fig. 1.17, are imposed

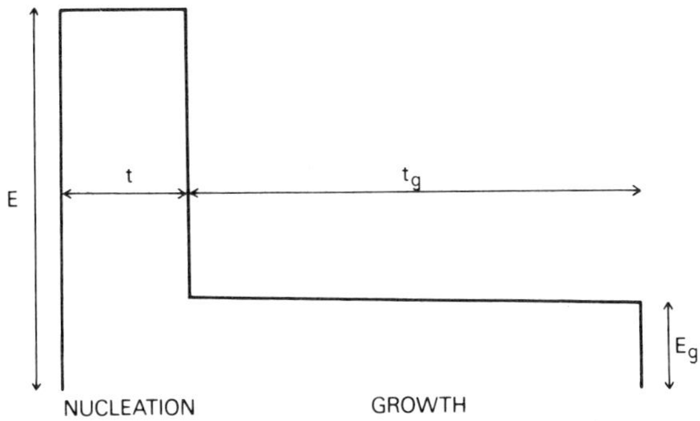

Fig. 1.17. Principle of the double pulse potentiostatic technique for studying the electrolytic nucleation of metals.

on an electrolytic cell consisting of two electrodes immersed in an electrolytic solution. The cathode is usually platinum, while the anode is of the metal whose ions are present in the solution. By suitable adjustment of the parameters E, E_g, t and t_g a certain number of invisible nuclei can be deposited during the first pulse, the second pulse serving only for their further growth up to visible size. The number of the nuclei produced can be directly determined by counting. In this way nucleation and growth can be separately studied under well defined conditions. The height of the first pulse E is a measure of the supersaturation as (per ion)

$$\Delta\mu = z\,eE\,,$$

where z denotes the valence of the neutralizing ions and e the elementary electric charge.

Fig. 1.18 shows a series of $N(t)$ plots in the electrodeposition of *mercury* nuclei at various overpotentials [73]. The transient character of nucleation kinetics is clearly demonstrated. The slopes of the linear parts of the curves presented depend exponentially on the overpotential in accordance with equation (1.50), permitting thus the evaluation of ΔG^* for each value of E (fig. 1.19). The applicability of eq. (1.59) is tested by plotting the values of $N/J_0\tau = f(x)$ against the dimensionless parameter $x = t/\tau$ (fig. 1.20). As is seen the points fall close to the theoretical curve

$$f(x) = x - \tfrac{1}{6}\pi^2 - 2\sum_{s=1}^{\infty} \frac{(-1)^s}{s^2} \exp(-s^2 x)\,. \tag{1.75}$$

The probabilistic approach can be also applied to the study of electrolytic nucleation because of the possibility of reproducing the experiments many times. The statistical distribution of the number of the nuclei formed under otherwise constant conditions obeys the Poisson law (fig. 1.21) which justifies a further examination of the probabilistic aspects of electrolytic nucleation. Fig. 1.22 shows experimental plots of the probability of formation of at least one *cadmium* nucleus against the time. The plots are S-shaped (in contrast to the integral distribution following from fig. 1.16), which again con-

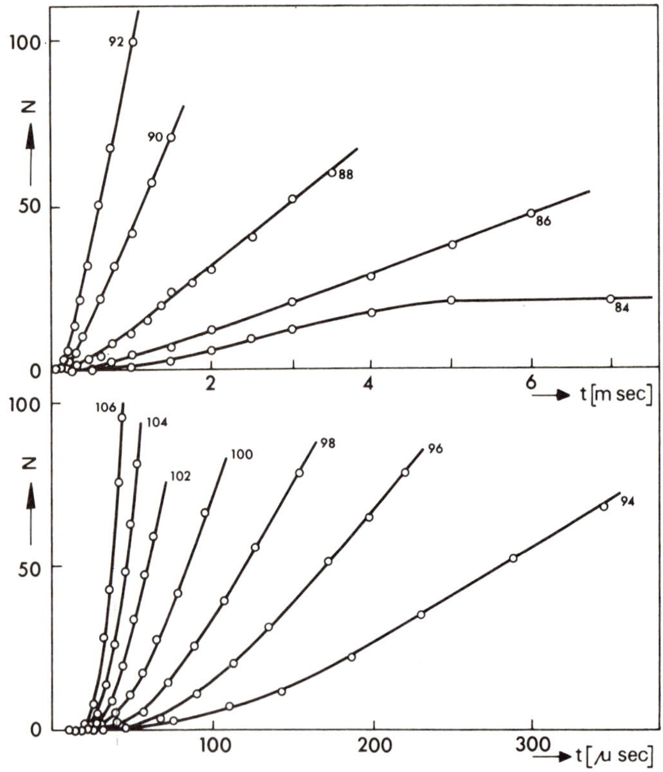

Fig. 1.18. Experimental plots of the number of nuclei vs time in the electrodeposition of mercury on platinum [73]. The figures denote the corresponding overvoltage in millivolts. (From Ber. Bunsenges. phys. Chem.).

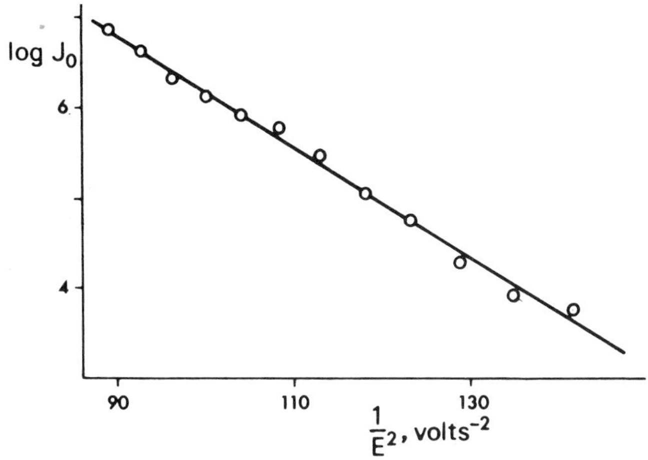

Fig. 1.19. Plot of the logarithm of the steady state nucleation rate vs the reciprocal of the square of the overvoltage in the electrodeposition of mercury on platinum [73]. (From Ber. Bunsenges. phys. Chem.)

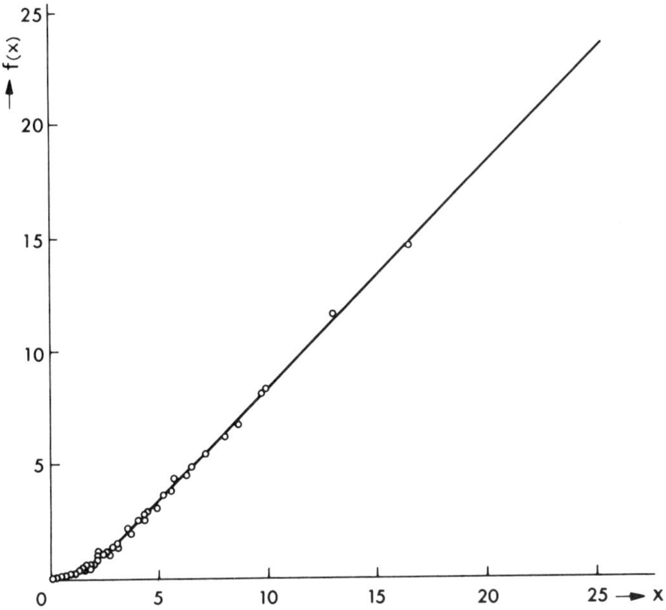

Fig. 1.20. Test of eq. (1.61) or (1.75) respectively based on experimental data on the nucleation kinetics of mercury on a platinum electrode [49].

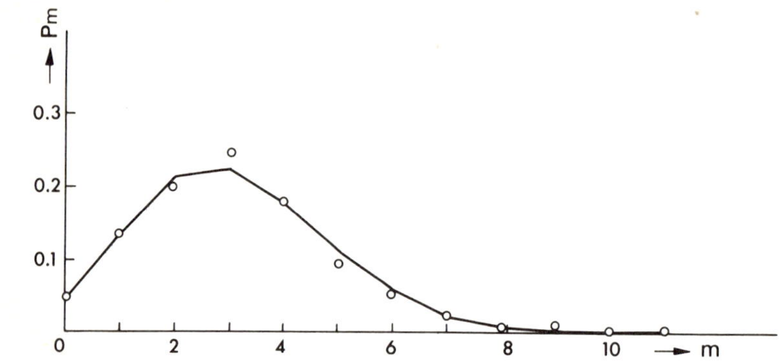

Fig. 1.21. Random distribution of the number of mercury droplets electrodeposited on platinum [50]
points – experimental data based on 500 identical experiments at E = 90 mV and $t = 1.68 \times 10^{-4}$ sec
discontinuous line – Poisson distribution law constructed by means of the experimental average value N = 3.14.

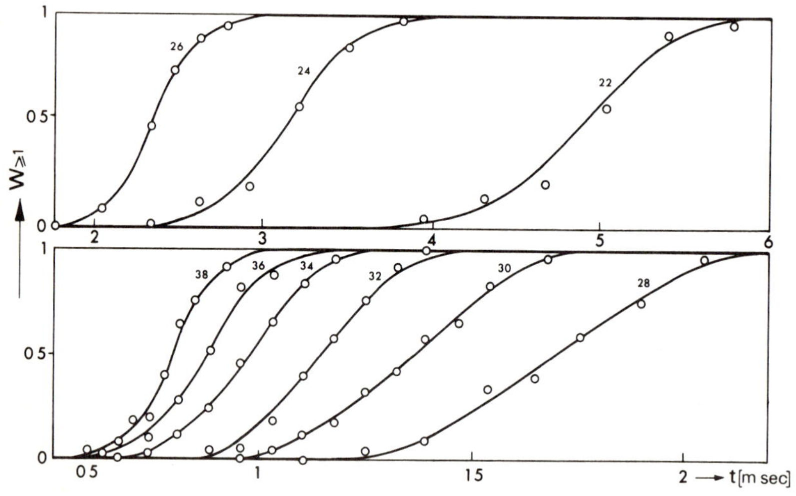

Fig. 1.22. The probability of formation of at least one nucleus in the electrodeposition of cadmium on platinum [71].

firms the finding that transient nucleation effects play an important role in the early stages of electrodeposition. The parameters of the process can be evaluated from the relations

$$\tilde{W} = 1 - \exp[-R(\tilde{x})], \qquad \tilde{x} = \tilde{t}/\tau , \tag{1.76}$$

$$J_0 = -\frac{2.3\tilde{x}\log(1-\tilde{W})}{\tilde{t}f(\tilde{x})}, \tag{1.77}$$

wherein $\tilde{t}$ and $\tilde{W}$ are the coordinates of the inflexion points of the $W_{\geqslant 1}(t)$ plots and $R(\tilde{x})$ is a known tabulated function [49]. It is interesting that if $J_0\tau > 50$ the $W_{\geqslant 1}(t)$ equation resulting from the combination of (1.72) and (1.61) can be very well approximated by the Gaussian distribution law. This is the case in the experiments described, in agreement with the calculations based on (1.76) and (1.77) yielding $J_0\tau = 135$ irrespective of the overvoltage applied.

*1.3.2. Nucleation in Vapours**

From the various techniques available to produce supersaturated vapours [75], three, in the main, have been employed to study condensation phenomena, namely adiabatic expansion in classical *cloud chambers, jet and nozzle techniques* and stationary mass and heat transfer in *diffusion chambers.* A common feature of these methods is that critical supersaturations are measured at different temperatures. Experiments in cloud chambers were carried out long ago**, but Volmer and Flood [1,77] were the first to use this device for testing nucleation theory. They determined the critical supersaturation for condensation of water and several organic substances and compared the experimental data with the values predicted by relation (1.74) assuming $N = 1$ nucleus per cm^3, $t = 1$ sec and calculating K_1 from an expression practically identical with (1.50′). The results are listed in table 1.2. As is seen, despite the approximations involved both in the theory and in the elaboration of the experimental procedure, the agreement is remarkably good. This situation has been regarded as a firm basis of the classical nucleation theory for many years. It was not until recently that Hirth and Pound

* The present state of the investigations on this subject is reviewed by Wegener and Parlange [74].

** Older cloud chamber work is surveyed in [17, 76].

Table 1.2
Data of Volmer and Flood (after [18])

Substance	Temperature °K	Critical supersaturations	
		calculated	observed
Water	261	5.0	5.0
Water	275	4.2	4.2
Methanol	270	1.8	3.2
Ethanol	273	2.3	2.3
n-Propanol	270	3.2	3.0
Isopropanol	265	2.9	2.8
n-Butanol	270	4.5	4.6
Nitromethane	252	6.2	6.1

[17] rejected the significance of these results, their main argument being that Volmer and Flood observed nucleation on gaseous ions. On the other hand, if the rate equation (1.50) is to be additionally multiplied by the Lothe-Pound factor $\Gamma = 10^{17}$ accounting for statistical mechanical effects, the expected critical threshold for water condensation should be by 30–40 per cent lower than the experimental finding. Hence, Hirth and Pound concluded that the observed agreement of theory and experiment was due to a fortuitous combination of a wrong theory with inadequate measurements. This criticism is perhaps rather severe, particularly in view of the still disputable value of Γ for the liquids studied, but it stimulated further work in this area. A certain preference is shown nowadays for nozzle and diffusion chamber techniques as in contrast to the Wilson chamber the former ensure continuous operation under stationary conditions. There is sufficient evidence to indicate that some liquids follow the classical rate equation, while the critical supersaturations observed with other liquids can be explained by much higher pre-exponential factors. Under the first class of substances fall H_2O, CH_3OH, C_2H_5OH, the n-alkanes C_6H_{14}, C_7H_{16}, C_8H_{18}, C_9H_{20} [78, 79]; under the second NH_3, C_6H_6, $CHCl_3$, CCl_3F [80, 81]. Attempts have been made to explain the different behaviour of the liquids studied based on the argument that the Lothe-Pound treatment is the correct one and that the discrepancy between theory and experiment can be removed if the increase in the surface tension of small clusters with respect to bulk liquids constituted of polar molecules is taken into account [82]. However, in view of the evidence with the nonpolar n-alkanes the matter is still unclear. It is interesting to note that with the exception of water and ethanol the two groups of liquids indicated have been investigated by different methods.

1.4.3. Nucleation in Melts

Unlike supersaturated vapours, melts can never be entirely free from extraneous particles catalyzing the nucleation process. However, if the bulk sample is dispersed in such a way that the number of the droplets produced exceeds the number of the catalyzing particles in the initial volume, some of the droplets will contain no active centers at all. The completely "purified" droplets exhibit some maximum undercooling ΔT_{max} as compared with the remaining droplets which solidify at different temperatures between the melting point T_0 and T_{max}, depending on the nucleation potency of the cores retained. The maximum undercooling observed represents then the critical threshold for homogeneous nucleation. This principle was applied by Turnbull and Cech [83] for studying the onset of nucleation of a large number of metals by means of a microscopic technique. Correlating the experimentally found undercoolings with relation (1.74) under the assumption that $K_1 = 10^{33}\ \text{sec}^{-1}\text{cm}^{-3}$ and $J_0 = N/t = 10^{-(1\pm 1)}\ \text{sec}^{-1}$ per 50 micron particles, Turnbull [84] calculated the specific surface energies at the crystal/melt interface. His results are summarized in table 1.3. It is seen that both the ratios $\Delta T_{max}/T_0$ and σ_m/Q_m, where σ_m is the gramatomic interfacial tension defined by

Table 1.3
Homogeneous nucleation of metals after Turnbull (from [18])

Metal	ΔT_{max}	$\frac{\Delta T_{max}}{T_0}$	σ erg.cm^{-2}	σ_m cal. gatom^{-1}	$\frac{\sigma_m}{Q_m}$
Mercury	58	0.247	24.4	296	0.53
Gallium	76	0.250	55.9	581	0.436
Tin	105	0.208	54.5	720	0.418
Bismuth	90	0.166	54.4	825	0.33
Lead	80	0.133	33.3	479	0.386
Antimony	135	0.150	101	1430	0.302
Aluminium	130	0.140	93	932	0.364
Germanium	227	0.184	181	2120	0.348
Silver	227	0.184	126	1240	0.457
Gold	230	0.172	132	1320	0.436
Copper	236	0.174	177	1360	0.439
Manganese	308	0.206	206	1660	0.480
Nickel	319	0.185	255	1860	0.444
Cobalt	330	0.187	234	1800	0.490
Iron	295	0.164	204	1580	0.445
Palladium	332	0.182	209	1850	0.450
Platinum	370	0.181	240	2140	0.455

$$\sigma_m \approx N_A^{1/3} v^{2/3} \sigma$$

are nearly constant over the whole range of the metals investigated. This evidence is demonstrated again in fig. 1.23; the slope of the continuous straight line, being 0.46, can be regarded as a further confirmation of nucleation theory as such a relationship between σ_m and Q_m could be in principle expected by analogy with Stefan's rule in the liquid/vapour case. However, since the melt/crystal interfacial tensions are not available by other means, an absolute verification of the theory seems hardly possible in this case.

Critical undercoolings have been also measured during the crystallization of small droplets of molecular compounds [85, 86] and alkali halides [87, 88]. An intermediary stage in these experiments is the formation of supersaturated vapours of the corresponding substances producing crystals via liquid droplets. The observed times required for the alkali halide droplets to solidify ($t \approx 10$ sec) are treated as non-steady state time lags which on the basis of expression (1.66) would invoke extremely low β' values. Generally speaking, the experience with non-metallic substances also confirms the existence of a nearly constant relationship between ΔT_m and T_o and σ_m and Q_m.

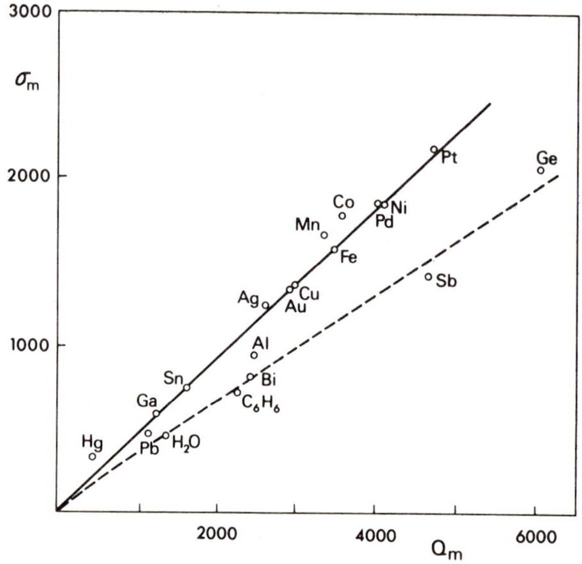

Fig. 1.23. Plot of the gram-atomic interfacial tension σ_m vs the molar heat of fusion Q_m for various metals (after Turnbull [84]).

1.4.4. Nucleation in Solutions

The relative stability of supersaturated aqueous solutions below some critical limit was observed long ago [16, 89, 90]. It is believed that heterogeneous nucleation can be avoided by very rapid mixing of solutions producing a sparingly soluble salt. Due to the very high supersaturations attained, homogeneous nucleation becomes the predominant process. Using a *turbidimetric method* Walton [91] was able to record the onset of nucleation for several substances and to evaluate interfacial energies. Nielsen [70, 92] investigated the nucleation of a large number of salts in a specially designed mixing chamber, measuring the number of the particles formed and the time needed for visible precipitation to take place. It must be noted that Nielsen's definition of the induction time is not the same as that considered so far, since he assumes that the delay in the appearance of the new phase is due to the slow growth of the particles in the solution. His discussion of the experimental results is based on the assumption of steady state nucleation kinetics. Other authors [14, 46, 68] interpret the delay times as non-stationary time lags, accentuating even more the nucleation stage in precipitation experiments. Summarizing experimental data for six substances Kahlweit [93] established that

$$\sigma_m/H_m \approx 0.18 ,$$

where H_m is the molar enthalpy of dissolution. In a recent paper Nielsen and Söhnel find a simple correlation between the interfacial tensions of 29 ionic substances obtained from nucleation experiments and the solubility of the corresponding salts in water [94]. This is explained by assuming that wetting of the surface of an ionic crystal and hydration can be described by the same interfacial tension.

1.5. Conclusion

The current theory of homogeneous nucleation is based on some fundamental ideas formulated several decades ago mainly by Gibbs, Volmer, Becker-Döring, Zeldovich and Frenkel. Despite the ever increasing interest in this field of science, no basically new concepts have been introduced since then, thus showing that the theoretical description of the nucleation process is very sound in principle. The major shortcoming arises from the thermodynamic treatment which assigns macroscopic properties such as vapour pressure and surface tension to very small clusters. This is a good approximation at relatively low supersaturations but the situation becomes very complicated when the critical nucleus contains only a few atoms. In such a case neither the use of the bulk surface tension, nor the division of the free energy ΔG^* into one volume and one surface term can be justified. The introduction of the curvature dependence of σ into the theory does not help very much because the treatment is confined to classical thermodynamic concepts which precludes extrapolation to very small dimensions. Similar difficulties are encountered in the kinetic treatment of nucleation as far as the decay constants D^- are calculated by means of an essentially thermodynamic relationship. The continuity approach to nucleation kinetics avoids operation with D^- assuming only that the free energy change $\Delta G(i)$ goes through a maximum at some critical size. Eq. (1.61) is believed therefore to be valid for relatively small clusters, provided that the mathematics of the derivation is not disturbed. Again this does not hold for the time lag which is defined in terms of the macroscopic surface tension (see (1.64)). At the present moment it is not even possible to predict the smallest size up to which the application of the theory is correct. It seems that progress in this field depends on further developments of a very fundamental nature linking together the microscopic properties of molecular aggregates with those of bulk phases.* Among the other approximations made the most essential are the assumption of a bimolecular chain mechanism, the preservation of the geometrical form of the clusters as well as approximations of a purely mathematical nature. More complicated cases involving compositional and structural changes have also found a satisfactory interpretation [97, 98].

Referring to the experimental evidence on homogeneous nucleation we note that it is based primarily on critical supersaturation measurements or on other indirect experiments. It can be said that the theory provides a good semiquantitative description of the available data. Moreover, nucleation experiments are a valuable source of information concerning interfacial tensions which in some cases can hardly be obtained by other methods. However, for many reasons of a both theoretical and experimental nature a complete and unambiguous quantitative verification of the theory is still lacking with the exception perhaps of the case of bubble formation which seems to be the ideal example for such studies. Nevertheless, the existing theory proved to be a reliable guide in predicting the basic features and trends of the early stages of phase transitions.

* Some interesting contributions in this area can be found in [95, 96].

Acknowledgment

The author is indebted to Professor R. Kaischew and Dr. D. Kashchiev for valuable discussions during the preparation of this manuscript.

References

[1] M. Volmer, Kinetik der Phasenbildung (Th. Steinkopf Verlag, Leipzig-Dresden, 1939).
[2] R. Defay, J. Prigogine, A. Bellemans and D.H. Everett, Surface Tension and Adsorption (Longmans Green and Co Ltd., London, 1966).
[3] A.I. Russanov, Fazovie Ravnovessiya i Poverkhnostnie Yavleniya (Izd. Khimia, Leningrad, 1967).
[4] R. Kaischew, unpublished lectures.
[5] R. Kaischew, Bull. l'Acad. Bulg. Sci. (phys.) 2 (1951) 191.
[6] D. Turnbull, in: Solid State Physics, ed. by F. Seitz and D. Turnbull, Vol. 3 (Academic Press Inc., New York, 1956) p. 224–306.
[7] S. Ono and S. Kondo, in: Handbuch der Physik, ed. by S. Flügge, Vol. 10 (Springer Verlag, Berlin-Göttingen-Heidelberg, 1960).
[8] R.C. Tolman, J. Chem. Phys. 17 (1949) 338.
[9] F.F. Abraham and G.M. Pound, J. Cryst. Growth 6 (1970) 309.
[10] J.Y. Parlange, J. Cryst. Growth 6 (1970) 311.
[11] R. Kaischew and I.N. Stranski, Z. physik. Chem. B26 (1934) 317.
[12] I. Gutzow, I. Konstantinov and R. Kaischew, Comm. Dept. Chem. Bulg. Acad. Sci. 2 (2) (1969) 249.
[13] Ya.I. Frenkel, Kinetic Theory of Liquids (Dover, New York, 1955).
[14] W.J. Dunning, in: Chemistry of the Solid State, ed. by W.E. Garner (Butterworths, London, 1955) p. 159–183.
[15] J.H. Hollomon and D. Turnbull, in: Progress in Metal Physics, ed. by B. Chalmers, Vol. 4 (Pergamon Press, London, 1953) p. 333–388.
[16] A.E. Nielsen, Kinetics of Precipitation (Pergamon Press, London, 1964).
[17] J.P. Hirth and G.M. Pound, Condensation and Evaporation (Pergamon Press, London, 1963).
[18] R.F. Strickland-Constable, Kinetics and Mechanism of Crystallization (Academic Press, London-New York, 1968).
[19] A.G. Sutugin, Uspekhi Khimii 38 (1969) 166.
[20] G. Zinsmeister, Kristall u. Technik 5 (1970) 207.
[21] F.H. Ree, T.S. Ree, T. Ree and H. Eyring, in: Advances in Chemical Physics, ed. by I. Prigogine, Vol. 4 (Interscience, New York, 1962) p. 1–66.
[22] R. Becker and W. Döring, Ann. Phys. 24 (1935) 719.
[23] R. Kaischew, Z. Elektrochem. 61 (1957) 35.
[24] D. Turnbull, Trans. AIME 175 (1948) 774.
[25] Ya.B. Zeldovich, Acta Physicochim. USSR 18 (1943) 1.
[26] B.Ya. Lyubov and A.L. Roitburd, in: Problemi Metallovedeniya i Fiziki Metallov (Metallurgizdat, Moskva, 1958) p. 91.

[27] M. Volmer and A. Weber, Z. physik. Chem. 119 (1926) 227.
[28] D. Turnbull and J.C. Fisher, J. Chem. Phys. 17 (1949) 71.
[29] I. Gutzow, S. Toschev, M. Marinov and E. Popov, Kristall u. Technik 3 (1968) 337.
[30] I. Gutzow and S. Toschev, Kristall u. Technik 3 (1968) 485.
[31] A. Kantrowitz, J. Chem. Phys. 19 (1951) 1097.
[32] R.F. Probstein, J. Chem. Phys. 19 (1951) 619.
[33] H. Wakeshima, J. Chem. Phys. 22 (1954) 1614.
[34] F.C. Collins, Z. Elektrochem. 59 (1955) 404.
[35] B.K. Chakraverty, Surface Science 4 (1966) 205.
[36] F.J. Farley, Proc. Roy. Soc. 212 (1952) 530.
[37] I.N. Stranski and D. Totomanow, Z. physik. Chem. A163 (1933) 399.
[38] D. Kashchiev, Surface Science 14 (1969) 209.
[39] J.C. Fisher, J.H. Hollomon and D. Turnbull, J. Appl. Phys. 19 (1948) 775.
[40] H.L. Frisch, J. Chem. Phys. 27 (1957) 90.
[41] E.R. Buckle, Proc. Roy. Soc. A261 (1961) 189.
[42] A. Ziabicki, J. Chem. Phys. 48 (1968) 4368.
[43] A. Ziabicki, J. Chem. Phys. 48 (1968) 4374.
[44] D. Kashchiev, Surface Science, 18 (1969) 389.
[45] D. Kashchiev and R. Kaischew, Comm. Dept. Chem. Bulg. Acad. Sci. 2 (1969) 263.
[46] R. Gindt and R. Kern, Ber. Bunsenges. physik. Chem. 72 (1968) 459.
[47] S. Toschev and I. Gutzow, Kristall u. Technik 7 (1972) 43.
[48] S. Toschev and S. Stoyanov, Comm. Dept. Chem. Bulg. Acad. Sci. 3 (1970) 295.
[49] S. Toschev and A. Milchev, Comm. Dept. Chem. Bulg. Acad. Sci. 3 (1970) 755.
[50] S. Toschev, A. Milchev and S. Stoyanov, J. Crystal Growth 13/14 (1972) 123.
[51] W.H. Rodebush, Chem. Revs. 44 (1949) 269.
[52] W.H. Rodebush, Ind. Eng. Chem. 44 (1952) 1289.
[53] F. Kuhrt, Z. Physik. 131 (1952) 205.
[54] J. Lothe and G.M. Pound, J. Chem. Phys. 36 (1962) 2080.
[55] J. Feder, K.C. Russell, J. Lothe and G.M. Pound, Adv. Phys. 15 (1966) 111.
[56] K.C. Russell, J. Lothe and G.M. Pound in: Condensation and Evaporation of Solids (Gordon and Breach, New York, 1965) p. 504.
[57] J.P. Hirth, Ann. New York Acad. Sci. 101 (1963) 805.
[58] W.J. Dunning, Proc. Case Inst. Technol. Symp. Nucleation (Cleveland, Ohio, 1965) p. 1.
[59] H. Reiss and J.L. Katz, J. Chem. Phys. 46 (1967) 2496.
[60] J. Lothe and G.M. Pound, J. Chem. Phys. 45 (1966) 630.
[61] F.M. Kuni and A.I. Russanov, Teoreticheskaya i Matematicheskaya Fizika 2 (1970) 265.
[62] S.M. Ohlberg and J.J. Hammel, Proc. 7th Intern. Congr. on Glass (Brussels, 1965) p. 32.
[63] V.I. Danilov, Stroenie i Kristallizatziya Ghidkostei (Izd. Akad. Nauk Ukr. SSR, Kiev, 1956).
[64] V.D. Kusnetsov, Kristalli i Kristallizatsiya (Goss. Izd. Techn. Teoret. Lit., Moskva, 1954).
[65] E.N. Sinitzin and V.P. Skripov, Ukr. Fiz. Ghurnal 12 (1967) 99.
[66] V.P. Skripov, V.P. Koverda and G.T. Butorin, Kristallografiya 15 (1970) 1219.
[67] D. Turnbull, J. Chem. Phys. 20 (1952) 411.
[68] R. Gindt and R. Kern, Ber. Bunsenges. physik. Chem. 69 (1965) 124.

[69] S. Toschev and I. Gutzow, Phys. Stat. Sol. 24 (1967) 349.
[70] A.E. Nielsen, Kristall u. Technik 4 (1969) 17.
[71] R. Kaischew, S. Toschev and I. Markov, Comm. Dept. Chem. Bulg. Acad. Sci. 2 (1969) 463.
[72] R. Kaischew, Chem. Ing. Technik 39 (1967) 554.
[73] S. Toschev and I. Markov, Ber. Bunsenges. physik. Chem. 73 (1969) 184.
[74] P.P. Wegener and J.Y. Parlange, Naturwiss. 57 (1970) 525.
[75] A.G. Amelin, Teoreticheskie Osnovi Obrazovaniya Tumana pri Kondenzatsii Para (Izd. Khimiya, Moskva, 1966).
[76] B.J. Mason, The Physics of Clouds (The Clarendon Press, Oxford, 1957).
[77] M. Volmer and H. Flood, Z. physik. Chem. A170 (1934) 273.
[78] J.L. Katz and B.J. Ostemier, J. Chem. Phys. 47 (1967) 478.
[79] J.L. Katz, J. Chem. Phys. 52 (1970) 4733.
[80] H.L. Jaeger, E.J. Willson and P.G. Hill, J. Chem. Phys. 51 (1969) 5389.
[81] D.B. Dawson, E.J. Willson, P.G. Hill and K.C. Russell, J. Chem. Phys. 51 (1969) 5389.
[82] F.F. Abraham, J. Chem. Phys. 50 (1969) 3977.
[83] D. Turnbull and R.E. Cech, J. Appl. Phys. 21 (1950) 804.
[84] D. Turnbull, J. Appl. Phys. 21 (1950) 1022.
[85] D.G. Thomas and L.A.K. Staveley, J. Chem. Soc. 1952, 4569.
[86] H.J. de Nordwall and L.A.K. Staveley, J. Chem. Soc. 1954, 224.
[87] E.R. Buckle and A.R. Ubbelohde, Proc. Roy. Soc. A259 (1960) 325.
[88] E.R. Buckle and A.R. Ubbelohde, Proc. Roy. Soc. A261 (1961) 196.
[89] A. van Hook, Crystallization: Theory and Practice (Reinhold Publ. Corp., New York, 1961).
[90] E.V. Khamskii, Kristallizatsiya iz Raztvorov (Izd. Nauka, Leningrad, 1967).
[91] A.G. Walton, Mikcrochim. Acta 3 (1963) 422.
[92] A.E. Nielsen, J. Phys. Chem. Solids, Suppl. I, Crystal Growth 1967, p. 419.
[93] M. Kahlweit, Z. physik. Chem. NF 28 (1961) 245.
[94] A.E. Nielsen and O. Söhnel, J. Crystal Growth 11 (1971) 233.
[95] J.J. Burton, Surface Science 26 (1971) 1.
[96] A. Bonissent, Thesis, Université de Provence, Marseille (1971).
[97] M. Bever, in: Energetics in Metallurgical Phenomena, Vol. 1 (Gordon and Breach, New York, 1965) p. 113.
[98] B.Ya. Lyubov, Kineticheskaya Teoriya Fazovikh Prevrashchenii (Izd. Metallurgiya, Moskva, 1969).

P. Hartman, ed., Crystal growth: an introduction. © (1973) North-Holland Publ. Co.

HETEROGENEOUS NUCLEATION AND CONDENSATION ON SUBSTRATES

B.K. CHAKRAVERTY

2.1. Introduction

Fundamental aspects of heterogeneous nucleation and condensation from a vapor phase on substrates form the subject matter of this chapter. We shall not try to make an extensive review of the subject, as one already exists in the excellent monograph by Pound and Hirth [1]. A more recent review of the ensemble of nucleation problems as treated by Pound, Lothe, Sigsbee, Walton and others, appears in the volume edited by Zettlemoyer [2]. Another recent extensive review is the one by Frankl and Venables [3]. Neither shall we attempt to give detailed mathematical formalisms, for which one can consult the published articles cited in the references.

Our aim is to provide a concise, coherent and relatively simple treatment of the heterogeneous nucleation phenomenon, with special appeal to some of the more recent, rather precise, quantitative experimental results. In this latter domain, the novel use of the mass spectrometer has truly opened up a vast new capability for experimental measurement that will certainly challenge many of the established ideas of nucleation theory, in the years to come.

2.2. Basis of heterogeneous nucleation theory

2.2.1. Introduction

The free energetic aspects of the nucleation phenomenon in homogeneous systems, has already been treated in ch. 1 by Toschev. If we are considering vapor-liquid condensation in a homogeneous system (i.e. in the absence of a foreign surface, on which condensation can catalyse), the phase transition is a first order process, and is initiated by intense localised density fluctuations in the supersaturated medium. These fluctuations lead to the formation of liquid-like droplets (clusters, in thermodynamic language) and form the starting point for the theoretical treatment of imperfect gases by the cluster integral method of Mayer. Statistical mechanics, as well as the free energy of formation of these clusters in a homogeneous nucleation system, has been the subject, more recently, of an extensive series of studies by Pound, Lothe, et al., and these will not be treated here.

In the case of heterogeneous nucleation, the clusters, if they do form, do not form in the gas phase but, by necessity, on the substrate within the matrix of an adsorbed phase of the supersaturated vapor. There is, in reality, very little experimental evidence as to whether these clusters of the condensed phase are liquid-like or solid-like. This leads to a certain amount of error and ambiguity in the calculation of the free energy of formation of the cluster (or nucleus, as it is often called).

In the subsequent treatment we shall make no attempt to remove this ambiguity and will treat the heterogeneous nucleation as a temporal sequence of two processes:

(A) Formation of an adsorbed layer on the substrate from the vapor phase.

(B) Formation of a critical nucleus (when it can be defined) from the adsorbed atoms and its growth to supercritical size.

These two processes, i.e. nucleation and growth, can be juxtaposed in time and the amount of the condensed phase at any time will be naturally determined by some sort of convolution.

2.2.2. Adsorbed layer formation

Vapor atoms impinging on the substrate are captured temporarily by the local potential wells. After a life time τ_s on the substrate they re-evaporate. At steady state, a vapor source operating under a pressure p establishes a steady state concentration of ad-atoms $c(1)$ per unit area on the substrate. Depending on the temperature T of the substrate and the strength of the attractive potential, these adsorbed atoms could be completely localised each in their potential well, or quite mobile along the substrate; in the latter case they may be considered to behave as a gas. For each substrate temperature the relation between p and $c(1)$ is a very complex one and can be, in the extreme limit, of two types – a localised isotherm of the Fowler type [4] or a mobile isotherm of the Hill-de Boer type [5].

Let us consider the mobile adsorbed gas-atoms, behaving as a two-dimensional gas with pressure φ, to exist as a two-dimensional analogue of the Van der Waals equation of state for an imperfect gas

$$\left(\varphi + \frac{a' n'^2}{A^2}\right)(A - n'b) = n'kT, \tag{2.1}$$

where: A is the area of the substrate on which n' atoms are adsorbed, b is the area per adsorbed atom, and a' is a parameter, which can be determined by the Virial theorem of Clausius.

Following Halsey et al. [6], we can use a hard-sphere interaction potential like

$$\begin{aligned} \epsilon(\rho') &= -\epsilon_0 \text{ for } \rho' = \rho_0' \\ \epsilon(\rho') &= \infty \quad \text{for } \rho' < \rho_0' \\ \epsilon(\rho') &= 0 \quad \text{for } \rho' > \rho_0' \end{aligned}$$

where ρ_0' is the distance of minimum approach between two adsorbed atoms, and obtain a where:

$$a = \frac{a'}{kT} = \frac{1}{2} \int_{\rho_0'}^{\infty} \frac{\epsilon(\rho')}{kT} 2\pi\rho' \mathrm{d}\rho' = \frac{\pi(\rho')^2 \epsilon_0}{4kT}$$

and

$$b = \tfrac{1}{2}(\rho_0')^2\sqrt{3}.$$

The relationship between the number of adsorbed atoms per unit area and over-all gas pressure p is obtained from the Gibbs adsorption equation which has the form:

$$\frac{\mathrm{d}\varphi}{\mathrm{d}\ln p} = kTc(1), \tag{2.2}$$

where $c(1)$ is the number of adsorbed atoms/unit area.
Using eq. (2.2) in (2.1) and integrating gives us:

$$\frac{p}{p_0} = k_2 \frac{\theta}{1-\theta} \mathrm{e}^{\theta/(1-\theta)} \mathrm{e}^{-k_1\theta}, \tag{2.3}$$

where $k_1 = \dfrac{2a}{b} = \dfrac{\pi\epsilon_0}{\sqrt{3}kT}$; $\qquad \theta = \dfrac{n'b}{A}$ = fractional coverage

The constant k_1 depends only on the adsorbed atoms and their interatomic attraction and is independent of the substrate material.

The constant k_2 on the other hand measures mainly the adsorption strength of the surface and is seen to be, for small θ,

$$\frac{p}{p_0} \approx k_2\theta \text{ or } \left[\frac{\mathrm{d}\theta}{\mathrm{d}(p/p_0)}\right]_{\theta\to 0} = \frac{1}{k_2} \approx \left[\frac{\theta}{p/p_0}\right]_{\theta\to 0}.$$

Thus the smaller is k_2, the larger is the amount of adsorption for a given increment of external pressure. p_0 is the equilibrium vapor pressure of the adsorbed substance. Eq. (2.3) is the Mobile Adsorption Isotherm. Some typical isotherms (taken from de Boer [5]) are shown in fig. 2.1. These same isotherms can also be obtained by means of the Bragg-Williams approximation of nearest neighbour pairs.

We note that k_1 is a direct measure of the sublimation energy of the condensed phase and k_2 is inversely proportional to the adsorption strength. Let us use $k_2 = 5$ and $k_1 = 1, 6, 6.5, 7, 7.5$ and 10. We see for a given ϵ_0, that as we decrease the temperature, i.e. k_1 increases, a region of instability appears in the $\theta - p$ diagram, which represents the onset of condensation into a

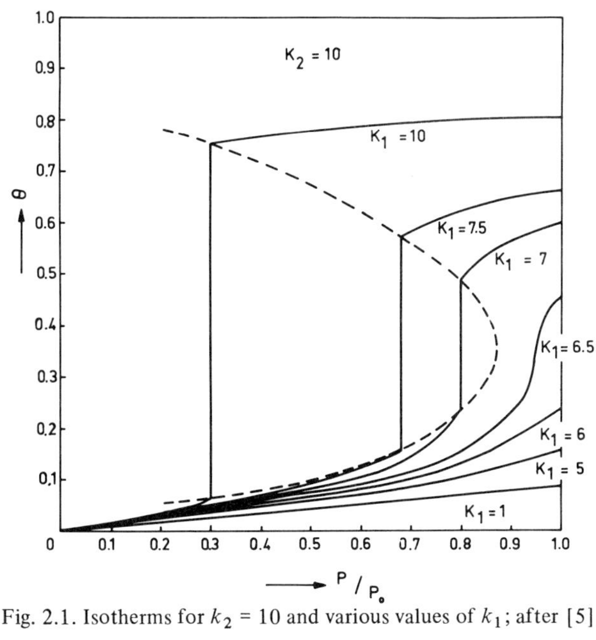

Fig. 2.1. Isotherms for $k_2 = 10$ and various values of k_1; after [5].

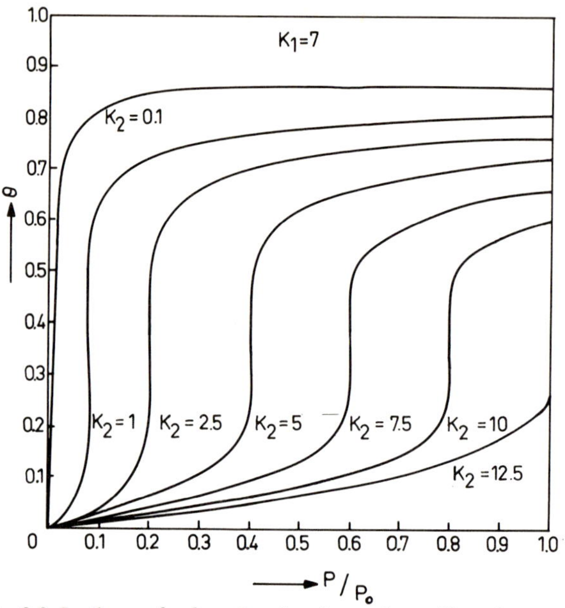

Fig. 2.2. Isotherms for $k_1 = 7$ and various values of k_2; after [5].

liquid phase at around $k_1 = 6.5$. The most interesting feature is that the liquid phase can form at values of p/p_0 less than unity, i.e. when the gas is undersaturated.

These isotherms do not give the gas-liquid co-existence lines, but only the Van der Waals loops – a result to be anticipated, because our starting point was the Van der Waals equation of state. If instead one uses the Yang-Lee [7] theory of condensation, one obtains directly the successive co-existence lines in the plots.

Equally we can keep k_1 fixed at say 7, and vary k_2 (fig. 2.2). We see once again that as the adsorption strength increases, i.e. k_2 decreases, liquid phase condensation can start at undersaturation. It is a layer by layer condensation mechanism. On the other hand if $k_2 = 100$, the value of p/p_0 at which liquid condensation sets in is ≈ 3.0, as seen in the fig. 2.3. In this case definite supersaturation is necessary before condensation can begin. A supersaturated vapor is a prerequisite for nucleation of the condensed phase to occur, because work is necessary to create new interfaces and this is furnished by the free energy stored in a supersatured vapor. Following the relative values of k_1 and k_2, one can anticipate three distinctly different condensation modes:

1) High desorption energy i.e. small k_2 compared to the heat of sublimation

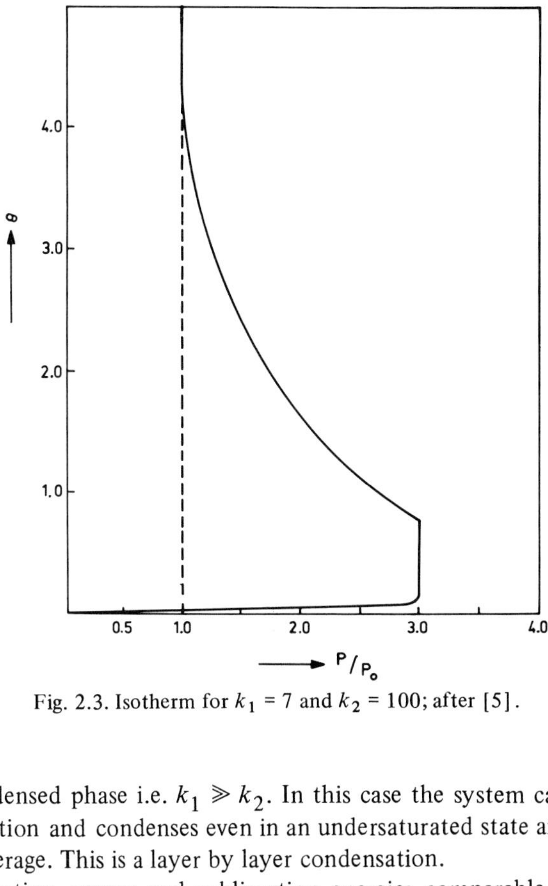

Fig. 2.3. Isotherm for $k_1 = 7$ and $k_2 = 100$; after [5].

of the condensed phase i.e. $k_1 \gg k_2$. In this case the system cannot achieve supersaturation and condenses even in an undersaturated state and at relatively high coverage. This is a layer by layer condensation.

2) Desorption energy and sublimation energies comparable i.e. $k_1 \approx k_2$. Here the coverage is less than a monolayer and the supersaturation is moderate. This is the regime of classical nucleation.

3) When the desorption energy is very low compared to the sublimation energy, the coverage is extremely low ($\approx 10^{-6}$), the supersaturation is extremely high and heterogeneous nucleation theory, which is a macroscopic thermodynamic theory, is difficult to apply. One should use some form of atomistic theory: Zinsmeister [8], Lewis-Campbell [9, 10], Walton-Rhodin [11], Kikuchi [12]. The modes (1) and (3), although very interesting, will not be discussed here and we shall only review the mode (2) of condensation which is by heterogeneous nucleation. The atomistic nucleation theory is discussed in ch. 2 §3.3.

2.2.3. Kinetics of the nucleation process

We have just seen that with an increase of supersaturation less than a monolayer of adsorbed atoms builds up on the substrate. It is natural that these adsorbed atoms will interact with each other and form atom clusters which, in their turn, will break up or grow bigger in a dynamic equilibrium. This equilibrium is represented most simply by the chemical equation: $iA \rightleftharpoons (A_i)$

$$K_i = \frac{c(i)}{\Sigma_{i=1} c(i)} \approx \frac{c(1)}{c(1)} \text{ for } c(1) \gg c(i) \tag{2.4}$$

Here K_i is an equilibrium constant. If the free energy of formation of a cluster of size i is ΔG_i we must have:

$$K_i = \exp(-\Delta G_i/kT)$$
$$\text{or } c(i) = c(1)\exp(-\Delta G_i/kT).$$

This free energy of formation is generally written as a sum of two terms:

$$\Delta G_i = \Delta G_{\text{supersaturation}} + \Delta G_{\text{surface}}, \tag{2.5}$$

where, in the process of formation of a cluster of size i from i single atoms, the cluster gains free energy, so to speak by the amount the gas is supersaturated and loses it in the form of work necessary to create the new surface. We assume that the cluster is a segment of a sphere (a spherical cap*) on a substrate

$$\Delta G_{\text{supersaturation}} = \Delta G_\text{V} \tfrac{4}{3}\pi r^3 f_1(\theta), \tag{2.6}$$

where ΔG_V is the supersaturation free energy available per unit volume of the condensed phase given by:

$$\Delta G_\text{V} = \frac{-RT}{V} \ln\left(\frac{p}{p_0}\right),$$

* The artifice of the spherical cap has been succesfully used by Pound to derive the three-dimensional nucleation kinetics for vapor deposition on a substrate. For later use we write down expressions for: the volume of the cap = $\frac{4}{3}\pi r^3 f_1(\theta)$, the surface area of the cap = $4\pi r^2 f_2(\theta)$ and the periphery of the cap = $2\pi r \sin\theta$, where $f_1(\theta)$ and $f_2(\theta)$ are functions of the contact angle θ, being namely: $f_1(\theta) = \frac{1}{4}(2 - 3\cos\theta + \cos^3\theta)$ and $f_2(\theta) = \frac{1}{2}(1 - \cos\theta)$, cf. fig. 2.22.

where V is the molar volume of the condensed phase. We note that ΔG_V is negative.

$$\Delta G_{\text{surface}} = 4\pi r^2 f_2(\theta)\sigma_{c-v} + \pi r^2 \sin^2\theta(\sigma_{c-s} - \sigma_{s-v}), \tag{2.7}$$

where σ_{c-v} is the cluster-vapor interfacial energy per unit area of the new surface created. σ_{c-s} and σ_{s-v} are the corresponding cluster-substrate and substrate-vapor interfacial free energy per unit area of substrate and $f_2(\theta)$ is the geometric function of the contact angle defined in the note on the spherical cap.

We note that σ_{c-v}, σ_{c-s} and σ_{s-v} are all positive quantities and are interrelated by:

$$\sigma_{s-v} = \sigma_{c-s} + \sigma_{c-v}\cos\theta \; .$$

We note also that $\Delta G_{\text{surface}}$ is positive.
Using (2.6) and (2.7), we obtain:

$$\Delta G_i = \Delta G_V \frac{\pi r^3}{3}(2 - 3\cos\theta + \cos^3\theta) + \sigma_{c-v}\pi r^2(2 - 3\cos\theta + \cos^3\theta), \tag{2.8}$$

where $i = \dfrac{\pi r^3}{3\Omega}(2 - 3\cos\theta + \cos^3\theta)$,

Ω being the atomic volume in the condensed phase.

Extension of this expression to nucleation on substrate imperfections and ledges has also been considered (Chakraverty and Pound [13], Fletcher [14]).

Eq. (2.8) gives us the free energy of formation of a cluster of size i. If this free energy of formation is negative, such a cluster of size i would be spontaneously created in the system and would have an infinite life-time. If it is positive such a cluster will be created only by some thermally-activated or fluctuation process and will have a finite life-time after which it will break up. When we trace ΔG_i as a function of its size, we see that up to a certain $r = r^*$ (i.e. $i = i^*$), the cluster free energy increases and reaches a maximum (fig. 2.4a), after which it rapidly drops.

For a single adsorbed atom to form a microscopic cluster, which in turn could grow bigger to give a macroscopic phase, it is evident that, at some stage of its evolution, it will have to go through this maximum energy barrier of formation and it is finally this difficulty that limits the rate of formation of a new stable phase. The cluster of size $i = i^*$ is called the critical cluster or the critical nucleus.

In any nucleation process each cluster, in order to become stable and grow further, must necessarily attain the critical nucleus size by some thermally-activated process before the cluster can become super-critical and hence become a growth center. The frequency of nucleation is the rate at which critical nuclei become super-critical per unit time and per unit area of the substrate and one measures it simply by counting the rate of appearance of stable particles on the substrate. The necessity to achieve $i = i^*$, is the true bottleneck of the phase transformation.

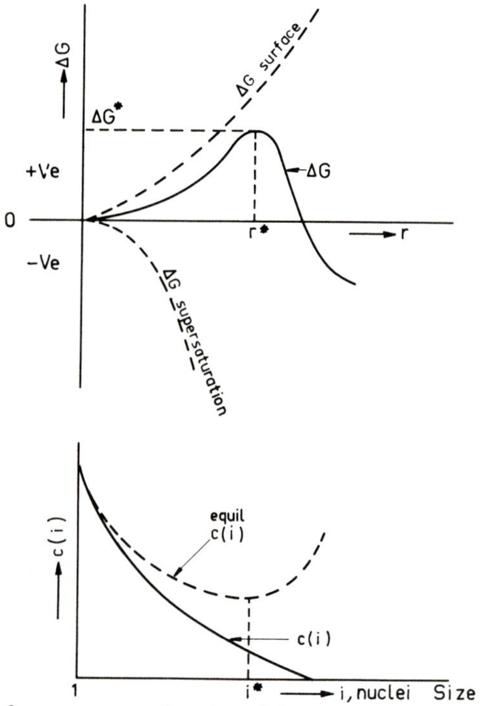

(a)

(b)

Fig. 2.4. a) Cluster free energy as a function of cluster radius; b) number of clusters $c(i)$ as a function of nucleus size.

We consider the stable nuclei to appear from the supersaturated ad-atoms by a process of bi-molecular reaction. We can represent the dynamics of the process, as a diffusion phenomenon (Chakraverty [15]) on the i-axis and write the appropriate continuity equation as:

$$\left[\frac{\partial c(i,t)}{\partial t}\right]_{\text{diffusion}} = \frac{-\partial}{\partial i}\left[-D_i \frac{\partial c(i,t)}{\partial i} + \frac{D_i}{kT} c(i,t) \frac{\mathrm{d}(\Delta G_i)}{\mathrm{d}i}\right], \tag{2.9}$$

where the terms within the bracket on the right hand side give the net flux of particle points flowing along the i-axis i.e.:

$$J_i = -D_i \frac{\partial c(i,t)}{\partial i} + \frac{D_i}{kT} c(i,t) \frac{\mathrm{d}(\Delta G_i)}{\mathrm{d}i} . \tag{2.10}$$

Here $c(i, t)$ is the number of clusters of size i at time t; D_i is a diffusion coefficient.

The flux J_i through the bottleneck i^*, per cm^2 per sec is the rate of nucleation i.e. the number of stable nuclei forming per cm^2 per sec. If one assumes a steady state nucleation frequency (fig. 2.4b),

$$J_i = J = J_{i^*} = \left[-D_i \frac{\partial c(i,t)}{\partial i} + \frac{D_i}{kT} c(i,t) \frac{\mathrm{d}(\Delta G_i)}{\mathrm{d}i} \right]_{i=i^*} \tag{2.11}$$

and one obtains from eq. (2.10)

$$J = Z\, \omega\, c(i^*), \tag{2.12}$$

where Z is the Zeldovitch factor, ω is the probability of addition of a single ad-atom to a critical nucleus so as to make it super-critical and hence stable. Eq. (2.12) is the central result of all nucleation theories. There is one characteristic of eq. (2.12) that is of the most fundamental importance. This has to do with a strong dependence of nucleation rate on the supersaturation ratio p/p_0.

Although for every p/p_0 a critical nucleus can be found, the rate of nucleation is not necessarily appreciable and no phase change occurs until a critical threshold value of supersaturation p/p_0 is attained. This behavior is clearly seen if we trace the adsorbed atom population c (1) with time (fig. 2.5a and b) remembering that each value of $c(1)$ is associated with a supersaturation p/p_0. When the value of c (1) remains below $c(1)_{\mathrm{cr}}$, no appreciable nucleation occurs and the adsorbed atom population attains a steady state value governed by desorption. However when $c(1)$ reaches $c(1)_{\mathrm{cr}}$ (hence $(p/p_0)_{\mathrm{cr}} = 1$), rapid nucleation begins, the ad-atom population reaches a maximum and, because numerous attachment centres for ad-atoms are now available, the continual growth of nuclei begins to deplete the ad-atoms population until it falls below the threshold value for nucleation when all fresh nucleation stops. The value $c(1)_{\mathrm{cr}}$ can be experimentally measured.

In order to complete the development of the classical nucleation theory one must consider both the nucleation frequency, as well as the growth of

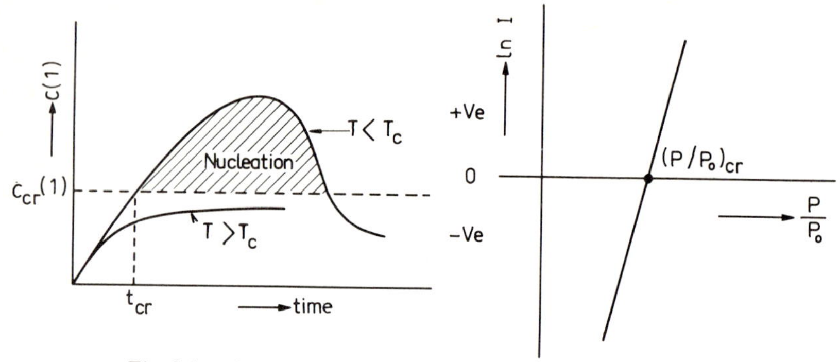

Fig. 2.5. Adsorbed atom population $c(1)$ as a function of time.

individual nuclei as time-dependent processes. We shall see, in ch. 2 §5, the profound implication of this time-dependence on the physics of coalescence phenomena. In general, however, the "observable" in a classical nucleation experiment is often the mass deposited as a function of the fraction of the surface covered with nuclei. The free surface remaining at time t can be written as an integral equation

$$1 - \theta(t) = \int_0^t J(\tau)\, \theta(t-\tau)\, g(t-\tau)\, \mathrm{d}\tau, \tag{2.13}$$

where $\theta(t)$ is the surface covered at time t, $J(t)$ is the time-dependent nucleation frequency, $g(t)$ is the size of the nucleus at t, and τ is a dummy time variable.

The required function $\theta(t)$ can be obtained analytically for a variety of simple $J(t)$ and $g(t)$, by the method of Laplace transformation, and the theorem that the Laplace transformation of a convolution is equal to a product of Laplace transforms. To give examples, we solve it for two physically distinct cases:

Case (a), growth is controlled by the surface diffusion of ad-atoms only:

$$\frac{\mathrm{d}r^2}{\mathrm{d}t} = \text{constant i.e. } g(t) = \alpha_1 t \text{ and assume } J(t) = J_0.$$

If we define

$$\theta(s) = \int_0^\infty \theta(t)\, \mathrm{e}^{-st}\, \mathrm{d}t,$$

we get

$$\theta(s) = \frac{s}{s^2 + \alpha_1 J_0}.$$

This gives

$$\theta(t) = \cos\beta_1 t,$$

where $\beta_1 = \sqrt{\alpha_1 J_0}$.

Case (b), growth of the nucleus is controlled by interface attachment kinetics:

$$\frac{dr}{dt} = \text{constant, i.e. } g(t) = \alpha_2 t^2.$$

Assuming $J(t) = J_0$,

we have $\theta(s) = \dfrac{s^2}{\beta_2^3 + s^3}$

and we get

$$\theta(t) = \frac{1}{3}\left[e^{-\beta_2 t} + 2 e^{\beta_2 t} \cos\left(\frac{\sqrt{3}}{2}\beta_2 t\right)\right],$$

which at $t \to 0$, behaves as

$$\theta(t) = 1 - \frac{\beta_2^3 t^3}{6}.$$

Low energy electron diffraction measurements (Dufayard and Chakraverty [16]) on condensation of a metal on a metal can often be used to obtain $\theta(t)$ which can be compared with the analytically expected time development.

2.3. Some experimental results and difficulties of classical nucleation theory

2.3.1. Experimental results

One can distinguish three distinct regions of supersaturations and substrate-ad-atom interactions:

a) moderate supersaturation	Classical theory works
b) high supersaturation	Atomistic theory works
c) very low supersaturation	no adequate theory exists

Classical nucleation theory, as developed in ch.2 §2 predicts two things; a threshold nucleation frequency, (eq. (2.12)) and a critical ad-atom concentration $c(1)_{\mathrm{cr}}$.

The expression for the nucleation frequency, when written out simply, is, for the three dimensional cap-shaped nucleus (Pound et al. [1]),

$$J = \beta \exp \frac{\Delta G_{\mathrm{des}} - \Delta G_{\mathrm{diff}}}{kT} \exp \frac{-\Delta G_{i*}}{kT}, \tag{2.14}$$

where

$$\Delta G_{i*} = \frac{16\pi\sigma_{\mathrm{c-v}} f_1(\theta)}{3(\Delta G_{\mathrm{v}})^2} \quad \text{(cf. eq. 1.22).}$$

ΔG_{des} and ΔG_{diff} are the activation free energies for desorption from and diffusion of an ad-atom on the surface, and β is a temperature independent constant. ΔG_{i*} is obtained from eq. (2.8) by putting $\partial \Delta G_i / \partial r = 0$, which

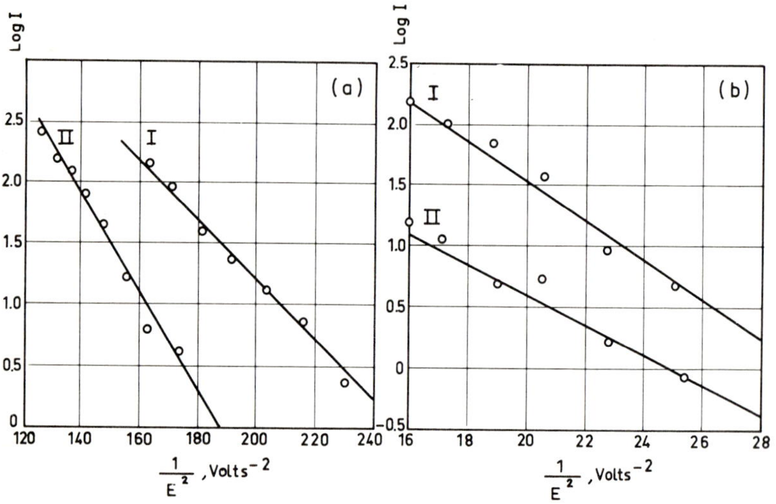

Fig. 2.6. Plots of eq. (2.18).

gives a value for r^* the radius of the critical nucleus, and by calculating ΔG_{i^*} at $r = r^*$. To test the validity of this equation one can either keep the substrate temperature constant and vary the supersaturation or vary both.

In the first case,

$$J = \beta' \exp\left(-\frac{16\pi\sigma_{\text{c-v}}^3 f_1(\theta)}{3(\Delta G_{\text{v}})^2 kT}\right). \tag{2.15}$$

In the second case one usually constructs a "Pound Plot" by rewriting equation (2.14) in the form:

$$\frac{1}{(\Delta G_{\text{v}})^2} = \frac{3}{16\pi\sigma_{\text{c-v}}^3 f_1(\theta)} kT \ln\left(\frac{\beta}{J}\right) + \frac{3}{16\pi\sigma_{\text{c-v}}^3 f_1(\theta)} (\Delta G_{\text{des}} - \Delta G_{\text{diff}}) \tag{2.16}$$

In the "Pound Plot" (see fig. 2.7) one plots $1/(\Delta G_{\text{v}})^2$ against $k\,T \ln(\beta/J)$, both being experimentally measurable. Linearity of this plot confirms the validity of the nucleation equation.

Both of these methods have been verified. The simplest experiment one can do is to electrodeposit the metallic ions contained in an electrolyte onto a cathode of a different metal. This was done in an extremely elegant experiment by Mutaftschiev and Kaischew [17] where they electrodeposited Hg and Ag on platinum. This is one of the most quantitative nucleation experiments to have been performed in electrolysis. The driving force ΔG_{v} is proportional to the externally applied electric potential E and can be written as:

$$\Delta G_{\text{v}} = \beta E. \tag{2.17}$$

For a constant substrate temperature eq. (2.15) gives:

$$\ln J = \ln \beta' - A/E^2. \tag{2.18}$$

Mutaftschiev measured J as a function of E, and the results are shown in fig. 2.6. We see that the nucleation equation is completely verified.

In the last few years most heterogeneous nucleation experiments have been performed in ultra-high vacuum on extremely clean substrates. Kenty [18, 19] and Ruth, Moazed and Hirth [20] have reconstructed a whole mass of data on the basis of Pound Plots. Kenty, for example, using the nucleation experiments of Walton, Rhodin and Collins, for Ag on NaCl, has shown that classical nucleation theory quite adequately describes the data. In these experiments, critical supersaturation was measured as soon as a deposit was visible optically.

A typical set of results, from Kenty's work, is shown in fig. 2.7. Once again the classical nucleation expression is quite well obeyed. Pound Plots are difficult to construct but they are the only valid way of interpreting the variation of J with substrate temperature. A linear variation of the logarithm of the nucleation frequency with reciprocal temperature is more often than not quite fortuitous.

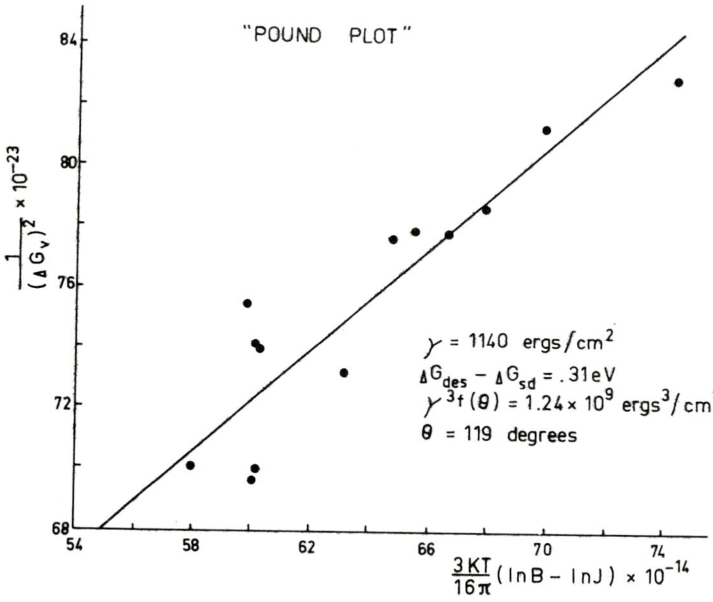

Fig. 2.7. Pound plot for nucleation of Ag on NaCl; after Kenty [18, 19].

Table 2.1

Calculation of various quantities related to the critical nucleus on the basis of classical nucleation theory

substrate temperature	S	ΔG_v erg/cm^3	r^* Ångström	ΔG^* (90°) eV.	ΔG^* (45°) eV.	n^* (90°) /cm^2	n^* (45°) /cm^2
400°C	2×10^6	7.8×10^{10}	2.46	0.86	0.10	3.55×10^8	1.78×10^4
425°C	5×10^5	7.42×10^{10}	2.7	0.95	0.11	1.38×10^8	1.6×10^4
	5×10^4	6.13×10^{10}	3.26	1.39	0.161	8.5×10^4	6.85×10^{13}
475°C	1.46×10^4	5.82×10^{10}	3.44	1.54	0.179	3.8×10^4	6.25×10^{13}
	4.2×10^3	5.06×10^{10}	3.95	2.05	0.238	14	2.5×10^{13}
533°C	725	4.31×10^{10}	4.64	2.8	0.325	4.3×10^{-3}	9.25×10^{12}
	138	3.22×10^{10}	6.2	5.05	0.58	10^{-13}	2.38×10^{11}
592°C	36	2.51×10^{10}	8	8.3	0.96	ϵ	2.5×10^9
	4.3	1.02×10^{10}	19.6	50	5.8	ϵ	ϵ
640°C	4	0.74×10^{10}	27	96	11.1	ϵ	ϵ

2.3.2. Critical ad-atom concentration

We have mentioned that nucleation begins when the ad-atom concentration on the substrate reaches a critical value. The best series of experiments demonstrating this fact was performed by Gretz [21] using a field emission microscope. In a typical field emission experiment one exposes a tungsten tip to a beam of metal vapor atoms which adsorb on the tip. One then applies an electric field to the tip which emits electrons. If a condensed phase forms on the tip, one sees the electron emission, due to this phase, as a bright spot on the fluorescent screen of the collector. Gretz evaporated Zn, Au, Cd and Ni on tungsten. In each experiment, for a constant substrate temperature and for a given flux J_v, he noted the time at which the bright spot appeared. Then he varied the beam flux without varying the substrate temperature and he obtained a curve of beam flux J_v versus reciprocal time necessary for a bright spot, due to a condensed phase, to appear. The curve is shown in fig. 2.8. Thus the critical value of $c(1)$, at which nucleation begins, $c(1)_{cr} = J_v t =$ constant,

$$\text{or } J_v = \frac{\text{const}}{t}. \tag{2.19}$$

Although these experimental results partly confirm the validity of the classical nucleation theory, more recent results on condensation in ultra-high vacuum cast doubt on the wisdom of indiscriminately using the classical capillarity idea for all supersaturation ranges.

This is most strikingly seen in the recent extensive series of measurements on the condensation of Ag on fused quartz using a microbalance technique, by Cinti [22]. In a typical experiment one can measure, with a high degree of accuracy, the condensation coefficient as a function of time, substrate temperature and impingement flux (in fig. 2.8, the impingement flux J_v is plotted against reciprocal nucleation time). It is seen that condensation is a monotonic function of time and that there is neither a critical supersaturation nor a critical flux density for nucleation to begin. For the experimental substrate temperatures and for the assumed contact angles of 90° or 45° with $\sigma \approx 1000$ erg/cm^2, table 2.1 shows the possible values of $\Delta G_v{}^*$, r^* and n^*, where n^* is the number of critical nuclei/cm^2, calculated on the basis of classical nucleation theory. It is evident from this table that classical nucleation theory can predict little or no nucleation for this system in complete contradiction to the experimentally-observed abundant three-dimensional nuclei seen on the substrates even at the lowest supersaturation. One way out, at least at the

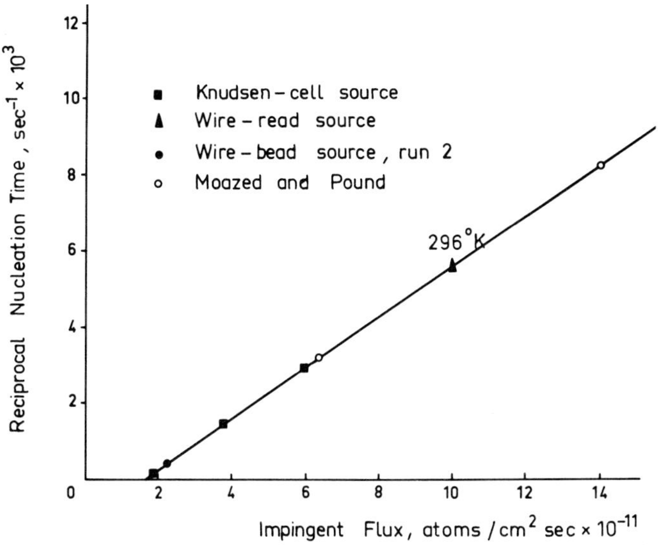

Fig. 2.8. Nucleation of metal vapours on a tungsten tip in a field emission microscope; cf. eq. (2.19); after Gretz [21].

high supersaturation end, is to use the atomistic theory, which we shall outline next. For the very low supersaturation range the theoretical situation is far from clear and it is discussed in ch. 2 §4.

2.3.3. Atomistic nucleation theory

The rather elaborate development of the atomistic nucleation theory in the last few years derives from the original work of Walton and Rhodin [11] that was subsequently woven into a coherent framework by Zinsmeister [8] and by Lewis and Campbell [10], amongst others. Walton and Rhodin define, for a cluster of i-atoms, *a binding energy* E_i necessary to dissociate such a cluster, *instead of defining a free energy* ΔG_i as is done in the classical theory. The number of such clusters $c(i)$ per unit area is then

$$\frac{c(i)}{N_0} = \left(\frac{c(1)}{N_0}\right)^i \exp\left(\frac{E_i}{kT}\right), \tag{2.20}$$

where N_0 is the number of sites/cm^2 and $c(1)$ is the number of ad-atoms/cm^2.

The nucleation frequency expression (2.12) is then used to obtain an analogous expression for the atomistic theory:

$$J = J_v\, \alpha_0^2\, N_0 \left(\frac{J_v}{\nu N_0}\right)^i \exp\left[\frac{(i^*+1)Q_{ad} + E_{i^*} - \Delta G_{diff}}{kT}\right], \tag{2.21}$$

where α_0 is a typical atomic dimension, ν is the Debye vibration frequency $\approx 10^{12}$Hz and Q_{ad} and ΔG_{diff} are the adsorption energy and the diffusion activation energy of the ad-atom respectively and i^* is the size of the critical nucleus.

Using this equation, one can show that for an incidence rate, J_v, of one monolayer per second and for $Q_{ad} \leqslant 3E_i/i$ the nucleation frequency is negligible for a critical nucleus containing seven atoms.

In the limit of high supersaturation the critical nucleus should be a single atom. In other words each time a pair is formed it will, on average, grow rather than decay into two single atoms. In that case, the nucleation rate is:

$$J_1 = J_v \left(\frac{J_v}{\nu N_0}\right) \exp\left[\frac{2Q_{ad} - \Delta G_{diff}}{kT}\right]. \tag{2.22}$$

As the substrate temperature is increased, a temperature is reached at which the pair is no longer stable. The next smallest stable cluster is one that has a minimum of two bonds per atom. One way of achieving a minimum of

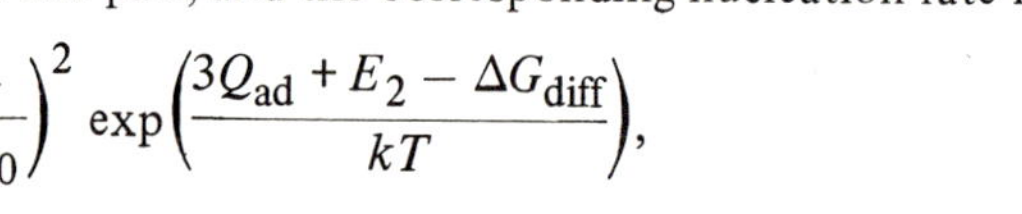

Fig. 2.9. Schematic models of two-dimensional critical nuclei.

two bonds per atom is to have three atoms in a triangle (fig. 2.9b). The critical nucleus is a pair, and the corresponding nucleation rate is

$$J_2 = J_v \left(\frac{J_v}{\nu N_0}\right)^2 \exp\left(\frac{3Q_{ad} + E_2 - \Delta G_{diff}}{kT}\right), \tag{2.23}$$

where E_2 is the dissociation energy of the pair.

Another way of achieving a minimum of two bonds per atom is shown in fig. 2.9d, with four atoms in a square. The critical nucleus for this arrangement can be taken as the configuration of three atoms, shown in fig. 2.9c. Its nucleation rate is

$$J_3 = J_v \left(\frac{J_v}{\nu N_0}\right)^3 \exp\left(\frac{4Q_{ad} + E_3 - \Delta G_{diff}}{kT}\right), \tag{2.24}$$

where E_3 is the dissociation energy of the three atoms shown in fig. 2.9c.

The temperatures T_1 and T_2 at which transition occurs from a one-atom to a two-atom critical nucleus (a) or from a two- to three-atom critical nucleus (c), can be obtained by equating J_1 and J_2, and J_2 and J_3, respectively, to obtain

$$T_1 = -\frac{(Q_{\text{ad}} + E_2)}{k \ln(J_v/\nu N_0)} \tag{2.25}$$

and

$$T_2 \approx -\frac{(Q_{\text{ad}} + \frac{1}{2}E_3)}{k \ln(J_v/\nu N_0)}. \tag{2.26}$$

These transition temperatures can be determined from experimentally observed discontinuities in the nucleation rate versus $1/T$ plot and can give some quantitative information about the relevant activation energies.

In the formulation of the atomistic theory of Zinsmeister and others, the idea of the critical nucleus is altogether abandoned and the condensation process is visualised essentially as a kinetic polymerisation process. Atom clusters form and decay by bi- or multiatomic collisions giving rise to the formation of pairs and higher aggregates. These processes are described by the following set of equations:

$$\frac{dn_1}{dt} = q' + \sum_{i>1} n_i\beta_i - \frac{n_1}{\tau_s} - n_1 \sum_{i\geqslant 1} w_{12}n_i, \tag{2.27}$$

$$\frac{dn_2}{dt} = \tfrac{1}{2} w_{11}n_1^2 + \sum_{i>2} n_i\beta_i - w_{12}n_1n_2 - n_2\alpha_2.$$

where q' is the rate of impingement from the vapor, β_i is the rate of decay of cluster i, $\Sigma n_i\, \beta_i$ the number of single atoms formed by decay of clusters of i atoms, τ_s is the desorption time of single atoms from the substrate, $\Sigma w_{12}n_i$ the number of single atoms disappearing by collision with i-clusters and α_2 is the addition rate to clusters with $i = 2$. In this, we have neglected direct hits from the vapor phase. The collision factor w_{12} is governed by surface diffusion of ad-atoms, using the normal collision formula from the kinetic theory of gases ($w_{12} \approx \frac{1}{2}w_{11}(\sqrt{i} + 1)$, $w_{11} \approx 10^{-3}$). In the decay process of the aggregates, evaporation will be governed by the relevant dissociation energy of the clusters. The possibility of re-evaporation of clusters as a single unit (for $i > 2$) is neglected. For the calculation of Zinsmeister *the binding energy E_2 of a pair of atoms is of fundamental importance for the condensation*

Table 2.2

Binding energy of atom pairs

		Hg	Cd	K	Na	Zn	Ag	Cu	Au	Ge	SiO_2	ZnS	LiF	H_2O
Boiling point	°C	357	765	776	883	906	1897	2322	2600	2700				
Heat of sublimation														
Q_{subl}	kcal/mol	15.4	26.9	21.7	25.9	31.1	68.3	80.8	87.3	89.3	136	62	57	11.2
	eV	0.67	1.17	0.94	1.12	1.35	2.96	3.50	3.88	3.97	5.88	2.68	2.47	0.49
$\frac{1}{6}Q_{subl}$	eV	0.11	0.19	0.16	0.19	0.23	0.49	0.58	0.63	0.65	0.98	0.45	0.41	0.08
E_2 measured[8]	eV	0.06	0.09	0.52	0.76	0.26	1.6	1.9	2.23	2.82				

process. The value of E_2 can be estimated as one sixth of the heat of sublimation for the known metal and can be compared with values measured directly by mass-spectroscopic methods. Table 2.2 shows a list of values compiled for some common metals. Comparison of the figures in the two bottom rows of table 2.2 show that atom-pairs of the more common metals are considerably more stable than might have been expected on the basis of the simple estimate $\frac{1}{6}\,Q_{\text{subl}}$. The dissociation energies are in fact so high (> 1.5 eV) that the dissociation of such pairs at normal substrate temperatures of less than 400°C can be totally neglected. In other words, an atom-pair can serve as a stable nucleus, although in a non-thermodynamic sense. In the contrast, the metals like Hg and Cd that are generally difficult to condense, exhibit precisely the opposite behavior i.e. the actual binding energy is far lower than the theoretical estimate. For cases, where a pair is a very stable cluster one can neglect the dissociation of higher aggregates and cast the kinetic equation for the formation of pairs in the form

$$\frac{\mathrm{d}n_2}{\mathrm{d}t} = \tfrac{1}{2} w_{11} n_1^2 - w_{12} n_1 n_2 - \alpha_2 n_2, \text{ with } \alpha_2 \approx \frac{1}{\tau_2}. \tag{2.28}$$

At steady state (with $\mathrm{d}n_2/\mathrm{d}t = 0$), one therefore has

$$n_2 = \frac{\tfrac{1}{2} w_{11} n_1^2}{\alpha_2 + w_{12} n_1}. \tag{2.29}$$

Here

$$\frac{1}{\tau_2} = \nu \exp \frac{-E_2}{kT}.$$

The simplified set of equations are

$$\frac{\mathrm{d}n_1}{\mathrm{d}t} = q' - \frac{n_1}{\tau_s} - \sum_{i \geqslant 1} w_{12} n_1 n_i$$

$$\frac{\mathrm{d}n_2}{\mathrm{d}t} = \tfrac{1}{2} w_{11}\, n_1^2 - w_{12} n_1 n_2$$

$$\frac{\mathrm{d}n_i}{\mathrm{d}t} = w_{1,i-1}\, n_1 n_{i-1} - w_{12} n_1 n_i$$

This simplified set of equations can be directly integrated for the special case $n_1 = N_1$ = constant and $\Sigma w_{12} n_1 n_i = 0$. The accompanying fig. 2.10 shows the time-development of n_2, n_3, etc. An aggregate n_k reaches equilibrium

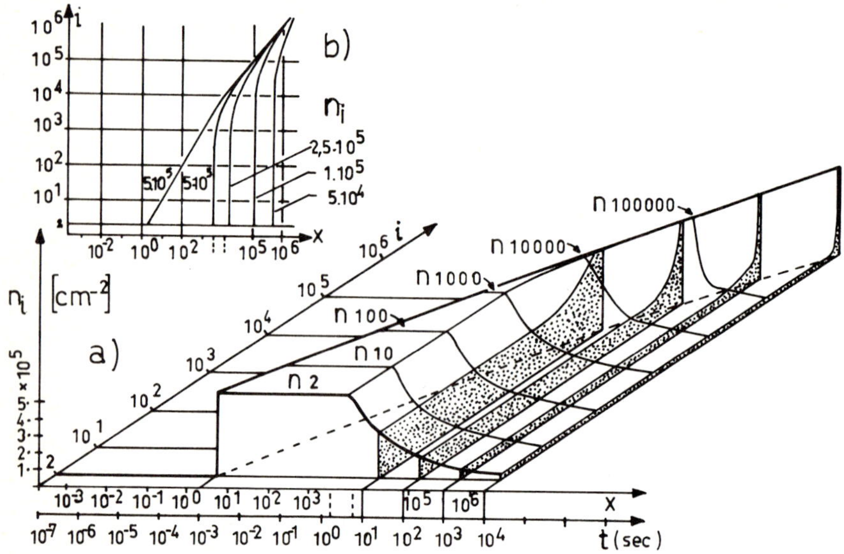

Fig. 2.10. A condensation process with $q = 10^{13}$, $\tau_A = 10^{-7}$, $w = 2 \times 10^{-3}$. a) Concentration n_i of the aggregates as a function of real time t or synthetic time x and the number i of atoms in an aggregate; b) top view of the aggregate distribution with aggregate concentrations as contour lines. After Zinsmeister [31].

only when all the preceding aggregates n_i $(i < k)$ have reached their equilibrium concentration. The differential condensation coefficient can also be calculated to give

$$\alpha_{\text{cond}} = \tfrac{1}{2}\, q' \tau_s^2 w_{11} \gamma_i, \tag{2.30}$$

where γ is a kinetic factor for the process and is of the order of unity for $\tau_s \approx 10^{-7}$. We note that the simplified atomistic theory here presented takes into account only the first stage of condensation.

2.4. New experiments using a mass spectrometric technique

2.4.1. Introduction

We have seen in the previous two sections that there is a basic sequence of molecular kinetic steps involved in the nucleation process. Initially molecules from the vapor phase impinge on the bare substrate. These molecules may either be reflected back to the vapor phase, with some memory of their previous momentum and energy, or may reach energetic equilibrium with the substrate. Those which are equilibrated will stay at the surface for some finite stay-time in an adsorbed state. As impingement of still more molecules occurs, an equilibrium ad-layer population will build up on the surface until the desorption rate is equal to the impingement rate. The adsorbed molecules, if the substrate temperature is not too low, are, in general, in motion over the surface during the time that they are adsorbed. In the course of this motion they encounter other adsorbed molecules and thus form clusters of various sizes which may be stable or unstable depending on the strength of bonding in the clusters. The net rate of formation of clusters of a given size will be dictated by the balance between the rate of formation, given by the rate of impingement of single molecules on clusters one molecule smaller than the size under consideration, and the rate of decomposition of clusters of this given size. When conditions on the surface are such that there is a net flux of material to clusters of ever increasing size, then nucleation is said to have taken place and growth can proceed without further impediment due to kinetic barriers until the substrate is completely covered with the condensed phase of the depositing material.

What kind of precise measurements can one do to check these different steps in the condensation process? Consider first the question of energy and momentum accommodation for molecules impinging on the surface. The degree to which this process occurs can be characterized in terms of the thermal accommodation coefficient (*TAC*) which is defined as

$$TAC = (T_a - T_s)/(T_v - T_s), \tag{2.31}$$

in which T_a is the effective temperature of the material leaving the surface, T_s is the surface temperature and T_v the temperature of the impinging vapor.

The adsorption process can be characterized in terms of τ_s, the mean stay-time for adsorption and n_1, the ad-layer coverage. The stay-time can further be related to the heat of desorption (ΔH_{des}) of single molecules by the relation due to De Boer

$$\tau_s = \tau_0 \exp(\Delta H_{des}/RT),$$

in which τ_0 is essentially the inverse of the ad-atom vibrational frequency normal to the surface. At any time the desorption rate from the surface will be given by

$$\dot{E} = \frac{n_1}{\tau_s}.$$

When adsorptive equilibrium is reached, this desorption rate will be equal to the molecular impingement rate.

Hudson [23, 24] was the first to use the mass spectrometry technique extensively; it has also been used very recently by Nguyen-Anh [25, 26].

2.4.2. Experimental technique

The basic technique is conceptually very simple. All that is required is to set up the surface to be studied in a vacuum system, provide a source of impinging molecules and set up a mass spectrometer so that there is a line-of-sight path between the surface under study and the ionization region of the mass spectrometer.

In such a system the useful signal consists of mass-analyzed ions formed from neutral molecules which have interacted with the surface and then followed a line-of-sight path from the surface. Successful operation of the system depends primarily on separating this signal from any other electrical signal coming out of the back end of the mass spectrometer, which is noise, whether it is purely electrical in origin, or results from the collection of ions formed from material other than that described above as signal. This means essentially that, in so far as is possible, the only material of the species under study reaching the mass spectrometer must be material that has followed the line-of-sight path from the surface to the detector and scattering within the detector must be scrupulously avoided.

The detector, mounted so that there is a line of sight between it and the substrate, is a monopole mass spectrometer. It is used for analysis of the residual gases in the system and to monitor the flux of beam material desorbed or evaporated from the substrate. Existing detectors can detect static partial pressures in the system in the mid 10^{-12} Torr range. They are capable of detecting an evaporation rate from the substrate equivalent to 10^{-8} Torr vapor pressure, or a desorption rate of 10^{13} molecule/cm^2 sec from the sample. The basic experimental set up is shown in fig. (2.11).

The basic measurement that one makes with the experimental arrangement is the response of the system to a step change in the impingement rate at constant surface temperature. One can do this by using the shutter to interrupt the flow of molecules from the vapor source to the substrate and observing the resultant transient in the desorption flux. If one begins with the shutter closed, then opens the shutter, waits for equilibrium and then closes

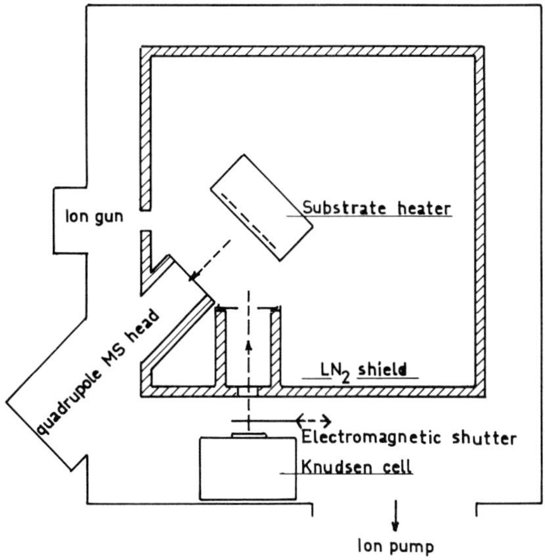

Fig. 2.11. Experimental set up for nucleation studies with a mass spectroscope.

the shutter, the result looks like fig. 2.12. This shows an oscilloscope trace obtained by Nguyen-Anh for indium on a silicon surface. When the shutter is closed the desorption signal decreases, again at a very-nearly exponential rate. Normally, for low coverage values, the time constant for the exponential rise is almost the same as that for the exponential decay. This time constant naturally depends sensitively on the substrate temperature.

Two sets of results obtained by this method are now presented:

A) Condensation of cadmium on tungsten – studied by Hudson [23, 24];

B) Condensation of indium on silicon – studied by Nguyen-Anh et al. [26].

2.4.3. *Cadmium on tungsten*

Cadmium on tungsten shows two different sets of activation energies for the mean stay-time of ad-atoms, depending on the substrate temperature. Thus at a high substrate temperature the activation energy of desorption is high ≈ 41 kcal/mol and a tightly bound adsorbed layer of cadmium occurs *without condensation.* One can get as much as two monolayers adsorbed in this way. The ad-layer coverage at any time can be unambiguously determined by graphical integration of the shaded area of fig. 2.12. If one now lowers the substrate temperature a completely different activation energy is found (≈ 21 kcal/mol) and the adsorbed phase is loosely bound. This loosely

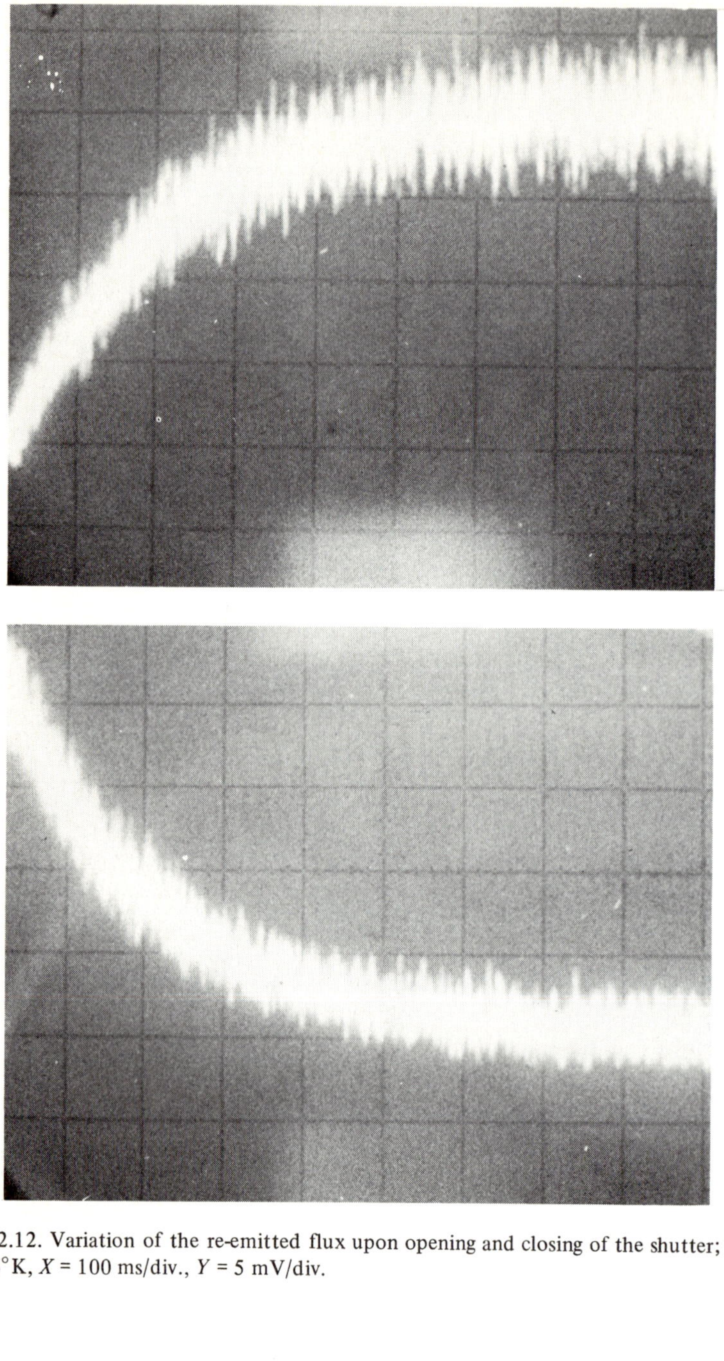

Fig. 2.12. Variation of the re-emitted flux upon opening and closing of the shutter; T_S = 1085°K, X = 100 ms/div., Y = 5 mV/div.

bound phase may extend to tens of monolayers before true condensation occurs. The stay-time data is shown in fig. 2.13.

One can now measure exactly the critical supersaturation for the onset of nucleation. This is done as follows: the substrate is heated to a high temperature and then exposed to the molecular beam. The substrate is then slowly cooled past the point at which the desorption rate shows a sharp decrease from the value typical of the desorption equilibrium and begins its descent to the value typical of the evaporation equilibrium. Fig. 2.14 shows a typical nucleation determination for Cd on W. The critical supersaturation is given by

$$S^* = p^*/p_0 = J_i\,(2\pi mkT_s)^{\frac{1}{2}}/p_0,$$

where p^* is the vapor pressure on the substrate corresponding to the incident vapor flux J_i and p_0 is the equilibrium vapor pressure at T_s. Results plotted in

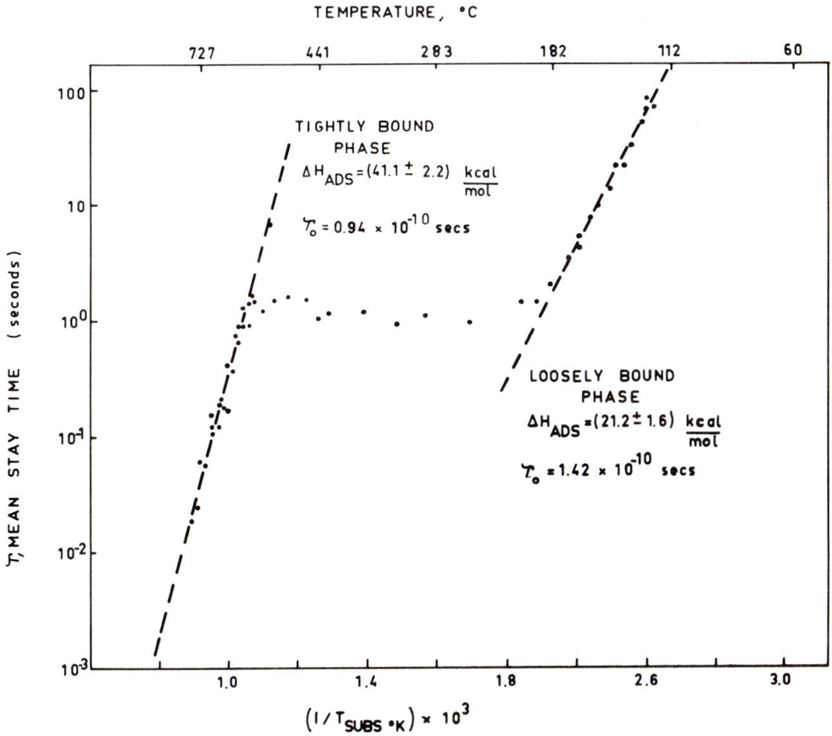

Fig. 2.13. Summary of data of stay-time for adsorption of cadmium on tungsten; after Hudson [23, 24].

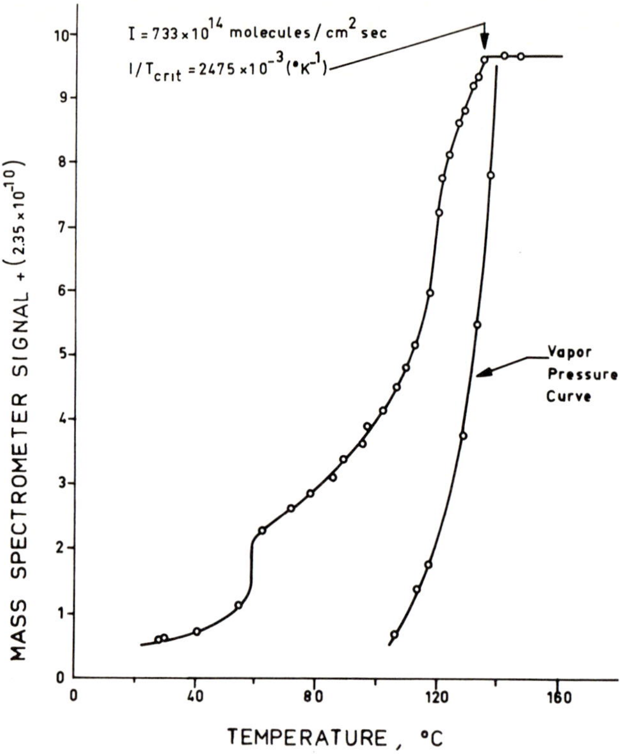

Fig. 2.14. Typical nucleation determination; after Hudson [23, 24].

the form of S^* against substrate temperature T_s are shown in fig. 2.15. Fig. 2.16 shows the critical ad-layer coverage at the onset of nucleation. The significance of these results and the results for In on Si are discussed later.

2.4.4. Indium on silicon

Mean lifetimes τ are measured for fluxes J_i ranging from 6.3×10^{12} to 9.4×10^{13} atoms cm^{-2} s^{-1} and for temperatures T_s from about 800°K to 1250°K. The incident and reemitted (i.e. desorbed) fluxes will be denoted by J_i and J_r. At high temperature τ is small and J_r is viewed on an oscilloscope. A typical example of these oscilloscope traces is shown in Fig. 2.12, corresponding to the opening and the closing of the shutter for $T_s = 1085$°K. These traces are pure exponentials: log i_s is a linear function of time.

As the temperature T_s is gradually decreased, the traces of J_r deviate more

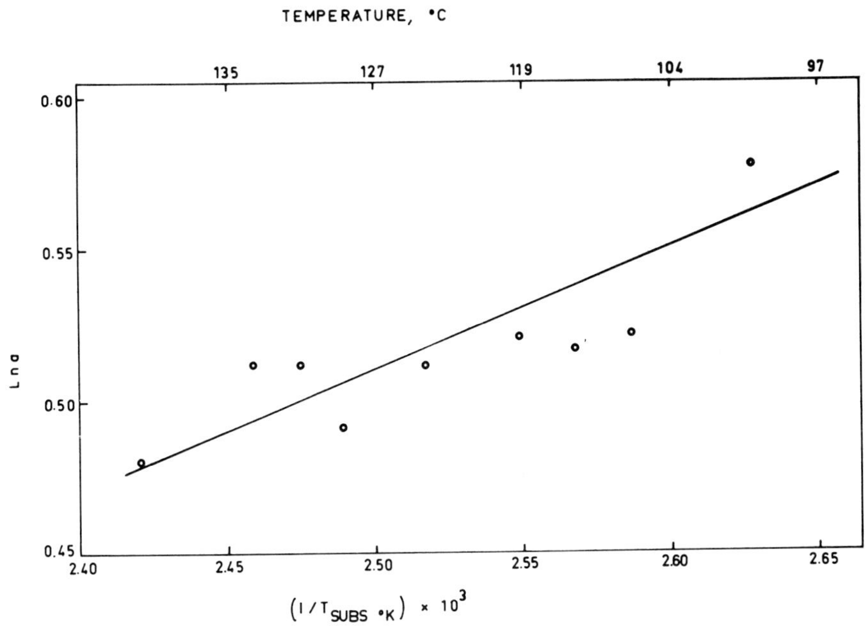

Fig. 2.15. Critical supersaturation for nucleation; ln a is proportional to S^*; after Hudson [23, 24].

and more from exponential form; the increasing and the decreasing curves are no longer symmetrical. The measurements of τ are unambiguous at high temperature but are more difficult in the low temperature range. In the latter case, two values of τ are obtained for every temperature, the first $(\tau_{\theta \approx 0})$ from the slope at the beginning of the increasing curves and which is related to small surface coverage, the other $(\tau_{\theta = \theta_0})$ from the slope at the origins of decreasing curves and which corresponds to equilibrium surface coverage θ_0. The values of $\tau_{\theta \approx 0}$ for $J_i = 6.3 \times 10^{12}$ atoms cm^{-2} s^{-1} and for $J_i = 1.7 \times 10^{13}$ atoms cm^{-2} s^{-1} are of the same order and the mean lifetime τ for a virgin surface is found to be:

$$\tau_p = 8 \times 10^{-13} \exp \frac{57\,000 \pm 3\,000}{RT}\,.$$

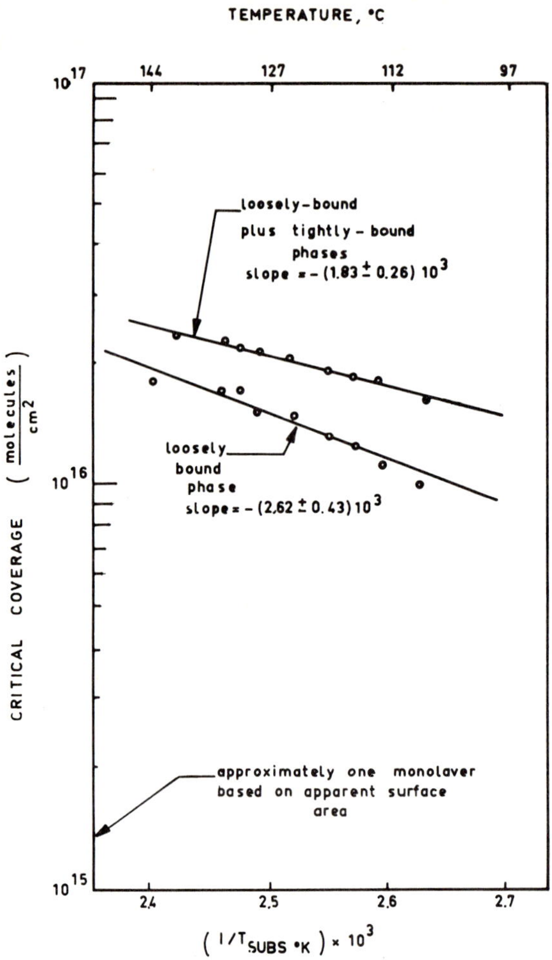

Fig. 2.16. Adlayer coverage at the onset of nucleation; after Hudson [23, 24].

Populations of ad-atoms at adsorption equilibrium are deduced from graphical integration of the curve J_r as a function of time (fig. 2.12).

The curves of log τ versus $1/T_s$ for a group of measurements are shown in fig. 2.17. The corresponding ad-atom population as well as the dependence of τ on n are shown in figs. 2.18 and 2.19 respectively. Both the sticking coefficient and the thermal accommodation coefficient were measured and were found to be unity.

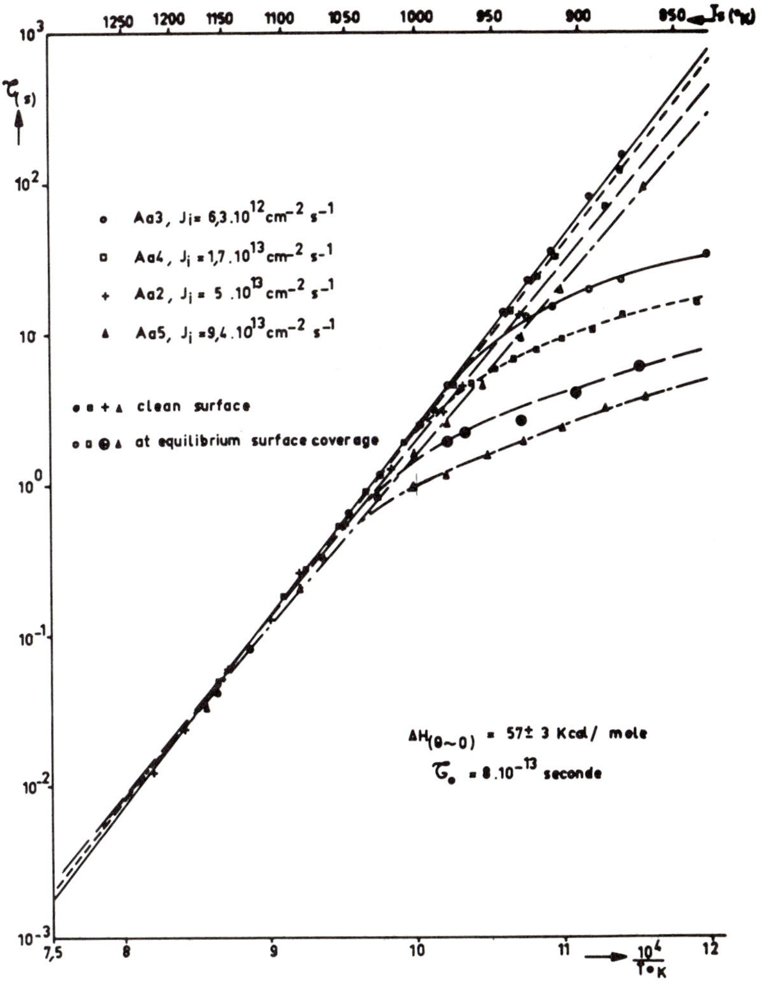

Fig. 2.17. Mean life-time of indium on silicon as a function of reciprocal temperature for various fluxes.

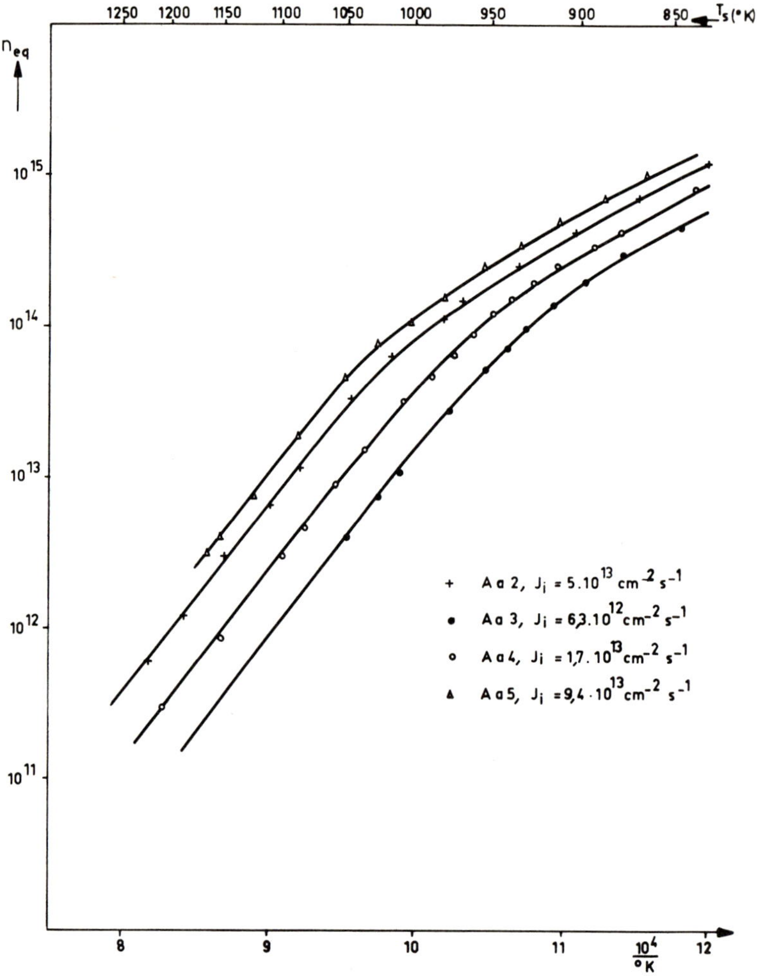

Fig. 2.18. Ad-atom populations corresponding to fig. 2.17.

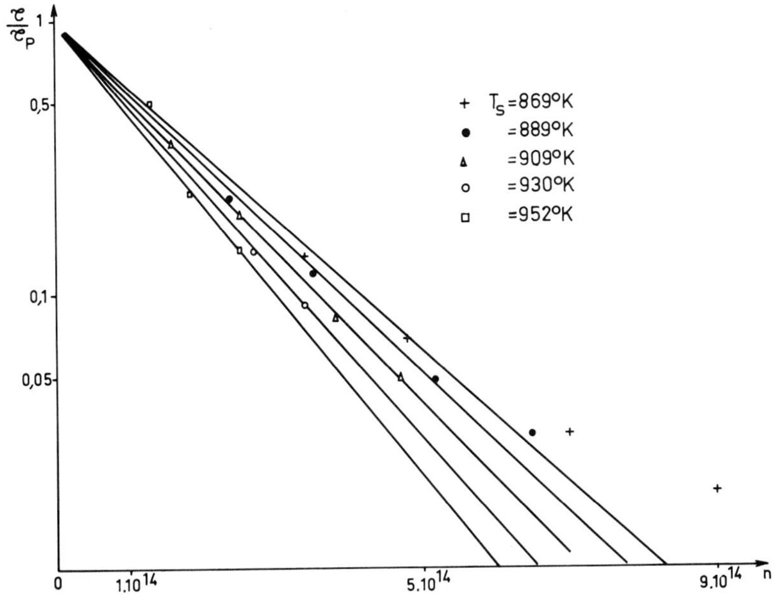

Fig. 2.19. Dependence of mean life-time on ad-atom populations for the data in fig. 2.17.

2.4.5. Critical supersaturation for nucleation

In these series of experiments a constant impingment flux was maintained. The substrate temperature was slowly decreased and mass spectrometric signals i_s were recorded for each temperature. The temperature T_s was decreased until i_s was zero, all incident atoms having by then condensed on the substrate. A deposit of a minimum of about one hundred monolayers was obtained thereby. After that the substrate temperature was slowly increased and the corresponding mass spectrometric signals were recorded. For each experiment, the re-emitted flux curve and the vapor pressure curve were obtained in this manner. Critical supersaturations for nucleation were determined by comparing these two curves. Some typical results are plotted in figs. 2.20 and 2.21, for impingment fluxes $J_i = 1.6 \times 10^{13}$ and 4.47×10^{14} atoms cm^{-2} s^{-1}. The re-emitted flux curves are all below the equilibrium vapor pressure curves. This indicates that condensation begins when the equivalent pressure of the impingment flux is still lower than the vapor pressure of the bulk phase of indium. There is an apparent undersaturation. This undersaturation is about 0.9 for the four different extensive series of experiments performed with four different impingment fluxes.

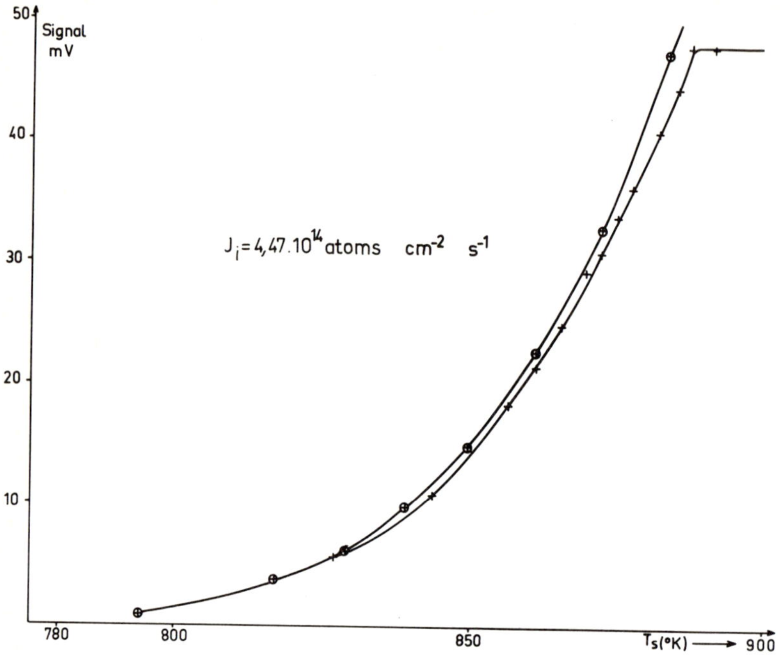

Fig. 2.20. Mass-spectrometric signal of re-emitted fluxes for an impingement flux of 4.47 $\times\ 10^{14}$ atoms cm^{-2} s^{-1}. (○ vapor pressure, + re-emitted flux)

2.4.6. Discussion

What do these results imply? It is clear from the results of thermal accommodation measurements (for both Cd on W and In on Si, it is unity) that there is no effective barrier to energy transfer between vapor and substrate, at least in the case of metals on metals or semiconductors, at these temperatures.

The case of Cd on W implies, as far as the nucleation process is concerned, that the high coverages observed in the ad-layer prior to nucleation rule out the direct application of any of the models of the nucleation process discussed earlier. Thus a new model, which treats the primary ad-layer as the substrate and which considers the formation of a bulk cadmium nucleus within the secondary phase, is required if we are to adequately describe the nucleation process in this case. The critical supersaturations for nucleation also show marked deviation from almost all the values indicated so far in the classical nucleation studies (1.6 to 1.8 compared to $10^6 - 10^{19}$), and thus rule out any utilisation of atomistic theory.

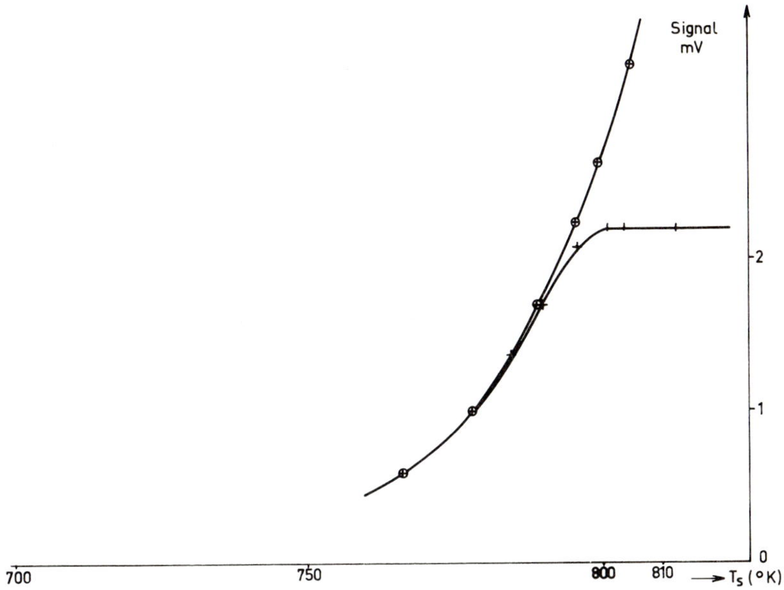

Fig. 2.21. As fig. 2.20 for an impingement flux of 1.6×10^{13} atoms cm^{-2} s^{-1}.

The case of In on Si is even more revealing and disturbing. Here we have seen that there is absolutely no critical supersaturation necessary for the condensation to begin. This had been previously indicated by the ultrasensitive mass-balance results of Cinti [22, 27]. But now, we not only do not have a supersaturation, but condensation starts visibly at *undersaturation*. Currently there is no theory which can account for this and the phenomenon is intimately related to phase transition in two-dimensional systems, as well as to localised surface states on silicon which probably serve as strong nucleating centers.

In conclusion, the mass spectroscopic technique has given all the quantitative data the nucleation theorists have, for years, been asking for and in the process has almost destroyed the basis of current nucleation theory. However it has also opened new doors for extremely precise measurements and has made nucleation study a very exciting field for both theoretical and experimental physicists and chemists.

2.5. Coalescence in thin films

2.5.1. Introduction

In the three previous sections we have been mainly concerned with the first stage of condensation on a substrate. In this stage, the condensing centres grow radially outwards at the expense of surrounding ad-atoms. Very soon however, the second stage of condensation sets in. This stage is outwardly characterized by a gradual and monotonic redistribution of the size classes amongst the nuclei themselves. Inwardly, the system makes an all-out drive to minimize the large interfacial energy of the system of discrete nuclei. Thus large nuclei grow larger at the expense of the small ones, which eventually disappear. At every instant a size distribution of nuclei occurs, weighted towards the largest possible size. This distribution is by no means gaussian. In this section, we shall try to understand this coalescence process (note the word "coalescence" by no means signifies actual impingement of one nucleus on another, which happens in the so-called third stage of condensation) and try to analyse the underlying mechanism. Two distinct cases are considered:

A) Coalescence, driven by capillary forces

B) Coalescence, on a purely atomistic basis, as a kinetic polymerisation phenomenon.

2.5.2. Coalescence driven by capillary forces

2.5.2.1. Growth rate of nuclei on a substrate.

This form of coalescence, often known as "Oswald Ripening", was first analysed in detail by Chakraverty [28]. Because the hydrodynamic method used gave some exact analytical solutions to the problem and illustrated a general and powerful method, common to all systems involving precipitation from a supersaturated phase, we shall set down the analysis in some detail.

The redistribution of the sizes of the nuclei is brought about mainly by the singly-adsorbed depositing atoms (called ad-atoms) attaching themselves to the larger nuclei, having transferred from a small nucleus by surface diffusion on the substrate. Because of the extreme mobility of these ad-atoms (generally the activation energy for surface diffusion is low) a considerable rearrangement of the nuclei sizes is possible even at room temperature. As the redistribution proceeds, the ad-atom population progressively diminishes and approaches the equilibrium value characteristic of the film material at the substrate temperature.

We consider unit area of the substrate at temperature T, containing a distribution of nucleus sizes and a certain average ad-atom population $\bar{n}$

(which is a monotonically diminishing function of time). We suppose each nucleus to be three-dimensional and of the shape of a spherical cap (cf. ch. 2 §2.3). This presupposes that the nucleus-vapor interfacial energy per unit area σ is isotropic (it also neglects the line tension at the base) and, in such a case, a nucleus of radius r can be in thermodynamic equilibrium only with an ad-atom population given by the Gibbs-Thomson relationship,

$$n_{\mathrm{r}} = n_{\mathrm{eq}} \exp \frac{2\sigma v_{\mathrm{m}}}{kT}, \tag{2.32}$$

where v_{m} = volume of an ad-atom,
k = Boltzmann constant,
T = the absolute temperature,
n_{eq} = ad-atom concentration corresponding to the equilibrium vapour pressure p_{eq} of the film material at T. It can be obtained as follows.
At steady state equality of the impingment and desorption rates gives

$$\frac{p_{\mathrm{eq}}}{\sqrt{(2\pi m_0 kT)}} = n_{\mathrm{eq}}\, \nu \exp \frac{-\Delta G_{\mathrm{des}}}{kT},$$

where m_0 = mass of an ad-atom,
ν = vibration frequency of an ad-atom on the substrate,
ΔG_{des} = the activation energy for desorption of an ad-atom from the substrate.

From this expression n_{eq} can be obtained.

The extent to which n_{r} is greater or less than the average ad-atom population $\bar{n}$ determines whether a particular nucleus of radius r evaporates or grows.

Let $f(r,t)$ be the required size distribution function for nuclei at time t, i.e. the number of particles of size r per unit area. Following Lifshitz and Slezhov, we write the continuity equation in the particle size space

$$\frac{\partial f(r,t)}{\partial t} + \frac{\partial}{\partial t}[f(r,t)\dot{r}] = 0, \tag{2.33}$$

where $\dot{r} = \mathrm{d}r/\mathrm{d}t$ is the velocity with which each particle travels on the r-axis.

The determination of $f(r,t)$ must be subject to the condition that the total amount of film deposited is conserved on annealing which is, for the three-dimensional nuclei we are considering,

$$\varphi = \alpha_0 \int_0^\infty f(r,t) r^3 \, \mathrm{d}r, \tag{2.34}$$

where $\alpha_0 = \frac{4\pi}{3} f_1(\theta) m$; m is the density of the film-material. The total number of discrete nuclei or particles is given by

$$n(t) = \int_0^\infty f(r,t) \, \mathrm{d}r. \tag{2.35}$$

The solution of the continuity eq. (2.33) depends on the proper choice of $\dot{r}$, which we shall now determine.

The mechanism of particle growth or dissolution is, as has already been pointed out, surface diffusion of ad-atoms and their eventual transfer at the interface of the nucleus. Thus it is a series process and the rate $\dot{r}$ will be determined by the slowest step (surface diffusion or interface transfer).

Considering surface diffusion first, Fick's second law, in polar coordinates, at steady state is

$$\frac{1}{R} \frac{\mathrm{d}}{\mathrm{d}R} \left[R D_\mathrm{s} \frac{\mathrm{d}n}{\mathrm{d}R} \right] = 0, \tag{2.36}$$

where R = distance from the centre of the nucleus of radius $r \sin\theta$, considered to be the coordinate origin,
$n(R)$ = local concentration of ad-atoms,
D_s = surface diffusion coefficient of ad-atoms.

The solution of eq. (2.36) is

$$n(R) = K_1 \ln R + K_2. \tag{2.37}$$

We use the boundary condition (cf. fig. 2.22)
$n(R) = n'$ at $R = r \sin\theta$, the edge of the nucleus
$n(R) = \bar{n}$ at $R = lr \sin\theta$,

where l is a "screening" distance at which the ad-atom population recovers its average value. l is taken to be a constant. Using (2.37), eliminating K_2, one obtains,

$$n(R) = \frac{\bar{n} - n'}{\ln l} \ln R$$

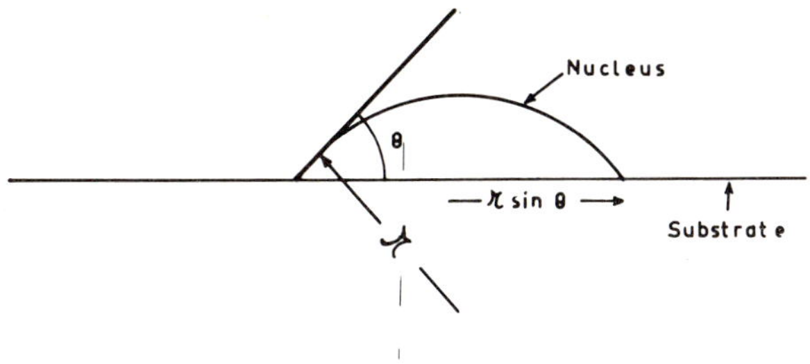

Fig. 2.22. Spherical cap model.

The number of atoms attaching themselves to the nucleus per second by surface diffusion is then

$$J_s = (2\pi r \sin\theta) \times \left[D_s \frac{dn(R)}{dR} \right]_{R=r\sin\theta} .$$

This gives

$$J_s = \frac{2\pi D_s}{\ln l} \left[\bar{n} - n' \right] . \tag{2.38}$$

For the mechanism of interface transfer on the other hand, the probability of atom addition or dissolution is proportional to the surface area and the local ad-atom concentration.
Thus,
Number of atoms being added per unit of time = $4\pi r^2 f_2(\theta) \times \beta_0 \times n'$.
Number of atoms detaching per unit time = $4\pi r^2 f_2(\theta) \times \beta_0 \times n_r$,
where β_0 is a probability constant assumed to be dependent only on temperature.

The last equality follows from the principle of microscopic reversibility for the nucleus in a hypothetic equilibrium. This gives the net rate of atom attachment by interface transfer as

$$J_t = 4\pi r^2 f_2(\theta) \beta_0 [n' - n_r] . \tag{2.39}$$

The two mechanisms of transfer (2.38) and (2.39) must give equal rates at steady state, viz.:

$$J = J_s = J_t \tag{2.40}$$

This we can use to eliminate n', the unknown and obtain

$$J = \frac{(2\pi D_s/\ln l)\,\beta_0 4\pi r^2 f_2(\theta)}{(2\pi D_s/\ln l) + \beta_0 4\pi r^2 f_2(\theta)}\,[\bar{n} - n_r] \tag{2.41}$$

We obtain $\dot{r}$ (= $\mathrm{d}r/\mathrm{d}t$) from the condition that

$$\frac{\mathrm{d}}{\mathrm{d}t}\left[\frac{4}{3}\pi r^3 f_1(\theta)\right] = J v_\mathrm{m}\,,$$

which gives

$$\frac{\mathrm{d}r}{\mathrm{d}t} = \frac{J v_\mathrm{m}}{4\pi r^2 f_1(\theta)}\,. \tag{2.42}$$

We define two quantities: the supersaturation of the ad-atoms on the substrate, s and the critical nucleus size r^*:

$$s = \bar{n}/n_\mathrm{eq}\,, \tag{2.43}$$

and r^* is obtained from eqs. (2.32) and (2.43) by observing that $n_{r^*} = \bar{n}$:

$$r^* = \frac{2\sigma v_\mathrm{m}}{kT \ln s}\,. \tag{2.44}$$

With (2.43) and (2.44) eq. (2.32) can be written in the form:

$$n_r = n_\mathrm{eq}\, s^{r^*/r}\,. \tag{2.32$'$}$$

Substitution of (2.32′), (2.43) and (2.41) into (2.42) gives:

$$\frac{\mathrm{d}r}{\mathrm{d}t} = \frac{v_\mathrm{m} n_\mathrm{eq}}{4\pi r^2 f_1(\theta)}\,\frac{(2\pi D_s/\ln l)\,\beta_0 4\pi r^2 f_2(\theta)}{\{(2\pi D_s/\ln l) + \beta_0 4\pi r^2 f_2(\theta)\}}\,[s - s^{r^*/r}]\,. \tag{2.45}$$

The equation (2.45) so far is quite general.

One now makes a central supposition of the whole theory, namely that the supersaturation s is small. Writing $s = 1 + h$ ($h \ll 1$), $\ln s \approx h$, and

$$s^{r^*/r} = (1+h)^x = 1 + xh + \tfrac{1}{2}x(x-1)h^2 + \ldots\ldots, \tag{2.46}$$

where $x = r^*/r$.

Retaining only the second term of the expansion (valid for $hx < 1$), eq. (2.45) becomes, after substituting for r^* from (2.44), using (2.46):

$$\frac{\mathrm{d}r}{\mathrm{d}t} = \frac{n_{\mathrm{eq}}\,\sigma\, v_{\mathrm{m}}^2}{kT\, 2\pi r^2 f_1(\theta) r^*} \frac{\dfrac{2\pi D_{\mathrm{s}}}{\ln l}\beta_0\, 4\pi r^2 f_2(\theta)}{\dfrac{2\pi D_{\mathrm{s}}}{\ln l} + \beta_0\, 4\pi r^2 f_2(\theta)} \left[1 - \frac{r^*}{r}\right]. \tag{2.47}$$

The rate expression (2.47) considerably simplifies for two limiting cases.

Case A

Interface transfer is much faster than surface diffusion. This implies that

$$\beta_0\, 4\pi r^2 f_2(\theta) \gg \frac{2\pi D_{\mathrm{s}}}{\ln l}.$$

Surface diffusion is then rate-controlling and (2.47) becomes

$$\frac{\mathrm{d}r}{\mathrm{d}t} = \frac{n_{\mathrm{eq}}\,\sigma v_{\mathrm{m}}^2 D_{\mathrm{s}}}{(\ln l) f_1(\theta) kT r^* r^2} \left[1 - \frac{r^*}{r}\right]. \tag{2.47$'$}$$

Case B

Surface diffusion is much faster than interface transfer. This implies

$$\frac{2\pi D_{\mathrm{s}}}{\ln l} \gg \beta_0\, 4\pi r^2 f_2(\theta),$$

and eq. (2.47) becomes

$$\frac{\mathrm{d}r}{\mathrm{d}t} = \frac{2 f_2(\theta) n_{\mathrm{eq}}\,\sigma v_{\mathrm{m}}^2 \beta_0}{kT f_1(\theta) r^*} \left[1 - \frac{r^*}{r}\right]. \tag{2.47$''$}$$

It is with the two expressions (2.47$'$) and (2.47$''$) that we shall determine the particle-size distributions for the limiting cases A and B.

2.5.2.2. Particle-size distribution for the surface diffusion mechanism.

A great simplification results by introducing two dimensionless variables ρ and τ_{s}, which replace r and t:

$$\rho = r/r^*(t), \text{ and}$$

$$\tau_{\mathrm{s}} = \ln\left[\frac{r^{*4}(t)}{r^{*4}(0)}\right]. \tag{2.48}$$

The second of the substitutions (2.48) introduces τ_s as a pseudo-time variable due to the fact that $\bar{n}(t) \to 0$ as $t \to \infty$ causing $r^*(t) \to \infty$, and thus $\tau_s \to \infty$, as $t \to \infty$. We now consider $d\rho/d\tau_s$ instead of dr/dt as given by (2.47′).

$$\frac{d\rho}{d\tau_s} = \frac{d\rho}{dt}\frac{dt}{d\tau_s} = \frac{r^{*3}(dr/dt) - r.r^{*2}(dr^*/dt)}{dr^{*4}/dt},$$

which becomes, using (2.48):

$$\frac{d\rho}{d\tau_s} = \frac{\lambda_s(\rho - 1) - \rho^4}{4\rho^3}, \tag{2.49}$$

where

$$\lambda_s = \frac{4A_s}{dr^{*4}/dt} \text{ and } A_s = \frac{n_{eq}\,\sigma v_m^2\, D_s}{kTf_1(\theta)\ln I}. \tag{2.49′}$$

Similarly the continuity equation (2.33) now takes on the canonical form

$$\frac{\partial f_s(\rho,\tau_s)}{\partial \tau_s} - \frac{\partial}{\partial \rho}\left[f_s(\rho,\tau_s)g_s(\rho,\lambda_s)\right] = 0, \tag{2.50}$$

where the subscript s on the distribution function refers to the surface diffusion case and

$$g_s(\rho,\lambda_s) = -\frac{d\rho}{d\tau_s} = \frac{\rho^4 - \lambda_s(\rho - 1)}{4\rho^3}. \tag{2.50′}$$

The conservation equation (2.34) becomes

$$\varphi = \alpha_0\, r^{*3}(0)e^{3\tau_s/4}\int_0^\infty f_s(\rho,\tau_s)\rho^3\, d\rho. \tag{2.51}$$

Let us suppose that the solution of (2.50) can be written in the variable separable form

$$f_s(\rho,\tau_s) = \psi_s(\rho)T_s(\tau_s). \tag{2.52}$$

It is evident on substitution of (2.52) into the conservation equation (2.51) that, for the left-hand side φ to be conserved, $T_s(\tau_s)$ must be of the form

$$T_s(\tau_s) = \beta_s e^{-3\tau_s/4}, \tag{2.52'}$$

where

$$\beta_s = \frac{\varphi}{\alpha_0 r^{*3}(0)\int_0^\infty \psi_s(\rho)\rho^3\,d\rho},$$

is a constant. Utilising (2.52′) and (2.52) in (2.50) gives, for $\psi_s(\rho)$

$$-\ln \psi_s(\rho) = \frac{3}{4}\int \frac{d\rho}{g_s(\rho,\lambda_s)} + \frac{dg_s(\rho,\lambda_s)}{g_s(\rho,\lambda_s)}. \tag{2.53}$$

In order that the solution $\psi_s(\rho)$ from (2.53) be a valid solution, it is necessary that the integrands on the right shall have no singularities. This is directly dependent on the behaviour of $g_s(\rho,\lambda_s)$ in the denominator, which as one sees in (2.50′), is a polynomial of fourth degree with four roots in general.

Let us consider the numerator of $g_s(\rho,\lambda_s)$ i.e. the function

$$y_s = \frac{d(\rho^4)}{d\tau_s} = \lambda_s(\rho-1) - \rho^4. \tag{2.54}$$

The three possible types of behaviour of y_s, depending on the value of λ_s are shown in fig. 2.23. One can have

a) $\lambda_s = \lambda_s^0$ (two roots real and equal, two imaginary)

b) $\lambda_s > \lambda_s^0$ (two roots real and unequal, two imaginary)

c) $\lambda_s < \lambda_s^0$ (all four roots imaginary).

Of these three possibilities, it can be shown that cases (b) and (c) are incompatible with the basic conservation condition (Chakraverty [28]).

The only stable situation the system can have is case (a), with $\lambda_s = \lambda_s^0$. This is obtained when both y_s and its first derivative with ρ touches the y-axis at $\rho = \rho_s^0$. We obtain from

$$\left.\begin{aligned} y_s &= \rho^4 - \lambda_s^0(\rho-1) = 0 \\ y_s' &= 4\rho^3 - \lambda_s^0 = 0 \end{aligned}\right\} \begin{aligned} &\text{at } \rho = \rho_s^0 \\ &\text{at } \rho = \rho_s^0 \end{aligned}$$

and (2.54′)

$$\rho_s^0 = 1.333,$$
$$\lambda_s^0 = 9.5.$$

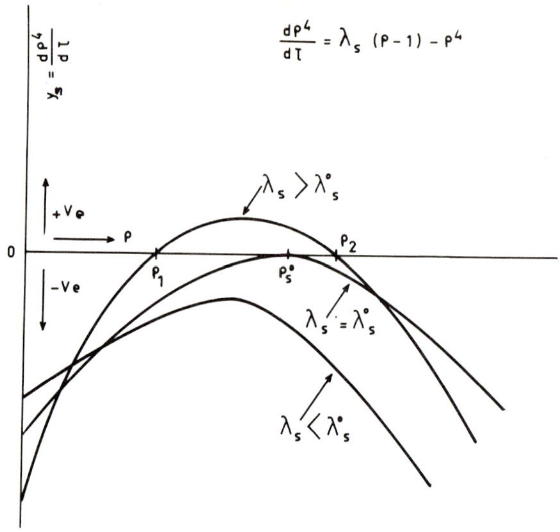

Fig. 2.23. Behaviour of the function $y_{s'}$ (eq. 2.54).

Because of the equality $\lambda_s = \lambda_s^0$, the required unique function $g_s(\rho,\lambda_s^0)$ causes the solution $\psi_s(\rho)$ in (2.53) to be also quite unique in that $\psi_s(\rho)$, as well as its derivatives, disappears at and beyond $\rho = \rho_s^0 = 1.33$. For the surface-diffusion control process we are considering, there can be no particle larger than $\rho = 1.33$, no matter what the initial distribution function of particles is in the thin film. The solution of (2.53), with $g_s(\rho,\lambda_s^0)$ is then

$$\psi_s(\rho) = \frac{4\rho^3 \exp\left[\dfrac{-2}{3(\rho_s^0-\rho)}\right]\exp\left[-\dfrac{1}{6\sqrt{2}}\tan^{-1}\dfrac{\rho+\rho_s^0}{\rho_s^0\sqrt{2}}\right]}{(\rho_s^0-\rho)^{19/6}(\rho^2+2\rho_s^0\rho+4\rho_s^0)^{23/12}}, \quad \text{for } \rho < \rho_s^0 = 1.33$$
$$= 0, \qquad \text{for } \rho \geqslant \rho_s^0 = 1.33 \tag{2.55}$$

This solution is plotted in fig. 2.24.

From $\lambda_s = \lambda_s^0 = 9.5$, one obtains by the definition of λ_s in (2.49′)

$$r^*(t) = r^*(0)\left[1+\frac{t}{t_s}\right]^{1/4} = r^*(0)\left[\frac{t}{t_s}\right]^{1/4}, \qquad \text{for } t > t_s \tag{2.56}$$

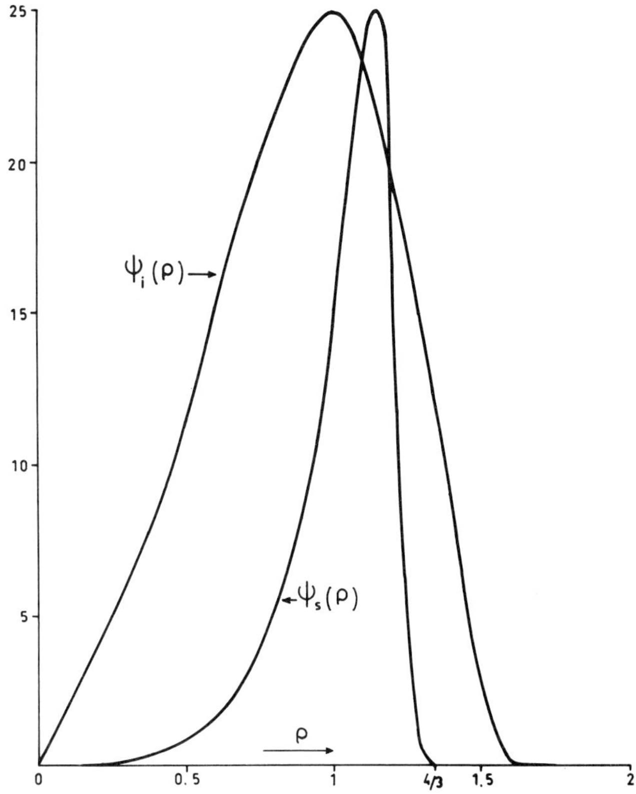

Fig. 2.24. The functions $\psi_s(\rho)$ (eq. 2.55) and $\psi_i(\rho)$ (eq. 2.71).

where

$$\frac{1}{t_s} = \frac{4A_s}{9.5} = \frac{4D_s v_m^2 \sigma n_{eq}}{9.5(\ln l) f_1(\theta) kT} \,.$$

We note that the screening distance is $l \approx 2$ or 3 and that the surface diffusion coefficient D_s in the transition state theory is given by

$$D_s = \nu a_0^2 \exp\left(\frac{-\Delta G_{diff}}{kT}\right)$$

where ν = Debye frequency $\approx 10^{13}$ Hz
a_0 = lattice parameter of the substrate
ΔG_{diff} = activation energy per atom for surface diffusion of an adatom on the substrate.

From (2.52), (2.52′) and (2.55), and replacing τ by t through $r^*(t)$, we get

$$f_s(\rho,t) = \frac{\text{const.}}{\left(1+\dfrac{t}{t_s}\right)^{3/4}}\psi_s(\rho) \quad \text{for } \rho < 1.33$$
$$= 0, \qquad \text{for } \rho \geqslant 1.33 \tag{2.57}$$

The total number of particles is obtained from

$$n(t) = \int_0^\infty f_s(\rho,t)\,\mathrm{d}\rho = \frac{t_s^{3/4}\ \text{const.}}{t^{3/4}}, \text{ for } (t > t_s). \tag{2.58}$$

Since

$$f_s(\rho,t)\,\mathrm{d}\rho = f_s(r,t)\,\mathrm{d}r, \text{ one has} \tag{2.59}$$

$$f_s(r,t) = \frac{\text{const.}}{\left(1+\dfrac{t}{t_s}\right)}\psi_s(\rho) \quad \text{for } \rho < 1.33$$
$$= 0, \qquad \text{for } \rho \geqslant 1.33. \tag{2.60}$$

2.5.2.3. Particle-size distribution for the interface transfer mechanism. Using the transformations

$$\rho = \frac{r}{r^*(t)} \text{ and } \tau_i = \ln\left[\frac{r^{*2}(t)}{r^{*2}(0)}\right], \tag{2.61}$$

eq. (2.47″) becomes

$$\frac{\mathrm{d}\rho}{\mathrm{d}\tau_i} = \frac{\lambda_i(\rho-1)-\rho^2}{2\rho}, \text{ where} \tag{2.62}$$

$$\lambda_i = \frac{2A_i}{\mathrm{d}r^{*2}/\mathrm{d}t} \text{ and } A_i = \frac{2f_2(\theta)n_{eq}\sigma v_m^2\beta_0}{kTf_1(\theta)}. \tag{2.63}$$

The continuity equation retains its canonical form (2.50)

$$\frac{\partial f_i(\rho,\tau_i)}{\partial \tau_i} - \frac{\partial}{\partial \rho}\,[f_i(\rho,\tau_i)g_i(\rho,\lambda_i)] = 0, \tag{2.64}$$

where the subcript i refers to interface transfer and

$$g_i(\rho,\lambda_i) = -\frac{\mathrm{d}\rho}{\mathrm{d}r} = \frac{\rho^2 - \lambda_i(\rho-1)}{2\rho}. \tag{2.64'}$$

The basic conservation equation (2.34) becomes

$$\varphi = \alpha_0 r^{*3}(0)\mathrm{e}^{3\tau_i/2}\int_0^\infty f_i(\rho,\tau_i)\rho^3\,\mathrm{d}\rho. \tag{2.65}$$

We expect the solution to be

$$f_i(\rho,\tau_i) = \beta_i \mathrm{e}^{-3\tau_i/2}\,\psi_i(\rho), \tag{2.66}$$

where

$$\beta_i = \frac{\varphi}{\alpha_0 r^{*3}(0)\int_0^\infty \psi_i(\rho)\rho^3\,\mathrm{d}\rho}.$$

Substituting (2.66) in (2.64) gives

$$\ln\psi_i(\rho) = -\frac{3}{2}\int\frac{\mathrm{d}\rho}{g_i(\rho,\lambda_i)} - \int\frac{\mathrm{d}g_i(\rho,\lambda_i)}{g(\rho,\lambda_i)}, \tag{2.67}$$

in complete analogy with the surface diffusion case. It can be shown that the only stable value of λ_i is λ_i^0 for which

$$y_i = \mathrm{d}\rho^2/\mathrm{d}\tau = \lambda_i^0(\rho-1) - \rho^2 = 0,\ \text{at}\ \rho = \rho_i^0,$$

and (2.68)

$$y_i' = \lambda_i^0 - 2\rho = 0\ \text{at}\ \rho = \rho_i^0.$$

This gives $\lambda_i^0 = 4$

and $\rho_i^0 = 2$ (2.69)

Now quite analogously to the surface diffusion case, the distribution function as well as all its derivatives vanish at and beyond $\rho = \rho_i^0 = 2$.
We write down the solution from (2.67),

$$\psi_i(\rho) = \frac{\text{const}\ \rho}{(2-\rho)^5} \exp\left[-\frac{6}{2-\rho}\right] \text{ for } \rho < \rho_i^0 = 2$$
$$= 0, \qquad \text{for } \rho \geqslant \rho_i^0 = 2. \tag{2.70}$$

This is identical to Wagner's equation VIII. 16a [29].

$\psi_i(\rho)$ is plotted in fig. 2.24 and has a maximum at $\rho = 1$. The particle size r^* at this maximum is obtained from (2.63), for $\lambda_i = \lambda_i^0 = 4$,

$$r^{*2}(t) = r^{*2}(0)\left[1 + \frac{t}{t_i}\right], \tag{2.71}$$

where $1/t_i = \frac{1}{2} A_i$.
The complete distribution function (2.66) is then, substituting for τ_i from (2.62),

$$f(\rho,t) = \frac{\beta_i}{\left(1+\frac{t}{t_i}\right)^{3/2}} \frac{\rho}{(2-\rho)^5} \exp\left[\frac{-6}{2-\rho}\right] \text{ for } \rho < 2$$
$$= 0, \qquad \text{for } \rho \geqslant 2. \tag{2.72}$$

The total number of particles diminishes with time according to

$$n(t) = \frac{\text{const}}{\left(1+\frac{t}{t_i}\right)^{3/2}}. \tag{2.73}$$

We can also write

$$f(r,t) = \frac{\text{const}}{\left(1+\frac{t}{t_i}\right)^2} \psi_i(\rho) \text{ for } \rho < 2$$
$$= 0, \qquad \text{for } \rho \geqslant 2. \tag{2.74}$$

This completes the two distribution functions that we sought at the beginning of the section.

More recently, Robertson [30] has extended this analysis by covering the phenomenon at the high supersaturation end, where it is not possible to obtain an analytical solution. Their numerical solution indicates a similarly sharply peaked distribution, without any cut-off.

2.5.3. Coalescence in the atomistic theory

As in the nucleation theory this does away altogether with the idea of a surface energy or critical nucleus. The analysis is due principally to Zinsmeister [31], who writes the governing kinetic equations as:

$$\frac{\mathrm{d}n_2}{\mathrm{d}t} = \frac{w}{2} n_1^2 - w n_1 n_2$$

$$\frac{\mathrm{d}n_3}{\mathrm{d}t} = w n_1 n_2 - w n_1 n_3$$

$$- - - - - - - - - - - - - - - - - -$$

$$\frac{\mathrm{d}n_i}{\mathrm{d}t} = w n_1 n_{i-1} - w n_1 n_i$$

By introducing a pseudo-time scale $x = w \int_0^t n_1(t)\mathrm{d}t$, one can rearrange the previous set of equations to

$$\dot{n}_2 + n_2 = \tfrac{1}{2} n_1$$

$$\dot{n}_3 + n_3 = n_2$$

$$- - - - - - - - - - - -$$

$$\dot{n}_i + n_i = n_{i-1},$$

where $\dot{n}_i = \mathrm{d}n_i/\mathrm{d}t$.

Analytic solutions to these equations were not possible. Instead the very simple approximate solution:

$$n_i(x) \approx \frac{1}{2} n_1(x - i), \qquad x > i,$$

$$n_i(x) = 0, \qquad x < i.$$

was used.

This solution permitted the following two calculations

a) The time dependence of the concentration $n_i(x) = f(x)$ of an aggregate consisting of i atoms.

b) Size distribution $n_i(x_1) = g(i)$ of all aggregates present on the surface at a time x_1.

A contour view of Zinsmeister's calculations is shown in fig. 2.10. As Zinsmeister points out, these distributions carry over quite well to those obtained by Chakraverty.

Space does not permit a discussion of the recent series of very interesting experiments, undertaken by Masson, Metois and Kern [32], on the brownian motion of nuclei (or crystallites) of gold on KCl substrates. As the subject falls outside the domain of coalescence (in their brownian motion, no coalescence or size redistribution was observed), we shall simply conclude by saying that it is quite probable that for a great number of condensing systems, (especially where the condensate does not wet or only slightly wets a substrate), the nuclei actually suffer physical displacement on the substrate and, by this process, can bring about a large amount of structural change including epitaxy. This phenomenon is relatively new, and no adequate explanation of such Brownian motions has yet been given.

2.5.4. Experimental results and discussion

Although there is a great wealth of electron microscopic evidence on the granular nature of thin films, there is not available a large number of quantitative measurements of the particle-size distribution under precisely stated conditions. Unless precisely stated most such distributions determined by electron microscopy (unless measured in situ) should be considered as having resulted from some annealing of the film during carbon replication, as well as unintentional ageing at room temperature. To some of the observed distributions, the results of the last section should be applicable. In the original paper by Chakraverty [28] the following experimental data points have been plotted so, that they are normalized to match at their respective maximum:

(1) Palatnik [33] et al., on nickel deposited on NaCl, at room temperature, film thickness 200 Å, fig. 4 of ref. [28];

(2) Herman and Rhodin [34], on annealing of Au on NaCl at room temperature and of Hampe [35] on annealing of Au in a matrix of SiO_2 on glass, at room temperature and at 400°C; (the data of Herman and Rhodin as well as those of Hampe are used in fig. 5; it is to be noted that of the four sets of data reported by Herman and Rhodin, only three sets are represented in fig. 5, while the other set is shown in fig. 6 of ref. [28];

(3) Herman and Rhodin [34], on Au on NaCl (one set of 10 Å points) and

the results of Blackman and Curzon [36] for Sn on amorphous carbon film, at room temperature.

It is seen that:

(a) The data of Palatnik et al., Herman and Rhodin (except one set of points) as well as those of Hampe can be quite satisfactorily described by the distribution function due to interface transfer. Of these the agreement with Palatnik's data is certainly striking.

(b) The data of Blackman and Curzon and the remaining set of Herman and Rhodin points seem to follow quite clearly the distribution function $\psi_s(\rho)$ due to surface diffusion control.

(c) On the high ρ-side, the experimental data points seem to lie above the theoretical distribution.

It is very interesting that only one 10 Å set of data of Herman and Rhodin prefers not to follow the interface controlled distribution. However, this also happens to be the set of points whose average particle-size is almost double that of the other three sets and looking at eq. (2.47), because of this larger r, one might possibly have

$$\beta_0 4\pi r^2 f_2(\theta) > 2\pi D_s/\ln l,$$

thus leading to a surface-diffusion-control distribution, as revealed by analysis. Certainly films with large growth in particle-sizes (i.e. grown under small supersaturation) on annealing will tend to obey surface diffusion kinetics. The reason for the discrepancies on the high ρ-side can be thought to originate simply from insufficient time or temperature of annealing, resulting in all $\rho > \rho_s^o$ or ρ_i^o not having had time to dissolve. This thus reflects the tail of the initial distribution which, given enough time and diffusive mobility of ad-atoms, should approach the distribution functions here presented and which are unique and universal in the sense of being completely independent of the initial distribution of particle-size.

The secondary maxima observed by Hampe on the films annealed at high temperatures probably reflect parallel atom-transfer processes, not covered by the theory. It could also be due to superposition of the two distributions here presented due to equal contribution from both at high temperatures. One interesting feature that emerges from all the experimental results here presented is the importance of the interface transfer as often being rate determining. This raises the question as to the meaning of β_0, the kinetic coefficient determining the interface rate. One theoretical estimate of β_0 can be obtained by stealth, if one considers the hypothetical equilibrium of a nucleus of radius r, with ad-atom population n_r. By microscopic reversibility one must have

$$n_r \beta_0 f_2(\theta) 4\pi r^2 = \frac{f_2(\theta)}{a^2} 4\pi r^2 \nu \exp \frac{-\Delta G_1(r)}{kT}$$

where a is a lattice parameter of the nucleus, $\Delta G_1(r)$ is the activation energy to desorb an atom from the nucleus surface and which is a function of r.

Letting $r \to \infty$, one gets

$$\beta_0 = \frac{\nu}{a^2 n_{\mathrm{eq}}} \exp \frac{-\Delta G_1^0}{kT}$$

where ΔG_1^0 refers to a flat surface.

A great amount of insight into the physics of thin film formation could be gained if measurements were obtained on the isothermal time development of particle-size giving maximum distribution or total number of particles, during annealing of as-grown film. The predictions of eqs. (2.57) and (2.58), as well as of (2.72) and (2.73) should be easily verifiable. However no such data seem to be available, except in a qualitative form.

We have omitted totally the effect of impingement on one particle by another and this is intended for a future publication. Also implications of such an effect would be very difficult to verify experimentally. It is interesting however to note, that the first theory of impingement, as applied to colloids, due to Smoluchowski [38], gives the same rate law of number of particles with time as that of Lifshitz and Slezhov [39].

Of the non-conservative systems, the pertinent experiments with which to compare the results of the preceding section are in-situ observations of condensation, and these are rare. One excellent set of in-situ distribution functions has been obtained however by Poppa [40] for Ag-deposition on amorphous carbon at 425°K. His results show good agreement with an interface-transfer-controlled theoretical distribution, from $\rho = 1$, to $\rho = 2$. It is also seen as a curious fact, that, irrespective of what the over-all vapour phase supersaturation is (in Poppa's case it was $\approx 10^3$), the surface supersaturation of ad-atoms, after the end of the nucleation period (in this case 70 sec after the cessation of nucleation) remains quite low (not very much larger than 1). Also the mean particle-size in Poppa's experiments increases as the square root of time, substantiating the interface transfer model hereby proposed.

It is evident that, during the nucleation period, the surface supersaturation of the ad-atoms may actually be constant; in such a case $dr^*(t)/dt$ is 0, $\lambda = \infty$, leading to constant average particle-size. All the time nucleation is continuing, the total number of particles would continue to increase. The end of the nucleation is signalled by the drop of supersaturation below the threshold value. From this time onwards, nuclei density monotonically decreases with an accompanied increase of particle-size with time. This is actually the case in the experiments with thin films of Si, reported by Law [41]. We have re-plotted the data of Law (supposing the number of silicon atoms deposited to be proportional to time) in fig. 3 of ref. [37] and we see that in the particle-growth period:

(a) r increases as $\sqrt{t}$;

(b) the total number of particles diminishes as $1/\sqrt{t}$, ($m = 1$). Thus equations (9) and (10) of Chakraverty [37] seem to be obeyed.

In conclusion, it is seen that the formalism originated by Lifshitz-Slezhov can be quite successfully used to treat the particle-size distribution in thin continuous films, for non-conservative systems as well as conservative systems. A number of experiments verify the theoretical findings. The time development of the surface supersaturation of ad-atoms plays the key role in the condensation process.

References

[1] J.P. Hirth and G.M. Pound, Condensation and evaporation (Pergamon Press, Oxford, 1963).

[2] Nucleation, ed. by A.C. Zettlemoyer (Marcel Dekker Inc., New York, 1969).

[3] D.R. Frankl and J.A. Venables, Adv. Phys. 19 (1970) 409.

[4] T.L. Hill, Introduction to statistical mechanics (Addison-Wesley, New York, 1962) ch. 14.

[5] J.H. de Boer, The dynamical character of adsorption (Clarendon Press, Oxford, 1968).

[6] F. Tsien and G.D. Halsey, J. phys. Chem. 71 (1967) 4012.

[7] C.N. Yang and T.D. Lee, Phys. Rev. 87 (1952) 404.

[8] G. Zinsmeister, in: Basic problems in thin film physics, ed. by R. Niedermayer and H. Mayer (Vandenhoek and Ruprecht, Göttingen, 1966) p. 33.

[9] B. Lewis, Thin solid Films 2 (1967) 85.

[10] B. Lewis and D.S. Campbell, J. Vac. Sci. Techn. 4 (1967) 209.

[11] T.N. Rhodin and D. Walton, Metal surfaces, ASM-AIME Monograph, (ASM, Ohio, 1963) p. 259.

[12] R. Kikuchi, J. chem. Phys. 47 (1967) 1653.

[13] B.K. Chakraverty and G.M. Pound, Acta Met. 12 (1964) 851.

[14] C.A. Sholl and N.H. Fletcher, Acta Met. 18 (1970) 1083.

[15] B.K. Chakraverty, in: Basic problems in thin film physics, ed. by R. Niedermayer and H. Mayer (Vandenhoek and Ruprecht, Göttingen, 1966) p. 43.

[16] D. Dufayard, Thesis, Grenoble, 1969.
[17] B. Mutaftschiev and R. Kaischev, Izv. Bulg. Akad. Nauk Ser. Fiz. 5 (1955) 77.
[18] J.L. Kenty, J. chem. Phys. 47 (1967) 3120.
[19] J.L. Kenty and J.P. Hirth, Trans. AIME 245 (1969) 2373.
[20] V. Ruth, K.L. Moazed and J.P. Hirth, J. chem. Phys. 44 (1966) 2093.
[21] R. Gretz, J. Phys. Chem. Solids 27 (1966) 1858.
[22] R. Cinti, Thesis, Grenoble, 1970.
[23] J.B. Hudson and J.S. Sandejas, J. Vac. Sci. Techn. 4/5 (1967) 230.
[24] J.B. Hudson and J.S. Sandejas, Surface Sci. 5 (1968) 125.
[25] A. Nguyen, Thesis, Grenoble, to be submitted.
[26] A. Nguyen, R. Cinti and B.K. Chakraverty, J. Crystal Growth 13/14 (1972) 174.
[27] R. Cinti and B.K. Chakraverty, Surface Sci. 30 (1972) 109, 125.
[28] B.K. Chakraverty, J. Phys. Chem. Solids 28 (1967) 2401.
[29] C. Wagner, Z. Electrochem. 65 (1961) 581.
[30] D. Robertson and G.M. Pound, to be published.
[31] G. Zinsmeister, Thin solid Films 4 (1969) 363.
[32] R. Kern, A. Masson and J.G. Métois, J. Crystal Growth 3/4 (1968) 196.
[33] L.S. Palatnik et al., in: Basic problems in thin film physics, ed. by R. Niedermayer and H. Mayer (Vandenhoek and Ruprecht, Gottingen, 1966), p. 110.
[34] D.S. Herman and T.N. Rhodin, J. appl. Phys. 37 (1966) 1594.
[35] W. Hampe, Z. Phys. 152 (1958) 470, 476.
[36] M. Blackman and A.E. Curzon, in: Structure and properties of thin films (Wiley, New York, 1959), p. 217.
[37] B.K. Chakraverty, J. Phys. Chem. Solids 28 (1967) 2413.
[38] M.V. Smoluchowsky, Z. Phys. 17 (1916) 557, 585.
[39] I.M. Lifshitz and V.V. Slezhov, J. Phys. Chem. Solids 19 (1961) 35.
[40] H. Poppa, J. Vac. Sci. Techn. 2 (1965) 42.
[41] T.J. Law, J. Phys. Chem. Solids 4 (1958) 91.

P. Hartman, ed., Crystal growth: an introduction. © (1973) North-Holland Publ. Co.

EPITAXY

M. GEBHARDT

3.0. Introduction

Epitaxy is a nucleation and growth relationship between two crystalline phases, which makes it possible for a crystalline phase "g" (guest crystal) to grow in a structure-dependent manner onto a given phase of given structure "h" (host crystal) developing, in general, an interfacial region which is chemically and structurally inhomogeneous. In principle, in the completed epitaxial reaction, there exists a two-phase system consisting of two adjacent heterochemical or isochemical epitaxial partners; – host and guest.

Epitaxial host-guest relationships occur frequently by the formation of crystalline solid layers on crystalline surfaces whether the crystalline layers are formed from solutions, from melts, or from the gas-vapour state.

As an example groups of such epitaxial systems are worth mentioning: systems used in modern thin-film physics and technology, many corrosion films, many corrosion protection layers on crystalline solid materials, but also in the living body "apatite-collagen" intergrowth systems and many kidney-, ureter- and bladder-stones. Epitaxy involves, in principle, nucleation and nucleus-growth on given cristalline host-surfaces. In consequence, understanding and regulation of epitaxial layer formation implies in each case a knowledge of the nucleation and nucleus-growth laws. Only with this knowledge is it possible, for example to grow an epitaxial single crystal film on a single crystal substrate surface (e.g. Ge on spinel or Ge(doped) on Ge) or to grow a polycrystalline film with defined grain size, grain distribution and orientation.

3.1. Epitaxial nucleation and growth

The orientation behavior of growing films has been under investigation for some time, and the problems are only just beginning to be recognized. Stowell [1] has recently pointed out that the orientation of a film on a suitable single crystal substrate may not only be influenced by the nucleation of the deposit but also by the subsequent growth. In fact, it is possible to

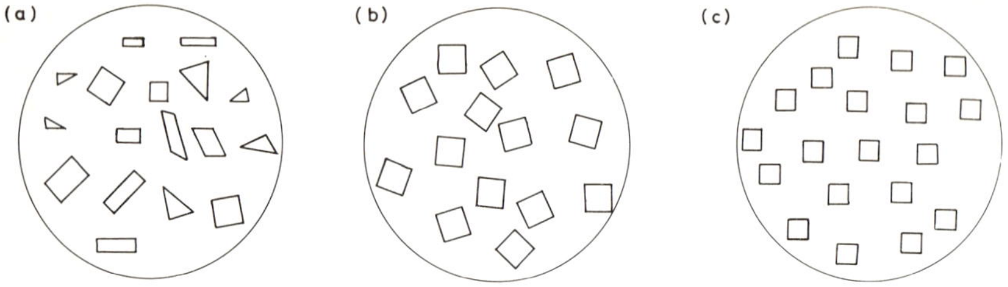

Fig. 3.1. Growth of a deposit (guest) on a substrate (host); a) fully non-oriented; b) texture orientation: a guest plane is parallel to the host surface; c) epitaxy: texture and azimuthal orientation.

divide epitaxial deposits into two groups, for the first of which the orientation is controlled by initial nuclei, and in the second of which the subsequent growth is the controlling parameter.

3.1.1. Epitaxial nucleation

Epitaxial nucleation is a special case of the heterogenous nucleation, treated by Chakraverty in ch. 2. It had been the nucleation of a *phase g* – that is the *guest phase* – on the surface of a *phase h* – the *host phase.* In fig. 3.1 the basically different kinds of intergrowth host/guest are schematically shown.

We see, that three types of orientation relationship between host and guest exist. According to fig. 3.1a host and guest crystallites are completely non-oriented with respect to one another. According to fig. 3.1b an orientation of texture may exist; that is, the guest crystallites all grow with the same low-index stable lattice plane on the substrate surface. They are not oriented with respect to one another. Finally, in Fig. 3.1c guest crystallites and the crystallites of the substrate show both, the textural and azimuthal orientation towards one another. Only these latter orientation relationships are designated as *epitaxial.*

As Bauer [2] has shown already, the deposition of a guest phase on the surface of a host phase is only possible, if the change of the free enthalpy* of the system, $\Delta G = G - G_0$, is negative. Epitaxial intergrowths host/guest are only to be expected, if additionally, for special crystal faces (hkl) the free interfacial energy host/guest (σ_{hg}) as a function of the azimutal orientation shows marked minima caused by the atomistic structure of the host surface. (fig. 3.2).

The *interfacial free energy* (σ_{hg}) increases with the *misorientation angle* φ. This is caused by elastic tensions in the substrate/nucleus interface and by the increase of the surface energy as a result of displacements, which appear by

(*otherwise known as the Gibbs free energy)

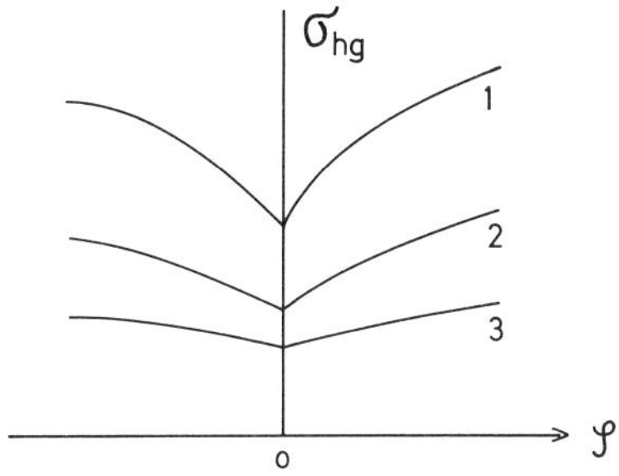

Fig. 3.2. Generalized σ_{hg}-curve vs. misorientation angle φ at low (1) and high (2, 3) supersaturations.

misfitting. These changes are important when the supersaturation is low (curve 1 in fig. 3.2.) The absolute value of σ_{hg} is reduced if the supersaturation is high (curve 2 and 3 in fig. 3.2). The same effect is caused by the relative change of σ_{hg} in relation to the misorientation. The difference between the nucleation probabilities for epitaxial and nonoriented growth gets smaller. The growth of the nonoriented nuclei leads to the formation of a polycrystalline layer. This change of the host/guest specific free interfacial energy can be quantitatively expressed, if the misorientation angle is low: (see Aleksandrov in ref. [3]):

$$\sigma_{hg} = \sigma_{hg}^0 + \frac{Eb\varphi}{4\pi(1-\nu)}\left(1 - \ln\frac{\varphi}{\varphi_{max}}\right), \tag{3.1}$$

σ_{hg}^0 = specific free interfacial energy host/guest for epitaxy,
E = elastic modulus (if host = guest),
ν = Poisson constant,
b = Burgers vector,
φ = misorientation angle.

Additional parameters are important for the realisation of an epitaxial layer formation.

a) *Contact angle.* Let us remember first the *nucleation frequency* expression used by Chakraverty (eq. 2.14):

$$J = \beta \exp\frac{\Delta G_{des} - \Delta G_{diff}}{kT} \cdot \exp\frac{-\Delta G_{i*}}{kT}, \tag{3.2}$$

in which $\beta = 2\pi r^* p Z(\sin\theta) n_0/(2\pi m kT)^{\frac{1}{2}}$,

$$\Delta G_{i^*} = \frac{16\pi\,\sigma_{gv}^3}{3\Delta G_v^2} \cdot f_1(\theta),$$

$f_1(\theta) = \frac{1}{4}(2 - 3\cos\theta + \cos^3\theta)$, see also ch. 2 §2.3,

ΔG_{des} = free desorption energy of ad-atoms from the host-surface,
ΔG_{diff} = free diffusion energy of ad-atoms on the host-surface,
ΔG_{i^*} = free formation energy for a critical nucleus containing i^* atoms,
ΔG_v = Gibbs free energy change per unit volume of guest material,
p = vapour pressure,
Z = Zeldovich non-equilibrium factor,
r^* = radius of critical nucleus,
n_0 = concentration of monomers,
m = molecular mass,
σ_{gv} = guest-vapour free interfacial energy.

If we combine, for the moment, irrelevant quantities in the nucleation frequency expression (3.2) in A and B, we get an expression for the critical nucleation frequency J^* as a function of the contact angle (θ) for the spherical cap model: (fig. 3.3a):

$$J^* = \exp[A] \cdot \sin\theta \cdot \exp[-Bf_1(\theta)], \tag{3.3}$$

or

$$\ln J^* = A + \ln\sin\theta - Bf_1(\theta), \tag{3.3a}$$

with $A = \ln[2\pi r^* p Z n_0/(2\pi mkT)^{\frac{1}{2}}] + (\Delta G_{des} - \Delta G_{diff})/kT$

and $B = \dfrac{16\pi\,\sigma_{gv}^3}{3\,kT(\Delta G_v)^2}$.

Let us define: J_1 as the nucleation rate appropriate to an epitaxial orientation and which is characterized by the contact angle θ; J_2 as the nucleation rate appropriate to an unfavourable orientation (also characterized by its own contact angle θ); then the quotient

$$\beta_r = (J_1/J_2) \gg 1$$

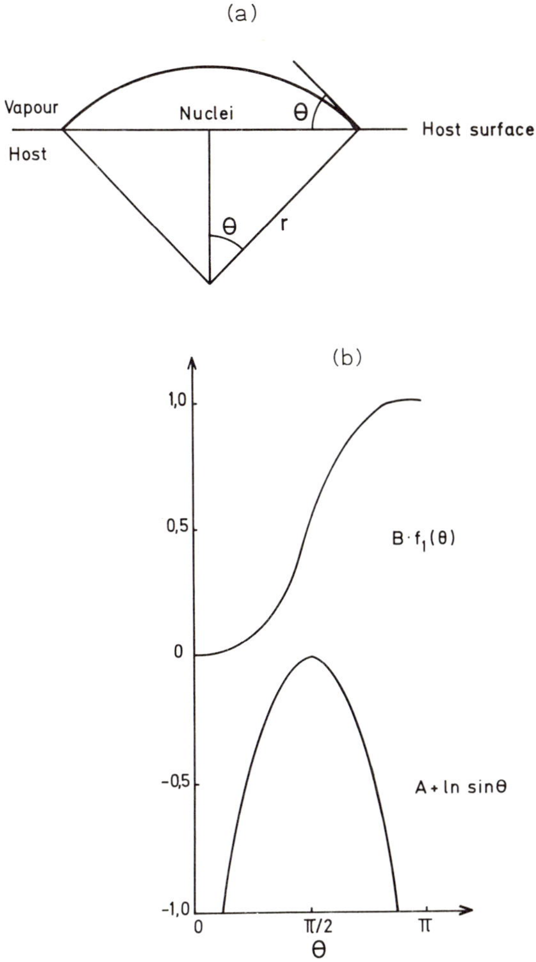

Fig. 3.3. a) spherical cap nuclei;
b) the functions $f_1(\theta)$ and $\ln \sin \theta$.

shows how much more favourable the epitaxial orientation for nucleation is than any other orientation. If the experimental conditions are given, all parameters, except the contact angle, are equal in both cases. Therefore β_r is according to eq. (3.3):

$$\beta_r = (\sin \theta_1/\sin \theta_2) \exp \{B[f_1(\theta_2) - f_1(\theta_1)]\} . \tag{3.4}$$

As σ_{hg} is smaller for the epitaxial orientation than for all other orientations then:

$$\theta_2 > \theta_1 \text{ and } f_1(\theta_2) > f_1(\theta_1).$$

From eq. (3.3) it can be deduced that not only very large contact angles lead to low values of A and B but also very small ones, which are necessary for the start of nucleation. Further for a given difference $(\theta_2 - \theta_1)$ minimal values of $f_1(\theta_2) - f_1(\theta_1)$ are to be expected, if $\theta \to 0°$ or $\theta \to 180°$ according to fig. 3.3b.

Therefore both factors in the exponent of eq. (3.4) assume low values for very small and for very large contact angles. As $(\sin\theta_1/\sin\theta_2) < 1$ it can be concluded that β is relatively small for small and large contact angles, too. This means that for these cases epitaxial nucleation is not significantly favoured in comparison with non-epitaxial nucleation. The conditions most favouring epitaxial nucleation seem to occur for a contact angle which is given by the upper curve of fig. 3.3b being tangent to the lower one, that is for

$$B = \frac{4\cos\theta}{3\sin^4\theta}.$$

It must be emphasised however that the results of this discussion are only conclusive for spherical nuclei with contact angles not smaller than about 5°; for very small contact angles the equations, based on this model, lose their validity. For example, with very strong adhesion forces between host and guest there is a perfect wetting of the substrate ($\theta = 0°$) and a perfect epitaxy is possible as the total process of layer formation is just a growth process.

b) *Substrate temperature.* Now the influence of the substrate temperature on the nucleation rate should be discussed. The factor A in eq. (3.3) for the critical nucleation rate increases with rising temperature. The main reason is the increase of the vapour density and the vibration frequency of the diffusing atoms on the surface which give rise to the critical nucleation rate. For the critical nucleation rate, the value of B has therefore to increase too with rising temperature. Consequently, acoording to eq. (3.4) the value of β_r has to rise with increasing substrate temperature.

As B of eq. (3.3) is an abreviation of

$$B = \frac{16\pi\sigma_{gv}^3}{3(\Delta G_v)^2 kT}, \tag{3.5}$$

B decreases if the supersaturation exceeds the critical value of ΔG_v, and so β_r decreases also. That means that, with increasing supersaturation, the epitaxial nucleation becomes less and less favoured.

How strongly β_r is connected with the supersaturation can be derived from eqs. (3.4) and (3.5).

$$\frac{\partial \ln \beta_r}{\partial \Delta G_v} = (f_1(\theta_2) - f_1(\theta_1)) \cdot (\partial B / \partial \Delta G_v), \tag{3.6}$$

$$\frac{\partial \ln \beta_r}{\partial \Delta G_v} = (f(\theta_2) - f(\theta_1)) \cdot \frac{-32\,\sigma_{gv}^3}{3kT(\Delta G_v)^3}. \tag{3.7}$$

Eq. (3.7) shows that the rate of decrease of β_r with increasing supersaturation is greater the lower the substrate temperature. This demonstrates that epitaxy is always favoured by increasing the substrate temperature whenever the nucleation process can be described by the classical nucleation model.

The conditions can change, however, if the nucleation is influenced by chemical reactions. If, in a special case, the free reaction enthalpy shows a considerable reduction in the concentration of adsorbed monomers with increasing temperature then it causes a decrease in A with increasing temperature.

In the case of the critical supersaturation the value of B must therefore be relatively small and β, too, decreases with increasing substrate temperature. In such a case the epitaxy is apparently favoured by the low substrate temperature. The influence of the substrate temperature on the epitaxial nucleation is highly dependent on the nucleation mechanism of the special case which is under consideration.

Although the above arguments are based on the spherical calotte model, it is to be expected that the temperature dependence of β_r, expressed by eq. (3.4), will be similar for all models. Only in the case of nucleation which is influenced by chemical reactions can epitaxy be supported at low substrate temperatures independently of the nucleus model.

Walton [4a, b] suggested a model for epitaxial nucleation, in which epitaxy can occur with critical nuclei containing only 2 or 3 atoms. According to equation (3.6) the value of B should be very small for such small nuclei. Such a situation, which might be possible for low substrate temperatures, would cause a low value of β_r according to equation (3.4). In consequence this would be unfavourable for epitaxial nucleation. In addition it is possible, that the special orientation of the small nuclei is indeed energetically favoured by

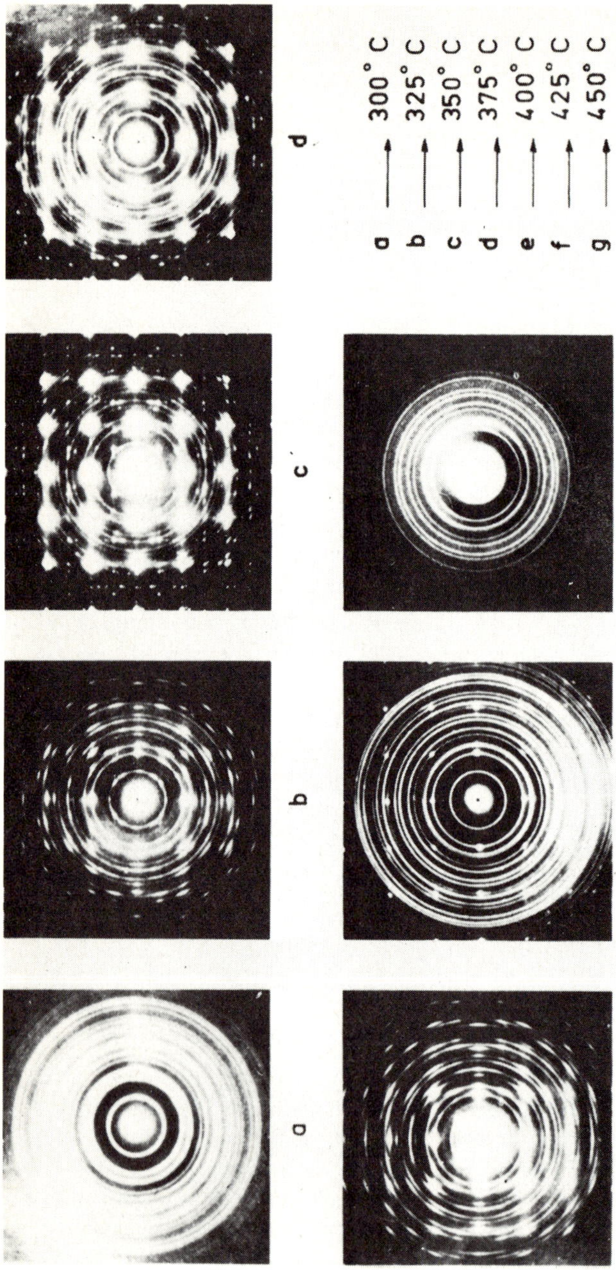

Fig. 3.4. Epitaxy of InSb on KI at various temperatures as shown by electron diffraction; after Lafourcade et al. [38].

comparison with any other orientation, so that it is still true that $\beta_r \gg 1$, although the value of B is relatively small. In these cases the Walton mechanism of the epitaxial nucleation might apply.

Fig. 3.4 gives an example: growing InSb on KI. It can be noticed, how the orientation of the InSb crystals (ca. 325°C) starts, improves and deteriorates again with rising temperature. The deterioration can be explained by the great liquid-like thermal motion of the K^+ and I^- ions on and from the host surface at the highest temperature. This leads in the end to an amorphous InSb-film above ca. 450°C.

Summarising, it can be stated that in normal cases epitaxial nucleation is favoured by high substrate temperatures, low supersaturation and special interactions between host and guest, characterized by θ. The contact angle is dependent on the various interfacial energies σ_{hg}, σ_{gv} and σ_{hv} (h = host; g = guest; v = vapour) which again are highly influenced by the coherence in the interface between nucleus and substrate, the adsorption of contaminants, the orientation of the substrate and the presence of structural faults in the substrate surface. We limit our discussion to the role of structural faults and contamination.

c) *Structural faults and contamination.* Usually the nuclei are not distributed regularly over the substrate surface; on the contrary some places of the surface seem to favour the nucleation event more than others. The first observations of this kind were probably made by Andrade and Martindale [5], who evaporated Au and Ag on quartz and glass. They observed a favoured nucleation on micro-cracks in the surface. Chakraverty and Pound [6] wrote about a favoured nucleation of gold on the macroscopic steps of NaCl and LiF. As the contact angle between nuclei and substrate was 95° these results are in good accord with the theoretical expectations of fig. 3.5. Nuclei, formed by condensing Cd, were found on the flat parts of the surface which again is in accordance with the theory, as the contact angle was 120°. Bassett [7] observed a decoration of monoatomic steps on gold surfaces caused by gold nuclei. His results suggest that the critical contact angle, which separates according to fig. 3.5 the cases of favoured nucleation on steps from those on flat parts, has nearly the same value for monoatomic and macroscopic steps. As the contact angle $\theta = 95°$ does not differ highly from the critical value 105° (fig. 3.5) some nuclei should be expected in the flat regions, too, according to the theory.

Contaminated layers can result in an increase or a decrease in the nucleation rate. Thus the nucleation rate of silver on copper in the temperature interval from 200° to 400°C in a vacuum of 10^{-9} Torr is higher than in a vacuum of 10^{-5} Torr. By contrast, Sears [8] reports that the adsorption of

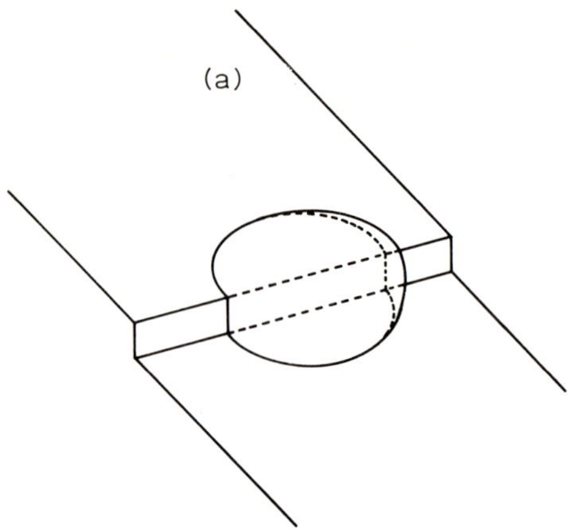

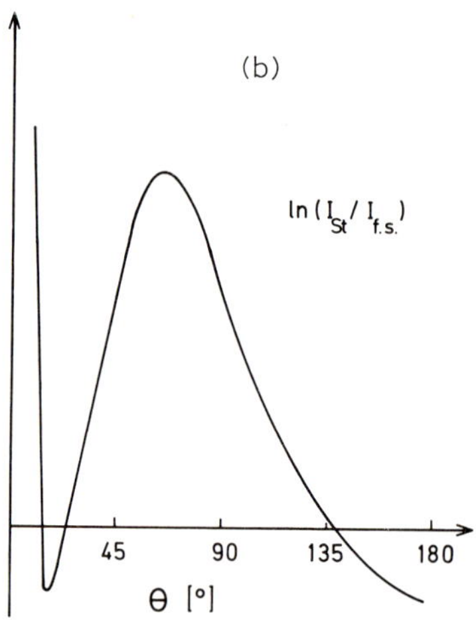

Fig. 3.5 a) Spherical cap nuclei on a mono-atomic step; b) ratio of nucleation rate on a step and on a flat surface as a function of the contact angle, according to Chakraverty and Pound [6].

Fe^{+++} cations increases the nucleation rate of solid LiF from an aqueous solution. It is to be expected that such a nucleation process is more or less equivalent to nucleation from the vapour phase.

Some recent papers which deal with both the influence of contaminants as well as structural faults demonstrate that a whole spectrum of surface sites with different catalytical activities exist, and that each type of surface site has its own special value of critical supersaturation. Of course each special supersaturation has its own effective contact angle. Koutsky et al. [9] demonstrated that adsorption by polar molecules can gradually change the catalytical efficiency of a substrate during a certain time. Hirth and Pound [10] inform us about a number of experiments, which show preferential nucleation on the structural faults and the role of the adsorption layers in a qualitative way.

3.1.2. Epitaxy during nucleus growth

Bassett [11, 12] and Pashley, Stowell et al. [1, 13–18] have developed the application of *transmission electron microscopy* for the direct observation of nucleation and nucleus growth. Pashley [15] has written a survey of the subject.

After the nucleation stages, the nuclei will sooner or later start to agglomerate. This agglomeration will increase as islands are formed. At a later stage, island chains join up until only channels are left. Eventually, even these will disappear and a continuous film will be obtained; the surface irregularities will be smoothed out by further growth.

Thin film growth experiments carried out inside an electron microscope have shown that the islands drift on the substrate during their growth, that they rotate and that they rapidly “flow together” after their contact, as it is schematically shown in fig. 3.6. This movement of the nuclei leads to a picture of liquid-like behaviour. However it could be shown that the nuclei are small single crystals during all stages of the growing process. Stowell et al. have shown that it is possible to apply sintering theory to the agglomeration process such that the mass transport mechanism is one of *surface diffusion*. One of the features of the *coalescence* phenomena is that the time taken for the coalescence of two islands is a function of islands size, the smaller the island the more rapid is the diffusion process. Given this, Stowell points out that there is a critical stage of growth at which the coalescence time of an island is greater than the time taken for the island to continue to grow and therefore touch and attempt to coalesce with the nearest neighbours.

Examination of growing films in the electron microscope has also shown that recrystallization by grain boundary migration occurs during growth.

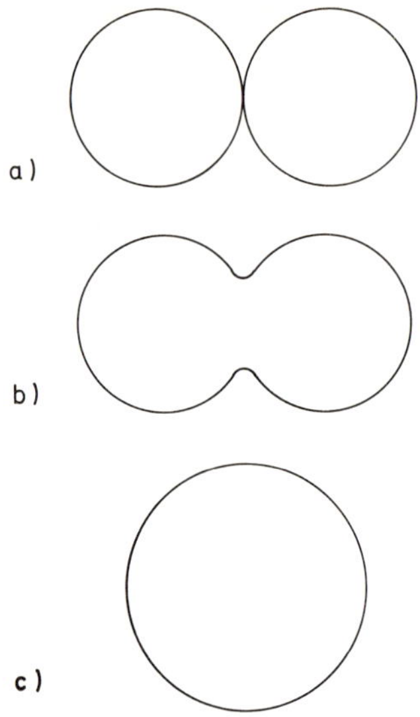

Fig. 3.6. Different coalescence stages of two spherical islands.

Grain boundaries are formed in the necks between islands of different orientations and stay in the islands until the neck is eliminated. This recrystallization is a rapid process in the early stages of growth when the islands are small, but after the critical growth stage is reached, no further recrystallization can occur by this process. Further grain growth can occur as surface irregularities are removed as the film grows and it is also possible to anneal the final film to increase the grain size. The effect of deposition parameters on the coalescence process has recently been analysed by Stowell [1], and he has shown theoretically that cases can exist in which growth is more important than nucleation in determining epitaxy. These results are in experimental agreement with those of other authors [19].

3.1.3. Influence of the host on the crystal structure of the guest

Layers growing on another material can crystallize in the structure of the substrate instead of the normal structure of the layer material. According to the research of Schulz [20] and Pashley [21] the Cs and Tl halides (which normally have the CsCl-structure) can accept the NaCl-structure, if they grow on Na and K halides.

Haase [22] and Jesser and Matthews [23, 24] grew cubic face centered γ-iron on (001)-surfaces of copper. Mariano and Chopra [25] grew SnS, SnSe and $PbSnS_2$ (usually having an orthorhombic structure) in a form having the NaCl-structure, by depositing the former compounds on NaCl substrates. Goddard and Wright [26] as well as Jesser and Matthews [27] have produced thin films of face-centered α-Co by growing the metal on Cu.

In order to make clear the substrate influence on the deposit structure a closer look at the growth of α-Co on Cu must now be taken. In ultra high vacuum (10^{-7}–10^{-9} Torr) thin Co films were deposited on cold (20°C) Cu surfaces, cleaned in the evacuated apparatus [27]. Either a diaphragm was permanently installed between the vapour source and the substrate or it was put between them during the experiment so that Co layers with a gradient of thickness resulted. A closer look to the thinner parts of the deposited layers led to the following conclusions:

a) the films grew as nearly monoatomic layers
b) the first deposited layers had a face-centred cubic structure (α-Co)
c) the layers were strained so far that their structure fitted the structure of the substrate.

When the thickness of the layers reached critical values (for Co on Ni ca. 50 Å) long and straight misfit dislocations were formed. The dislocation lines ran exactly parallel to the [110] and [$1\bar{1}0$] directions in the (001) plane of the grown layer. With increasing thickness of the Co film the total length of the dislocation lines grew and it became apparent that stacking faults connected most of the dislocation lines with the free Co surface. Although the forming of partial dislocations has not yet been observed, it is possible that they occur according to a mechanism schematically shown in fig. 3.7. The Thompson notation [28] is used for the Burgers vector of the dislocation. The positive direction of the dislocation line reaches from the substrate into the layer if the unit-cell parameter of the substrate is smaller than that of the deposit. Alternatively, it reaches from the layer to the substrate if the unit-cell parameter of the substrate is bigger than that of the deposit (a_0Cu > a_0α-Co > a_0Ni). The dashed plane in fig. 3.7 represents the stacking faults. The length of the stacking fault in fig. 3.7 grows with increasing thickness of the Co film. At last a thickness will be reached for which the equilibrium distance between the partial dislocations becomes infinite. It is useful to define this thickness as the critical thickness which is decisive for the creation of the partial dislocations. This thickness is reached when the force influencing the partial dislocation is equal to the line tension of the partial dislocation. It is caused by the elastic strain in the Co layer and by the negative stacking fault energy in f.c.c. Co.

It is known that the α phase, which is stable above 430°C, transforms into β-Co by plastic deformation only, and not by temperature lowering alone.

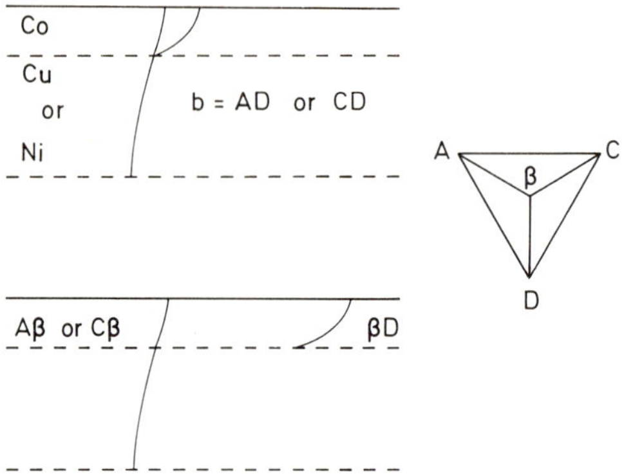

Fig. 3.7. Formation of a stacking fault and of partial adaption dislocation lines by dissociation of a dislocation.

The deposition of Co onto Cu- and Ni-surfaces at 20°C leads however to β-Co in a thicker layer, because of the elastic strain in the Cu/Co- and Ni/Co-interfaces. One method of retarding this transformation of α-Co in a thinner layer into β-Co in a thicker layer is to deposit the Co layers on substrates which are free from stacking faults. In this case no misfit dislocations can be produced by the mechanism shown in fig. 3.7. The phase transformation must now occur by nucleation and the subsequent formation of partial dislocation lines. The film thickness which is necessary for starting such a nucleation is dependent on the elastic strain in the Co film. An other possibility is that it is dependent on the smoothness of the Co surface or on the presence of point defect agglomerations. If the elastic strain is small, the surface smooth with no point defect agglomerations there, then it seems to be possible to avoid the occurrence of a transformation altogether.

3.1.4. Summary

We have seen that orientation effects can be produced at all stages in film growth. It is now recognized that both nucleation and growth stages can be important in obtaining epitaxy. High supersaturation associated with growth on substances with low adsorption energies precludes perfect epitaxy in the resultant guest deposit. Epitaxial relationships can lead to the formation of metastable or "host-like" crystal structures of the guests.

3.2. Epitaxy and its applications *)

3.2.0. Introduction

Epitaxial layers are important factors in science and industry.

They are interesting as: –

a. natural and artificial corrosion layers,
b. anticorrosion layers for protecting underlying materials,
c. intermediate layers between metal and silicates, for example Co, Ni or Fe oxide layers for enamelling
d. layers to improve the chemical or abrasive properties; for example metal borides, carbides or nitrides.

The fastest growing applications are epitaxial semiconducting layers (for example Si and Ge) on insulator surfaces or on single crystal surfaces of nearly the same composition. These examples may be sufficient to demonstrate the extensive application of epitaxial layers. Let us discuss some of these in detail.

3.2.1. Epitaxial corrosion layers on iron

The corrosion of metals takes place, in general, spontaneously with the release of free energy. In the primary contact of a "pure" metal surface with the atmosphere, phases are naturally formed which result from the high atmospheric oxygen and water content. Pronounced oxidation phases are formed. The coating thickness of these, generally very thin ($\leqslant 0.1$ micron), primary corrosion films may be very different, depending on the growth velocity and the nature of the reaction species. Here, we are interested only in the ideal case of an essentially impervious coating. This is generally feasible only in a precisely controlled experiment, but it is occasionally also observed in corrosion, which takes place freely. For the case of the impervious crystallite coating of the initial corrosion layer, it follows that further growth in thickness is caused almost exclusively through diffusive transport (topotaxial transport) of the carrier metal through the initial corrosion film up to the particular "coating-atmosphere" phase boundary. As corrosion progresses, the reaction film thickens and the path of diffusion of the carrier metal to the coating-atmosphere phase boundary is lengthened more and more until, after achievement of a certain thickness of the layer, growth kinetics and the growth law change their character [29, 30].

With progressive thickening of the primary layer, the oxygen potential

* This section is based on a discussion contributed to the General Motors Symposium in 1967 [29].

drops more and more at the "metal-oxide film" phase boundary. With increase of temperature (hydrothermal or high-temperature corrosion) there is then formed, after the redox reactions have occurred in the solid state between the metal and the primary oxide, a non-uniform, thick layer of less highly oxidized phases which is designated as the secondary layer. The growth mechanism of the primary layer changes almost simultaneously owing to the complete exhaustion of the iron supply to the "layer-atmosphere" phase boundary. Also, the growth mechanism of the primary layer changes because of the atmosphere which results in the formation of higher oxide and eventually hydrated terminal phases.

In the oxidation corrosion of iron-surfaces, there result collectively very complex-structured corrosion layers with a rigid sequence of phases with respect to both time and space. These phases are not always built one after the other with an increasing degree of oxidation from the carrier to the atmosphere. Also, when enough water is present, we have to consider the degree of hydration (table 3.1).

In the special case of the slow oxidation of iron at moderately elevated temperatures (for example, 400°C) and strongly reduced O_2 pressure (for example 10^{-4} Torr), a very thin film of Fe_3O_4 forms first. With progressive

Table 3.1
Oxide and Hydroxide Corrosion Products of Pure Iron under Various Environmental Conditions

Increasing Degree of Oxidation	Corrosion Products on Iron		
	Atmospheric Corrosion $-50°C < T < +50°C$	Hydrothermal Corrosion $T > 50°C$	High-Temperature Corrosion $T > 600°C, P \leqslant 1$ atm (dry oxides)
↓	FeO wüstite (metast.) $Fe(OH)_2$	$Fe(OH)_2.2H_2O$ $Fe(OH)_2$ hex.	FeO (stab.)
	$Fe_3(O,OH)_4$ cub. Fe_3O_4 magnetite	$Fe_3(O,OH)_4$ Fe_3O_4 cub.	Fe_3O_4
	γ-Fe_2O_3 maghemite α-FeOOH goethite β-FeOOH γ-FeOOH lepidocrocite	γ-Fe_2O_3 tetr. α-FeOOH o.rh. β-FeOOH tetr. γ-FeOOH o.rh. δ-FeOOH hex.	γ-Fe_2O_3 α-Fe_2O_3

growth in thickness of this primary oxide layer, a metastable FeO layer is formed. While redox reactions in the solid phase proceed, γ-Fe_2O_3 is formed at the boundary between the unchanged carrier lattice and the atmosphere. In the case of a layer thickness of about 2000 Å, there is finally a change of the reaction mechanism whereby γ-Fe_2O_3 is converted into α-Fe_2O_3. Accordingly, the oxidation of iron at elevated temperatures leads to a layered sequence of phases with increasing oxygen content from the inside to the outside in steps and therefore leads to optimum stability with respect both to the free atmosphere and to the iron substrate. One then has the following sequence of structures (written as stoichiometric phases): Fe-FeO-Fe_3O_4-γ-Fe_2O_3-α-Fe_2O_3.

All these phases are connected with one another through orientation relations, and they form reaction zones which are more or less deep-seated between carrier crystallites and guest phases. Therefore, in a rigid sense they are not only epitaxial (rigidly two-dimensional relationship), but also, through successively repeated epitaxial reactions and through topotaxial events, as fig. 3.8 shows, they form whole chains of epitaxial intergrowing guest phases (orientation chains). In table 3.2 are assembled the known epitaxies and topotaxies between iron metal, iron oxides, and iron hydroxides [29, 30].

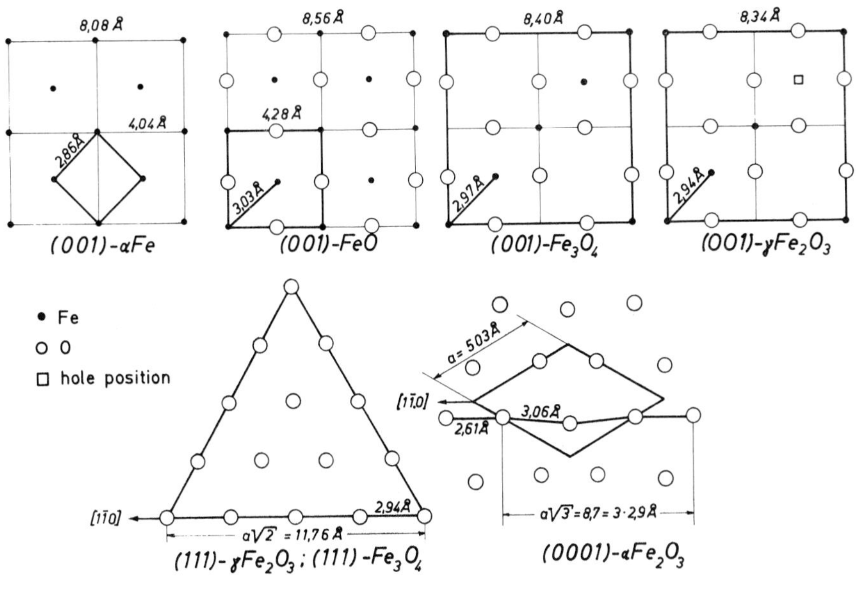

Fig. 3.8. Affinitive lattice planes of iron and iron oxides.

Table 3.2.
Orientation Relationships of Iron Metal, Iron Oxide, and Iron Hydroxide

Epitaxies	Fe-FeO Fe-Fe_3O_4 Fe_3O_4-γ-Fe_2O_3	Fe-$Fe(OH)_2$ Fe-γ-Fe_2O_3 Fe_3O_4-α-Fe_2O_3	Fe-$Fe_3(O, OH)_4$ Fe_3O_4-FeO
Topotaxies	Fe_3O_4-γ-Fe_2O_3 γ-Fe_2O_3-α-Fe_2O_3 δ-FeOOH-α-FeOOH	Fe_3O_4-α-Fe_2O_3 γ-FeOOH-γ-Fe_2O_3 δ-FeOOH-α-Fe_2O_3	Fe_3O_4-FeO α-FeOOH-α-Fe_2O_3 β-FeOOH-γ-Fe_2O_3
Orientation and nuclei chains	FeO-Fe_3O_4-γ-Fe_2O_3-α-Fe_2O_3 / Fe– \\ $Fe(OH)_2$-$Fe_3(O, OH)_4$-Fe_3O_4-γ-Fe_2O_3-α-Fe_2O_3 γ-FeOOH δ-FeOOH \\ / -γ-Fe_2O_3-α-Fe_2O_3- / \\ β-FeOOH α-FeOOH		
Hereditary orientation	α-Fe-(Fe_3O_4)-α-Fe_2O_3 (001) (11$\bar{2}$4)		

3.2.2. Protective layers and epitaxy

3.2.2.1. Burnishing layers on iron alloys. Burnishing is defined as the production of thin, brown-black to deep-black oxide films on iron and iron alloys through the oxidation of carefully cleaned metal surfaces. The three following very different processes are used in their production: 1) reaction with high-pressure steam, 2) oxidation in hot electrolyte solutions, and 3) reactions with oxidizing molten salts. In our own experiments [30], we chiefly used the third method of production. If in this case one selects a melt of sodium hydroxide + sodium nitrate + sodium nitrite as the burnishing bath, then the following polishing phases are formed on α-Fe: γ-Fe_2O_3, Fe_3O_4, γ-Fe_4N, ε-Fe_3N and ε-Fe_2N. In addition, the following other phases in the burnishing layers have been detected on several steels: (a) on V2A steel: $NiFe_2O_4$, $Fe(Cr,Fe)_2O_4$ and $NiCr_2O_4$; (b) on Co-W steel of the approximate composition 50% Fe, 40% Co, 7% W and 2.5% Cr: Co_3O_4, $CoFe_2O_4$ and Co_2N; (c) on Fe-Mn steel: MnO and $MnFe_2O_4$. All these layer phases are oxides or nitrides throughout at the selected reaction temperatures and reaction conditions. The systematic removal of these burnishing layers of different formation periods and thickness through ionic-radiation etching revealed the following sequence of phases: α-Fe–Fe(N)–γ-Fe_4N–ϵ-Fe_3N–ϵ-Fe_2N–Fe_3O_4–γ-Fe_2O_3. The following layer-formation mechanism results.

For short-time action of the burnishing melt on the pure iron surface, there is formed first, by analogy with the oxide corrosion layers [31], a very thin Fe_3O_4 film. However, in deviation from the simple oxide corrosion, active nitrogen (N*) is always present as a reaction species (formed by the reciprocal effect of iron with NO_3^- and NO_2^-). This N* diffuses through the already formed primary film of magnetite and is first absorbed in the regions of the iron lattice adjacent to the surface in the form of a solid solution, Fe(N). In the further course of the burnishing reaction, the oxide layer thickens outward in the direction of the melt, and then topotaxial γ-Fe_2O_3 is formed in the outer parts. At the same time, however, the diffusion of N* occurs in the direction of the remaining pure iron lattice with formation of other nitride phases.

The former layer formations correspond to a great extent to oxide corrosion and therefore do not require any further explanation. However, the epitaxial growth of nitride secondary layers at the phase boundary of the oxide primary layer with the pure iron lattice which results in the thickening of the oxide corrosion layer is novel. The nitride formation originally consists solely in a statistical absorption of N* in the octahedral holes of the metal lattice. It leads rather rapidly, with increasing N* supply, to the formation of genuine nitride phases with a nitrogen content increasing from the inside outward. With the advancing growth of the thickness of this nitride layer and simultaneously advancing growth in thickness of the oxide layer, the diffusion path for N* increases successively and rapidly. This finally does not advance up to the phase boundary of the unchanged carrier lattice. This gives the sequence of nitride phases: Fe(N)–Fe_4N–Fe_3N–Fe_2N. All these secondary phase formations naturally are based on the reactions in the solid state. Thus, with the burnishing layers on α-Fe, there exists a sequence of phases with a rising degree of nitride content from the inside to the outside, followed by layers of increasing degrees of oxidation: Fe–Fe(N)–γ'-Fe_4N–ϵ-Fe_3N–ϵ-Fe_2N–(FeO)–Fe_3O_4–γ-Fe_2O_3–(α-Fe_2O_3).

The epitaxies between the particular adjacent phases of this sequence of phases are listed in table 3.3. Fig. 3.9 gives in addition a representation of the lattice geometric relationships between α-Fe and the nitride phases. As can be seen from both, the host-guest lattice comparisons are excellent throughout.

In the replacement of the pure iron carrier, through iron-, cobalt-, and nitrogen containing hosts, there are formed additional oxides such as $CoFe_2O_4$ and Co_3O_4 as well as nitrides – for example, Co_2N – which, on

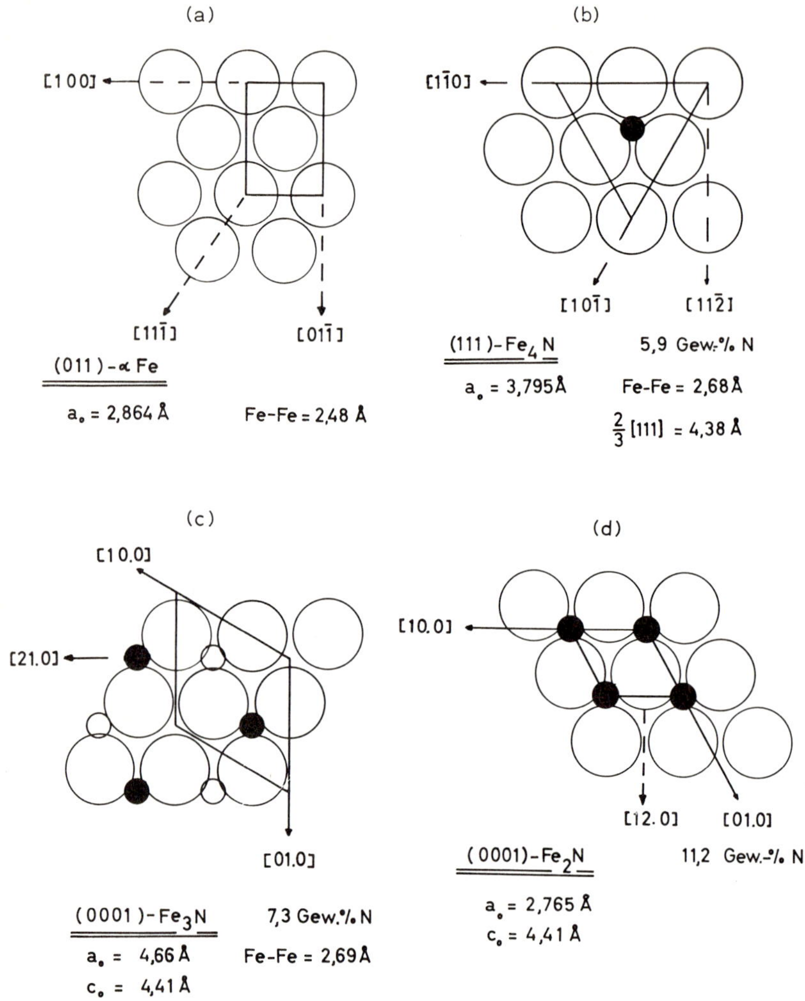

Fig. 3.9. Affinitive lattice planes of iron and iron nitrides; a) (011)–α-Fe; b) (111)–Fe_4N; c) (0001)–Fe_3N; d) (0001)–Fe_2N.

the basis of their lattice parameters, can be arranged readily in the epitaxial chains.

Table 3.3
Topotaxy Chain in Burnishing Layers on Iron

Phases		Epitaxies		Periods (Å)		Misfit
Host	Guest	Host	Guest	Host	Guest	(%)
α-Fe	Fe_4N	(011) [100]	(111) $[1\bar{1}0]$	2.86	2.68	− 6
Fe_4N	Fe_3N	(111) $[1\bar{1}0]$	(0001) [21.0]	2.68	2.69	< 1
Fe_3N	Fe_2N	(0001) [21.0]	(0001) [10.0]	2.69	2.77	+ 3
Fe_2N	Fe_3O_4	(0001) [10.0]	(111) $[1\bar{1}0]$	2.77	2.97	+ 7
Fe_3O_4	γ-Fe_2O_3	(111) $[1\bar{1}0]$	(111) $[1\bar{1}0]$	2.97	2.94	− 1

3.2.2.2. Carbide, nitride and boride layers on metal substrates. In many applications of metals, their bulk properties are less important than their surface properties. For many applications, a surface improvement is sufficient on some materials through appropriate coating layers. Thus, iron or steel surfaces are improved with the best results by employing thin chromium, vanadium, niobium, tungsten or molybdenum layers, which subsequently can be improved further through formation of carbides, nitrides or borides.

a) Formation of carbide layers on chromed steel. As a first example for an improvement of a steel surface, we use galvanic chroming with subsequent gas carburisation (process for carbide formation). The galvanic chroming of iron or steel leads to a chromed layer alloyed metallurgically with the substrate metal, which is not connected directly with the pure iron lattice but via an iron-chromium mixed crystal layer. In the subsequent carburisation, for example, with CCl_4 and H_2 at higher temperatures, a heterogeneous single-phase layer is formed in the previously described burnishing layers which consists of a sequence of layers built up from phases with a carbon content increasing from the inside to the outside [32]. This sequence of layers is made up of (1) a thin outer layer consisting of Cr_3C_2, (2) a thicker middle layer of Cr_7C_3 and finally, (3) in contact with the metallic chrome, a thin layer which is built up from the phase $Cr_{23}C_6$. The orientation relations to be expected and partly found, namely epitaxies, topotaxies and nuclei chains, are recorded in table 3.4.

Table 3.4
Epitaxies in Carbide–Chrome Layers

Phase:	Cr	$Cr_{23}C_6$	Cr_7C_3	Cr_3C_2
Symmetry:	Cubic-Body Centered	Cubic	Hexagonal	Orthorhombic
Intergrowth plane	(011)	(001)	(0001)	(001)
Direction	[1$\bar{1}$1]	[110]	[10.0]	[010]
Periods (Å)	15.02	15.07	13.98	11.46
Misfit (%)	← + 0.3 →	← – 6.9 →	← – 10.9 →	

b) *Formation of nitrides of tinanium.* In the preparation of titanium nitrides with pure nitrogen at temperatures of about 950°C, there is formed, in 24 hours, a nitride layer about 80 microns thick and of a golden colour, which, in itself, is again built up of several zones [32]. The outermost film, only 0.1 micron thick, consists of the cubic δ-phase TiN. It is preceded by an 8- to 10-micron-thick layer of the hexagonal ϵ-phase Ti_3N, which gives the high microhardness of 1500 kg/mm^2. After this follows the innermost layer with a solid solution of nitrogen in titanium with a decreasing lattice period and microhardness (1300–700 kg/mm^2) toward the pure carrier lattice.

The formation of the nitride layer is diffusion-controlled with transportation of N* toward the pure titanium lattice. The correlation between Ti, Ti(N), Ti_3N and TiN compounds (table 3.5) indicates once again that topotaxial phases are formed.

Table 3.5
Epitaxies in Nitride Layers on Titanium

Phase:	α-Ti	Ti(+N)	Ti_3N	TiN
Symmetry:	Hexagonal	Hexagonal	Hexagonal	Cubic
Intergrowth plane	(0001)	(0001)	(0001)	(111)
Direction	[10.0]	[10.0]	[12.0]	[1$\bar{1}$0]
Periods (Å)	2.95	2.96	2.84	3.00
Misfit (%)	← + 0.3 →	← – 3.6 →	← + 5.6 →	
	← (+ 1.7) (α-Ti to TiN) →			

Table 3.6
Epitaxies in Boride Layers on Tungsten

Phase: Symmetry:	W Cubic-Body Centered	W_2B Cubic	WB Orthorhombic	W_2B_5 Hexagonal
Intergrowth plane	(011)	(011)	(010)	(0001)
First direction	[100]	[1$\bar{1}$1]	[100]	[10.0]
Periods (Å)	3.16	3.23	3.19	3.00
Misfit (%)		+ 2.2	− 1.2	− 6.0
Second direction	[1$\bar{1}$1]	[100]	[101]	[12.0]
Periods (Å)	5.48	5.58	4.43	5.19
Misfit (%)		+ 1.8	− 20.6	+ 17.1
		→ good adhesive bond	→ poor adhesive bond	

c) Formation of borides on tungsten. The preparation of tungsten borides takes place in the temperature range of 1100°C to 1500°C by means of the gas reaction with BCl_3 and H_2. A layer about 100 microns thick is formed, for example, after a reaction period of two hours, which again consists of a phase sequence with the metal lattice having a decreasing degree of boride content [32]: W_2B_5–WB–W_2B–W. The corresponding microhardnesses in this connection decrease from 2800 (WB + W_2B_5) to 2000 (W_2B) to 300 (W-metal) kg/mm^2. The boride formation on tungsten therefore effects an enormous increase in surface hardness, abrasion and corrosion resistance. For the quality of the surface layer to act as an improvement layer, the adhesive bond between metal carrier and layer phases is naturally essential. After the excellent correlation (table 3.6), one can expect that the adhesive bond between tungsten and the innermost boride layer (W_2B) will be excellent, while the outermost layer zone in this respect is far less favorable.

3.2.2.3. Oxalation layers on iron, zinc, manganese, magnesium and aluminium. Oxalation is defined as the production of crystalline protective layers from baths which consists mainly of oxalic acid or alkali oxalate with varying additions of heavy metal oxalates.

Table 3.7
Phases in Oxalation Layers on Iron, Manganese, Zinc, Magnesium, and Aluminium

Metal Substrate	Baths		
	Alkali Oxalate	Iron Oxalate	Manganese Oxalate
Fe	$FeC_2O_4.2H_2O$	$FeC_2O_4.2H_2O$	$FeC_2O_4.2H_2O$ $(Mn, Fe)C_2O_4.2H_2O$
Mn	$MnC_2O_4.2H_2O$	$FeC_2O_4.2H_2O$ $(Fe, Mn)C_2O_4.2H_2O$	$MnC_2O_4.2H_2O$
Zn	$ZnC_2O_4.2H_2O$	$ZnC_2O_4.2H_2O$ $(Fe, Zn)C_2O_4.2H_2O$	$ZnC_2O_4.2H_2O$ $(Mn, Zn)C_2O_4.2H_2O$
Mg	$MgC_2O_4.2H_2O$	$MgC_2O_4.2H_2O$ $(Fe, Mg)C_2O_4.2H_2O$	$MgC_2O_4.2H_2O$ $(Mn, Mg)C_2O_4.2H_2O$
Al	–	$FeC_2O_4.2H_2O$	$MnC_2O_4.2H_2O$

a) Phase stability of the oxalation layers. In the normal pickling operation using an acid bath of alkali oxalate on a metal substrate A (A = Fe, Mn, Zn, Mg), oxalation layers are formed which are built up from a homogeneous phase of A-oxalate dihydrate. If the initial bath contains heavy metal ions (B), such as Fe^{++} and Mn^{++}, there is formed once more only a monophase layer built up of mixed crystals of A-oxalate and B-oxalate, written as $(B,A)C_2O_4.2H_2O$. In this case, the portion of metal A (metal substrate) is about 10 atom %. If a very strong pickling reaction occurs which results in the enrichment of the ions of the substrate metal in the immediate vicinity of the metal surface (A) to be oxalated, primary oxalation layers are formed each time. These thin layers almost always consist of the pure carrier metal-oxalate, which gradually is transformed into an oxalate layer with a predominant portion of primary bath metal ion (B) (table 3.7).

b) Textures, epitaxies and adhesive bonds: "Metal-oxalate". Oxalation layers on rolled or drawn textured metals were always oriented and therefore, at the intergrowth surface one always observed the monoclinic (010) surface of the oxalate. This led to the epitaxies recorded in table 3.8 (fig. 3.10).

The intergrowths listed in table 3.8 can be understood if one considers the structure of the oxalate dihydrates. $FeC_2O_4.2H_2O$ (humboldtine) is monoclinic, possesses the space group C2/c, has four formula units per unit cell and has the following lattice constants: a_0 = 12.060 Å; b_0 = 5.550 Å; c_0 =

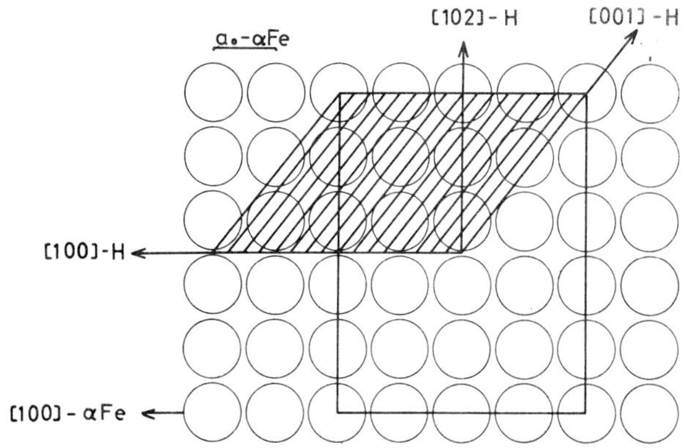

Fig. 3.10. Affinitive lattice planes: (010)-humboldtine on (001)–α-Fe.

9.085 Å and β = 127°58′. Every iron ion in humboldtine is surrounded octahedrally by four oxygen atoms of two planar C_2O_4 groups and two H_2O molecules (fig. 3.11). Since every C_2O_4 ion is connected with two iron ions, chains are formed, ... Fe ... C_2O_4 ... Fe ... C_2O_4 ... and these are positioned in parallel with the crystallographic b axis, [010], while the direction H_2O ... Fe ... H_2O of the crystallographic a axis is parallel with [100].

The dihydrates of manganese (II), zinc and magnesium oxalate have naturally somewhat different lattice constants and thus chain periods. Magnesium oxalate has the larger cell constants, and zinc oxalate the smaller. The adhesive bond according to the above law is characterized by a perpendicular

Table 3.8
Epitaxies in Oxalation Layers

Phases		Epitaxial Laws		Periods (Å)		Misfit
Host	Guest	Host	Guest	Host	Guest	(%)
α-Fe	$FeC_2O_4.2H_2O$	(001)	(010)			
		[100]	[100]	11.46	12.06	+ 5
Zn	$FeC_2O_4.2H_2O$	(0001)	(010)			
		[12.0]	[010]	9.34	9.09	– 3
Mg	$FeC_2O_4.2H_2O$	(0001)	(010)			
		[10.0]	[001]	9.60	9.09	– 6
Al	$FeC_2O_4.2H_2O$	(001)	(010)			
		[100]	[100]	12.15	12.06	– 1

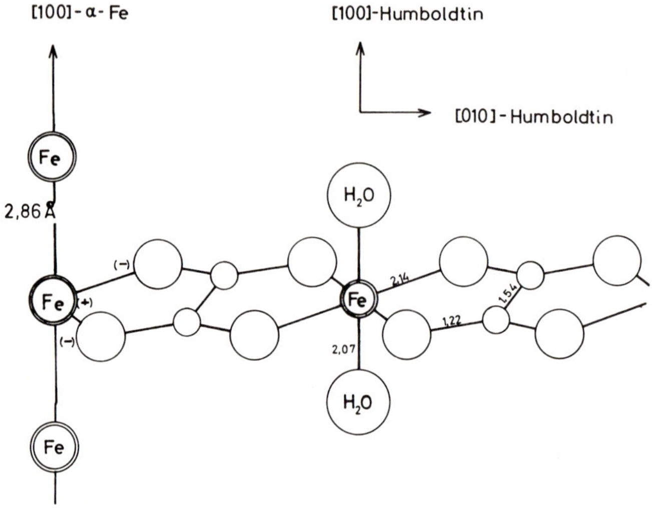

Fig. 3.11. Bonding at the interface α-Fe-humboldtine.

orientation of the individual oxalate chains of humboldtine to the metallic surface. From this, there results, for the individual oxalate chain, an atomic adhesive and valence bond according to fig. 3.11. Both terminal-positioned oxygen ions of the oxalate chains lie in the potential wells of iron structural units, assumed to be polarized, in the metal surface. This results in a main valence connection $Fe^{+}\cdots{}^{-}O$ of the chain ends with the polarized iron atoms, and thus a very strong adhesive bond between the individual oxalate chain and the metal surface. The latter is supported by good correlation of the intergrowing host-guest lattice sections, which effect a solid lateral dipole and van der Waals' connection of all chains with one another to the oxalate lattice, with the formation of a densely packed oxalate layer in the solid state.

3.2.2.4. Phosphating layers on iron, zinc, copper, brass and aluminium. Phosphating is defined as the production of crystalline protection coatings from baths, which consist predominantly of phosphoric acid or alkali phosphate with varying additions of heavy metal phosphates. So far we have investigated iron, zinc, manganese, calcium, (calcium + zinc) and alkali phosphate layers on the metals iron, zinc, copper, brass and aluminium. Two of the findings made were verified [33, 34], and, instead of the data of the latter reference, additional studies have led to a large number of new phases.

a) Construction of layers. The layer construction of the phosphating layers investigated will be demonstrated on the basis of two examples with other results of the phase analysis for the metals iron and zinc given in table 3.9.

In phosphating from the "bath which does not build a layer" – that is, a primary heavy-metal-free bath – on iron a very dense layer of $Fe_3(PO_4)_2.8H_2O$ (vivianite) forms, the growth of which practically stops at a layer thickness of approximately 0.1 micron. With the addition of the required heavy metal to such a non-layer-forming bath the formation of a phosphate layer only occurs from a substrate metal with the iron in the form of Fe^{2+}, by the diffusion of two electrons from the iron lattice through the phosphate

Table 3.9
Composition of Phosphate Coatings on Metallic Iron and Zinc

Heavy Metal Portion in Bath	Metallic Substrate	
	Fe	Zn
Alkali	$Fe_3(PO_4)_2.8H_2O$ (vivianite-monoclinic)	$Zn_3(PO_4)_2.4H_2O$
Fe	$Fe_5H_2(PO_4)_4.4H_2O$ (Fe-hureaulite-monoclinic) $FePO_4.2H_2O$* (strengite-monoclinic)	$Zn_3(PO_4)_2.4H_2O$ $Zn_2Fe(PO_4)_2.4H_2O$ $Fe_5H_2(PO_4)_4.4H_2O$
Mn	$(Mn, Fe)_5H_2(PO_4)_4.4H_2O$ (hureaulite-monoclinic) (with Mn-Fe-diadochy)	$Zn_3(PO_4)_2.4H_2O$ $Mn_5H_2(PO_4)_4.4H_2O$ (Mn-hureaulite-monoclinic)
Zn	$Zn_2Fe(PO_4)_2.4H_2O$ (phosphophyllite-monoclinic) $Zn_3(PO_4)_2.4H_2O$ (hopeite-rhombic)	$Zn_3(PO_4)_2.4H_2O$
Zn + Ca	$Zn_2Fe(PO_4)_2.4H_2O$ $Zn_2Ca(PO_4)_2.2H_2O$ (scholzite-rhombic) $Zn_3(PO_4)_2.4H_2O$	$Zn_3(PO_4)_2.4H_2O$ $Zn_2Ca(PO_4)_2.2H_2O$
Ca	$CaHPO_4.2H_2O$ (brushite-monoclinic) $CaHPO_4$ (monetite-tetragonal)	$Zn_2Ca(PO_4)_2.2H_2O$ $CaHPO_4.2H_2O$ $CaHPO_4$

* Only from HNO_3-H_3PO_4 baths.

layer to the particular "solid-fluid" phase boundary. The cause for the termination of growth then is obviously the result of the rapidly increasing hindrance to the diffusion of iron. This also holds for the alkali phosphating of other metals.

In phosphating baths which contain iron or (iron + zinc), there is formed, on zinc, one of the crystallite structures in the case of the higher layer portions of three-phase layer portions: $Zn_3(PO_4)_2.4H_2O$ (hopeite), $Zn_2Fe(PO_4)_2.4H_2O$ (phosphophyllite) and $Fe_5H_2(PO_4)_4.4H_2O$ (iron-hureaulite). This structure is transformed, depending on the substrate (zinc), to a more and more pure zinc phosphate layer as the primary layer. On the contrary, in the case of very iron-rich baths the primary layer appears also to be of two phases, consisting of hopeite and phosphophyllite. The bath metal in this case has already, at the moment of the primary layer formation, diffused to the metal surface. With progressive growth of the layer, the portion of the substrate in the phosphate layer diminishes more and more so that the external portions of the layer reflect the conditions in the phosphating solution. The diffusion secondary supply of the substrate material to the "solid-fluid" phase boundary with increasing thickness of the layer is involved in phosphate layer formation and this process slows down and finally ceases.

b) Textures, epitaxies, nucleation and adhesive bonds: metal-layer phases. All the phosphating layers shown in table 3.9, as well as the phosphating layers on copper, brass and aluminium, are known to develop texture under appropriate experimental conditions, and indeed the quality of the orientation of the phosphate crystals increases markedly with the quality of the orientation of the metal crystallites. It was necessary to investigate texture correlations between metal and phosphating crystallites. The texture correlations demonstrated so far in phosphating layers are the following: hopeite grows with (010) to (001)–α-Fe, (0001)–Zn, (001)–Cu, (011)–α-brass, (001)–A1 and (011)–A1; phosphophyllite with (100) to (001)–α-Fe and (0001)–Zn; scholzite with (100) to (001)–α-Fe, (0001)–Zn, (001)–Cu, (011)–α-brass, (001)–A1 and (011)–A1; and hureaulite with (100) to α-Fe, (0001)–Zn, (001)–Cu, (001)–α-brass, (001)–A1 and (011)–A1. (010)–Vivianite intergrows with (011)–α-Fe. This means that the phosphating systems employed are characterized through nonequivocal texture correlations of the phosphate crystallites (guest phase) to the metal crystallites (host phase). To seek out first the genuine crystal structures – that is, the epitaxial ones out of the many textural correlations of host and guest crystallites – the corresponding host-guest surface pairs must be fixed at least with a pair of corresponding host-guest lattice directions within these intergrowth levels.

Table 3.10
Epitaxies in Phosphate Coatings on Iron and Zinc

Phases		Epitaxial Laws		Periods (Å)		Misfit (%)
Host	Guest	Host	Guest	Host	Guest	
α-Fe	Vivianite	(011)	(010)			
		[1$\bar{1}$1]	[001]	4.97	4.70	– 5
		[11$\bar{1}$]	[100]	9.94	10.08	+ 1
α-Fe	Hopeite	(001)	(010)			
		[100]	[001]	11.46	10.64	– 7
α-Fe	Phosphophyllite	(001)	(100)			
		[100]	[001]	11.46	10.49	– 8
α-Fe	Scholzite	(001)	(100)			
		[210]	[001]	6.42	6.61	+ 3
α-Fe	Brushite	(001)	(010)			
		[100]	[100]	5.73	5.89	+ 3
α-Fe	Hureaulite	(001)	(100)			
		[210]	[001]	9.63	9.51	– 1
Zn	Hopeite	(0001)	(010)			
		[10.0]	[100]	10.66	10.64	< 1
Zn	Phosphophyllite	(0001)	(100)			
		[10.0]	[001]	10.66	10.49	– 1
Zn	Scholzite	(0001)	(100)			
		[12.0]	[001]	6.95	6.61	– 5
Zn	Brushite	(0001)	(010)			
		[12.0]	[001]	6.95	6.38	– 8
Zn	Hureaulite	(0001)	(100)			
		[12.0]	[010]	9.26	9.11	– 2

Table 3.10 and figs. 3.12 and 3.13 show directly that the corresponding guest-host lattice surfaces of vivianite, hopeite, phosphophyllite, scholzite, brushite and hureaulite on the one hand, and of iron and zinc (similar to copper, brass and aluminium) on the other hand, possess to a high degree the mono- and two-dimensional lattice correlations generally required for epitaxies [35, 37]. There only remained to prove that the corresponding lattice points of the metal-phosphate intergrowth planes are also suitably distributed with respect to valence. This proof is obtained through direct structural comparison of the intergrowing host-guest lattice (contact planes). Since, up till now, only the structures of hopeite, phosphophyllite and vivianite have been determined, this could be done only for these phosphates.

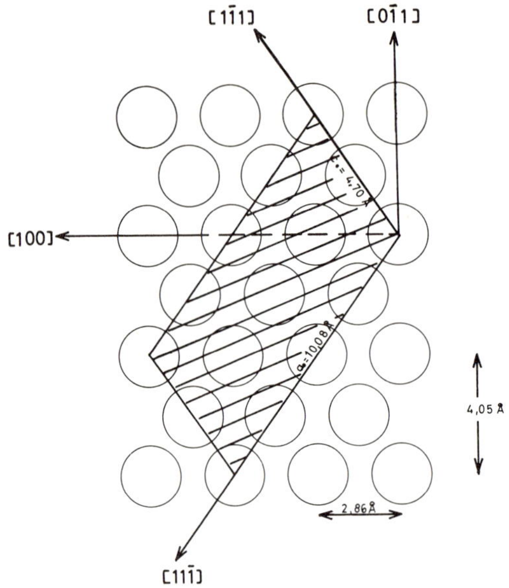

Fig. 3.12. Affinitive lattice planes: (011)-vivianite on (011)–α-Fe.

The intergrowth planes of the participating host lattices form extremely simple lattices packed with metal atoms. The lattice planes of hopeite, phosphophyllite and vivianite intergrowing with it form with the corresponding host lattices good "polymer analogue" lattice samples with exclusive PO_4 distribution, whereby in every case a corner of the PO_4 tetrahedron points to an Me^+ lattice building stone of the host lattice (fig. 3, 14). The following "host-guest" structures result:

$$Me^+ \ldots {}^-0 - P \begin{matrix} \diagup 0 \\ - 0 \\ \diagdown 0 \end{matrix} \quad \text{or} \quad Me - 0^- \ldots + P \begin{matrix} \diagup 0 \\ - 0 \\ \diagdown 0 \end{matrix}$$

or a transition form of both valence states.

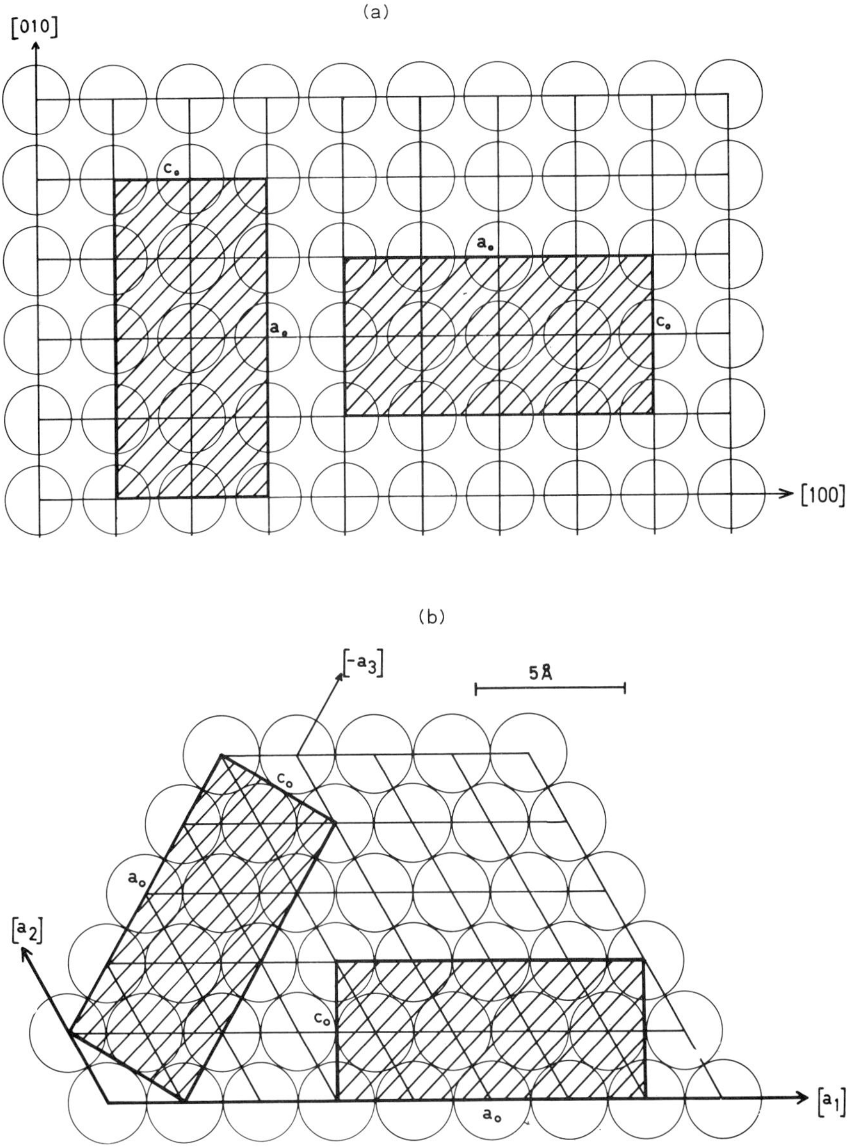

Fig. 3.13. Affinitive lattice planes. (Top) (010)-hopeite or (100)-phosphophyllite on (100)–α-Fe; (bottom) (010)-hopeite or (100)-phosphophyllite on (0001)-Zn.

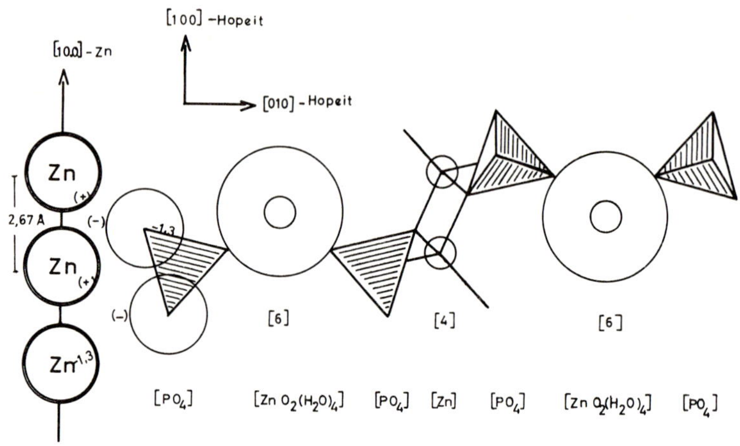

Fig. 3.14. Bonding at the interface zinc-hopeite.

3.2.3. Epitaxial layer growth

All discussions of epitaxial layer growth to be found in the literature were carried out on systems which are chemically (elements or AB-compounds) and structurally (cubic or hexagonal sphere packing) very simple. In contrast to this, the example of phosphating of Fe and Zn in a Zn-phosphate bath is now discussed.

3.2.3.1. General processes of phosphate layer formation. The classical prescription for the composition of Zn-phosphate baths is Zn-phosphate + free phosphoric acid, for which the relative proportions of Zn^{2+}, H^+, $H_2PO_4^-$ and HPO_4^{2-} ions are adjusted in such a manner that the solubility products of the secondary and tertiary Zn-phosphates are not exceeded. The equilibrium of the reaction, which is important for layer formation,

$$2\,H_2PO_4^- \rightarrow HPO_4^{2-} + PO_4^{3-} + 3\,H^+$$

is strongly temperature dependent and shifts to the right with rising temperature; i.e. according to the working temperature the proportion of Zn-phosphate to phosphoric acid must be chosen differently.

During the treatment of Zn-objects in such a bath, etching is the initial reaction:

$$Zn^0 + 2\,H^+ \rightarrow Zn^{2+} + H_2$$

at which the attack takes place preferentially at grain boundaries and on those faces which do not belong to the equilibrium form. The diminution of the H^+ ion concentration, connected with this reaction, results in the solubility products of the secondary and tertiary phosphates being rapidly exceeded.

Therefore nucleation and crystallization is produced in the proximity of the metal/solution phase-boundary and in the main it remains limited to the immediate boundary-layer since H^+ ions are formed during crystallization, which repress the dissociation to insoluble phosphates. Therefore, the original equilibrium condition of the solution will be approximately regained.

In addition to the ratio Zn-phosphate/phosphoric acid, the absolute concentration in the bath is also important. Thus, for small absolute concentrations, the change in pH takes place very rapidly and does not remain limited to the immediate metal/solution boundary-layer. The consequence of this is that, after exceeding the solubility products, the crystallization of Zn-phosphates takes place on a large scale in the phosphating bath itself. In contrast, solutions with very high absolute concentrations cause the formation of a great quantity of Zn^{2+} ions. Since the absolute concentration of H^+ ions is very high, the pH value shifts only slowly, whereby the layer formation is strongly retarded and the Zn-enrichment takes place in the solution far from the metal. This results in strong etching, a considerable formation of reaction products in the form of an incoherent mass, and only very little layer formation.

In a phosphating bath with the appropriate absolute concentrations, the supersaturation, necessary for nucleation, is generated only in the vicinity of the Zn-surface. According to the theory (ch. 3 §1.1) the free energy of formation of a critical Zn-phosphate nucleus (ΔG^*) for heterogeneous nucleation is half the value for homogeneous nucleation, if a contact angle $\theta = 90°$ is assumed. Consequently the crystallization of Zn-phosphate implies layer formation and not the formation of an incoherent mass.

Because of the crystalline character of the bath phases, the specific host/guest free interfacial energy σ_{hg} is certainly anisotropic, i.e. there are minimal values for certain pairs of faces and directions. Therefore the nucleation energy for definite orientations of the nucleus is distinctly small and correspondingly the nucleation frequency is distinctly large; this is the necessary and ideal condition for epitaxial attachment.

Just as reaching the value of the critical supersaturation leads to a very rapid rise in the nucleation rate, particularly for the special orientations, so epitaxial Zn-phosphate nuclei were formed upon exceeding the solubility product. These nuclei grow to micro-crystallites until the supersaturation is destroyed. Further growth of the micro-crystallites is possible only if, by the etching attack on those parts of the metal surface that are still free, Zn is dissolved and therewith the solubility product of Zn-phosphate is exceeded again. This leads to tangential and normal growth of the primary Zn-phosphate crystallites and to the formation of new nuclei on the metal surface. So the amount of uncovered Zn surface becomes progressively smaller and the thickness of the layer greater, i.e. on the metal surface there is not a formation of an absolutely closed layer, but a layer with asymptotically decreasing porosity. The pore-level is, e.g. after phosphating for 2 min, about 5%, after 4 min 0.7% and after 6 min still 0.5%.

In the phosphating of Zn in Zn-phosphate baths, naturally only Zn-phosphates can be formed. Therefore it is not surprising that by X-ray and electron diffraction studies of all parts of the layer only $Zn_3(PO_4)_2.4H_2O$ (as mineral known as hopeite) could be found. A different situation occurs with the phosphating of Fe. Here Fe^{2+} ions are formed by the etching attack, so that after exceeding the solubility products in addition to Zn-phosphate, Fe-phosphates and (Zn,Fe)-phosphates can also be formed. As the solubility product of Zn-phosphate is essentially smaller than that of Fe(II)-phosphate, the primary phosphate-layer immediately adjacent to the Fe substrate should consist of $Zn_3(PO_4).4H_2O$. Only the subsequent crystallization layers can partly consist of Fe-phosphate, because the etching attack through the pores which are present must supply the necessary Fe^{2+} concentration. Consequently, further tangential growth of the phosphate crystallites occurs and this results in a decrease of the pore cross-section.

Electron diffraction photographs were made from layers which were successively thinned by ion bombardment-etching and from layers in their initial state. These photographs show however, that the primary phosphate layer consists of the compound $Zn_2Fe(PO_4)_2.4H_2O$ (as mineral known as phosphophyllite), on which follows an epitaxial secondary layer of $Zn_3(PO_4)_2.4H_2O$. Only in essentially later crystallization layers do both phosphates $Zn_3(PO_4)_2.4H_2O$ and $Zn_2Fe(PO_4)_2.4H_2O$ participate in layer formation side by side; in this case phosphophyllite is sunk like islands between the hopeite crystal aggregates.

The determination of the Fe-content of the solution as well as the amount of hopeite and phosphophyllite in the layer shows that after phosphating for 6 min, 0.1 μm is carried off on the average, and that the proportion Zn/Fe in the layer is about 20 : 1 (that means hopeite/phosphophyllite = 6 : 1) and that about 2.2 mol of tertiary phosphate is separated per mol dissolved Fe. Formally this can approximately be described by an equation for the layer formation:

$$20\,Zn^{2+} + 14\,PO_4^{3-} + 6\,H^+ + 3\,Fe^0 \rightarrow$$
$$Zn_2Fe(PO_4)_2 + 6\,Zn_3(PO_4)_2 + 3\,H_2 + 2\,Fe^{2+}$$

This equation shows that only the smaller part of the dissolved Fe is built into the layer as $Zn_2Fe(PO_4)_2$; the phosphating bath therefore becomes more and more rich in Fe. An explanation of the fact that phosphophyllite is the primary layer-forming phase can only be given if the solubility product of phosphophyllite is smaller than that of hopeite. In this case nuclei of phosphophyllite are formed first on the Fe substrate, which grow further into micro-crystallites until the local supersaturation of Fe is consumed. With this a local supersaturation of Zn^{2+} is quickly reached, so that epitaxially nucleation of hopeite on phosphophyllite takes place, while the first formed phosphophyllite crystallites grow further laterally and new phosphophyllite nuclei are formed on the still uncovered Fe surface, where the local supersaturation of Fe^{2+} is reached later. When a largely closed layer is formed (about 20% free pore-level), the diffusive Fe transport to the bath takes place only through the pores, while the main layer growth takes place on the crystallite faces, turned towards the bath, which will not be reached by the "Fe^{2+}-stream". Such a mechanism explains additionally the observation that the phosphophyllite crystallites in the outer parts of the layer are sunk like islands between the hopeite crystal aggregates.

3.2.3.2. Elementary processes. If one considers as the simplest case an ideal (001)-surface network plane of α-Fe, then each of its building blocks is surrounded by the four nearest ($\frac{1}{2}a_0\sqrt{3}$) and five next nearest (a_0) neighbours (in the structure 8 or 6). The undisturbed potential field over this surface has the same rigorous periodic construction as the surface network plane itself [36]. This is expressed by curve a (fig. 3.15) in the shape of the potential profile of (001)-α-Fe in the direction [100]. When an oxygen ion of a PO_4 group in the phosphating process moves into one of these potential troughs, an "ad-atom" is formed and therewith a locally limited disturbance of the potential field (curve c in fig. 3.15). Within the range of this disturbance, the potential field therefore loses its strictly periodic structure. The potential troughs in the vicinity of the "ad-atom" become deeper than normal and at the same time change their local positions. They coincide within a certain range about the "ad-atom" and not any more with the lattice positions of the potential minima between, in each case, four iron atoms of the pure host surface. With this distortion of the potential field through the first "ad-atom", under equilibrium conditions, the position of the second "ad-atom" is also fixed. In the deepest trough (fig. 3.15) immediately adjacent to

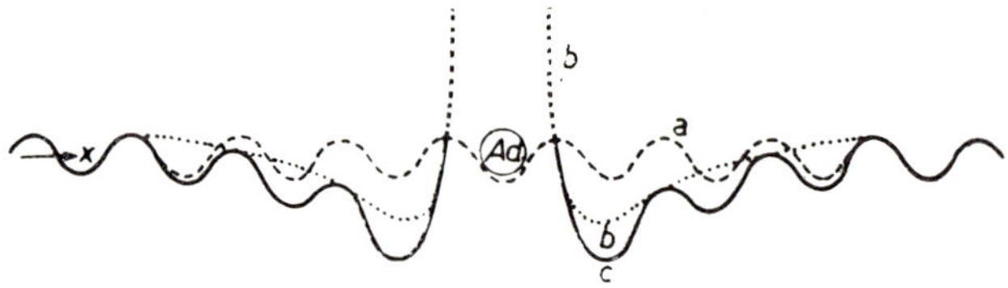

Fig. 3.15. Potential of a (001)–α-Fe surface in the direction [100] according to Mayer [36]; a) undisturbed potential field directly above the surface of the host; b) potential of a single adsorbed atom; c) potential distribution resulting from (a) and (b).

the first "ad-atom" or in one voluminous "ad-atom", as in the case of PO_4 ions, the adjacent, somewhat more shallow, trough is occupied and therewith the symmetry of the overall field is deformed further.

Through this, rigorously ordered ranges are formed of PO_4 ions on the α-Fe surface, from which (in the case of zinc phosphate layers), through addition of zinc ions, H_2O molecules and more PO_4 ions form hopeite nuclei, which are necessarily auto-oriented. The tangential growth of these hopeite nuclei into microcrystallites and thus the active host-guest growing together follow to an extent such that the deviation of the PO_4 positions of the potential minima of the host network becomes so great that a new primary "ad-atom" is formed. This generates a new chain section of equal orientation which is energetically more favorable than continuation of the old chain. These chain breaks should, in general, occur after a few chain periods. Naturally they

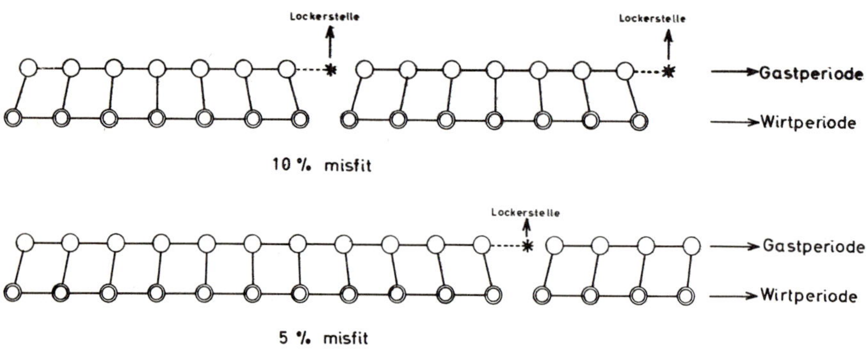

Fig. 3.16. Rupture of an active intergrowth chain in the case of distinct differences of corresponding host-guest periods.

require a correspondingly dense arrangement of loose positions, according to fig. 3.16.

Thus, an ultra-dispersed primary layer results, strongly oriented to the carrier lattice but not necessarily oriented in parallel. Further growth of this primary layer then ensues in the shape of a short period sequence of epitaxial nuclei formation which is also diffusion-controlled.

The regularly repeated, epitaxially directed nuclei formation, growth and growth-retarding mechanism assure the transmittal of the substrate orientation to the primary phosphate layer and to all the subsequent crystallite deposits. Thus, there results at the same time for the layer a properly adherent, fine-grained and about equally sized crystallite structure.

3.2.4. Conclusion

All the protective layers discussed above such as oxide, nitride, boride, carbide, oxalate and phosphate coatings are built up from phases which under normal pressure-temperature-concentration conditions are either thermodynamically stable or extremely inert. The layer crystallites are linked epitaxially or topotaxially in all cases with the host crystallites via strong reciprocal reaction forces.

References

[1] M.J. Stowell, Thin films 1 (1968) 55.
[2] E. Bauer, Z. Krist. 110 (1958) 372.
[3] H.G. Scheider, Ed., Epitaxie–Endotaxie (VEB Deutscher Verlag für Grundstoffindustrie, Leipzig, 1969).
[4a] D. Walton, J. Chem. Phys. 37 (1962) 2182.
[4b] D. Walton, Phil. Mag. 7 (1962) 1671.
[5] E.N. Andrade and J.G. Martindade, Phil. Trans. Roy. Soc. (London) 235 A (1935) 69.
[6] B.K. Chakraverty and G.M. Pound, Acta Met. 12 (1964) 851; in: Condensation and Evaporation of Solids, ed. by Rutner et al. (Gordon Breach, New York, 1964) p.533.
[7] G.A. Bassett, Phil. Mag. 3 (1958) 72.
[8] G.W. Sears, J. Chem. Phys. 33 (1960) 1068.
[9] J.A. Koutsky, A.G. Walton and E. Bauer, Surface Science 3 (1965) 165.
[10] J.P. Hirth and G.M. Pound, Condensation and Evaporation: Nucleation and Growth Processes (Pergamon, Oxford, 1963).
[11] G.A. Bassett, Proc. Eur. Reg. Conf. Elektron Microscopy. Ned. Ver. Electronenmicros., Delft (1961) 270.
[12] G.A. Bassett, in: Condensation and Evaporation of Solids, ed. by Rutner et al. (Gordon and Breach, New York, 1964) p.599.

[13] D.W. Pashley and M.J. Stowell, in: Proc. 5th Int. Conf. Electron Microscopy, ed. by Breese (Academic Press, New York, 1962).
[14] D.W. Pashley, M.J. Stowell, M.H. Jacobs and T.J. Law, Phil. Mag. 10 (1964) 127.
[15] D.W. Pashley, Adv. in Phys. 14 (1965) 327.
[16] D.W. Pashley and M.J. Stowell, J. Vac. Sci. and Techn. 3 (1966) 156.
[17] M.J. Stowell and T.J. Law, Phys. Stat. Solid. 16 (1966) 117.
[18] M.J. Stowell and T.J. Law, Phys. Stat. Solid. 16 (1966) 479.
[19] B.W. Sloope and C.O. Tiller, J. Appl. Phys. 32 (1961) 1331.
[20] L.G. Schulz, Acta Cryst. 4 (1961) 487.
[21] D.W. Pashley, Proc. Phys. Soc. A 65 (1952) 33.
[22] O. Haase, Z. Naturforsch. A 14 (1959) 920.
[23] W.A. Jesser and J.W. Matthews, Phil. Mag. 15 (1967) 1097.
[24] W.A. Jesser and J.W. Matthews, Phil. Mag. 17 (1968) 595.
[25] A.N. Mariano and K.L. Chopra, Appl. Phys. Letters 10 (1967) 282.
[26] J. Goddard and J.G. Wright, Brit. J. Appl. Phys. 15 (1964) 807.
[27] W.A. Jesser and J.W. Matthews, Phil. Mag. 17 (1968) 461.
[28] N. Thompson, Proc. Roy. Soc. B 66 (1953) 481.
[29] A. Neuhaus and M. Gebhardt, in: *Interface Conversion for Polymer Coatings* (American Elsevier Publishing Company, Inc., 1969) p.91.
[30] A. Neuhaus and M. Gebhardt, Werkstoffe Korrosion, 17 (1966) 567.
[31] M. Gebhardt, *Protection against Corrosion by Metal Finishing* (Förster-Verlag, Zürich, 1967) pp. 323–330.
[32] G.V. Samsonov and A.P. Epik, *Coatings of High-Temperature Materials,* Part 1 (Plenum, New York, 1966).
[33] A. Durer and E. Schmid, Korrosion Metallschutz 20 (1944) 161.
[34] J. Saison, Compt. Rend., 248 (1959) 2586.
[35] A. Neuhaus, Fortschr. Mineral., 29–30 (1950–1951) 136.
[36] M. Mayer, Physik dünner Schichten (Stuttgart, 1955).
[37] M. Gebhardt and A. Neuhaus, *Epitaxie-Daten für anorganische und organische Kristalle,* Landolt-Bornstein, Bd. Kristallstrukturdaten (III/8) (Springer-Verlag, Berlin – Heidelberg – New York, 1972).
[38] L. Lafourcade, Ch. Bonnafous, P. Larroque and B. Legros, J. Microscopie 6 (1967) 431.

P. Hartman, ed., Crystal growth: an introduction. © (1973) North-Holland Publ. Co.

CHEMICAL TRANSPORT REACTIONS

H. SCHÄFER

4.1. The mobility via the gas phase of a solid in a temperature gradient

The transport of a solid from a region of higher temperature T_2 to a region of lower temperature T_1 may occur for any one of five different reasons; only one of these involves chemical transport. ((e) below and see [1]).

a) *Sublimation,* e.g.
$I_{2,s} = I_{2,g}$
$LiF_{,s} = LiF_{,g}$ ($Li_2F_{2,g}$; $Li_3F_{3,g}$)

b) *Sublimation with suppression of thermal decomposition,* e.g.
$TiCl_{3,s} = TiCl_{3,g}$; in presence of $TiCl_{4,g}$
$NbCl_{4,s} = NbCl_{4,g}$; in presence of $NbCl_{5,g}$

c) *Catalytic sublimation,* e.g.
$4\ P_{,red} = P_{4,g}$; in presence of I_2, PI_3, P_2I_4 [2]

An investigation has shown that the important concentration gradient is determined by $P_{4,g}$. The reaction is therefore not one of chemical transport.

d) *Thermal decomposition and recombination (dissociative sublimation),* e.g.
$NH_4Cl_{,s} = NH_3 + HCl$
$3\ SbSI_{,s} = Sb_2S_{3,g} + SbI_{3,g}$ [3]
$CdO_{,s} = Cd_{,g} + 0.5\ O_2$

e) *Chemical transport,* e.g.
$Fe_2O_{3,s} + 6\ HCl = 2\ FeCl_{3,g} + 3\ H_2O_{,g}$
$ZnS_{,s} + I_{2,g} = ZnI_{2,g} + 0.5\ S_{2,g}$
$OsO_{2,s} + O_2 = OsO_{4,g}$
$Ni_{,s} + 4\ CO = Ni(CO)_{4,g}$

How can one find out which of the possibilities (a) – (e) occurs in a given case? All of the processes (a) – (e) are reversible, but when carried out in a high vacuum only the processes (a) and (d) are possible. The other processes need the presence of foreign gas molecules. The processes (a) – (d) are endothermic in each case. On the other hand chemical transport (e) can be endothermic or exothermic. Therefore when a solid moves up the temperature gradient as in the Ni/CO case it must occur clearly as a result of chemical transport. The nature of the different possibilities can of course be analysed using thermodynamic calculations if suitable data is available.

4.2. Chemical transport – the rate determining step

Consider as an example the chemical transport of Fe_2O_3 with HCl as transport agent ($1000° \rightarrow 800°$), see §4.1(e). Very roughly we can distinguish the following steps, (i) the chlorination reaction involving the consumption of Fe_2O_3, (ii) gas motion involving diffusion, convection and streaming and (iii) the back reaction involving the deposition of Fe_2O_3.

When a low total pressure (e.g. $< 10^{-2}$ atm) is used, as a rule the heterogeneous reaction is relatively slow, but the gas motion is relatively fast. The transport rate is thus controlled by the chemical reaction and the surface of the solid and its reactivity become important.

When working in a closed tube with total pressures of $\geqslant 10^{-2}$ atm the gas motion is generally slow compared with the velocity of the chemical reaction. The gas motion is therefore rate determining. This is the usual situation in chemical transport. At both ends of the closed tube local equilibria between the solid and the gas phase are established and thus transport may be described in terms of gas diffusion between regions of local equilibrium having temperatures T_2 and T_1. In such a case the transport rate is independent of the amount of solid starting material, proportional to the cross section of the tube (diffusion path) and inversely proportional to the length of the tube [1].

Even when the equilibrium at the surface of the solid material at T_2 and T_1 is established, the calculation of the transport rate in an exact theoretical sense is impossible. The difficulties arise from thermodiffusion, thermal convection, an inexact knowledge of the temperature dependence of the diffusion coefficients and the requirement for the use of individual diffusion coefficients for gas species in a multi-component gas mixture.

Fortunately for practical purposes simplifications which neglect thermodiffusion and use an average value for the diffusion coefficients are possible and useful. (See [1]).

It should be noted that more important errors in calculations of transport rates are caused as a rule by uncertainties in the thermodynamic data used for the calculation of the concentration gradient rather than by uncertainties in respect of the gas motion.

4.3. Chemical systems, suitable for chemical transport

It is conventional to write the solid material on the left hand side of the chemical equation, e.g.

$$A_{,s} + B_{,g} = C_{,g} + D_{,g} \cdots g;\ \Delta H^{\circ}_{\text{react}}$$

The basic process of chemical transport is only possible when no more than *one* condensed phase is involved (more complicated cases are considered in [1]). Clearly the reversibility of the reaction is also necessary for chemical transport.

The direction of transport of a solid in a temperature gradient ($T_1 < T_2$) is given by the sign of $\Delta H^{\circ}_{\text{react}}$:

Endothermic reaction: $T_2 \rightarrow T_1$,
Exothermic reaction : $T_1 \rightarrow T_2$.

Table 4.1 is a selection of the large number of solids which have been used in chemical transport experiments. Table 4.2 shows a comparable selection of

Table 4.1
Examples of solids which have been chemically transported

elements,	halides (F, Cl, Br, I),
oxides,	chalcogenides (S, Se, Te),
Me_nX_m	X = N, P, As, Sb; Si, Ge, Sn; B
borates, silicates, germanates, phosphates, tungstates, ternary oxides, ternary and quaternary sulfides, oxyhalides, sulfide halides.	

Table 4.2
Examples of transporting agents

F_2	Cl_2	Br_2	I_2
HF	HCl	HBr	HI
H_2	H_2O	CO	CO_2
O_2	S_2	Se_2	Te_2
$HgCl_2$	$AlCl_3$	$SiCl_4$	$NbCl_5$

transporting agents. The amount of transporting agent used is usually about a mg per ml of tube volume. However chemical transport can sometimes be observed with traces of transporting agent, for example in the cases of Ni/HCl or Cu_2O/HCl [1]. On the other hand transport reactions with pressures of some hundred atmospheres are well known ("hydrothermal" reactions). The inclusion of small amounts of transporting agent in growing crystals are considered in [4].

The following equations give an impression of the very different reaction types, applicable to chemical transport. References are given to those reactions which are not treated in [1].

$Cu + 1/2\ I_2 = CuI_{,g}\ (1/3\ Cu_3I_{3,g})$;	$T_1 \rightarrow T_2$	
$Zr + 2\ I_2 = ZrI_{4,g}$;	$T_1 \rightarrow T_2$	
$Ni + 2\ HCl = NiCl_{2,g} + H_2$;	$T_1 \leftarrow T_2$	
$Pt + 2\ CO + Br_2 = Pt(CO)_2Br_{2,g}$;	$T_1 \rightarrow T_2$	[5]
$W + 2\ H_2O + 6\ I_1(3\ I_2) = WO_2I_{2,g} + 4\ HI$;	$T_1 \rightarrow T_2$	[6]
$(Nb_6F_{12})F_3 + 9\ NbF_{5,g} = 15\ NbF_{4,g}$;	$T_1 \leftarrow T_2$	[7]
$CrCl_3 + 1/2\ Cl_2 = CrCl_{4,g}$;	$T_1 \leftarrow T_2$	
$Nb_3Cl_8 + 4\ NbCl_{5,g} = 7\ NbCl_{4,g}$;	$T_1 \leftarrow T_2$	
$WO_2 + I_2 = WO_2I_{2,g}$;	$T_1 \leftarrow T_2$	[8]
$SiO_2 + 4\ HF = SiF_4 + 2\ H_2O$	$T_1 \rightarrow T_2$	[9]
$Nb_2O_5 + 3\ NbCl_{5,g} = 5\ NbOCl_{3,g}$;	$T_1 \leftarrow T_2$	
$NbO + 3/2\ I_2 = NbOI_{3,g}$;	$T_1 \rightarrow T_2$	
$CdS + I_2 = CdI_{2.g} + 1/2\ S_{2,g}$;	$T_1 \leftarrow T_2$	
$Al_2S_3 + 3\ I_2 = 2\ AlI_{3,g} + 3/2\ S_2$;	$T_1 \leftarrow T_2$	
$2\ GaAs + GaI_{3,g} = 3\ GaI_{,g} + 1/2\ As_{4,g}$;	$T_1 \leftarrow T_2$	

(The arrows indicate the transport direction where $T_1 < T_2$)

4.4. Techniques

Closed tubes. Tube materials are usually glass and silica (quartz-glass). The water content of nearly all types of silica is especially important when transport occurs in closed tubes. At temperatures $> 700°C$ the out-diffusion of water from the walls of silica tubes becomes significant [10]. Consequently there is a build up of water content in the reaction tube, which may be either helpful or disruptive. It is recommended that the water content of the silica be controlled by monitoring the $\approx 3600\ cm^{-1}$ IR absorption band. The majority of the water may be removed by outgassing the quartz at 1000°C for one or more days [11].

It should also be remembered that the diffusion rate of hydrogen through the wall of the quartz glass tube can be high. At temperatures $\geqslant 600°C$ this effect must be taken into account. For example at 600°C in an ampoule containing the equilibrium mixture $HI/H_2/I_2$, the change of the composition caused by hydrogen diffusion can be observed by following the increase in the iodine pressure (absorption spectrum) [12]. The loss of hydrogen can be prevented by establishing the same hydrogen pressure on the outside as on the inside of the reaction tube. This arrangement also allows the possibility of changing the H_2-pressure during the transport experiment.

It is worth noting that, for HF containing systems, tubes may be fabricated in Ni or Pt. Discussion of the use and sealing of porcelain tubes is given in [13]. Also the use of Mo foil as a transport enclosure additionally permits heating by means of an electric current [14]. The important van Arkel method, which is concerned with the exothermic transport of metals onto a glowing wire by means of iodine, is described (together with its new developments) by Rolsten [15]. Special arrangements for the transport of material over short distances have been developed by Sirtl ("sandwich method") [16] and by Lydtin [17].

The *introduction of the transporting agent* can be achieved in a variety of different ways. In each case one starts with an ampoule containing the solid to be transported which is open at one or both ends.

(a) Admit the gaseous transport agent (Cl_2, HCl, HBr, CO ...) at room temperature – usually with a pressure of 1 atmosphere and seal the tube.

(b) Admit known amounts of I_2, Br_2, NH_4Cl ...; cool, when necessary with dry ice, evacuate and seal.

(c) Seal small amounts of transporting agent in glass capillaries. The transport ampoule 1 is connected via a seal capillary with an other ampoule 2. The capillary is brought into the ampoule 2, and the whole system is sealed under high vacuum. Afterwards the capillary is opened by a magnetic hammer or by strong heating at one point of the silica wall and the transporting agent is condensed in ampoule 1. Finally the connection between ampoules 1 and 2 is sealed.

(d) The transporting agent is prepared by quantitative chemical reactions and condensed within the ampoules. Usually we use an inert gas or the reacting gas (O_2, H_2) as a carrier. Examples [18] are:

Cl_2 : thermal decomposition of
$PdCl_2$, $PtCl_2$, K_2PtCl_6, K_2PbCl_6

Br_2 : thermal decomposition of
$PtBr_2$, K_2PtBr_6
$2\ CuBr + O_2 = 2\ CuO + Br_2$

I_2 : thermal decomposition of
PdI_2, WO_2I_2
$2\ CuI + O_2 = 2\ CuO + I_2$
$2\ AgI + 5\ V_2O_5 = 2\ AgV_5O_{12.5} + I_2$

H_2O : thermal decomposition of
$BaCl_2 \cdot 2\ H_2O$

HF, HCl, HBr, HI: reduction with H_2 of PbF_2, AgCl, K_2PtCl_6, AgBr, AgI respectively

4.5. Supersaturation

Supersaturation is an important factor in controlling crystal growth (see chs. 5, 7 and 10 §2). As an example one may quote experiments [12] which have been carried out on the nucleation of several materials (listed in table

Table 4.3
Supersaturation on quartz glass;
$\Delta T = T_2 - T_1$; T_2 = source temperature; T_1 = nucleation temperature (°C) Supersaturation Kp_2/Kp_1. The Kp-values are expressed for reactions involving 1 mol MeX or 0.5 mol Al_2S_3 or 0.5 mol Nb_2O_5 respectively. Closed tube experiments.

solid substance	transporting agent mg/ml	ΔT	T_1	Kp_2/Kp_1
1 ZnS	0.02 H_2	13	913	1.38
2 ZnS	0.25 HCl	28	810	1.29
3 ZnS	2 I_2	43	800	1.64
4 ZnSe	2 I_2	54	800	1.41
5 CdS	2 I_2	32	790	1.37
6 CdSe	2 I_2	20	715	1.21
7 0.5 Al_2S_3	2 I_2	37	820	1.8
8 0.5 Nb_2O_5 N-modification	2.90 Cl_2 + 8.8 $NbCl_5$	51	840	1.7

Kp_1 and Kp_2 are the equilibrium constants for the reactions at temperatures T_1 and T_2 respectively.

4.3) on clean silica which was etched with a mixture of diluted HF and strong HNO_3 and afterwards degassed at about 900°C for 24 h on a vacuum line. Experiments with ZnS/I_2 (+ N_2 carrier) were performed in a flow system. The gas equilibrated with ZnS at 850°C and then passed along a tube having a temperature gradient where the highest temperature where nucleation occurred within some hours was observed. The same results with ZnS/I_2 were found using a closed ampoule, situated in a transparent silica oven. The results obtained with ZnS/I_2 and other systems are given in table 4.3.

It was sometimes found that values of ΔT as large as 50°C were necessary for nucleation; a favourable situation for crystal growth. It seems remarkable that the K_{p2}/K_{p1} values for the different reactions have approximately the same value. But for a given K_{p2}/K_{p1} the practically important ΔT can vary widely for different reactions depending on the absolute value of K_p. As we will see in ch. 4. §6, the transport rate will be large, when the value of $\log K_p$ for the transport reaction is about zero. It can be shown that, under the same condition and for a given supersaturation K_{p2}/K_{p1}, the ΔT value is large. This may be an important criterion if one is searching for transport reactions suitable for the growth of crystals of a given solid.

4.6. Thermodynamic considerations

It was explained that usually the transport rate can be understood with reference to the model of gas motion (diffusion) between spaces in which the chemical equilibrium between the solid and the gas phase is established. The important concentration gradient is therefore calculable by chemical thermodynamics. From a knowledge of the large fund of thermodynamic data and chemical experience it is possible to find new transport reactions and to understand old ones.

Calculation shows that in a general reaction

$$A_{,s} + B_{,g} = C_{,g}$$

the gradient $P_{c,T2} - P_{c,T1} = \Delta P_c$ is close to its highest possible value when $\log K_p$ is about zero. Qualitatively it is easy to see that reactions for which the equilibrium position is far to the left or far to the right are unsuitable for chemical transport, because the ΔP_c-value is close to zero. Besides these considerations, chemical thermodynamics allow us to compose some rules; (1) The transport rate is proportional to $\Delta P_c/\Sigma P$, when the process is diffusion controlled, (ΣP = total pressure) (2) the sign of ΔH°_{react} determines the transport direction in a temperature gradient, (3) with reactions proceeding with a change in the number of gaseous molecules a measureable transport rate can not be expected when ΔS°_{react} and ΔH°_{react} have different signs. For a fuller discussion of these rules see [1]. ΔS°_{react} is the entropy change for the reaction.

Qualitative thermodynamic considerations can also be very helpful. As an example we refer to the connection between the chemical transport of metals and their sulfides by means of iodine. This leads to the result, that when the metal is transportable, with iodine, one may also expect the sulfide, selenide and telluride to transport. Discussion in detail may be found in [3].

Exothermic equilibria can be used for the transport of a solid onto a glowing filament. This is the basis of the classical van Arkel method for purification of metals [15]. As another example of the applicability of equilibrium considerations, even in complicated gas mixtures, one may quote the transport of tungsten at about $\geqslant 2000°C$ between neighbouring turns of a tungsten spiral by means of O_2 (ppm amounts of O_2 in N_2) which acts as a transporting agent. The observations show that the transport direction in the temperature gradient depends on the O_2-concentration and the absolute temperature [19]. This is in quantitative agreement with the thermodynamic expectations [20].

The interplay of endothermic and exothermic transport reactions for the same solid, resulting in the inversion of the transport direction, is a well known phenomenon [1]. Such cases, and in each system involving more than one equilibrium, a total concentration which can be understood as the reversible solubility of the solid in the gas phase must be introduced into the calculation.

The separation of two solids occurs when the transport equations for both have different signs of ΔH°_{react} [1, 3]. A new example is the separation of W and WO_2 by means of $I_2 + H_2O$ as transporting agents [6].

When the transport rate cannot be understood using known thermodynamic information new gaseous species are to be expected and information about such species and their thermodynamics can be obtained from chemical transport experiments. A recent example is the chemical transport of $CrCl_{3,s}$ with $Al_2Cl_{6,g}$. Quantitative transport experiments with a well defined diffusion path connecting equilibrium spaces showed the existence of the new species $(CrCl_3.3\,AlCl_3)_{,g}$. The equilibrium $CrCl_{3,s} + 1.5\,Al_2Cl_{6,g} = (CrCl_3.3\,AlCl_3)_{,g}$ is responsible for this $CrCl_3$-transport [21].

Another investigation showed, that a large number of metal sulfides are transportable by means of sulfur [22]. Since TaS_2 under special conditions moves up the temperature gradient there is no doubt that chemical transport via a yet unknown gas species is responsible.

References

[1] H. Schäfer, Chemische Transportreaktionen (Weinheim, 1962) Chemical Transport Reactions (New York, 1964).
[2] H. Schäfer and M. Trenkel, Z. anorg. allg. Chem. 391 (1972) 11.
[3] H. Schäfer, J. Crystal Growth 9 (1971) 17.
[4] H. Schäfer and H. Odenbach, Z. anorg. allg. Chem. 346 (1966) 127.
[5] H. Schäfer and U. Wiese, J. Less-common Metals, 24 (1971) 55.
[6] J.H. Dettingmeijer, J. Tillack and H. Schäfer, Z. anorg. allg. Chem. 369 (1969) 161.
[7] H. Schäfer, H.G. Schnering, K.J. Niehues and H.G. Nieder-Vahrenholz, J. Less-common Metals 9 (1965) 95.
[8] J.H. Dettingmeijer and B. Meinders, Z. anorg. allg. Chem. 357 (1968) 1.
[9] R. Gruehn, Münster 1967.
[10] A.J. Moulson and J.P. Roberts, Trans. Faraday Soc. 57 (1961) 1208.
[11] H. Schäfer and L. Zylka, unpublished.
[12] H. Schäfer, W. Jagusch and H. Wend-Erdel, unpublished.
[13] H. Klotz, Vakuum-Technik 15 (1965) 66.
[14] H. Schäfer and W. Fuhr, Z. anorg. allg. Chem. 319 (1962) 52.
[15] R.F. Rolsten, Iodide Metals and Metal Iodides (John Wiley and Sons, New York, London, 1961).
[16] E. Sirtl, Z. Naturforsch. 18a (1963) 884.
[17] H. Lydtin, Z. Naturforsch. 22a (1967) 571.
[18] H. Schäfer and Mitarb., Experimental experience during the last 10 years.
[19] R.J. Campbell, J. High Temp. Science 1 (1969) 303.
[20] H. Schäfer, Z. anorg. allg. Chem. 376 (1970) 11.
[21] K. Lascelles and H. Schäfer, Angew. Chem 83 (1971) 112, Angew. Chem. internat. Edition 10 (1971) 128, Z. anorg. allg. Chem. 382 (1971) 249.
[22] H. Schäfer, F. Wehmeier and M. Trenkel, J. Less-common Metals 16 (1968) 290.

P. Hartman, ed., Crystal growth: an introduction. © (1973) North-Holland Publ. Co.

CRYSTAL GROWTH BY MEANS OF TWO-PHASE AND MULTIPHASE EQUILIBRIA

A. RABENAU

5.1. Introduction

Vapour growth by chemical transport, solution growth – including growth under hydrothermal conditions – and flux growth are usually based on the temperature dependence of chemical equilibria. With chemical transport reactions these equilibria are often known and may be calculated, given sufficient thermodynamic data. In the simplest case two phases are involved, the gas phase and the substance to be transported as a solid phase. In the case of solution and flux growth we have to deal with the liquid (solution) phase and a solid phase *. The number of possibilities increases considerably if chemical equilibria in which three or more phases are involved are also considered [1].

For systems with three phases this leads to the combinations

I. one solid phase + one liquid phase + a gas phase
II. two solid phases + a gas phase
III. one solid phase + two liquid phases *
IV. two solid phases + one liquid phase *.

Examples of I and II and, in addition, two examples where four phases are involved, will be discussed and demonstrated with the growth of covellite, CuS, as a "model".

5.2. Experimental part

5.2.1. *General considerations*

The mineral covellite is a superconductor (1.67°K) and has a marked anisotropic structure (B-18 type). The choice of an appropriate method of crystal growth involves consideration of the phase relationships of CuS in the

* The gas phase also present is not actively engaged and can be ignored.

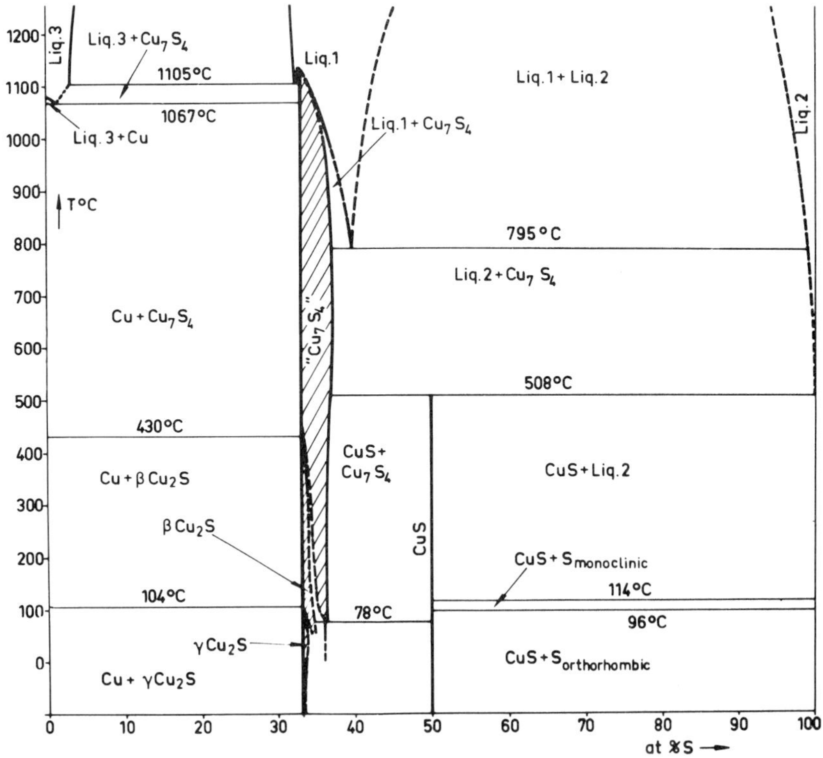

Fig. 5.1. Cu–S phase diagram.

system copper–sulphur (fig. 5.1). The phase diagram shows CuS to be a narrow phase melting peritectically at the rather low temperature of 508°C. For the preparation of single crystals, therefore, neither growth from its own melt nor sublimation can be used; growth from a sulphur melt on the other hand is not possible because the melt in equilibrium with CuS consists of almost pure sulphur, e.g., the solubility of CuS in liquid sulphur is negligibly small. Covellite crystals therefore cannot be obtained from the copper–sulphur system alone; ternary or higher order systems have to be considered.

5.2.2. *Chemical transport*

In similar cases chemical transport reactions (see ch. 4) have been applied with success in recent years. At the temperatures in question, only copper halides have a sufficient high partial pressure to make copper transport feas-

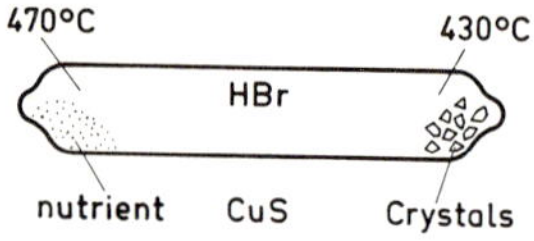

$$2\,CuS + 2HBr = \tfrac{2}{3}(CuBr)_3 + H_2S + \tfrac{1}{2}S_2 \mid 470 \rightarrow 430°C \mid$$

Fig. 5.2. Chemical transport of CuS with HBr.

ible and therefore halides have been used in a suitable chemical equilibrium. Best results were observed with HBr as the transporting agent as shown schematically in fig. 5.2. Transport rates increase with increasing HBr pressures in the range between 100 mm Hg and 30 atm.

Numerous, but small, crystals can be obtained in this way. In order to get crystals of sufficient size for physical measurements, chemical transport has to be applied in an arrangement which suppresses the influence of high nucleation rates by seed selection. Two different possibilities are discussed in §5.4. In our case the linear version of the temperature gradient reversal method (ch. 5, §4.2) has been applied, resulting in crystals up to 2 cm^2 and 0.1 mm thickness. (The arrangement is the same as in fig. 5.3 with omission of the CuCl/CuBr melt).

5.2.3. *One solid phase + one liquid phase + a gas phase*

Transport rates, for the method in 2.2, are rather slow due to the low partial pressure of copper bromide at the temperatures considered, namely of the order of only some tenths of a mm Hg. The crystal growth of CuS can be considerably improved, however, if there is a melt of copper bromide and copper chloride where crystal growth takes place (fig. 5.3). Only hydrogen bromide, hydrogen sulphide and sulphur are exchanged through the gas phase but copper is not. The copper necessary for the formation of the crystals is provided by the melt (an equivalent of anions is transported to the source). The growth rate proved to be almost independent of the pressure of HBr. This means that the growth rate is determined by what happens in the melt. This may be either a reaction or a transport (diffusion, convection) process in the melt. With this arrangement not only are the growth rates improved but the size of the crystals is also favourable (about 1 mm thick).

The combination of chemical transport through a gaseous phase with growth from a melt in which the metal is present in ion form, seems to embody an important principle which is applicable where no compounds of

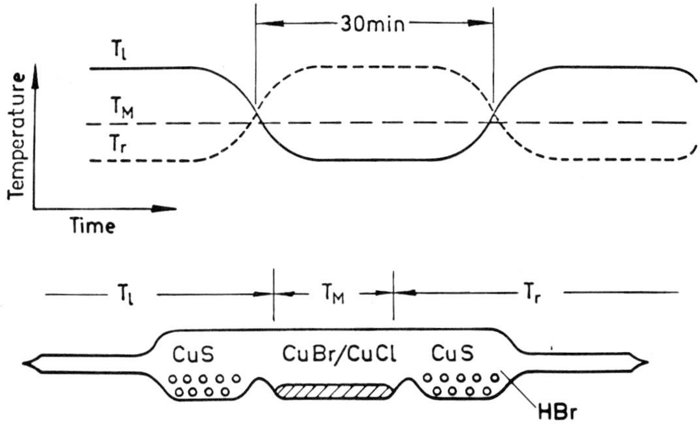

Fig. 5.3. Growth of covellite crystals on the basis of ch. 5 §2.3, making use of the temperature gradient reversal method (see ch. 5 §4.2). The CuS crystals grow on the bottom of the CuBr/CuCl melt in the middle of the ampoule. The temperature in the middle T_m must be below the average temperature of the right and left hand side $(T_r + T_l)/2$. Source: 2 g CuS powder on each side, melt: a mixture of 1.5 g CuBr and 2.5 g CuCl. Ampoule about 1/3 natural size. Temperature difference $(T_l - T_r)$ about 30°C, $T_m = 430°C$, $P_{HBr} \approx 1$ atm.). (From Z. physik. Chem. 53 (1967) 1).

the metal exist with sufficiently high partial pressures at the required temperatures, as in the case of copper, the rare earth elements and others.

5.2.4. *Two solid phases + a gas phase*

An example of this combination is the growth of CuS crystals based on the chemical equilibrium

$$CuS(s) + \tfrac{1}{2}I_2(g) = CuI(s) + \tfrac{1}{2}S_2(g) . \tag{5.1}$$

At 450°C the partial pressures of iodine and sulphur are almost equal. In addition, the equilibrium is strongly temperature dependent and consequently the conditions for crystal growth are favourable. CuS is placed on one side of an evacuated glass ampoule, and CuI on the other. The ampoule is heated so that the CuS is at 440°C and the CuI at 450°C. Within a few days, CuS and CuI crystals form at 450°C and 440°C, respectively.

In this way crystals of two substances can be grown simultaneously.

5.2.5. *Two solid phases + one liquid phase + a gas phase*

The crystals grown by the method described in the previous section were too small. The results can be improved if, with the same arrangement and on the basis of eq. (5.1), a 4-phase system is applied. Above the solid CuI a melt is formed by addition of a small amount of CuCl (eutectic temperature 284°C). The equilibrium (5.1) is not influenced because the activity of CuI in the melt is equal to unity (as long as solid CuI is present). Growth of the CuS crystals from the melt improves their size and quality. The method is similar to that in §2.3, with the advantage that in this case thermodynamic data can be calculated in principle on the basis of the pure components CuS, CuI, I_2 and S_2.

5.2.6. *Two solid phases + two liquid phases*

This method is based on the solubility of Cu_2S in a melt which contains CuCl. (Pure CuCl dissolves about 15 mol% Cu_2S at 500°C). The arrangement is shown in fig. 5.4. Above the solid Cu_2S there is a CuCl melt, where KCl

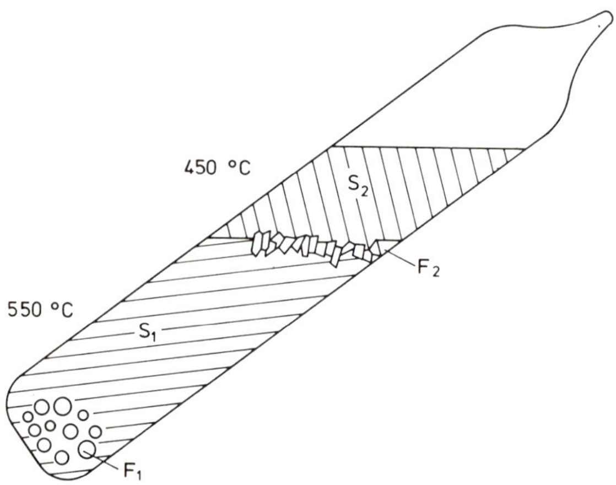

Fig. 5.4. Growth of CuS crystals according to ch. 5 §2.6. $F_1 = Cu_2S$, F_2 = CuS crystals. S_1 = melt of 20 g KCl, 13 g LiCl and 4.5 g CuCl, S_2 = liquid sulphur. (From Z. physik. Chem. 53 (1967) 1).

and LiCl are added in order to lower the melting point of CuCl. Liquid sulphur covers the salt melt. CuS crystals grow on the interface of the two melts, possibly according to

$$Cu_2S_{\text{dissolved}} + S_{\text{liquid}} = 2\ CuS_{\text{solid}}\ . \tag{5.2}$$

5.3. Discussion

Methods of crystal growth making use of third, or higher, order phase equilibria may be successfully applied where no suitable two-phase equilibria exist.

It should be possible, for a given substance AB, to find a suitable equilibrium of the form

$$AB(s) + CD(g) = AC(s) + BD(g) . \tag{5.3}$$

AB and AC are condensed phases which do not necessarily have to have a significant partial pressure at the temperatures involved. (In a proper chemical transport reaction AC is part of the gas phase, a condition limiting the choice considerably.) The mass action constant of eq. (5.3) must be temperature dependent and not deviate too much from unity. Under these conditions crystals of AB can be grown in a closed ampoule with an applied temperature gradient. On one side of the ampoule there is the source material, e.g., AB in a polycrystalline form, and on the other side AC and a suitable flux of sufficiently high solubility for AC. From the melt saturated with AC, crystals of AB can be grown as long as AC is present in a solid phase – in principle as long as the activity of AC is not below a critical value. Only the components B and C are transported via the gas phase.

The advantages of this method lie in the fact that the equilibrium can be calculated in principle from the thermodynamic data of the pure substances (AB, AC, BD, CD) although the crystals are grown from a melt. It should be mentioned, however, that the application may be limited by unwanted formation of solid solutions between AB and AC or between AB and the flux. The situation is less critical with hydrothermal synthesis. (See ch. 7.)

5.4. Chemical transport and seed selection

The growth of large single crystals via a chemical transport reaction is often hampered by a high nucleation rate resulting in numerous small crystals. Two examples of chemical transport arrangements are given which cope with these difficulties.

5.4.1. *Transport-pulling method* [2]

The principle is illustrated schematically in fig. 5.5. The substance to be transported is enclosed, together with the transporting agent, in a quartz glass

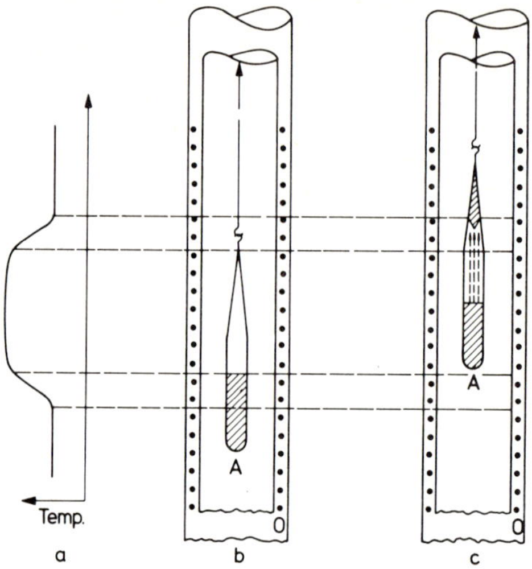

Fig. 5.5. a) Curve of temperature as a function of height in the furnace. b) Situation at the start of the pulling operation. c) A later stage of the transport process. (From Philips Technical Review.)

ampoule, the upper part of which is tapered to a point. The tapered end enters the zone of highest temperature first. This cleans the inner surface of the ampoule by the chemical transport of any crystallites deposited on the glass during the initial heating which would otherwise cause spurious nucleation. As the pulling operation progresses the lower end of the ampoule enters the maximum temperature zone, and at the same time the pointed tip passes into a zone of lower temperature; at a certain point the transported substance starts to deposit. The shape of the upper part of the ampoule is such that it favours seed selection, and consequently the number of crystals that grow laterally across the tip of the ampoule is limited. The method proved to be especially suitable for the growth of homogeneous mixed crystals, e.g., $GaAs_{1-x}P_x$.

The principle of the method is shown in fig. 5.5. Ampoule A is suspended in a furnace O consisting of two concentric quartz tubes, the heating element being wound round the smaller of the two. A section of the inside surface of the outer quartz tube carries a gold film which reflects radiant heat and so creates a high-temperature zone (see diagram a) in this part of the furnace. A vertical slit is left in the film, and so the ampoule can be observed during the pulling process.

Fig. 5.6. A battery of five crystal pulling furnaces.

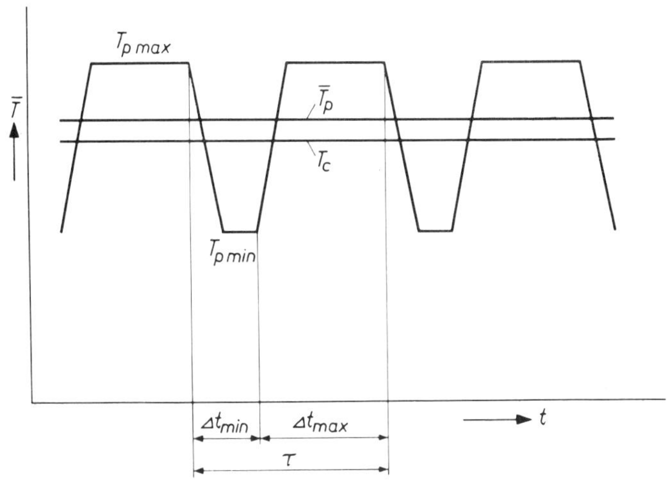

Fig. 5.7. Temperature–time curve for the axially symmetrical method. (From Philips Technical Review.)

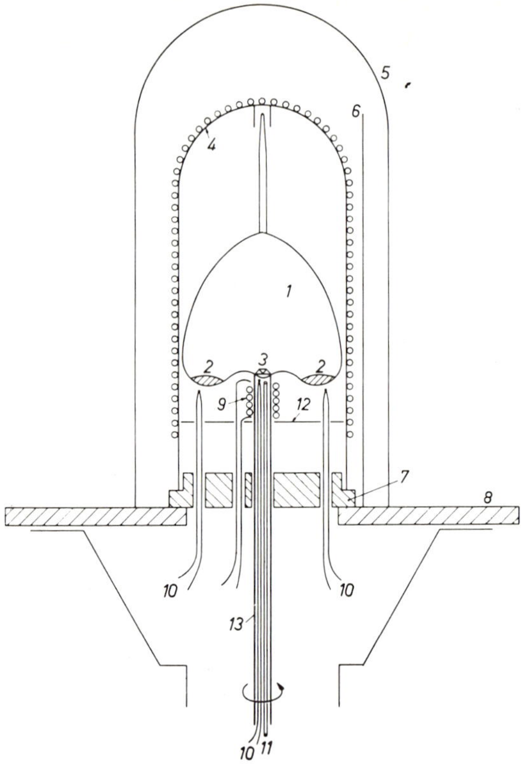

Fig. 5.8. Experimental arrangement for axially-symmetrical methods. 1 – reaction vessel, which is slowly rotated. 2 – source material at T_p. 3 – seed crystal at T_c. 4 – quartz glass dome, whose top and cylindrical parts can be heated by two separate * heater coils. 5 – heat-resistant glass dome to minimize heat losses due to convection. 6 – heat-reflecting screen which helps to set up a temperature gradient perpendicular to the axis *. 7 – ceramic support for quartz glass dome. 8 – baseplate of ceramic material. 9 – heating for base. 10 – three thermocouples. 11 – air-cooling tube. 12 – radiation screen. 13 – hollow drive shaft for reaction vessel; this shaft is rotated by a motor. (From Philips Technical Review.)
* Not relevant to this type of axial method.

5.4.2. *Temperature-gradient reversal methods* [3]

The special feature of this method is that the survival probability of spontaneously formed nuclei strongly diminishes. In temperature-gradient reversal methods, growing and etching follow one another, and therefore seed selection and the healing of faults on the growing crystal are possible. Material transport proceeds to the point of lowest average temperature $\overline{T}_c$. The con-

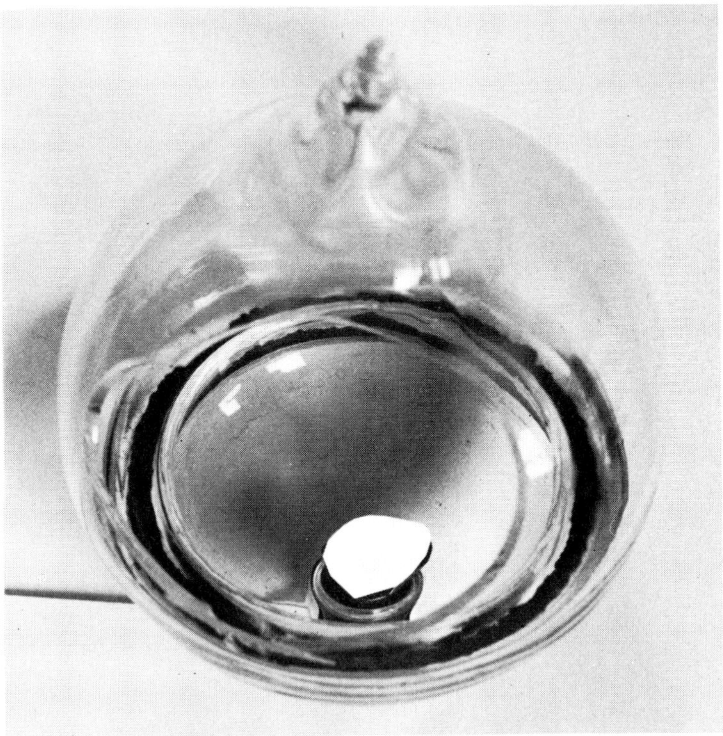

Fig. 5.9. The reaction vessel with a hematite α-Fe_2O_3 crystal grown by axial symmetrical method. The remainder of the source material can also be seen.

trolling factor for material transport is the difference between the average temperature $\overline{T}_p$ and $\overline{T}_c$. Alternation of temperatures is effected using a programmed power input. Rotation of the vessel here serves only to smooth out the inhomogeneities of the system.

References

[1] A. Rabenau and H. Rau, Z. physik. Chem., Neue Folge 53 (1967) 155.
[2] A. Rabenau, Philips Techn. Rev. 26 (1965) 117.
[3] H. Scholz, Philips Techn. Rev. 28 (1967) 316.

P. Hartman, ed., Crystal growth: an introduction. © *(1973) North-Holland Publ. Co.*

HYDROTHERMAL GROWTH

R. A. LAUDISE

6.1. Introduction – methods of crystal growth

Although some of the largest single crystals formed in nature (beryl crystals > 1000 kg) and the single crystalline material produced in the greatest volume by man (quartz world production ≈ 750,000 kg/yr) are both of hydrothermal origin, the technique is probably the least practiced on a laboratory scale of any of the major methods of crystal growth. If we use the criteria of at least three publications/yr over the past five years, there are less than a dozen groups active in hydrothermal crystal growth and less than fifty papers/yr being published. This is unfortunate. The technique is worthy of further study on the grounds of its technological efficacy, its importance in geochemistry, and the intrinsic interest of its diverse physical chemistry. The principal thrust of our review will be on the use of hydrothermal growth as a crystal growth technique. To appreciate this use, it is necessary to briefly review other crystal growth techniques.

The classification scheme for growth techniques which we will use is an abbreviated version of that of Laudise [1] and is summarized in table 6.1. The two broad classes of growth techniques are monocomponent and polycomponent methods. Component is used in the same sense as it is used in the phase rule. Crystals may be grown by perturbing an equilibrium toward the direction of a solid crystalline phase. The growth may be from the gas, the liquid, or the solid state:

(a) gas → solid crystal
(b) liquid → solid crystal
(c) solid → solid crystal

Thus, depending upon whether the growth is from a mono- or a polycomponent system, there are six broad classes of crystal growth methods. Solid-solid methods are quite specialized, and are not discussed further. The role of the added components in polycomponent growth may be considered to be to

Table 6.1
Growth methods.

I. Monocomponent

  A. Solid-solid
    1. Strain annealing
    2. Devitrification
    3. Polymorphic-phase change

  B. Liquid-solid
    1. Conservative
      (a) Directional solidification – (Bridgman-Stockbarger)
      (b) Cooled seed – (Kyropoulos)
      (c) Pulling (Czochralski)

    2. Nonconservative
      (a) Zoning (horizontal, vertical, float zone, growth on a pedestal)
      (b) Verneuil (flame fusion, plasma, arc image)

  C. Gas-solid
    1. Sublimation-condensation
    2. Sputtering

II. Polycomponent

  A. Solid-solid
    1. Precipitation from solid solution

  B. Liquid-solid
    1. Growth from solution (evaporation, slow cooling, and temperature differential)
      (a) Aqueous solvents
      (b) Organic solvents
      (c) Molten-salt solvents
      (d) Solvents under hydrothermal conditions
      (e) Other inorganic solvents

    2. Growth by reaction (media as above – temperature change, concentration change)
      (a) Chemical reaction
      (b) Electrochemical reaction

  C. Gas-solid
    1. Growth by reversible reaction (temperature change, concentration change)
      (a) Van Arkel (hot-wire) processes

    2. Growth by irreversible reaction
      (a) Epitaxial processes

allow growth at a lower temperature than would otherwise be possible. In gas → crystal growth the component complexes to form a new, more volatile species permitting growth at a lower temperature than in the monocomponent (sublimation) case. In liquid → crystal growth the additional component lowers the melting point of the material being crystallized (i.e., dissolves it so that growth is from solution). In both cases the additional component permits growth at a lower temperature. Among the important advantages of lower temperature growth are:

(1) It permits the direct growth of low temperature polymorphs. Of course, one may sometimes form such polymorphs by solid-solid transition from high temperature grown single crystals. Such processes are, however, difficult to control and the product is usually strained, often twinned and sometimes polycrystalline.

(2) It allows one to avoid decompositions, including incongruent melting, loss of volatile constituents (in melt growth) and oxidation reduction problems associated with higher temperature processes.

(3) It avoids strains, reduces vacancy concentration, and sometimes reduces dislocations and low angle grain boundaries associated with high temperature growth.

(4) It is experimentally more convenient. Higher temperature processes are often more demanding of equipment and more difficult to control, and harder to keep clean so that the product is pure.

Whether one would elect vapor phase growth or solution growth for a particular material depends principally upon whether reactions to increase volatility or liquid solvents seem more promising. At the present state of the art, vapor phase growth has had more practical use in the preparation of thin films, while solution growth has had its main successes in the preparation of bulk crystals. Our attention here will focus upon solution growth.

Lest one think that low temperature growth in general and solution growth in particular is a panacea, the disadvantages of polycomponent growth are enumerated:

(1) The additional component will be a contaminant and will have some solid solubility in the grown crystal.

(2) Elimination of the additional component at the growing interface will set up an upper limit on rate. Diffusion will be important in the process.

(3) Because of the concentration gradients at the growing interface, constitutional supercooling will often occur, facet effects, cellular growth, and dendritic growth can thus be problems.

The three main classes of solution growth are:

(1) Growth from aqueous solution at ambient conditions.

(2) Growth from nonaqueous solvents, especially molten salt (flux growth) and growth from liquid metals (e.g., GaP from Ga).

(3) Hydrothermal growth.

Hydrothermal growth is usually defined as the use of an aqueous solvent at elevated temperature and pressure to dissolve a solute which would ordinarily be virtually insoluble at ambient conditions. The principal advantages of hydrothermal growth over the other solution methods are:

(1) It combines a low temperature with high solvent power. Thus, it is an excellent method for the growth of low temperature polymorphs of refractory materials.

(2) It is a seeded process fairly easy to scale up. Thus, it is particularly suited to the growth of large quantities of single crystals.

(3) It usually produces crystals of good homogeneity and high quality. Its disadvantages are mainly those associated with high pressure – unwieldy equipment and inability to observe growth during the process. In addition, so little is known about hydrothermal chemistry and the systems used in hydrothermal growth usually involve at least three components so that a lengthy empirical study is usually required to find growth conditions. For this reason, hydrothermal growth is generally not the best technique to employ in a preliminary exploratory materials research program.

6.2. History of hydrothermal research

There are two different traditions in hydrothermal growth. One is the phase equilibria–geochemical tradition; the other is the crystal growth–electronic materials tradition. Since our present interest is crystal growth, we will emphasize the latter tradition. However, it is interesting to briefly trace the former. The first hydrothermal studies were atempts at mineral synthesis. Quartz, the feldspars, wollastonite, analcite, and many other minerals were prepared. Indeed, G.W. Morey [3] surveyed the literature and lists 80 mineral species for which hydrothermal synthesis is claimed before 1900. Unfortunately, the conditions were not defined, complex systems were investigated before simple systems were understood and all in all no reliable data on what minerals formed under what conditions was obtained. Morey [3] cites an experimenter who found that by reacting in glass for three months at 550°, a mixture of colloidal silicon, colloidal ferric hydroxide, colloidal ferrous oxide, lime water, magnesium hydroxide, and potassium hydroxide some crystals "resembling" hornblende were obtained! Morey and Niggli [4] summarized this work up to 1913. Indeed their paper is still a valuable key to references which are otherwise difficult to locate and which give qualitative guides to synthesis conditions. G.W. Morey of the Geophysical Laboratory of the Carnegie Institute of Washington can perhaps rightly be called the founder of systematic geochemistry-oriented hydrothermal phase equilibrium studies. H.S. Yoder, R. Roy and O.F. Tuttle have carried on these traditions, and their publications should be consulted for a further review [5–7]. It is interesting to point out that an important logical coupling of this laboratory work with field geological studies has finally been forged with the development by Roedder at the U.S. Geological Survey [8] and Yermakov [8a] in the Soviet Union of techniques for the careful assessment of liquid bubble inclusions in minerals.

The crystal growth–electronic materials tradition too had its beginnings in the pre-1900's mineral syntheses. In particular, we should mention the attempts by De Senarmont and G. Spezia to synthesize quartz. These attempts were successful, but conditions were unspecified and conditions for the sustained repetitive growth of large crystals were not obtained. In contrast to the geochemical activities which progressed steadily, to a large extent hydrothermal crystal growth fell into an inactive period following the early work on mineral syntheses. It was only during World War II when various nations were cut off from crystalline natural quartz that crystal growth activities began again. In particular, R. Nacken in Germany succeeded in preparing small quantities of quartz. Following the War, groups in the U.S. and England made sustained attempts to produce commercially useful synthetic quartz. The three main groups were the General Electric Co. Ltd. (Wembley, England) – Oxford group of Brown, Kell, Thomas, Wooster and Wooster; The Clevite Co. (Cleveland, Ohio) – group of Hale, Jeffe and others; and the Bell Telephone Laboratories group (Murray Hill, New Jersey) of Walker, Kohman, Buehler, and others. Literature references to this era can be obtained in Laudise [9]. Systematic studies of growth rate, solubilities and acoustic properties were carried out by Laudise, Ballman, and others [10], and the first description of pilot plant production was published by Sullivan and Laudise [11]. Presently in Western Europe, the Soviet Union, Japan, and the U.S., there are at least a dozen commercial quartz producers. The quartz success stimulated similar investigations of a variety of other electronic materials. It will be the purpose of this review to describe the quartz studies and the studies of the hydrothermal growth of other substances. The viewpoint will be tutorial, and no claim for complete coverage of all hydrothermal work is made. Indeed, for convenience sake most examples will be chosen from the author's own experience.

6.3. Hydrothermal equipment

Hydrothermal experiments in crystal growth are conducted at pressures up to about 3 kilobars; consequently, the main unique equipment requirement is high pressure autoclaves. Laudise and Nielsen [12] reviewed hydrothermal equipment, so that our descriptions here will be somewhat abbreviated and will emphasize recent trends. Table 6.2 lists the main autoclave types and their p–T ranges.

Glass vessels have the advantages of ease of observation and resistance to acid solutions (except HF), but with the disadvantages of low p–T capabilities and ease of attack by basic solutions. Their use has been generally neglected in hydrothermal work mainly because early interest centered on materials requiring basic media and p–T conditions beyond the limits of glass. When the solvent and p–T conditions permit, and when visual observation is an aid, glass vessels have much to recommend them and should be used more frequently in hydrothermal work.

The workhorse vessel for low pressure crystal growth where inert conditions are required is the autoclave designed by G.W. Morey [4], illustrated in fig. 6.1. Vessels are commonly lined with Pt, Au, or Ag. Liners are deep drawn to a wall thickness of ≈ 0.5 mm. It is common to make the seal out of a harder material than the body of the liner (e.g., coin silver in the case of Ag liners, or Pt–Ir in the case of Pt liners). The seal is made by the use of a metal disc, generally of the same material as the liner. The disc has a thickness of 100–200 μm. The seal is made by the pressure of the plunger on the disc which acts as a gasket against the noble metal covering of the shoulder. This pressure is applied by putting a torque on the cover with a large wrench. The gasket often pressure welds to the shoulder and must be cut or machined off following the run. The pressure limits are set by the available torques and by the creep of the steel at high temperatures. Tool steel has been used to about 500° at 330 bar; Inconel can probably be used to above 700° at nearly 500 bar. The vessel length can be short if small gradients or isothermal conditions are desired and larger for higher gradients or for the growth of more crystals. Common lengths are 10 cm or 20 cm. Diameters larger than 2.5 cm result in reductions in the useful pressure range.

Fig. 6.2 shows the modified Bridgman closure. This design is not easy to line so that growth of materials compatible with steel * or the use of an internal can is required. The wedge shaped seal ring initially makes a line

* Quartz grown from Na^+ containing solutions is the main material compatible with steel. Acmite, $Na_2O \cdot Fe_2O_3 \cdot 4SiO_2$, forms and protectively coats the vessel.

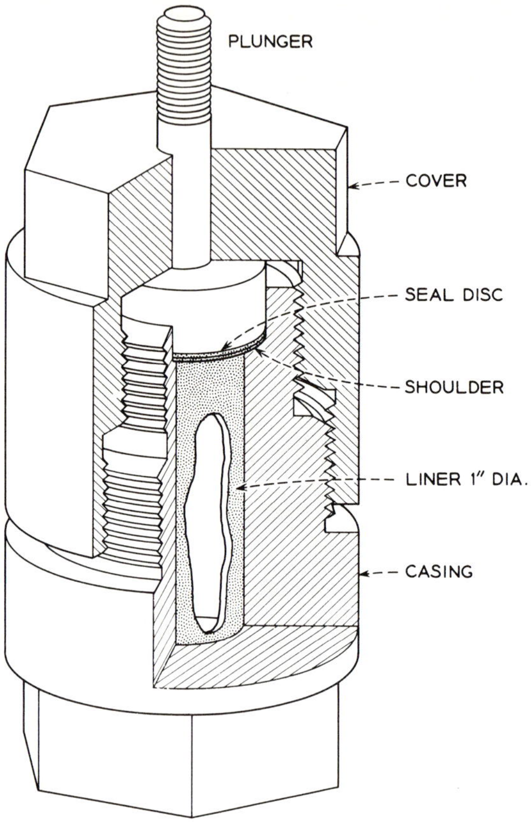

Fig. 6.1. Morey type autoclave (from [12]).

Table 6.2
Autoclaves.

Pyrex tube – 5 mm I.D. 2 mm wall thickness	6 bar at 250 °C
Quartz tube – 5 mm I.D. 2 mm wall thickness	6 bar at 300 °C
Flat plate seal – Morey type closure – 2.5 cm I.D. – easy to line with Pt, Au, or Ag	400 bar at 400 °C
* Modified Bridgman – unsupported area 2.5 cm I.D.	3,700 bar at 500 °C
** Cold seal-core – Tuttle-Roy type 5 mm I.D.	5,000 bar at 750 °C

* Ordinarily unlined; inert conditions obtained by use of welded internal can.
** Ordinarily unlined; reactions usually carried out in welded noble metal capsule.

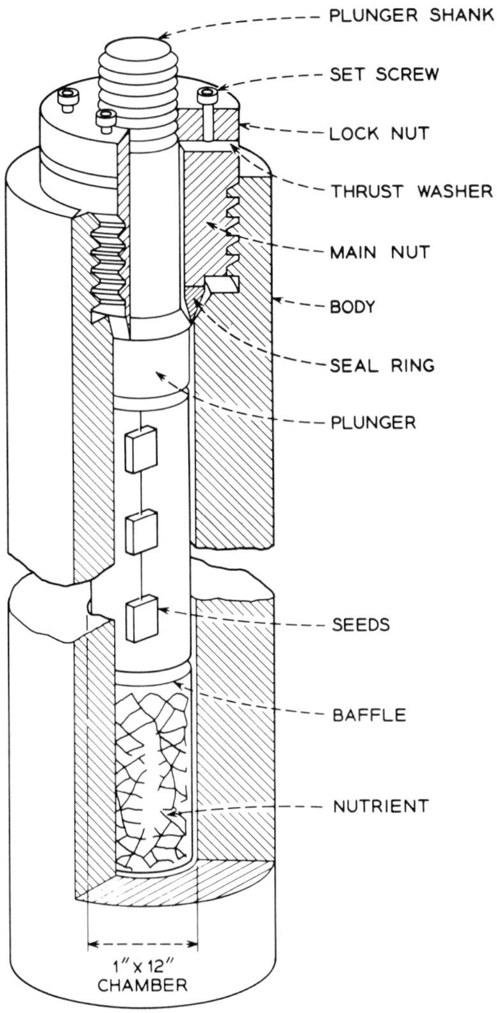

Fig. 6.2. Modified Bridgman seal autoclave (from [1]).

contact with the body and with the plunger. Slight internal pressure transmitted through the plunger will cause the ring to deform to produce a surface contact. Under all conditions the ratio (area plunger)/(area surface contact in seal) will be such that the pressure in the seal will always be larger than the pressure in the autoclave. Hence, the seal is *self-energized.* It is an *unsupported area* seal because the area of the shank of the plunger is not supported.

Careful machining of the surfaces including lapping of the internal seal surfaces with an iron lap and silicon carbide abrasive is generally required to prevent leaks. The closure is initially made by tightening the set screws sequentially (always moving to the screw as near as possible diametrically opposite in the tightening sequence) with a torque wrench. The material of construction can be a low carbon steel such as ASME 4140 for comparatively low temperature service (400–500°C max) or a more exotic alloy such as Inconel, Udimet or Stellite (with a considerable increase in initial and machining cost and somewhat reduced corrosion resistance dictating that noble metals can usually be employed) for high temperature service (600–800°C max). Since the seal is self-energized, the constraints on diameter are not as severe as for the Morey autoclave. A common laboratory size is 2.5 cm ID × 30 cm length; quartz production equipment 25 cm ID × 3.5 m length is common.

For phase equilibria, solubility and small scale crystal growth experiments, the Tuttle cold-seal-"test tube" autoclave [13] is generally used (fig. 6.3). Closure is effected by a cone seal; the seal region is not usually placed in the

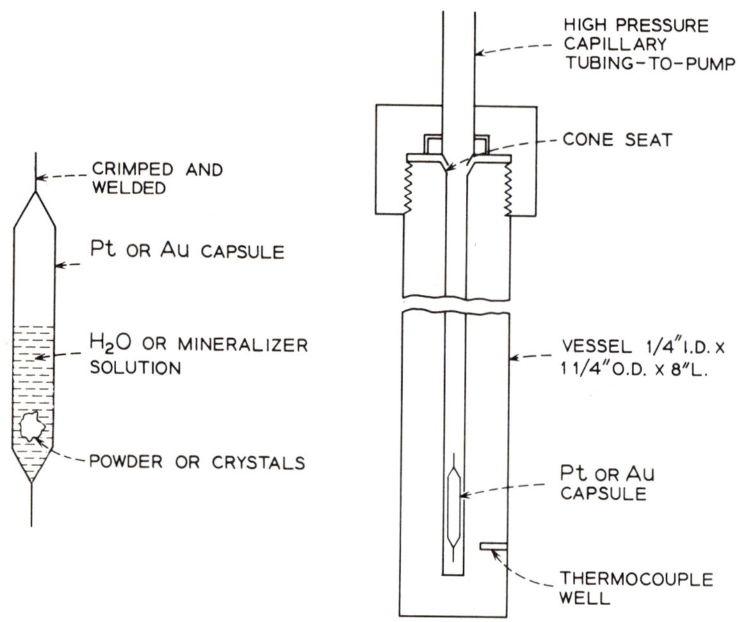

Fig. 6.3. Tuttle cold seal test tube autoclave (from [12]).

hot zone of the furnace. The sample to be studied is placed in a Pt, Ag, or Au capsule (solution is added with a hypodermic syringe or micro-buret) and the capsule is welded shut. Pressure is usually developed by pumping water into the system with an air driven intensifier. Stillite, Udimet or Inconel are the usual materials and p–T limits above those of table 6.1 are probably reasonable with some shortening of the autoclave life at higher ranges (e.g., 900°–500 bar; 700°–5000 bar).

To obtain intense conditions in the modified Bridgman apparatus it is useful to use a welded Pt or Ag can of the sort shown in fig. 6.4. The outer diameter of the can and its length are made close to the internal dimensions of the autoclave in which it will be used. This will allow small dimensional changes in the can (short of those causing leakage) to cause large changes in the volume between the outside of the can and the autoclave. These large volume changes can thus cause large pressure changes if needed, ensuring pressure balance between the fluid in the can and the balancing fluid outside the can. The covers on the can are welded with a "heli-arc" inert gas envelope carbon arc DC arc welder. A fixture for positioning the torch and the work piece and rotating the work piece is helpful to ensure a uniform weld. After the bottom cover is welded,

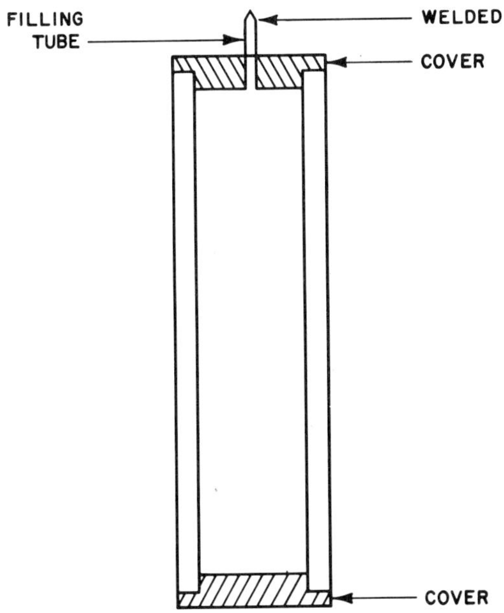

Fig. 6.4. Noble metal can.

solids are added to the can and the top cover (to which the small diameter filling tube has previously been welded) is welded on. Then the liquids are added with a syringe and the filling tube is welded while the can is kept in cold water to prevent vapor loss.

6.4. Phase equilibria studies

The first requirement for hydrothermal growth is the knowledge that the desired phase is stable at the p–T conditions of interest. Since it is common

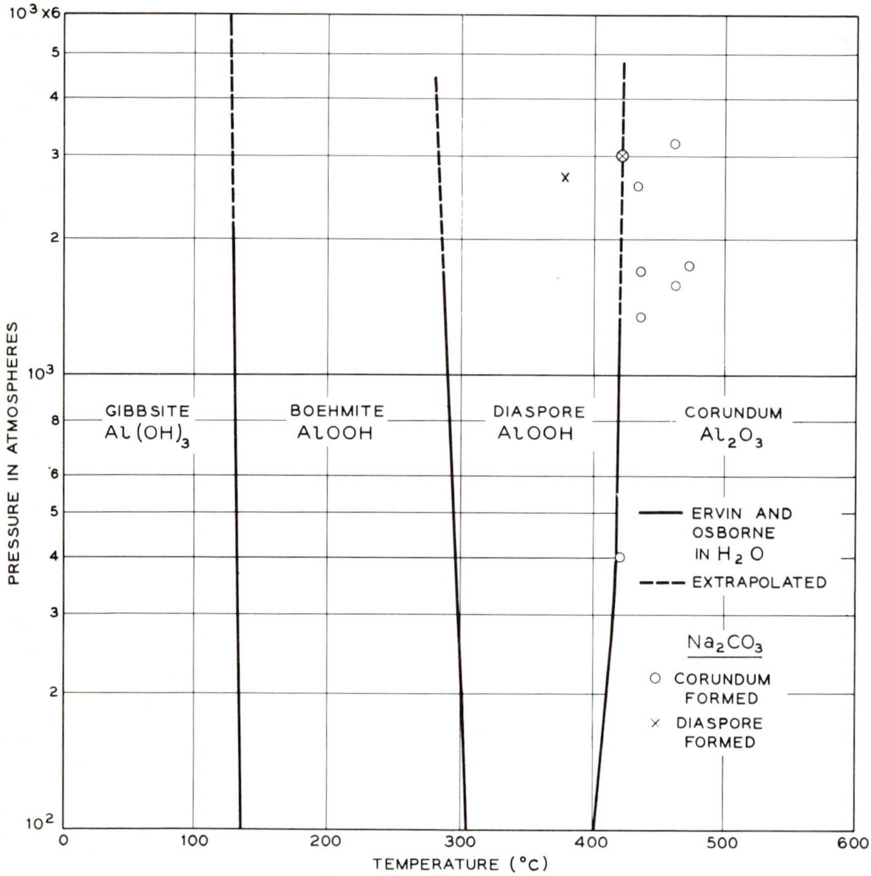

Fig. 6.5. Part of the Al_2O_3–H_2O diagram (from [9]).

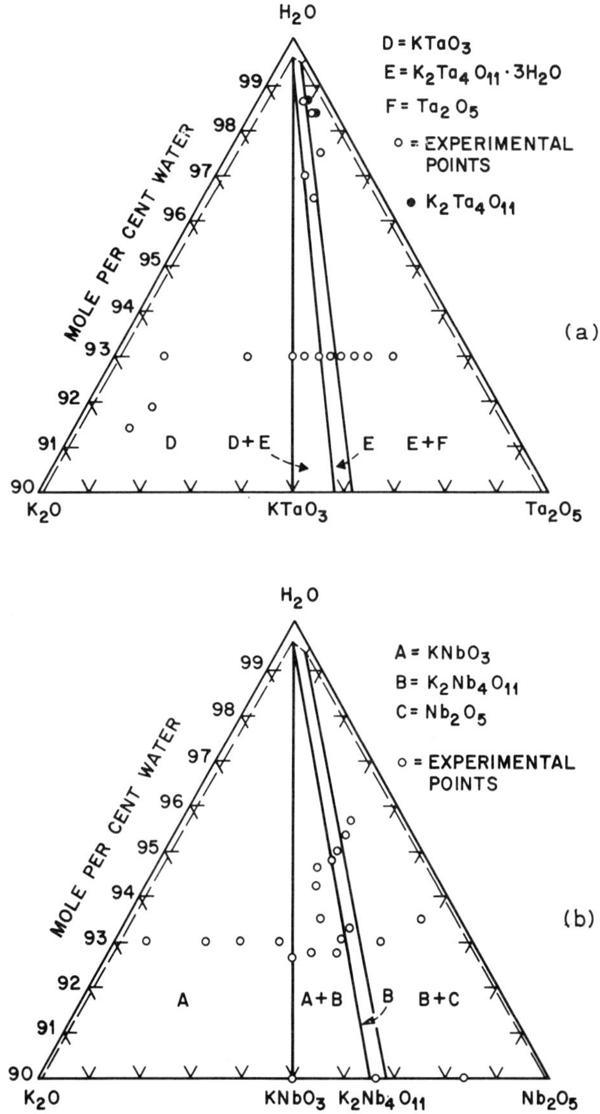

Fig. 6.6. Phase relationship involving K(TaNb)O_3(KTN) (from [24]).

that the solubility of materials of interest is too low for growth, it is usual to add an additional component (usually called a mineralizer or complexer) which complexes or reacts with the solute so as to form additional species not ordinarily present in appreciable concentrations in pure water and thus raise its solubility. Consequently, it is important to determine phase stability at various mineralizer concentrations. A variety of different techniques and apparatus have been used but that which we have found most convenient is the Tuttle autoclave. The components to be studied are placed in the capsule, and

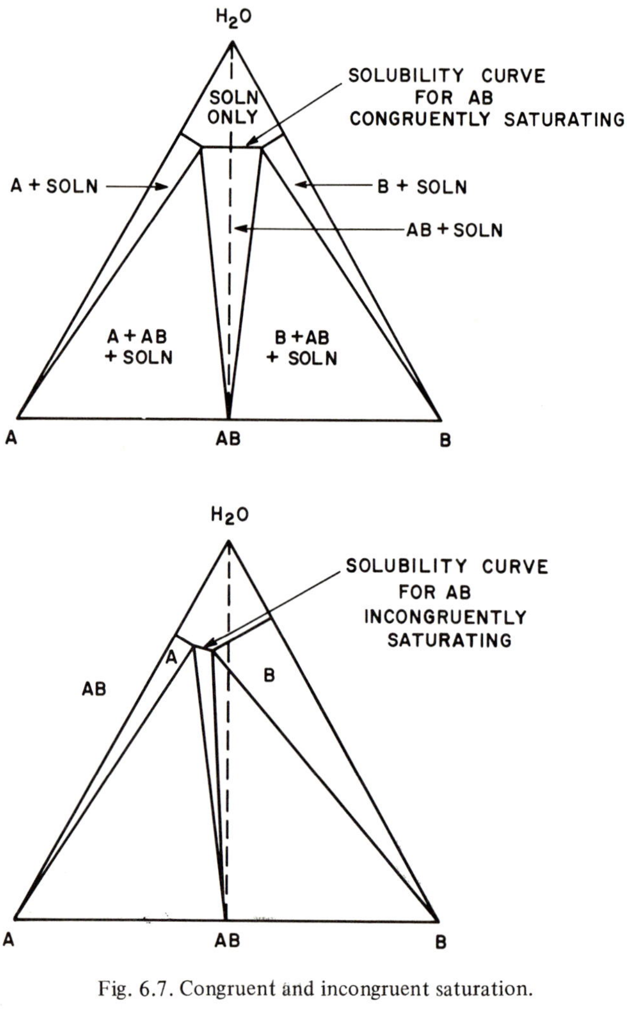

Fig. 6.7. Congruent and incongruent saturation.

the appropriate amount of the desired solution is added. Following welding, etc., the capsule is brought to the desired p–T conditions, run for a time longer than the equilibrium time (this time is found by making a series of runs as a function of time) and quenched rapidly to room temperature. The phase in the capsule is identified by X-ray diffraction and optical microscopy and one has obtained one point in p–T space. Typical of phase diagrams obtained are figs. 6.5 and 6.6. Fig. 6.5 shows the stability of the hydrates of Al_2O_3 as a function of p and T [14, 15], while fig. 6.6 shows the results of a ternary investigation [16].

The concept of incongruent and congruent materials is important for hydrothermal work and is illustrated by fig. 6.7. In congruent saturation a compound AB forms from a solution where the ratio A/B is identical to that in the compound. When A/B in the solution where AB is synthesized is different from A/B in the compound, it is incongruently saturating. The analogy to incongruent melting is obvious. Extensive reviews of hydrothermal phase equilibria studies have been published [5], and an excellent compilation of diagrams is available [17].

6.5. Solubility studies

For crystal growth by the temperature differential method a solubility, s, of at least a few weight percent is required and ds/dT, the temperature coefficient of solubility, should be 0.01–0.1 wgt percent/10°C to provide reasonable supersaturation. For almost all refractory materials this requires the addition of a mineralizer. For acidic and amphoteric materials (most of the materials so far grown) $(OH)^-$ has proven particularly effective. Solubilities were originally measured by removing samples of solution from high pressure systems and by the careful measurement of the loss in weight of solute exposed to fairly large volumes (25–500 cm^3) of the mineralizer solution for periods of time greater than the equilibrium time. When the weight solubility is greater than about 1% (the range of interest for crystal growth), it has been shown that acceptable accuracy in solubility measurements can be obtained using small volume (a few cm^3) capsules in Tuttle vessels. Weighed pieces of solute plus mineralizer solution are placed in the capsule and it is equilibrated in a Tuttle vessel. Following a quench, weight loss measurements of the remaining solute pieces allow the solubility to be determined. The results are valid provided (1) all solute pieces are recovered (it is therefore advantageous to use one large piece), (2) no deposition occurs during quench (microscopic examination and varying quench rates are useful for tests), and

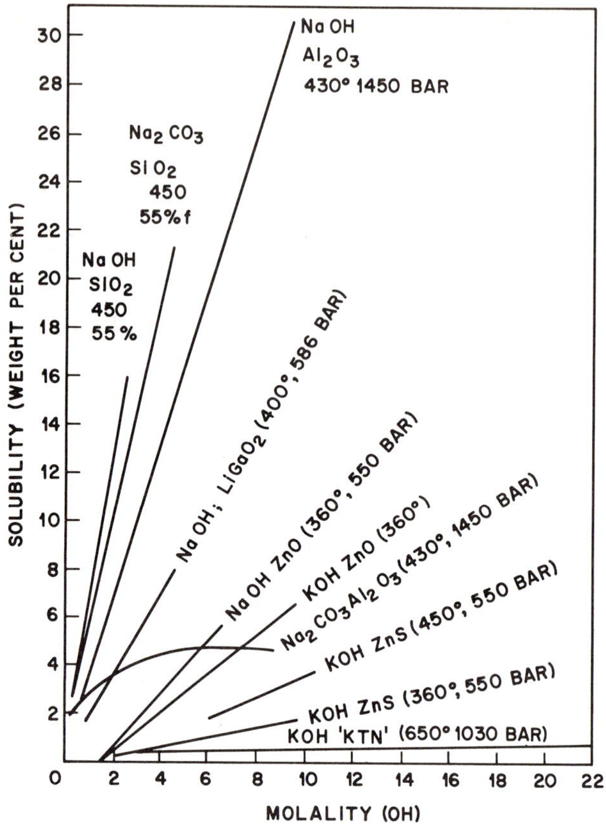

Fig. 6.8. Solubility data for various hydrothermal systems (from [18]).

(3) only one fluid phase is present in the system. We have applied this technique to many materials of hydrothermal interest. Fig. 6.8 [18] shows the solubility of α-quartz [19], sapphire [20], $LiGaO_2$ [21], zincite (ZnO) [22], sphalerite (ZnS) [23] and KTN $\{K(Ta_{0.65}Nb_{0.35})O_3\}$ [24] in $(OH)^-$ and $(CO_3)^{2-}$ solutions under conditions in which these phases are the stable solids. The noteworthy feature of fig. 6.8 is that in all cases (with the exception of Al_2O_3 in Na_2CO_3 where the strong concentration dependence of the ionization of $(CO_3)^{2-}$ is a complication) * solubility is a linear function of

* The linear dependence of SiO_2 solubility in $[CO_3]^{2-}$ is probably associated with the fact that these measurements were made at constant density while the Al_2O_3 and all other solubilities were made at constant pressure.

$(OH)^-$. The linear dependence of solubility on the value of $(OH)^-$ would not at first sight be expected, and its understanding in the wide variety of systems presented in fig. 6.8 could provide us with the beginnings of a general picture of hydrothermal dissolving reactions.

A generalized reaction for the dissolving of an oxide in $(OH)^-$ is

$$MO + nOH^- \leftrightarrows MO(OH)_n^{n-} , \tag{6.1}$$

where K, the equilibrium constant, is given by

$$K = s[OH]^{-n} , \tag{6.2}$$

with [OH] the hydroxide ion concentration and s the solubility of the material dissolving. * We assume here that the solubility in pure water is negligible. If $MO(OH)_n^{n-}$ gives up water by reactions such as

$$MO(OH)_2^{2-} \rightarrow MO_2^{2-} + H_2O \tag{6.3}$$

our formalism will not be affected.

Now,

$$K = s([OH]_i - sn)^{-n} \tag{6.4}$$

where $[OH]_i$ is the initial $(OH)^-$ concentration, and when eq. (6.1) proceeds essentially to completion,

$$s = n^{-1} [OH]_i . \tag{6.5}$$

According to eq. (6.5), s would be temperature-independent. If however eq. (6.1) proceeds hardly at all to the right

$$s = K [OH]_1^n \tag{6.6}$$

while for intermediate cases eq. (6.4) must be used.

* Owing to a lack of data on activity constant, we are forced to use concentrations. Some data on the ionization of hydroxide solutions under hydrothermal solutions are available [24] and have been used [24,25] in calculations of the sort described here, but its use has been neglected in the present discussion, since our conclusions are not altered if the partial ionization of bases is ignored.

Table 6.3
Summary of solubility dependence on mineralizer concentration.

Compound	Complexer	$A = \dfrac{\text{moles solute}}{\text{moles [OH]}_\text{i}}$	$\mathrm{d}s/\mathrm{d}T$	n
SiO_2	NaOH	0.67	1 wt %/100°	3/2
SiO_2	Na_2CO_3	2	1 wt %/100°	1/2
Al_2O_3	NaOH	0.38 – 0.42	1 wt %/100°	2 – 3
Al_2O_3	Na_2CO_3	not constant	–	–
$LiGaO_2$	NaOH	0.17	–	*
ZnO	NaOH	0.5	–	1
ZnO	KOH	0.03	–	1
ZnS	KOH	0.003	–	*
KTN	NaOH	0.017	–	*

* $n = 2$ fits the requirement of s linear with $[\mathrm{OH}]_\mathrm{i}$; in the case of ZnS, $n = 1$ also fits.

For s to be linear with $(OH)^-$ either eq. (6.5) or eq. (6.6) with $n = 1$ must be valid. An additional requirement is that n be constant over the $(OH)^-$ range studied. Table 6.3 gives data enabling one to judge whether eq. (6.1) goes far enough to the right to make eq. (6.5) valid. As can be seen, the quantity A of table 6.3 suggests that reaction (6.1) goes to completion for quartz (SiO_2) and essentially to completion for sapphire (Al_2O_3) in NaOH. For the remaining materials of table 6.3 the value of A suggests that reaction (6.1) does not go to completion. Thus, for the observed linearity of s with (OH) for these materials, eq. (6.6) must be valid either because $n = 1$ or because $n > 1$ with reaction products which raise the power dependence of s. When eq. (6.5) is valid in table 6.3 $n = 1/A$, otherwise, $n = 1$ or $n > 1$, with appropriate adjustment of the power dependence of s. The dissolving reactions suggested by the data of table 6.3 are:

$$3SiO_2 + 2(OH)^- \rightarrow [3SiO_2 \cdot 2(OH)]^{2-} \rightarrow Si_3O_7^{2-} + H_2O \text{ (for NaOH)}\,, \qquad (6.7)$$

$$SiO_2 + 2(OH)^- \rightarrow [SiO_2 \cdot 2(OH)]^{2-} \rightarrow SiO_3^{2-} + H_2O \text{ (for } Na_2CO_3)\,, \qquad (6.8)$$

$$\tfrac{1}{2}Al_2O_3 + (OH)^- \rightarrow [\tfrac{1}{2}Al_2O_3 \cdot (OH)]^- \rightarrow AlO_2^- + \tfrac{1}{2}H_2O \qquad (6.9)$$

(for NaOH; not calculable for Na_2CO_3) ,

$$ZnO + (OH)^- \rightarrow [ZnO(OH)]^- \rightarrow \tfrac{1}{2}Zn_2O_3^{2-} + \tfrac{1}{2}H_2O\ , \tag{6.10}$$

$$ZnS + (OH)^- \rightarrow [ZnS(OH)]^-, \tag{6.11}$$

or

$$ZnS + 2(OH)^- \rightarrow Zn_2O_3^{2-} + H_2O + HS^-\ , \tag{6.12}$$

$$KTa_{0.65}Nb_{0.35}O_3 + 2(OH)^- \rightarrow$$

$$Ta_{0.65}Nb_{0.35}O_4^{3-} + K^+ + H_2O, \tag{6.13}$$

$$LiGaO_2 + 2(OH)^- \rightarrow Li^+ + GaO_3^{3-} + H_2O\ . \tag{6.14}$$

Equations (6.12), (6.13), and (6.14) still lead to a linear dependence of s on (OH) since the general form of their equilibrium constants is

$$K = s^2\,[OH]_i^{-2} \tag{6.15}$$

because

$$[Zn_2O_3]^- = [HS]^- = s\ , \tag{6.16}$$

$$[Ta_{0.65}Nb_{0.35}O_4]^- = [K]^+ = s\ , \tag{6.17}$$

and

$$[Li]^+ = [GaO_3]^{3-} = s\ . \tag{6.18}$$

Thus, in all the cases so far studied, solubility increases linearly with (OH) either because the dissolving reaction does not proceed very far toward completion and reaction is with one hydroxide ion or the reaction goes entirely to completion. In the case of chalcogenides like ZnS, the possibility of $(HS)^-$ formation would allow complexing with two $(OH)^-$ ions. For salts like $KTa_{0.65}Nb_{0.35}O_3$ and $LiGaO_3$ complexing is with two (OH) ions with the dependence still linear on $[OH]_i$ because the alkali ion is a product of the dissolving reaction.

It has been shown elsewhere that the solubility depends approximately linearly on solution density and that the temperature dependence is often van't Hoff [9]. Table 6.4 tabulates some heats of solution. Recent studies of

Table 6.4
Heats of solution from Van't Hoff plots.

Substance	Solution	Percent fill	Pressure	Heat of solution
Quartz [19]	Water	71.5%	–	ΔE = 8.8 kcal/mol
Quartz	0.51 m NaOH	80%	–	ΔE = 1.03
Quartz	0.51 m NaOH	85%	–	ΔE = 1.82
Quartz	0.51 m NaOH	87%	–	ΔE = 2.68
Quartz	5 wgt % Na_2CO_3	80%	–	ΔE = 3.42
ZnO [21]	6.47 m KOH	–	550 bar	ΔH = 1.1
ZnO	9.07 m KOH	–	550 bar	0.8
ZnO	6.24 m NaOH	–	550 bar	0.4
Al_2O_3 [20]	10 m NaOH	–	430–2800 bar	ΔH = 0.2
Al_2O_3	3.4 m Na_2CO_3	–	430–2800 bar	ΔH = 1–5*
ZnS (cubic) [22]				
	6.47–11.9 m KOH	–	550 bar	ΔH = 7.2
ZnS	6.24–38.4 m NaOH	–	550 bar	ΔH = 7.2
ZnS	8 m CsOH	–	550 bar	ΔH = 7.2
$KTa_{.65}Nb_{.35}O_3$ [24]				
	14.7–15 m KOH	–	620–1720 bar	ΔH = 18
$LiGaO_2$ [21]	0.5 m NaOH	50–60–70%	–	ΔE = 4.5
Se [25]	1.25 m Na_2S	(20–80°C)	1 bar	ΔH = 3.88

* Van't Hoff Eq. obeyed poorly.

trigonal Se which is soluble and is the stable solid phase in $S^{=}$ solutions from ambient to hydrothermal conditions (200°C) show linear dependence on $S^{=}$ and van 't Hoff dependence on temperature. In addition, the ratio Se/S suggests that at temperatures near 50° the dissolving reaction [25] is

$$S^{=} + 2Se \rightleftarrows Se_2S^{=} . \tag{6.19}$$

6.6. Hydrothermal equations of state

In the simplest terms the role of water in hydrothermal crystal growth is to depress the melting point of the material being crystallized. To keep the water at a high enough density that its solvent power will be appreciable, a high pressure system is required. In most hydrothermal systems the pressure arises because a fixed volume of water is heated in a confined volume. Sometimes the pressure is generated by an external pump. In a typical hydrothermal system the density (and pressure at any temperature) is determined by the fraction of the free volume (percent of fill, filling factor) of the autoclave which is initially filled with water at room temperature. In an autoclave filled to greater than ≈ 32% fill, (≈ 0.32 g/cm^3 is the critical density) the liquid-gas meniscus moves up as the autoclave is heated and autoclaves fill with liquid at temperatures below 373°C (the critical temperature). The greater the initial fill, the lower the temperature at which the autoclave fills with the liquid phase. For fills less than 32% (seldom used in crystal growth because the solvent power is too low), the meniscus moves down and autoclaves "boil dry" containing only the gas phase at temperatures below 373°. The lower the fill, the lower the temperature at which the autoclave boils dry. In an autoclave filled to 32%*f*, the meniscus doesn't move during heating and disappears at 373°, above which temperature the fluid is by definition a gas regardless of its density. The pressure of solutions generally is lower than that of pure water and their critical point (actually critical end point for polycomponent systems) is elevated. Very little $p–v–T$ data on hydrothermal solutes and mineralizers has been published, so we are generally forced to make estimates based on the $p–v–T$ data for water [26] of fig. 6.9.

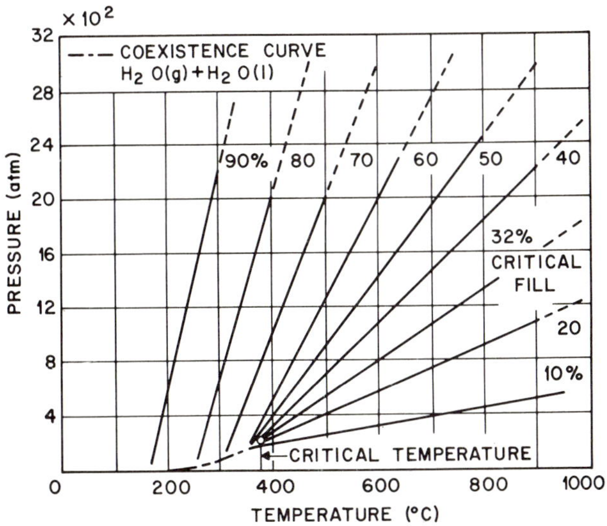

Fig. 6.9. $P–v–T$ relations for H_2O (from [27]).

6.7. Kinetics of crystallization

Quartz was the earliest [9, 10, 12, 27] and has been the most extensively studied hydrothermal material, mainly because of its commercial use as a piezoelectric. In order to interpret the kinetic studies, one must have in mind the geometry of the system ordinarily used for the growth of quartz (and most other materials). Relatively finely divided particles of α-quartz *nutrient* are placed in the bottom of the vessel and suitably oriented single crystal α-quartz seeds are suspended in the upper growth region (see fig. 6.10). Quartz growth usually takes place directly in steel vessels. For many other materials a liner or a can would be required. The vessel is filled to some fraction of its free volume, typically 0.80 (80% fill) with a basic solution of 0.5 M NaOH. With the long axis vertical, the vessel is placed in a furnace capable of heating the lower *dissolving* section isothermally hotter than the upper growth region which is also maintained isothermal. The furnaces incorporate appropriate temperature controllers, temperature sensors and suitable

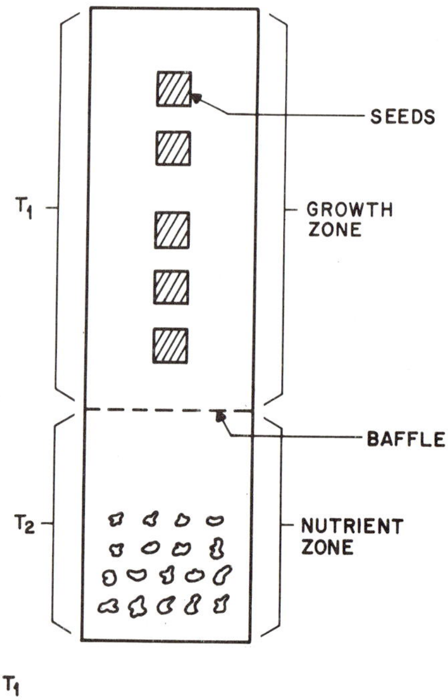

Fig. 6.10. Schematic of autoclave for quartz growth.

shielding against the not inconsiderable explosion hazard. A perforated metal disc called a "*baffle*" is often placed within the vessel separating the dissolving and growth zones to help in localizing the temperature differential, ΔT. If a can is used, an additional baffle is placed at the same level between the outer wall of the can and the autoclave. For a growth zone temperature of 400° and a dissolving temperature of 350°, the pressure is about 2 kbar, the solution saturates at 400° and moves by convection to the growth zone where rates on (0001) seeds of 2.0 mm/day (growth rate = seed thickness increase/time of experiment) are obtained.

For quartz growth the important variables which effect rate have been found to be:

1. Identity and concentration of mineralizer
2. Degree of fill of the vessel
3. Crystallization temperature
4. Temperature difference (ΔT) between dissolving and growth zones
5. Percent open area of the baffle
6. Seed orientation
7. Rate (dissolving surface area)/(growing surface area)

Quartz kinetic studies have shown that the variables (1)-(7) are interrelated so that

$$R_{hkl} = \alpha k_{hkl} \Delta S \,, \tag{6.20}$$

where R_{hkl} is the rate in a particular crystallographic direction, k_{hkl} is a velocity constant in that direction, ΔS is supersaturation and α is a dimensional conversion constant. It should be remembered that the fact that the rate can be fitted to a simple equation *does not* establish that the growth mechanism is simple. However, the dependence on variables (1)-(7) is easily understood in terms of the rate equation.

(1) The nature and concentration of the mineralizer establishes ΔS. In NaOH and probably KOH between about 0.5 and 5M the mineralizer does not markedly affect k_{hkl}. For chemically different mineralizers, k_{hkl} would probably be changed.

(2) Percent fill affects s and $\mathrm{d}S/\mathrm{d}T$ and thus its effect is mainly on ΔS. In NaOH between about 65% and 87% fill the effect is only on ΔS.

(3) The effect of crystallization temperature is to alter k_{hkl}. In $(OH)^-$ and $(CO)_3^=$ k_{hkl} obeys the Arrhenius equation as shown in fig. 6.11. Typical energies of activation, ΔE^*'s, are 20 kcal/mol (some dependence on direction observed) in NaOH and Na_2CO_3. Such values are too high for a pure diffusion process.

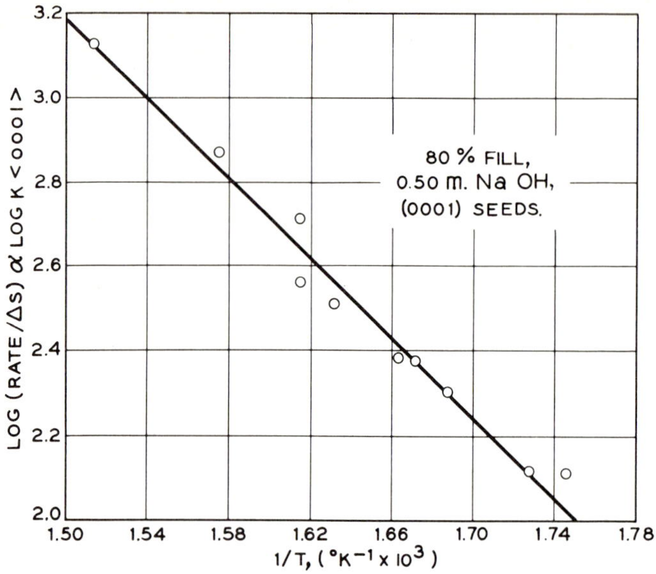

Fig. 6.11. Arrhenius plot – quartz (from [27]).

(4) ΔT affects only ΔS, thus as fig. 6.12 shows, R_{hkl} is linear with ΔT over regions where ΔS is linear with ΔT.

(5) The percent open area of the baffle between 2 and 50% changes the rate as shown in fig. 6.13. However, when internal temperatures are measured carefully, it can be shown that the observed R_{hkl} changes of fig. 6.13 can be accounted for by changes in T and ΔT. This indicates that convective circulation is rapid enough that it is not rate limiting. Typical baffle openings are 5–10%. An isothermal growth region aids in producing uniform crystals while an isothermal dissolving region prevents growth on the nutrient.

(6) Seed orientation in a noncentrosymmetric crystal like quartz has a great effect on k_{hkl} and hence R_{hkl}. See fig. 6.14.

(7) Provided the rate (area dissolving)/(area growing) > 5, the rate doesn't depend on the ratio. Thus, when the dissolving area is large enough, dissolution is not rate limiting.

Chernov and Khadzhi [28] have shown that the log of the sodium concentration is a linear function of $1/R$, and that colloidal inclusions occur as a result of the burial of fragments of the adsorbed film during growth. If the growth rate is greater than a critical value, burial occurs before desorption and colloidal inclusions result. The number of trapped inclusions observed as a function of temperature and growth rate agrees with this model.

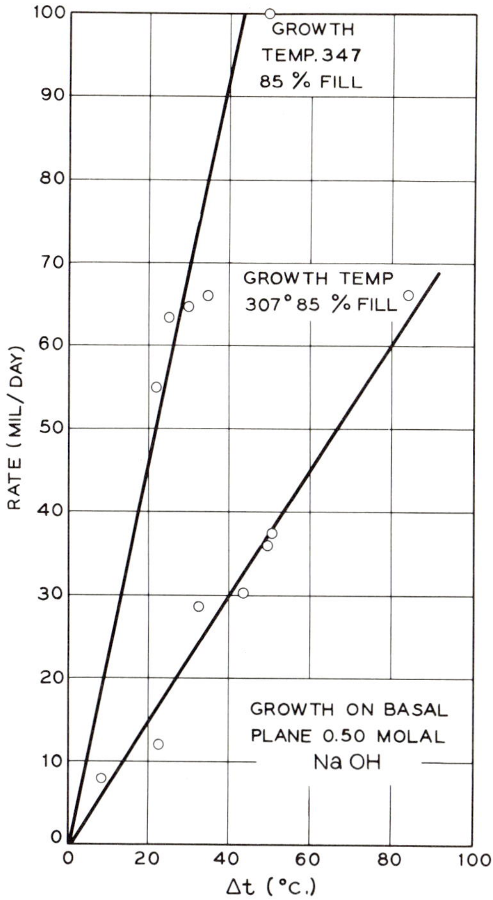

Fig. 6.12. Rate vs. ΔT – quartz (from [27]).

Where similar (less detailed studies) have been made on other materials (ZnO is probably the second most studied from the kinetics viewpoint), the results are similar. In summary, in quartz (and probably most other hydrothermal growth) the rate limiting process involves processes at the interface and the reaction is consecutive, so that more than one step may be involved in establishing the overall rate.

The very same effects are observed in large scale equipment of the sort shown in figs. 6.15–6.17. Indeed pilot plant studies [11] showed that a scaleup in process by a factor of over 300 did not significantly alter the kinetics.

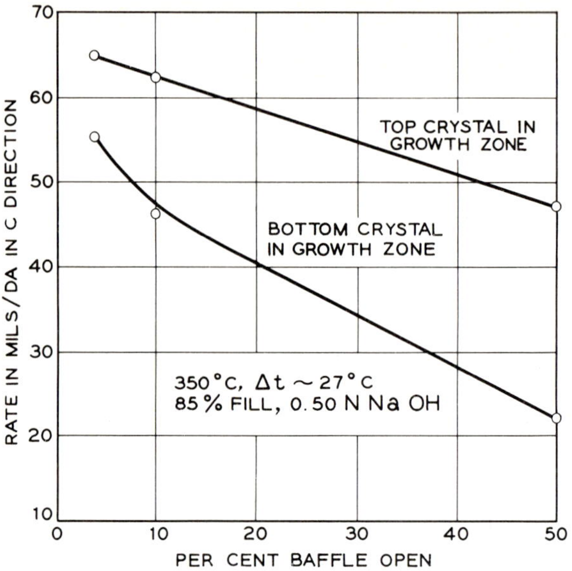

Fig. 6.13. Rate vs % open area of baffle – quartz (from [27]).

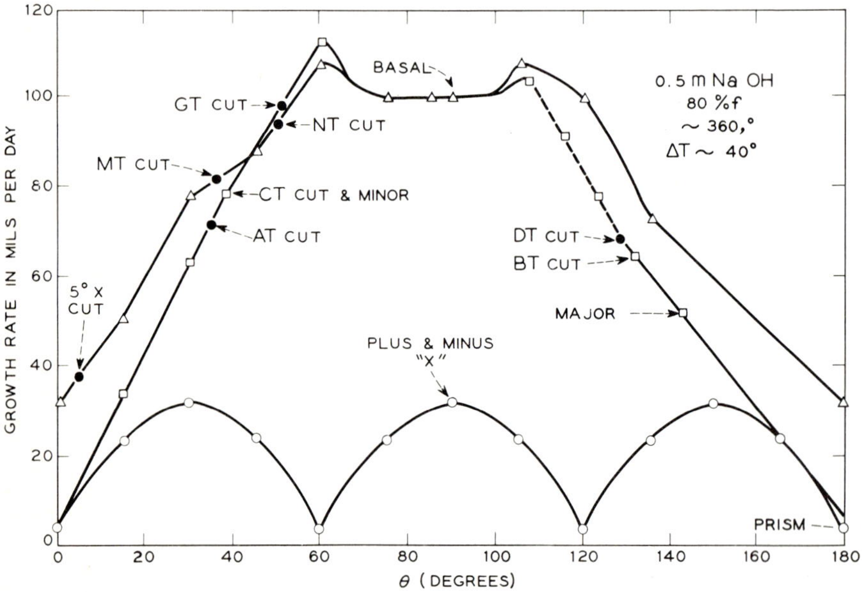

Fig. 6.14. Rate as a function of seed orientation (from [10]).

Fig. 6.15. Large scale quartz growth.

Fig. 6.16. Large scale quartz growth.

Fig. 6.17. Large scale quartz growth (courtesy of D.L. Rudd, Western Electric Co., Merrimack Valley Works, North Andover, Mass., U.S.A.).

6.8. Quartz perfection

The most noticeable imperfection in hydrothermal quartz has been variously called "crevice flawing", "fissure flawing", cellular growth and dendritic growth. It occurs most often on nonequilibrium faces and was the principal problem to be overcome in obtaining good growth on (0001) faces. (Cf. ch. 16 §4.4.4).

The cause of such growth is perfectly analogous to cellular and dendritic growth in metallic systems and probably the most accurate term for the imperfections is cellular growth.

When growth is rapid on a nonequilibrium face of high specific surface energy (or better yet high Jackson α) [29] so that nearby orientations can be stable (the facet effect) and when conditions are such that diffusion begins to play a dominant role, constitutional supercooling will occur, the peaks of the facets which protrude into regions of higher ΔS will begin to grow more rapidly and cellular-dendritic-crevice flawed growth will result. In the practical case p-T-ΔT-mineralizer space was mapped to find regions where $R_{(0001)}$ could be high but not severely diffusion controlled so as to eliminate the flaws under conditions of economically useful rates.

Acoustic Q has been shown to be associated with H^+ [30] (by noting a correlation with an IR absorption at 3μm corresponding to the OH stretching frequency). Lowering the growth rate, and growing in the presence of Li^+, decreases H^+ in the lattice and improves the Q [31, 32]. It would appear that interstitial H^+ charge compensates substitutional Al^{+3}, Fe^{+3} and Fe^{+2} so

Fig. 6.18. X-irradiated quartz producing color center showing inhomogeneities in growth.

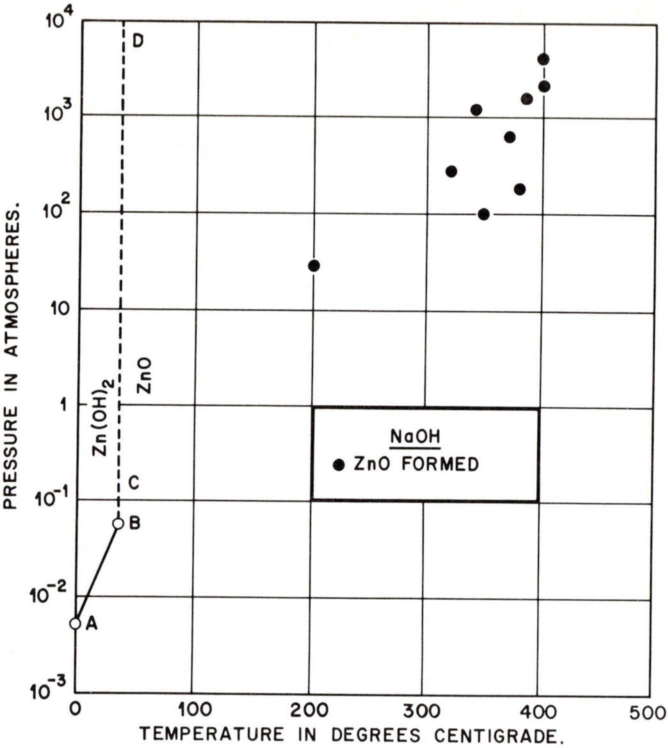

Fig. 6.19. Part of the ZnO–H_2O diagram (from [19]).

that measures which reduce their concentration greatly improve the Q of quartz. Li^+'s effect may be because it is more effective at charge compensating than H^+ or it may physically block the entry of H^+. Measures which improve Q also improve the optical quality of quartz [32]. Optical inhomogeneities in quartz are common as illustrated by fig. 6.18. Striagraphs also reveal index of refraction variations. Etching studies have also shown unusual etch pits [33], and recent reports of the growth of dislocation free quartz have been made.

6.9. Hydrothermal growth of other materials

6.9.1. ZnO

The common method of zincite (ZnO) growth is from the vapor phase, but this technique usually produces small c-axis needles or a-axis plates and controlled growth of large crystals is difficult to achieve reproducibly. Flux growth usually produces thin (0001)-$(000\bar{1})$ plates and is also difficult to control. Large ZnO crystals are quite easy to produce hydrothermally. The ZnO–H_2O phase diagram was redetermined and shown to be unaltered in the presence of (OH) [34]. See fig. 6.19. The solubility of ZnO in a variety of mineralizers has been studied [22] and the best growth seems to be achieved at the following conditions:

Solvent – 5 M KOH
Cryst. Temp. – 350°
ΔT - 10°
% Fill – 85%
Rate – ≈ 0.25 mm/day on (0001)-$(000\bar{1})$ seeds
[rate $(0001) \neq$ rate $(000\bar{1})$] .

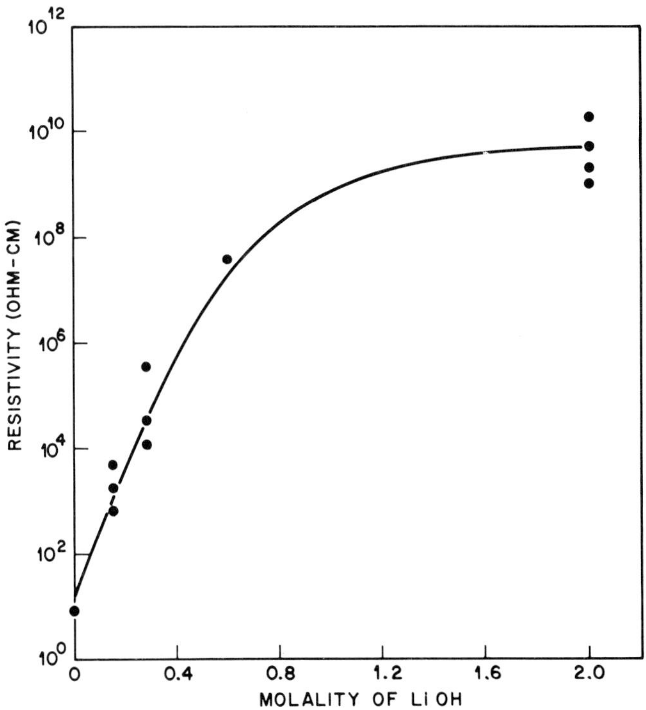

Fig. 6.20. Resistivity of ZnO as a function of Li^+ concentration in growth solution (from [35]).

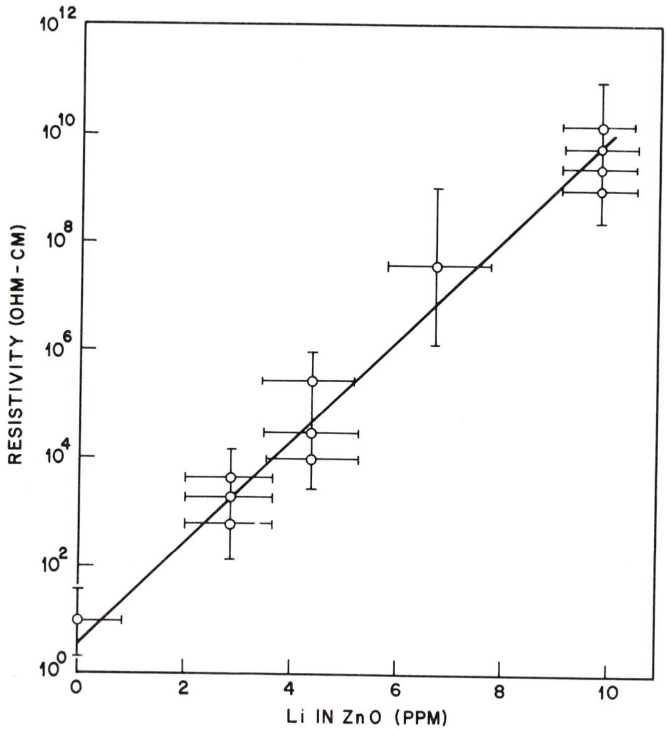

Fig. 6.21. Resistivity of ZnO as a function of lithium concentration in grown crystal (from [35]).

Growth on (0001)–(000$\bar{1}$) plates tends to be flawed. The addition of Li^+ ($\approx$ 0.1 M LiOH) salts seems to repress flawing. This can be interpreted as being due to a rate reduction or because adsorption on (0001) and (000$\bar{1}$) reduces the surface free energy so that cusps rather than lobes occur in the Wulff plot at ⟨0001⟩ and ⟨000$\bar{1}$⟩. Li^+ acts as an acceptor so that by adjusting Li^+ in the solution, the resistivity can be controlled [35] as shown in figs. 6.20 and 6.21. The data of figs. 6.20 and 6.21 provide us with the only quantitative data for the calculation of hydrothermal distribution coefficients. Inert conditions require the use of a lined Morey autoclave.

6.9.2. ZnS [33]

Generally similar conditions have been used for the growth of ZnS, where it is surprising that even in 5M KOH reactions of the sort

$$ZnS + OH^- \rightarrow ZnO + HS^- \tag{6.21}$$

do not proceed far enough to the right to be troublesome. The cubic polymorph of ZnS is produced in quite high quality hydrothermally. Other techniques of growth where the preparation temperature is closer to the cubic-hexagonal transition temperature usually produce ZnS with a high concentration of stacking faults. Preliminary investigations of hydrothermal ZnS revealed a much reduced stacking fault density. Further investigation of the perfection of hydrothermal ZnS would be worthwhile.

6.9.3. Sapphire–Al_2O_3 [36]

The second material to be grown following quartz was sapphire (α-corundum or Al_2O_3) (Laudise and Ballman, 1958). This work showed that a process above 400° would be required because of the reaction:

$$Al_2O_{3\,(\text{corundum})} + H_2O \xrightleftharpoons{\approx 400^\circ} 2AlOOH_{(\text{diaspore})} \tag{6.22}$$

at the growth pressures. The p–T boundary for corundum-diaspore was shown to be reasonably insensitive to $(OH)^-$. Later solubility studies (Barns et al., 1963) showed that the best supersaturation would occur in $CO_3^=$ mineralizers (fig. 6.22). Inert conditions at high pressure (to achieve reasonable rates) require the use of welded cans. Typical conditions are:

Mineralizer – 1 M K_2CO_3
Cryst. Temp. – 490°
ΔT – 50°
Fill – 80%
Rate – ≈0.25 mm/day [on (0001) seeds].

Chromium doping is possible. To a certain extent the ready availability of Czochralski grown sapphire and ruby has decreased interest in hydrothermal growth recently; but if very high perfection is desired, hydrothermal growth may have considerable advantage.

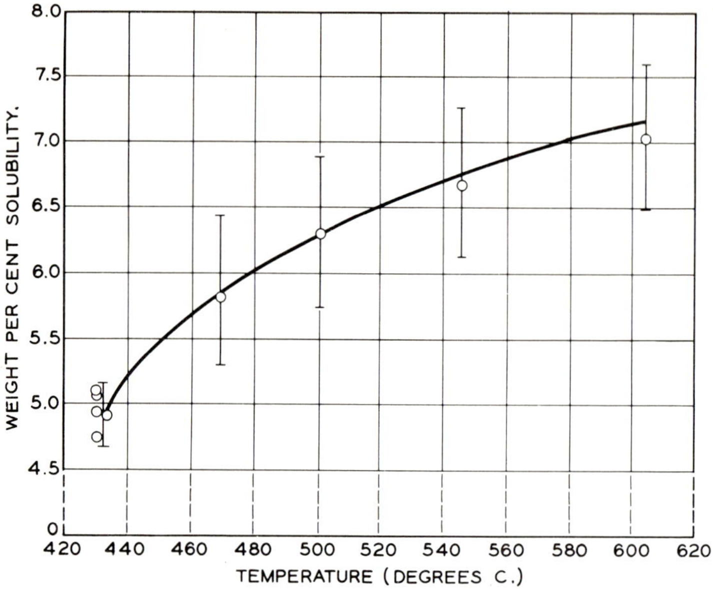

Fig. 6.22. Solubility of Al_2O_3 (from [20]).

6.9.4. Garnet [37, 38]

The phase diagram Fe_2O_3–Y_2O_3–H_2O was investigated with various mineralizers and some conditions where $Y_3Fe_5O_{12}$ (YIG) was congruently saturating were reported. Large high quality garnets whose magnetic and IR transmission properties are identical to the best flux grown garnet, can be grown at the following conditions:

Mineralizer – 20 M KOH
Cryst. Temp. – 350°
ΔT – 10°
Fill – 88%
Rate – ≈ 0.1–0.2 mm/day.

Inert low pressure lined Morey autoclaves are required. Recent work has shown that with some changes the substituted uniaxial garnets and heteroepitaxial garnet may also be grown. Similar conditions have been used for the growth of rare earth orthoferrites.

6.9.5. $LiGaO_2$ [21]

Difficulties in $LiGaO_2$ growth under hydrothermal conditions have come about because of attempts to grow in retrograde solubility regions. Solubility studies are required to deduce appropriate conditions for growth. By the use of such studies non-retrograde regions can be found and growth rates as high as 2.25 mm/day have been obtained. The best conditions of growth have been found to be:

Mineralizer – 3.5 M NaOH
Cryst. Temp. – 385°
ΔT – 35°
Fill – 70%

Hydrothermal growth is an attractive method for the crystallization of $LiGaO_2$ and good quality material grows faster than any other hydrothermal material at moderate conditions.

6.9.6. Selenium [39]

Previous mention was made of the solubility studies of Se in $S^=$ solutions. The best conditions for hydrothermal growth were found to be:

Mineralizer – 0.375 M Na_2S
Cryst. Temp. – 175°
ΔT – 25°
Fill – 95%

Growth tends to be dendritic, but the crystal quality and size are quite respectable, if the difficulties of obtaining high quality Se crystals by other procedures are borne in mind.

The solubility studies and growth of Se at ambient conditions from $S^=$ solutions, together with reports of CuCl–Cl^- (gel growth) and HgS–$S^=$ growth at ambient conditions suggest that the mineralizer concept might be more generally extended to ambient growth (see reference [39]).

6.10. General remarks

Hydrothermal growth has been used seriously for the preparation of no more than a few dozen materials. It has unique advantages especially those associated with low temperature growth, high solvent power, high product perfection and ease of scaleup. In addition, it has an intriguing physical chemistry. It is the author's hope that this brief review will result in the more wide spread use of this powerful and neglected method of crystal growth.

Acknowledgment

We would like to especially acknowledge many discussions with E.D. Kolb concerning hydrothermal experiments. The experimental work of A.J. Caporaso which has been essential in obtaining many of the results discussed here is most gratefully acknowledged.

References

[1] R.A. Laudise, The Growth of Single Crystals (Prentice-Hall, Englewood Cliffs, New Jersey, 1970).
[2] R.A. Laudise, J.R. Carruthers and K.A. Jackson, in: Annual Reviews of Materials Science, Vol. 1, 1971.
[3] G.W. Morey, J. Am. Ceram. Soc. 36 (1953) 279.
[4] G.W. Morey and P. Niggli, J. Am. Chem. Soc. 35 (1913) 1086.
[5] R. Roy and O.F. Tuttle, Physics and Chemistry of the Earth 1 (1956) 138.
[6] See 'Carnegie Institute of Washington – Year Book(s)' for the years 1959–1970, published by Carnegie Institution, Washington.
[7] R. Roy and W.B. White, J. Crystal Growth 3, 4 (1968) 33.
[8] E. Roedder, Scientific American 207, (4) Oct. (1962) 38.
[8a] N.P. Yermakov, Research on the Nature of Mineral Forming Solutions (Pergamon Press, New York, 1965)..
[9] R.A. Laudise in: Progress in Inorganic Chemistry, ed. by F.A. Cotton (Interscience, 1962) p. 1.
[10] See A.A. Ballman and R.A. Laudise in: The Art and Science of Crystal Growth, ed. by J.J. Gilman (Wiley, New York, 1963) p. 231, for a review of quartz kinetics.
[11] R.A. Laudise and R.A. Sullivan, Chem. Eng. Prog. 55 (1959) 55.
[12] R.A. Laudise and J.W. Nielsen, Solid State Phys. 12, ed. by F. Seitz and D. Turnbull (Academic, New York, 1961).
[13] O.F. Tuttle, Geol. Soc. Am. Bull. 60 (1949) 1727.
[14] G. Ervin and E.F. Osborn, J. Geol. 59 (1951) 385.
[15] R.A. Laudise and A.A. Ballman, J. Am. Chem. Soc. 80 (1958) 2655.
[16] R.A. Laudise, J.H. Crocket and A.A. Ballman, J. Phys. Chem. 65 (1961) 359.
[17] Phase Diagrams for Ceramists, ed. by E.M. Levin, C.R. Robbins and H.F. McMurdie (Am. Ceram. Soc., Columbus, Ohio, 1964).
[18] R.A. Laudise and E.D. Kolb, Endeavour 28 (1969) 114.
[19] R.A. Laudise and A.A. Ballman, J. Phys. Chem. Ithaca 65 (1961) 1396.
[20] R.L. Barns, R.A. Laudise and R.M. Shields, J. Phys. Chem. 67 (1963) 835.
[21] D.J. Marshall and R.A. Laudise, J. Crystal Growth 1 (1967) 88.

[22] R.A. Laudise and E.D. Kolb, Am. Miner. 48 (1963) 642.

[23] R.A. Laudise, E.D. Kolb and J.F. DeNeufville, Am. Miner. 50 (1965) 382.

[24] D.J. Marshall and R.A. Laudise, Crystal Growth, p. 557, supplement to J. Phys. Chem. Solids (Pergamon Press, New York, 1967). See also, S.P. Clark (Editor), Handbook of Physical Constants, (Geol. Soc. America, New York, 1966) p. 401, 418.

[25] E.D. Kolb and R.A. Laudise, J. Crystal Growth, in press.

[26] G.C. Kennedy, Am. J. Sci. 248 (1950) 540.

[27] R.A. Laudise, J. Am. Chem. Soc. 81 (1959) 562.

[28] A.A. Chernov and V.E. Khadzhi, J. Crystal Growth 3, 4 (1968) 641.

[29] K.A. Jackson, in: Crystal Growth, ed. by H.S. Peiser (Pergamon, New York, 1967) p. 17.

[30] D.M. Dodd and D.B. Fraser, J. Phys. Chem. Solids 26 (1965) 673.

[31] A.A. Ballman, R.A. Laudise and D.W. Rudd, Appl. Phys. Letters 8 (1966) 53.

[32] A.A. Ballman, D.M. Dodd, N.A. Kuebler, R.A. Laudise, D.L. Wood and D.W. Rudd, Appl. Opt. 7 (1968) 1387.

[33] J.W. Nielsen and F.G. Foster, Am. Miner. 45 (1960) 299.

[34] R.A. Laudise and A.A. Ballman, J. Phys. Chem. 64 (1960) 688.

[35] E.D. Kolb and R.A. Laudise, J. Am. Ceram. Soc. 48 (1965) 342, and 49 (1966) 302.

[36] R.A. Laudise and A.A. Ballman, J. Am. Chem. Soc. 80 (1960) 2655.

[37] E.D. Kolb, D.W. Wood, E.G. Spencer and R.A. Laudise, J. Appl. Phys. 38 (1967) 1027.

[38] E.D. Kolb, D.L. Wood and R.A. Laudise, J. Appl. Phys. 39 (1968) 1362.

[39] E.D. Kolb and R.A. Laudise, J. Crystal Growth, in press.

P. Hartman, ed., Crystal growth: an introduction. © (1973) North-Holland Publ. Co.

HYDROTHERMAL SYNTHESIS IN ACID SOLUTIONS

A. RABENAU

7.1. **Introduction**

The basic principles of the hydrothermal method of crystal growth are given in Chapter 6. In its conventional form it proved to be very suitable for oxidic compounds, in particular quartz and silicates, but not for metallic compounds of sulphur, selenium and tellurium, which are of particular interest in modern solid state research. This is rather odd, since there are many natural examples of sulphide minerals formed under hydrothermal conditions. These include sulphides, which are among the less soluble compounds. The usual method for increasing the solubility by working in an alkaline medium does not – apart from a few chalcogenides of zinc and cadmium (ch. 6 §9) – lead to the desired result, since hydrolysis takes place. An attempt to use strong sulphidic solutions, adding alkaline polysulphides, thiosulphides etc. in the hope of forming sulphidic complexes, was equally unsuccessful. Quite apart from the practical difficulties in these experiments, due to the extremely corrosive action of such solutions, the results have been disappointing.

We therefore investigated whether concentrated hydrohalic acids could be used as solvents [1]. There were two reasons for this: the tendency of sulphides to dissolve in acids with the formation of hydrogen sulphide and secondly, the tendency of heavy metals to form complexes with halogen ions. The first tendency can be understood from the fact that an increase in the hydrogen-ion concentration brings a decrease in the activity of the sulphide ions. (With reasonably soluble sulphides gaseous hydrogen sulphide may in fact be liberated, but this does not matter since the materials are part of a closed system and continue to take part in the equilibrium.) Moreover, it can be assumed that hydrogen sulphide will be miscible with super-critical water, just as had been observed with other systems which water forms with volatile compounds. Moreover, the existence of complexes of heavy metals in the presence of halogen ions under hydrothermal conditions has been proved, as shown by the work of Franck et al. [2].

7.2. The acid method

As hydrohalic acids solvents are too corrosive to be handled in the conventional way, a new technique had to be developed. It is based on the observation that within the range of these investigations, that is to say up to 500°C and 3000 bar, quartz glass remains almost unattacked by aqueous solutions of hydrogen halides even in the presence of free halogen or elements of the 5th or 6th group of the periodic system. All experiments can therefore be done in closed quartz glass ampoules. Such ampoules can withstand a high external pressure if diameters are not too large. Consequently, the danger of explosion due to the generally unknown internal pressure developed during the heating can be eliminated by simply ensuring that the ampoule is subjected to a sufficiently high external pressure; say 2500 bars at the working temperature with an ampoule of 15 mm diameter. However great the internal pressure then becomes, provided it remains lower than the external pressure, the ampoule will not explode. The equipment used is shown in fig. 7.5, the general procedure being described in detail in Ch7 §5. An autoclave for the temperatures and pressures quoted above can be made of low-alloy steels of the type DIN 24CrMoV55 (0.45 Mn; 1.35 Cr; 0.55 Mo; 0.20 V), which can be machined by conventional methods. The technique can quite easily be employed in small laboratories.

7.3. Results and discussion

Table 1, summarizing some characteristic results, gives an impression of the range of applicability of the method. Besides sulphides we find examples of selenides, chalcogenohalides, halides and elements. Some general observations which can be made from this table may be described in brief as follows:

(i) various compounds of a system can be grown from the same solvent as, for example, covellite (1) and digenite (2) in HBr.

(ii) The type of solvent plays a role. While CdS (5) shows little transport in HCl, in HBr (6) a high transport rate is observed, yielding well developed crystals.

(iii) A chemical reaction may occur. In 2 molar hydrochloric acid, cinnabar (mercury sulphide) is transported (7), while in 12-molar hydrochloric

Table 1
Results of hydrothermal syntheses in acid solutions

Compound	Starting materials	Solvent	Degree of fill (%)	Temperature (°C)	Time (days)	Remarks
1	2	3	4	5	6	7
1. CuS	18 g CuS	9 molar HBr	65	450 : 420	10	plates, 10×6×1 mm
2. Cu_9S_5	20 g Cu_9S_5	9 molar HBr	65	450 : 430	8	octahedra, 5×5×5 mm
3. $CuISe_3$	10 g CuI 5 g Se	5 molar HI	60	390→200	10	dark violet crystals, ca. 2.5 mm; CuI
4. $AuITe_2$	6.2 g Au 4 g Te	10 molar HI	65	450→150	10	silvery white spears, Ø 0.5 mm, l = 50 mm, Au
5. CdS	10 g CdS	6 molar HCl	57	450 : 430	12	small crystals, little transport
6. CdS	10 g CdS	9 molar HBr	65	450 : 430	11	dark red crystals, 4 × 4 × 8 mm
7. HgS	17 g HgS	2 molar HCl	58	330 : 320	8	cinnabar, platelets, 4×4×0.4 mm
8. γ-$Hg_3S_2Cl_2$	27 g HgS	12 molar HCl	55	450 : 400	11	amber crystals, 4 × 4 × 4 mm
9. SnSe	5 g SnSe	9 molar HBr	65	500→300	10	silvery plates, 5 mm
10. PbS	10 g PbS	12 molar HCl	55	450 : 430	11	cubes, 5×5×5 mm
11. $Pb_5S_2I_6$	10 g PbS	10 molar HI	65	480 : 465 480→100	6 10	carmine needles Ø 1.5 mm, $l \leqslant$ 50 mm
12. Sb_2S_3	10 g Sb_2S_3	9 molar HBr	65	455 : 440	7	spear shaped crystals 15 mm; also orange needles of SbSBr
13. SbSI	10 g Sb_2S_3	10 molar HI	65	490→250	10	dark red needles, $l \leqslant$ 50 mm
14. BiSI	10 g Bi_2S_3	10 molar HI	65	500→200	10	black needles, Ø 0.7 mm, $l \leqslant$ 25 mm
15. γ-CuI	10 g CuI	2 molar HI	57	365→170	10	yellow plates, 6×6×3 mm, sphalerite type
16. Te	10 g Te	10 molar HI	65	440→420	10	deformed cubes, 2×2×2 mm
17. As	10 g As	10 molar HI	65	450 : 450	10	$\leqslant$ 8 mm, AsI_3

Use of table 7.1.. The data given are referred to a quartz glass ampoule of the size stated in fig. 7.4. The inner volume of the ampoules will be between 10 and 15 ml. Legends of columns:

1. Material to be prepared.
2. Starting materials are usually of analytic grade and are added as powders. The purity of the crystals depends on the starting materials, although a purification effect has been observed.
3. The densities of the concentrated acids used were: 1.19 (12 molar HCl); 1.50 (9 molar HBr); 1.95 (10 molar HI).
4. The term "degree of fill" is explained in the General Procedure section.
5. 350 → 150 means the autoclave is heated to 350°C and cooled down to 150°C uniformly within the time given in column 6. 450 : 430 means the autoclave is heated to 450°C at the lower end (charge zone) and 430°C at the upper end and kept under these conditions for the time given in column 6. After the treatment the autoclave is cooled down to room temperature. In some cases a cooling programme follows the heat treatment.
7. The ampoule normally does not contain the elements in stoichiometric amounts referred to the composition in 1. Moreover we are not dealing with an isothermal reaction. Crystals of various phases may therefore be obtained at the same time.

The conditions given have been found empirically to be satisfactory. No attempts have been made to optimize the data.

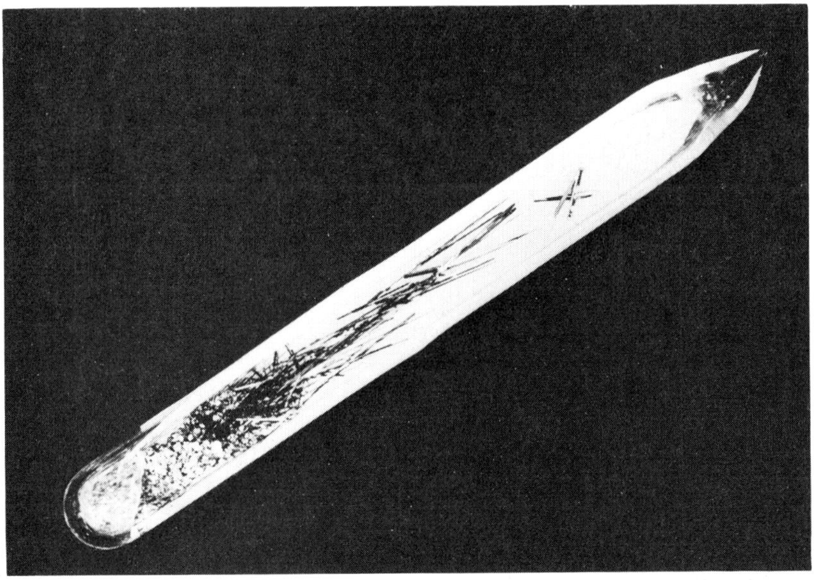

Fig. 7.1. Quartz-glass ampoule in which crystals of $AuITe_2$ together with gold have been grown. This result was obtained by bringing together gold and tellurium in equal numbers of gramme-atoms in 10-molar HI at a temperature of 450°C and then allowing it to cool for 10 days to 150°C. (2/3 actual scale.) (From Philips Technical Review.)

acid a mercury thiochloride is formed (8). Similar cases are the formation of SbSI, antimony thioiodide (13), and BiSI, bismuth thioiodide (14), starting from Sb_2S_3 and Bi_2S_3 respectively. In these cases liquid hydrogen sulphide as a separate phase can be observed at room temperature after the reaction.

(iv) Observation (iii) suggests that the method is especially suitable for the synthesis of chalcogenohalides. The special features of the method in this respect are that:

(1) subsolidus equilibria are easily attained, whereas by a solid state reaction, this is often hampered by low diffusion rates or the formation of meta-stable phases;

(2) the phases formed are usually obtained in the form of isolated small single crystals which can be subjected to chemical and X-ray analysis.

(Completely new series of compounds have in fact been synthesized with this method for the first time. Examples: 3, 4 (fig. 7.1) and 11).

7.4. The growth of gold crystals

We shall deal below in more detail with the growing of gold crystals. This brings us to a problem that has long preoccupied geologists. They have assumed that the greater part of the primary gold strata originated under hydrothermal conditions. In the laboratory, however, gold has previously been used only as a material for coating the autoclave to protect it against alkaline corrosion. This indicates that it is unlikely that the gold strata originated under such conditions.

At an earlier stage of our experiments we had already noticed that HI is an excellent solvent for gold. When 10 g of gold, in the form of wire or foil, are kept in 10-molar HI at a temperature of 500°C (with no temperature gradient), then in 10 days a group of gold crystals is formed. If a temperature gradient is applied, for example from 480 to 500°C, the gold is transported from cold to hot regions (fig. 7.3).

In our further investigations we tried to find an answer to the following two questions:

1) Can gold transport of this kind also take place under conditions actually found in nature?
2) Is transport from cold to hot regions a reasonable assumption for these conditions?

The second question is important because all previous discussions and experiments on this subject have started from the assumption that gold is a deposit

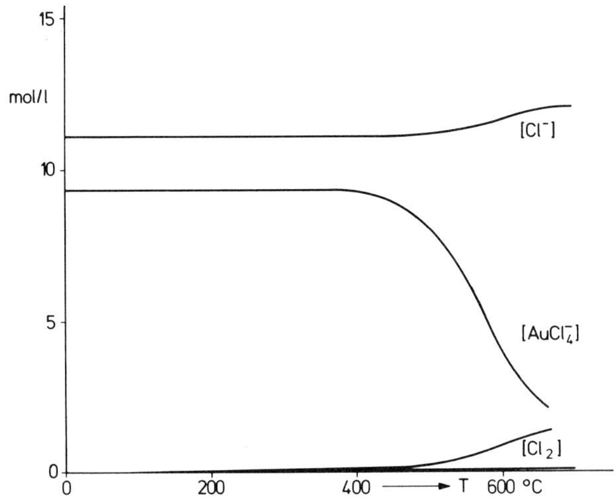

Fig. 7.2. Calculated concentration-temperature diagram, based on a solution equilibrium which we assume to occur in the hydrothermal growth of gold. The complex ion $AuCl_4^-$ is significant here (see text). Gold transport from cold to hot could take place in the temperature range in which the concentration of $AuCl_4^-$ shows marked variations. (For clarity the concentration of $AuCl_4^-$ is shown on a scale 10 times larger than the calculated value.) (From Philips Technical Review.)

formed in the course of time from a gradually cooling solution. It has been found that gold is also transported in solutions containing hydrogen chloride and bromide, provided an oxidizing agent is present, such as free halogen or oxygen. Where the solvent is HI, the free halogen is formed by dissociation from the solvent itself.

A current view of geologists is that natural gold deposits must have been deposited at temperatures between 50 and 550°C and at pressures up to 2000 bars. In natural conditions, weakly acid salt solutions are available as a solvent, with rock salt as the main constituent, in concentrations greater than 2-molar. It has been established by experiment that gold is in fact transported in weakly acid 10-molar NaCl solutions. The oxidizing agent in this case was oxygen, added in the form of H_2O_2.

We now assume that the transport of gold is based on the following equation:

$$Au(solid) + 1.5\ Cl_2(gas) + Cl^-(aq) \rightleftharpoons AuCl_4^-(aq).$$

$AuCl_4^-$ is a fairly stable complex and its thermodynamic data are known.

Fig. 7.3. "Gold diggers in the lab". Gold crystals, grown in 10-molar hydriodic acid at 500°C.

These data relate to normal conditions. The values applicable to hydrothermal conditions, obtained by extrapolation, must of course be regarded as a very rough approximation. Fig. 7.2 shows the result of a calculation based on extrapolated values. At low temperatures there is virtually no free halogen present. As the temperature rises, the complex dissociates. In the region in which the concentration of the complex varies strongly as a function of temperature we may expect a transport of gold from cold to hot. Considering the approximation used, the agreement between experiment and calculation may be regarded as satisfactory.

7.5. Hydrothermal synthesis in acid solutions: general procedure

Example: Growth of antimony thioiodide crystals (table 1 (13)): Sb_2S_3 + 2 HI = 2 SbSI + H_2S.

The volume of the quartz glass ampoule which serves as a reaction vessel (fig. 7.4) is measured up to the constriction, using a graduated burette. The

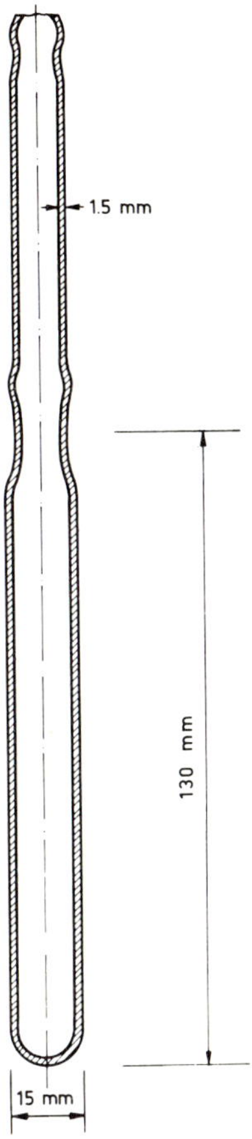

Fig. 7.4. Quartz glass ampoule.

degree of fill (see table 7.1) is referred to this volume after deducting the volume of the starting materials. For example: Volume of the ampoule 11.85 ml; volume of 10 g Sb_2S_3 about 2.15 ml; free volume 9.70 ml. 65% of 9.70 ml = 6.3 ml 10-molar HI are poured into the ampoule which is then dipped into liquid nitrogen almost up to the constriction to freeze the acid. The powdered starting materials, here 10 g Sb_2S_3, are loaded into the ampoule with the help of a suitable funnel. After cleaning the constriction on the inside with a cotton plug, the ampoule is connected to a vacuum system, evacuated and sealed off. During this operation the level of the liquid nitro-

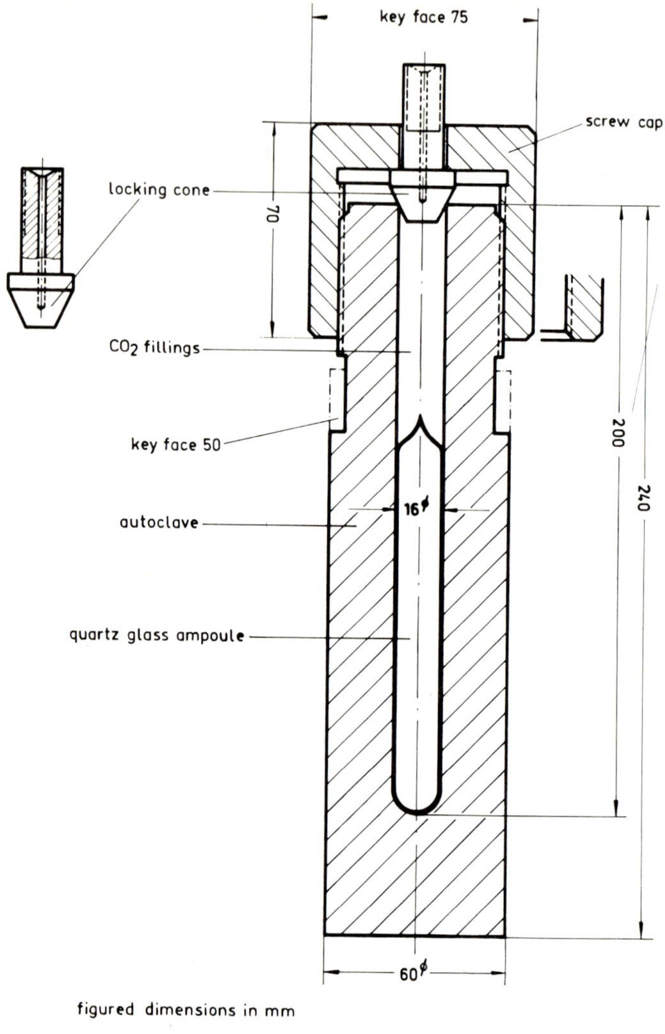

Fig. 7.5. Autoclave.

gen should be corrected from time to time to avoid thawing of the acid, which would otherwise immediately react with the starting materials. The sealed ampoule is transferred immediately from the liquid nitrogen into running hot water until a film of liquid acid has formed at the inner wall of the ampoule. In this way bursting of the ampoule by the increase in volume of the melting acid is avoided. The ampoule is then kept in a hood to allow slow thawing of the rest of the acid.

The ampoule is inserted into an autoclave (fig. 7.5) and the free volume remaining in the bore hole is calibrated using water from a burette, the water being afterwards removed. To counterbalance the pressure inside the ampoule developed under experimental conditions, carbon dioxide is used. The amount of CO_2 required is read off from the graph (fig. 7.6) taking into account the free volume in the autoclave, the temperature and the pressure of the experimental conditions.

As the experimental pressure is not known in most cases and as ampoules of the design given can support very high outside pressures, for safety reasons a load of 0.75 to 0.8 g CO_2 per ml of the free volume of the autoclave should be used. All experiments in table 7.1 were run under these conditions. A rod of dry ice (solid carbon dioxide) of suitable diameter is prepared by placing a piece of dry ice of sufficient size in the mould (fig. 7.7) which is compressed between the jaws of a vice. A piece is cut off the rod with a knife, the weight

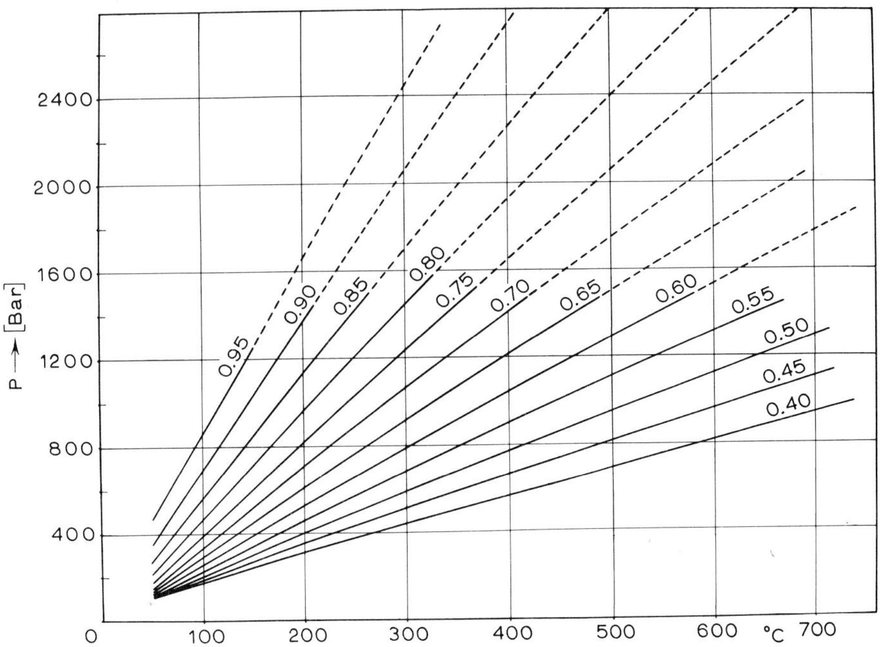

Fig. 7.6. P–T curves for CO_2. Parameter: g CO_2 per ml of free volume.

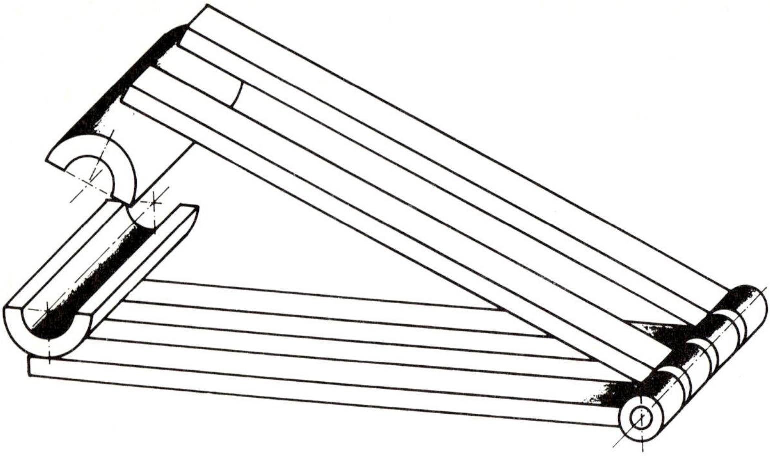

Fig. 7.7. Mould for pressing rods from dry ice.

of the dry ice being a little more than that calculated (plus 0.5 g to counterbalance the evaporation losses during the closing procedure). The rod is kept on a balance until the correct weight is obtained. The piece is transferred immediately into the autoclave above the ampoule. The locking cone is mounted and the screw cap bolted. Two elongated (about 1 m) wrenches are used for this purpose. The force applied should be sufficient to deform the locking cone a little bit in the sealing region in order to make the autoclave gas-tight. During all manipulations protective glasses should be worn.

The autoclave is mounted in a tube furnace with two independent filament windings to provide two temperature zones. The axis of the furnace should be at an angle of about 15° to the horizontal. This causes high convection when a temperature gradient is applied. The furnace is subjected to a temperature-time treatment according to the specifications given in table 7.1 for the respective compounds. For the preparation of SbSI the furnace is cooled down uniformly from 480°C to 250°C over a period of 10 days. If a temperature gradient is applied, the lower part (charge zone) is at the higher temperature. Temperatures are measured with thermoelements at the ends of the autoclave. After the heating period the autoclave is allowed to cool to room temperature. The screw cap is loosened carefully until carbon dioxide escapes.

Caution: High pressure may exist inside the ampoule (H_2S, in some cases H_2). The following operations should be done in a hood behind a protective shield, for example plexiglas. Heavy, protective gloves should be worn. The cap is removed and the ampoule transferred into liquid nitrogen in a Dewar vessel. The cooled-off ampoule is rolled in several layers of filter paper and broken open by tapping with a hammer. The contents of the ampoule are transferred to a porcelain dish and washed several times with methanol. The crystals are dried between layers of filter paper. The crystals of the respective compounds can be isolated mechanically from the accompanying material, e.g., quartz glass chips or other solid phases. The autoclave is ready for further experiments without any machining.

References

[1] A. Rabenau and H. Rau, Philips Techn. Rev. 30 (1969) 89.
[2] H.-D. Lüdemann and E.U. Franck, Ber. Bunsenges. physik. Chem. 72 (1968) 514.

P. Hartman, ed., Crystal growth: an introduction. © (1973) North-Holland Publ. Co.

MELT GROWTH

D.T.J. HURLE

8.0. Introduction

Growth from the melt is the most widely used method for the preparation of large single crystals. The technology associated with it has become quite sophisticated in recent years and the underlying science is proving to be complex. This chapter attempts to outline the main techniques of melt growth together with some of the phenomena which are important to the solidification process. Growth techniques are described in sections 1 and 2 whilst sections 3 and 4 deal with the segregation of solutes. Section 5 considers briefly the role played by convection in the melt. Space does not permit a detailed consideration of either the growth of specific materials or the many factors which control the perfection of grown crystals although the most important factors which give rise to chemical inhomogeneities are indicated in sections 4 and 5. It should perhaps be mentioned that it is now possible to grow large single crystals of some semiconducting materials (notably Si) which are entirely free from dislocations. However, even when this is done, unwanted imperfections, arising from the condensation of vacancies, remain [36].

The literature on the subject of melt growth is very extensive and the few references at the end of this chapter are to be considered as illustrative of the topics discussed. They are taken from the fields with which the author is most familiar and are not intended to provide a balanced review of the field.

8.1. Basic Principles and Techniques

8.1.1. Introduction

We are concerned in this chapter with the controlled *solidification* of a melt in such a manner as to promote the extension of a single nucleus without the introduction of new nuclei and with the minimum of chemical and structural disorder in the crystal. To avoid forming new nuclei and to avoid instability in the growth surface, which leads to chemical inhomogeneity in doped crystals, we must avoid having extensive zones of supercooling in the melt. To do this we must remove the latent heat generated by the solidification process by conduction into the solid rather than into the liquid. To achieve this we must have directional solidification in which growth occurs onto part of the crystal whilst a heat sink is attached to the remainder. Consider a molten bar solidified by passing a planar solid-liquid interface along it, normal to its axis at a velocity v (fig. 8.1). If the *temperature*

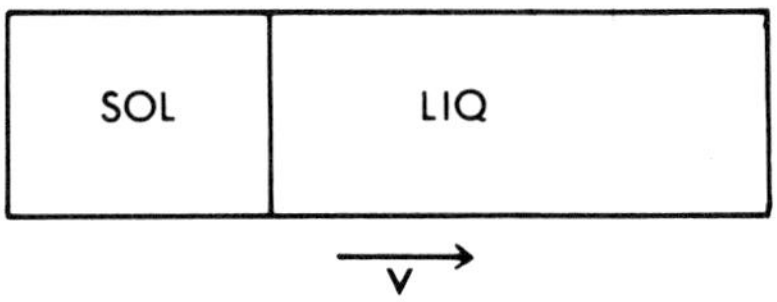

Fig. 8.1. Schematic representation of directional solidification.

gradients in solid and liquid at the interface are G_S and G_L then the continuity of heat flux requires: –

$$K_S G_S - K_L G_L = Lv, \tag{8.1}$$

where K_S and K_L are the *thermal conductivities* of solid and liquid and L is the *latent heat of solidification* per unit volume. If we require $G_L > 0$ then clearly for $v \geqslant 0$, we must have $G_S > 0$ i.e. we must extract heat from the growing crystal. This can occur by conduction down the crystal and radiation from its surface. In the melt in addition to these mechanisms, convection will be important.

If the material is neither a pure element nor a congruently-melting compound then additionally we must consider the segregation of components (either solutes or components of the compound) at the growth surface. Segregation products are removed from the vicinity of the solid-liquid interface by diffusion and convection. Mass diffusion occurs much more slowly than thermal diffusion and we shall see that often the rate at which we may perform the solidification is limited by the need to allow time for mass diffusion in the vicinity of the interface.

We classify the various techniques of melt growth under two headings: (a) normal freezing in which a molten charge is directionally solidified as in fig. 8.1 and (b) zone-melting in which a zone is melted in a solid ingot and then caused to pass along the ingot.

8.1.2. Normal freeze techniques

8.1.2.1. Crystal pulling. Undoubtedly the most important technique is the pulling method due originally to *Czochralski.* The elements of pulling are illustrated in fig. 8.2. The charge material is contained in a crucible which can be heated to above the melting point of the charge. A pull rod with a chuck

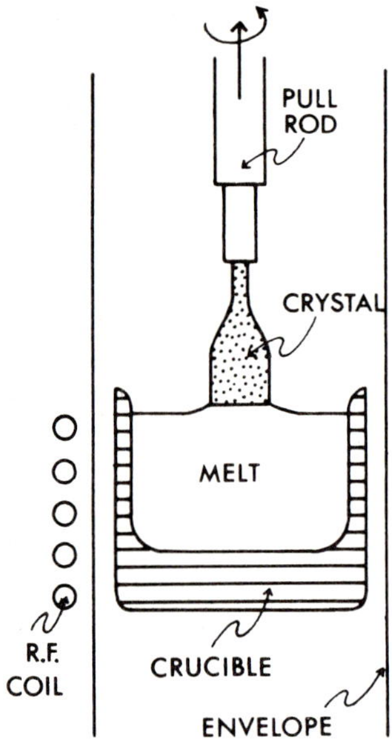

Fig. 8.2. Crystal pulling.

containing a seed crystal at its lower end is positioned above the crucible. The seed crystal is dipped into the melt whose temperature is adjusted until a meniscus can be supported by the seed crystal. The pull rod is then slowly lifted and rotated and by carefully adjusting the power supplied to the crucible a crystal of the desired diameter can be produced. The whole is maintained within an enclosure which permits control of the ambient gas and visibility of the growing crystal. The technique has been employed for an extremely wide range of materials from low melting point metals (e.g. Ga, MP. 28°C) up to spinel (MP. 2135°C). Its chief advantages are (i) the growing crystal is not in contact with a mould (ii) by using oriented seed crystals, crystals of a variety of orientations can be prepared (iii) suitable dopants can be added to the melt during growth (iv) the crystal is viewed as it is growing. Its disadvantages are (i) crucible materials must be found which do not contaminate the material in its molten state (ii) the shape of the resultant crystal is not exactly reproducible.

8.1.2.2. Pedestal growth. This method avoids the crucible contamination problem by pulling from a melt formed by melting the end of a large charge rod (fig. 8.3). (It could be argued that it is in fact a zone-melting technique). It has been found useful for the growth of Si for which no completely inert crucible material has been discovered (The commonly used SiO_2 crucible contaminates the melt with oxygen).

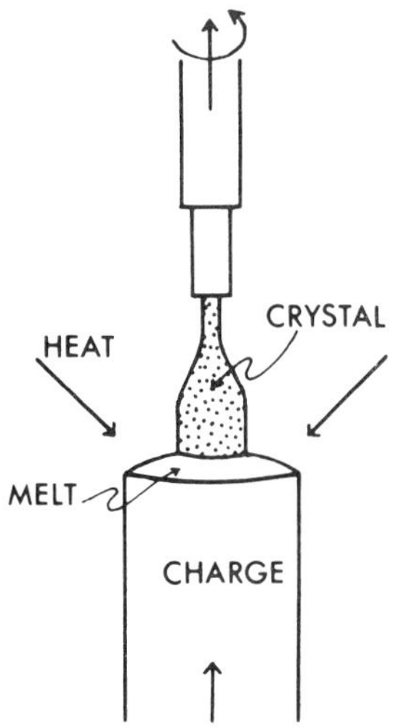

Fig. 8.3. Pedestal technique.

8.1.2.3. Vertical Bridgman/Stockbarger. In this technique the charge is completely melted in a pointed crucible within the upper furnace (fig. 8.4). The crucible is then lowered through the baffle into the lower furnace which is at a much lower temperature and progressive crystallization upwards from the tip of the crucible occurs.

8.1.2.4. Kyropoulos. The apparatus is basically that used for crystal-pulling, usually with forced cooling of the seed crystal except that here growth is achieved by slowly lowering the melt temperature rather than by pulling out a crystal.

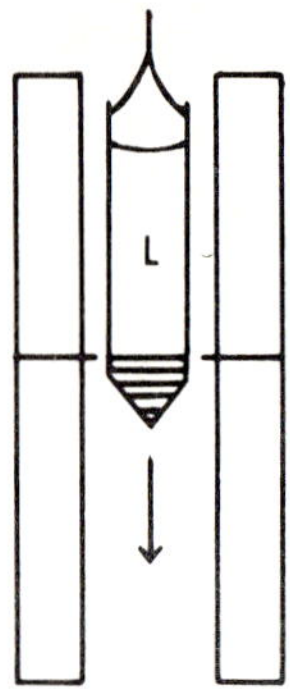

Fig. 8.4. Bridgman/Stockbarger technique.

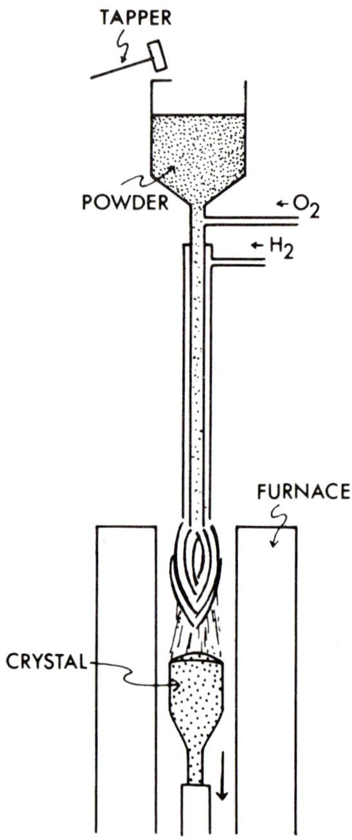

Fig. 8.5. Verneuil technique.

8.1.2.5. Verneuil (flame fusion). This is illustrated in fig. 8.5 and is a well established technique for the growth of *refractory oxide* crystals. Powdered charge material is fed at a controlled rate down a tube by a tapper. Hydrogen and oxygen are fed down outer, annular tubes and burnt. The flame melts the powder which rains down into a shallow pool of liquid on top of the seed crystal. As the crystal grows it is lowered down into the afterheater furnace at a rate commensurate with the feed rate. The crystal can be translated laterally and/or rotated to improve crystal homogeneity. A considerable amount of development effort has been put into this technique in recent years. Alternative heating methods such as induction-coupled plasmas have been used together with different geometries of seed crystal. By and large however the perfection of a crystal obtained by this technique is inferior to that currently obtainable by crystal pulling. The disadvantages of the technique include poor control of the thermal environment and limited freedom in the choice of gas ambient. The very small liquid volume predisposes towards inhomogeneities in the crystal.

8.1.2.6. Horizontal normal-freeze. (Variously known as *horizontal Bridgman* or *Chalmers technique*). The charge is contained in a boat which is placed within a horizontal furnace tube which can be filled with the required ambient gas (or evacuated). A muffle furnace is placed around the tube and serves to melt the charge. Directional solidification is obtained by slowly withdrawing the boat from within the furnace (or moving the furnace away from the boat).

8.1.3. Zone-melting techniques

8.1.3.1. Horizontal zone melting. This is as in section 1.2.6 above except that a short furnace (frequently an r.f. coil) is used to melt a zone which is then moved slowly along the bar. The advantages of this over the normal freeze process are (a) reduced contamination of the melt by the boat (b) less heater power required and (c) zone-refining for purification and zone levelling for uniform doping can be used. (See section 3).

8.1.3.2. Float-zone melting. In this technique (fig. 8.6) a molten zone is maintained between a vertically mounted seed crystal and a charge rod by surface tension forces. The principal advantage of this technique is that it does not require a crucible: the melt is in contact only with its own solid. The technique is of considerable industrial importance for the preparation of oxygen-free Si. Materials with high surface tension and low liquid density are required.

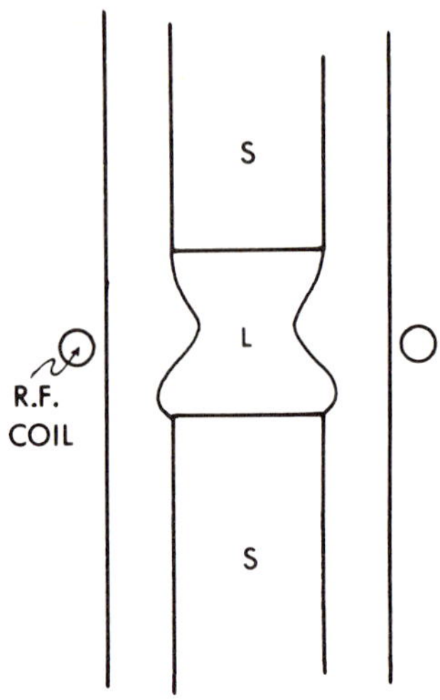

Fig. 8.6. Float-zone melting.

8.1.4. Experimental techniques

We conclude this section with a few general observations on experimental techniques. More detailed reviews are available in the literature [1–3].

8.1.4.1. Heating. Resistance heating and r.f. induction heating are the most commonly employed. Plasma heating, laser or electron beams and arc-image furnaces have all been used. The advantages of resistance heating are its low capital cost and, with high current/low voltage sources, the ability to make self-supporting heaters of complex shapes. R.F. induction heating is "clean" – that is it can apply the heat only where it is required. It is capable of accurate control with very short time constants but the capital and running costs are high.

8.1.4.2. Temperature (or power) control. Basically one needs a servo-system that compares recorded temperature or power with a requested value and adjusts the power source to obtain a null difference. Simple on/off devices are in general not adequate and servo-controls with proportioning action are necessary. In most applications servo-controlling for powder is the most satisfactory.

8.1.4.3. Crucibles and boats. The requirements of a crucible material are that it shall be: (a) chemically inert to the charge when it is in molten state (b) must have a smaller expansion coefficient than the crystal if the crystal is to be grown into it (c) its thermal conductivity shall be compatible with the heat flow requirements (d) it shall not contaminate the charge with foreign impurities.

8.2. Technology

In this section we discuss the specification of modern equipments for crystal-pulling, float-zoning and Stockbarger growth with reference to the basic design principles involved.

8.2.1. Crystal-pulling

A modern, versatile crystal pulling apparatus capable of producing highly perfect single crystals of a very wide range of materials is shown in fig. 8.7. Its essential features must include: –

(a) a rigid, vibration-free frame
(b) a precision lead screw and gearing driven by servo-controlled motors to provide a wide range of pulling and rotating speeds. Pulling speeds as low as 10^{-5} cm sec^{-1} and as high as 10^{-2} cm sec^{-1} are sometimes required. Rotation rates are commonly in the range 1–200 r.p.m.
(c) a stable power supply to heat the crucible to the required temperature which is very precisely servo-controlled for either constant power or constant temperature at some point on the crucible. (The melt temperature itself will fluctuate because of non-steady convection (see section 5)).
(d) a variety of work chambers, ranging down from a simple silica envelope to a water-cooled pressure vessel suitable for high pressure growth.
(e) facilities for the rapid removal of the work chamber for easy cleaning and loading of the puller.
(f) a range of crucibles.
(g) gas metering and purification systems and vacuum gear.

A typical procedure for operating such a puller might be as follows: –

(a) load crucible and fit seed crystal into chuck having first removed work damage by etching.
(b) assemble chamber, evacuate and then flush with ambient gas (if used).
(c) melt the charge and hold its temperature a little above the melting point.
(d) lower the seed to a position a little above the melt surface and allow to come to thermal equilibrium; apply seed rotation. Then lower it very slowly into the melt to minimise thermal shock.

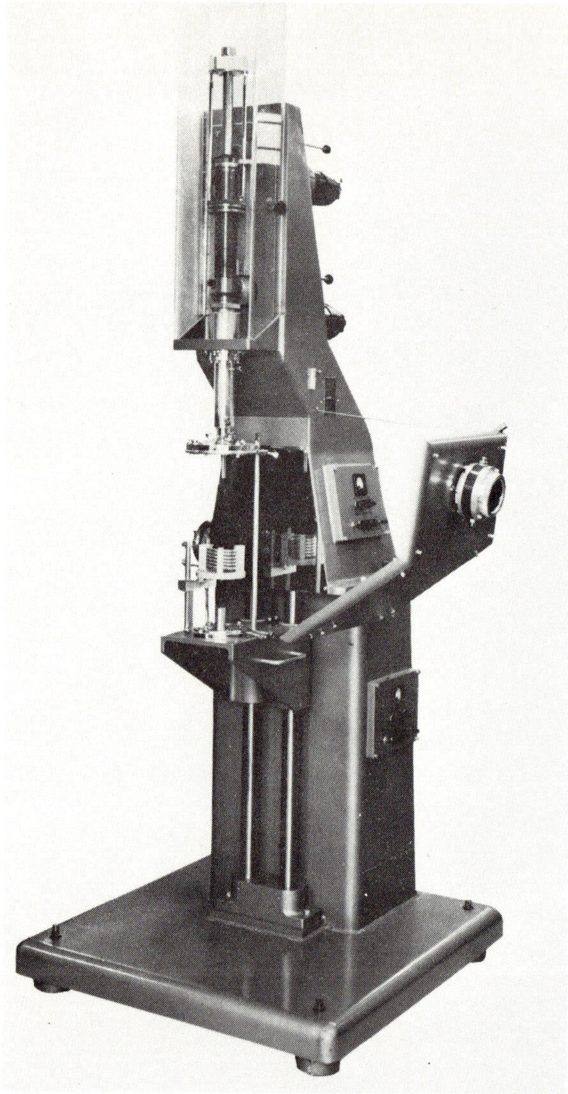

Fig. 8.7. Crystal puller (Courtesy of Metals Research Ltd England).

(e) make fine adjustments to the melt power until the seed supports a meniscus.

(f) commence pulling and adjust temperature (or pull rate) so that a thin neck is grown to the crystal. (This serves to minimise the transfer of imperfections in the seed into the crystal.)

(g) increase the crystal diameter slowly and smoothly (by reducing the power to the melt) until the growing crystal has the required diameter.

We now consider extensions or refinements of the basic technique required to grow specific classes of materials: –

8.2.1.1. Semiconductors. Crucible rotation: in addition to rotation of the seed, the crucible can also be rotated [4]. In the case of Si this can reduce the rate of uptake of oxygen into the molten Si from the silica crucible by reducing the rate of convective mixing. An additional advantage is that rotation of the crucible ensures symmetrical heating of the melt.

8.2.1.2. Oxides. Materials such as yttrium aluminium garnet, sapphire, ruby, calcium tungstate and spinel can be pulled from iridium crucibles [5]. If growing crystals are subjected directly to the cold flowing ambient gas they experience severe thermal stresses. To minimise this crystals are pulled into afterheaters typically consisting of an inverted iridium-lined ceramic crucible with a hole in its base through which the pull rod passes. Heating is provided by coupling the Ir lining into the r.f. field. By this means axial temperature gradients can be reduced to as low as $50°C\ cm^{-1}$. Under these conditions crystals with no low angle boundaries and dislocation densities of only $10^2 \rightarrow 10^3\ cm^{-2}$ can be prepared.

8.2.1.3. Volatile materials. Demands made by solid state physicists for large single crystals of such materials as GaAs and InAs raised severe problems for the crystal grower. For example the dissociation pressure of As over molten GaAs at the melting point is ≈ 1 atmosphere and melting GaAs in a conventional crystal puller results in most of the arsenic distilling off and condensing on the cold walls of the enclosure. Initial attempts to overcome this were based upon having As vapour in equilibrium with the melt by maintaining the whole enclosure at a temperature above the sublimation point of As and by having a reservoir of As whose temperature was adjusted to control the As pressure. The problem arises of how to provide the lift and rotation to the pull rod under these conditions. Two solutions were developed [1]. One called the syringe puller involved a bearing surface of boron nitride which provided a mechanical seal capable of working at temperatures $> 600°C$ and in an ambient of As_4 vapour. The other used magnetic coupling to a high Curie point alloy attached to the pull rod which was sealed within the heated enclosure.

These methods have been largely superseded by the development of the *liquid encapsulation technique* [6]. In this technique a layer of boric oxide is used to cover the melt and the enclosure, now at room temperature, is raised to a pressure greater than the dissocation pressure. Boric oxide is a glassy, viscous substance in which As is virtually insoluble, and it eliminates virtually completely the loss of As from the melt. Pulling can be performed by lowering the seed through the boric oxide layer and growth occurs at the interface of the two liquids. Moreover as the crystal rises through the encapsulant a thin coating remains on its surface which effectively prevents evaporation from the solid during cooling. More recently pressure vessel enclosures have been developed which permit the growth of GaP and InP whose dissociation pressures are respectively ≈ 35 and ≈ 21 atmospheres [7, 8]. A pressure chamber is shown in figure 8.8. Two side ports are provided. Illumination is supplied down one of these; a television camera views the growing crystal down the other.

Fig. 8.8. Pressure chamber for puller.

Liquid encapsulation techniques have been utilized for a range of other materials including II-VI and IV-VI compounds.

8.2.2. Float-zone refining [2]

A research apparatus for the float-zone refining of Si is shown in fig. 8.9. The requirement of two aligned chucks with independent rotation facilities and separate or coupled translation necessitates a rather complex mechanical

Fig. 8.9. Float-zone refining apparatus.

design. Theory shows that there is a maximum length of zone which can be supported. For thin rods this is $l = 2\sqrt{3}r$ when r is the rod radius; for large diameter rods it is $l = 2.8\,(\gamma/\rho g)^{1/2}$ where γ = interfacial tension, ρ = density and g = acceleration due to gravity. The need to melt a short zone limits the heating techniques which can be used. R.F. induction heating is used for Si and other semiconductors with coils specially wound to concentrate the r.f. field. The low electrical conductivity of Si at room temperature necessitates some auxilliary heating to raise the temperature to the point where the r.f. field can couple into the Si. Counter-rotation of the charge and crystal tends to straighten bent rods and maintain them straight.

The low value of γ/ρ for metals means that l and therefore, indirectly r, must be kept small. The high electrical conductivity of these materials make electron beam heating attractive. This of course requires a vacuum ambient which additionally serves to remove volatile impurities.

Neither of the above two methods are possible with oxides because of their low electrical conductivity. Here optical systems have been used. Initially carbon or xenon arc lamps focussed onto the molten zone were used. The development of high power CO_2 lasers has provided an improved alternative [9].

Another solution is the strip-heater technique [10] for oxides in which a resistively-heated strip of precious metal actually passes through the molten zone. This strip has several small holes in it through which molten oxide can flow when the zone is traversed through the charge.

The heating geometry required in all float-zone techniques leads inevitably to large radial temperature gradients and stresses. The use of fore and after heaters can reduce these but nonetheless, in general, the crystal perfection achieved by the floating-zone technique is inferior to that obtainable with the pulling technique.

8.2.3. Stockbarger technique

This technique has been extensively used for fluorides such as CaF_2, LiF etc. Large crystals up to 15 cm diameter and weighing 15 kg can be produced by this means [11]. By suitable design of heaters and radiation baffles axial temperature gradients can be reduced to quite low values and good quality crystals obtained. Two main features about the heat transport are to be noted: (a) the geometry minimises the degree of natural convection in the melt and (b) with the fluorides at the low rates of growth used ($\sim$ 1 mm/hr) the dominant heat transfer mechanism is by radiative exchange through the transparent melt. Power stabilisation is particularly important with the large crystals.

8.3. Macrosegregation

8.3.1. Introduction

We are here concerned with the macroscopic solute distributions which result from the application of directional solidification and zone-melting processes to initially uniform ingots. We define a *segregation coefficient* (sometimes called a *distribution constant*) $k\ (T) = C_S/C_L$ which is the ratio of the solute concentrations in solid and liquid which are in equilibrium at temperature T. This quantity is commonly less than unity (corresponding to downward sloping liquidus and solidus curves – fig. 8.10a) but can be greater than unity (corresponding to upward sloping solidus and liquidus curves – fig. 8.10b). Thus, for $k < 1$ for example, if we solidify a small portion of a melt of solute concentration C_L the resulting solid will have a concentration of $C_S = k\ C_L < C_L$ and the remaining liquid will now have a concentration greater than C_L. Progressive solidification will therefore result in a progressively in-

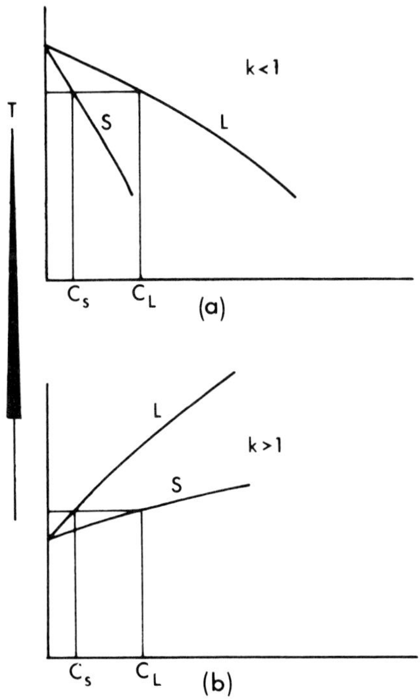

Fig. 8.10. Liquidus (L) and Solidus (S) curves for solutes with $k \lessgtr 1$.

creasing solid concentration. In the following sections we calculate analytical expressions for these distributions assuming the following

(i) that solute mixing in the liquid is complete
(ii) there is no diffusion in the solid
(iii) a steady growth rate is employed
(iv) the solid-liquid interface(s) is/are planar and normal to the macroscopic direction of solidification
(v) k is constant.

Consider the *zone melting* process exhibited schematically in fig. 8.11. A charge of unit cross section contains a molten zone whose length (in the z direction) at time t is l and whose freezing and melting interfaces are moving at velocities v_f and v_m respectively. The solidified material has a solute concentration $C_f(z)$, the liquid C_L and the charge to be melted C_m.

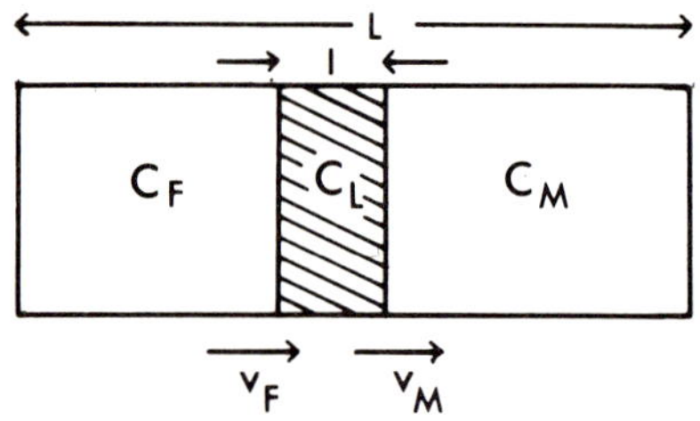

Fig. 8.11. Schematic representation of zone-melting.

The rate of change of solute concentration in the liquid zone is:

$$\frac{d}{dt}(lC_L) = C_m v_m - C_f v_f , \tag{8.2}$$

where

$$C_f = kC_L \tag{8.3}$$

and

$$\frac{dl}{dt} = v_m - v_f . \tag{8.4}$$

If v_m and v_f are constant then (8.4) integrates to give

$$l = l_0 + (v_m - v_f)t, \tag{8.5}$$

where l_0 is the zone length at $t = 0$. Integrating (8.2) one obtains

$$\int_{C_0}^{C_L} \frac{dC_L}{C_m v_m - [(k-1)v_f + v_m]\, C_L} = \int_0^t \frac{dt}{l_0 + (v_m - v_f)t}, \tag{8.6}$$

where C_0 is the initial concentration in the liquid zone at $t = 0$. We next evaluate (8.6) in the two cases of interest:

8.3.2. Normal freezing distribution

– i.e. where the whole charge is initially molten. In this case $v_m = 0$ and $l_0 = L$, the length of the ingot. Integrating (8.6) using (8.3), we obtain

$$C_f = kC_0(1 - g)^{k-1}, \tag{8.7}$$

where $g = v_f t/L$ is the fraction solidified at time t.

One may measure the segregation coefficient k by measuring the solute distribution in a directionally solidified ingot. Taking logarithms of (8.7):

$$\log(C_f/C_0) = \log k + (k-1)\log(1-g). \tag{8.8}$$

Thus, from a plot of $\log C_f/C_0$ versus $\log(1-g)$ one obtains an intercept of $\log k$ and a slope of $(k-1)$. If $k \approx 1$ or $k \gg 1$ the slope provides an accurate measure. However, for $k \ll 1$, its value must be obtained from the intercept.

8.3.3. Zone melting distributions

We first consider the distribution after the passage of one zone of length l_0. We set $v_f = v_m \equiv v = z/t$. Then from (8.6) and (8.3)

$$C_f = C_m - (C_m - kC_0)\exp[-kz/l_0]. \tag{8.9}$$

If we assume that the ingot was initially doped to a concentration C_0 (i.e. $C_m = C_0$) then, from (8.9) we get

$$C_f/C_0 = 1 - (1-k)\exp[-kz/l_0]. \tag{8.10}$$

Inspection of (8.10) shows that the first solid to form has a concentration of kC_0 and, for $kz/l_0 \ll 1$,

$$C_f/kC_0 = 1 + (1-k)\,z/l_0 \tag{8.11}$$

and when $kz/l_0 \gg 1$ then $C_f \to C_0$ resulting in the distribution shown, for k = 0.2, in fig. 8.12.

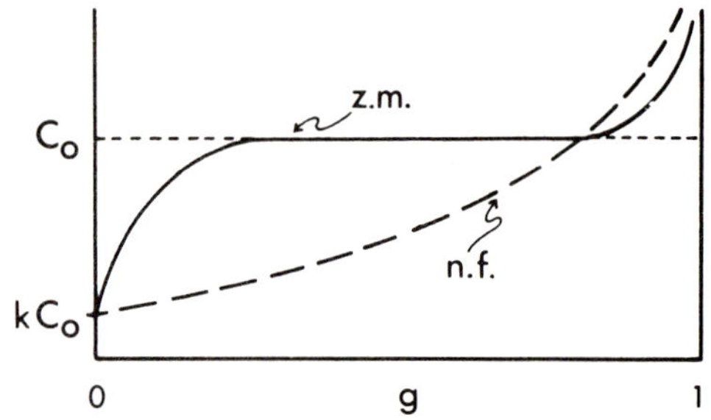

Fig. 8.12. Zone-melting and normal-freeze distributions.

When the melting interface reaches the end of the ingot the remaining portion of the ingot solidifies to give the normal-freezing distribution described in section 3.2 above. We see then that solutes with $k < 1$ are carried along in the molten zone and tend to accumulate at the trailing end of the ingot. On the other hand solutes with $k > 1$ accumulate at the leading end of the ingot. The maximum separation (purification) obtainable using a single zone pass, occurs when the exponent kz/l_0 is minimised i.e. for the maximum value of l_0. This value is clearly $l_0 = L$ and corresponds to a normal freeze of the ingot. A normal freeze curve, also for $k = 0.2$ is shown in fig. 8.12.

If a normal freeze achieves a greater degree of segregation than a zone pass, what then is the advantage of zone melting? The answer is to be found by considering what happens when the processes are repeated. In a normal freeze, melting the ingot re-mixes the solute so that the distribution after the second normal freeze is the same as that after the first. By contrast, the passage of a second molten zone further purifies the ingot. This is the basis of zone refining.

After the zone has moved a distance z on its nth pass, the amount of solute in the zone is:

$$\frac{1}{l_0}\int_0^{z+l} C_{(n-1)}(z)\mathrm{d}z - \frac{1}{l_0}\int_0^{z} C_{(n)}(z)\mathrm{d}z = \frac{1}{k}\,C_{(n)}(z)\;, \tag{8.12}$$

where $C_{(n-1)}(z)$ is the concentration in the ingot at position z after the $(n-1)$th pass. This holds for the region of the ingot $0 \leqslant z \leqslant L - nl_0$. For $z > L - nl_0$ the distribution is modified as a result of the "back reflection" of the effects of the normal freezing of the last zone length. The actual distribution after the nth pass $(C_{(n)}(z))$ is obtained by solving the n simultaneous integral equations (8.12): This is too complex to be considered here.

We confine ourselves here to the question – Is there an ultimate distribution, or does purification continue indefinitely as more zones are passed? The answer is that, for an ingot for which there is no "back reflection", there is an ultimate distribution. This is obtained by putting $C_{(n-1)}(z) = C_{(n)}(z)$ in (8.12). This yields

$$C_n(z) = A e^{Bz}, \tag{8.13}$$

where

$$A = C_0 BL / \{e^{BL} - 1\}$$

and

$$k = Bl / \{e^{Bl} - 1\}$$

This distribution is steeper the smaller is the quantity l_0/L. Hence long zones initially move the biggest amount of solute but short zones provide the best ultimate purification. In a zone-refining process it is therefore preferable to pass m long zones followed by n short zones rather than $m + n$ zones of intermediate length.

8.3.4. Zone levelling

If we start with an initially pure ingot ($C_\text{m} = 0$), but with a concentration C_0 in the molten zone, then, from (8.9)

$$C_\text{f}/C_0 = k \exp[-kz/l_0] \tag{8.14}$$

and for $kz/l_0 \ll 1$

$$C_\text{f}/C_0 \approx k(1 - kz/l_0) \approx k.$$

Thus, for solute with a very small k, we can get a uniform solute distribution by loading the zone with a concentration C_f/k. This process is known as zone levelling [2].

8.3.5. Discussion

Let us now re-examine the assumptions which have been made in deriving the above theory. In particular let us reconsider solute mixing in the melt. The rejection of solute into the melt at the freezing interface (for $k < 1$) leads to an enhanced concentration in this region which is transported to the bulk of the melt by diffusion and convection. Close to the interface diffusion will be the dominant mechanism because the convective velocities become zero at the surface; in the bulk melt convection dominates. We may suppose that there exists a layer of thickness δ_m adjacent to the interface within which diffusion is the only mechanism of solute transfer and beyond which convec-

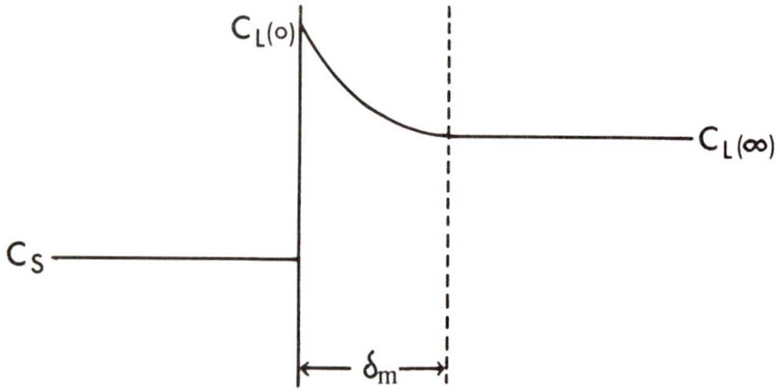

Fig. 8.13. Solute boundary layer ahead of a growing crystal.

tion ensures complete mixing (fig. 8.13). Our assumption made hitherto, of complete mixing is equivalent to $\delta_m \to 0$.

In a co-ordinate system located on the solid-liquid interface and moving with it at a steady velocity v, the diffusion equation is, in one dimension: –

$$\frac{\mathrm{d}}{\mathrm{d}z}\left(D\frac{\mathrm{d}C_L}{\mathrm{d}z}\right)+v\frac{\mathrm{d}C_L}{\mathrm{d}z}=\frac{\mathrm{d}C_L}{\mathrm{d}t}, \tag{8.15}$$

where D = mass diffusion coefficient.
After an initial transient $\mathrm{d}C_L/\mathrm{d}t = 0$.

We can integrate (8.15) subject to the boundary conditions:

$$C_L = C_L(\infty) \text{ at } z = \delta_m$$

and

$$D\left(\frac{\mathrm{d}C_L}{\mathrm{d}z}\right)_{z=0} = v(C_S - C_L(0)) = -v(1-k)C_L(0)\,,$$

to give the solute distribution in the boundary layer $0 \leqslant z \leqslant \delta_m$. The second boundary condition above expresses conservation of solute at the interface. Taking D to be constant one obtains:

$$C_L(0) = \frac{C_L(\infty)}{k^* + (1-k^*)\exp(-v\delta_m/D)} . \tag{8.16}$$

One can define an *effective segregation coefficient* $k_{eff} = C_S/C_L(\infty)$, which is experimentally measurable, in terms of the "interface" segregation coefficient $k^* = C_S/C_L(0)$. From (8.16):

$$k_{eff} = \frac{k^*}{k + (1-k)\exp(-v\delta_m/D)} . \tag{8.17}$$

This is the well-known Burton, Prim and Slichter equation [12].

In the absence of complete mixing in the liquid, the normal freeze and zone-melting distributions derived above may still be used, without serious error, if the quantity k is replaced by k_{eff}. A calculation of δ_m involves a knowledge of the convective flow conditions in the melt. In the case of crystal pulling with crystal rotation dominating the flow, the flow pattern is similar to that at a rotating disk [12] and under these conditions

$$\delta_m \sim 1.6\, D^{1/3}\, \nu^{1/6}\, \omega^{-1/2} , \tag{8.18}$$

where ν = kinematic viscosity

ω = angular velocity of the crystal.

We see, from (8.17) that increasing v and/or δ_m causes k_{eff} to approach unity. Hence for efficient zone refining action, good stirring of the molten zone is necessary. The choice of optimum growth rate lies between a large number of zone passes at a high growth rate with relatively inefficient segregation per pass or a small number of zone passes at a low growth rate with efficient segregation at each pass.

Finally we note that changes in k occur under certain condiditions when a planar solid-liquid interface becomes unstable; these effects are considered in section 4.

8.3.6. Eutectic solidification

Thus far we have considered only those situations in which the solidification product is a single phase. If we slowly solidify a concentrated alloy (α) then macrosegregation will take us down the liquidus curve until, finally, we may reach the eutectic composition C_E (see fig. 8.14). Provided the nucleation barrier can be overcome then a solid (β) phase will also be formed. The two-phase solid is called a eutectic and it should be noted that zone refining will not separate the components of a eutectic. The morphology of the α and β phase eutectic mixture depends on the growth conditions, and on the properties of the phases. Specifically, if facets do not occur on the solid-liquid interface of either phase then simple rod (fig. 8.15) or lamellar (fig.

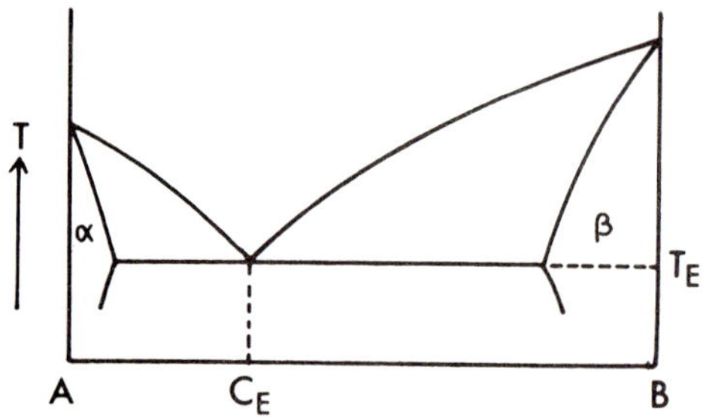

Fig. 8.14. Simple eutectic phase diagram.

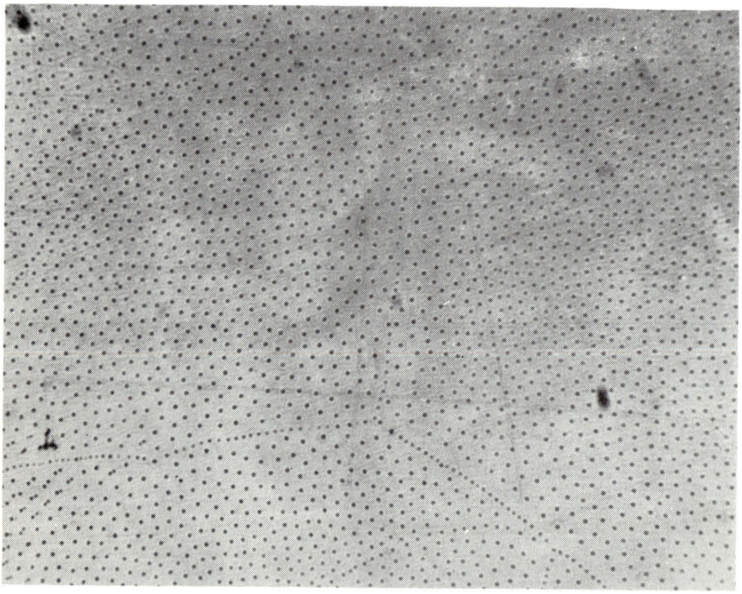

Fig. 8.15. Rod eutectic; rod axis and growth direction normal to surface. Rod spacing $\approx 20\mu m$.

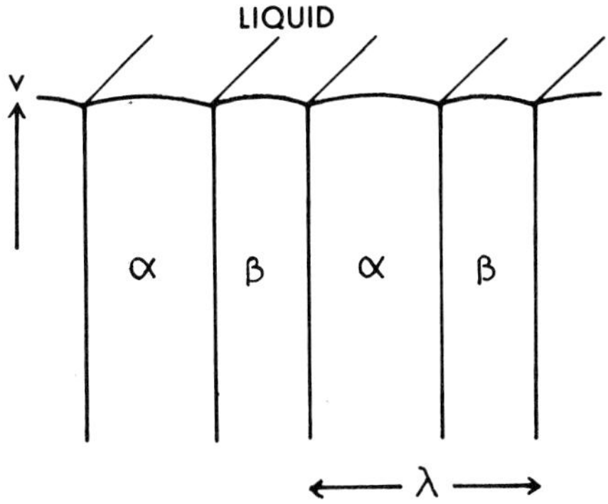

Fig. 8.16. Schematic representation of a lamellar eutectic.

8.16) structures will form depending on the value of C_E. This is characteristic of metal-metal systems [13]. If facets form on either interface more complicated or less regular structures occur [14]. We confine ourselves to non-facetting systems.

Consider the lamellar structure shown in fig. 8.16. Component B is rejected by the α phase and component A by the β phase. There is thus a lateral component of diffusion. These solute gradients imply an undercooling (ΔT_d) of the solid-liquid interfaces which varies spatially. A contribution to the undercooling (ΔT_c) below the eutectic temperature T_E also arises from the curvature of the solid-liquid interfaces (Gibbs-Thomson effect). If the interface is isothermal (a good approximation) then the total undercooling ($\Delta T = \Delta T_d + \Delta T_c$) is spatially constant. The interface adjusts its shape to achieve this condition. The magnitude of the total undercooling will depend on the periodicity of the structure (λ) for fixed growth velocity (ν). A coarse structure reduces ΔT_c but increases the lateral diffusion path and hence increases ΔT_d. Conversely a fine structure increases ΔT_c but decreases ΔT_d. There thus exists some value of $\lambda = \lambda_c$ for which ΔT is a minimum. By evaluating $(\partial \Delta T/\partial \lambda)_\nu = 0$, it has been shown that: –

$$\lambda_c^2 \nu = C_1$$

and

$$\Delta T_{min}^2/\nu = C_2,$$ where C_1 and C_2 are constants.

Similar results hold for rod eutectics where λ is now to be interpreted as the rod spacing [15].

Suppose now we change the composition of the bulk melt to C_∞. The liquid at the interface has to remain close to C_E if coupled growth of the eutectic is to occur [15]. A boundary layer therefore develops in the direction normal to the growth front whose height is approximately equal to $C_\infty - C_E$. Provided this is maintained, a eutectic structure can be grown from melts of composition differing from C_E [16]. The volume ratio of the solid phases changes so that the solid now has a mean composition equal to C_∞. As we shall see later, the existence of a boundary layer ahead of a growing crystal, whether single phase or eutectic, can lead to instability in the shape of the interface and this is considered by Sekerka in Chapter 15.

The directional solidification of eutectic and near-eutectic liquids has received considerable study in the last few years. Potential applications for such materials exist in the fibre-reinforcement field (e.g. for turbine blades) and in the electronics field [17, 18].

8.4. Segregation and the solid-liquid interface

In the previous section we assumed that segregation at a solid-liquid interface was characterised by a constant k which was a property of the alloy system. Incomplete mixing in the melt lead to the formation of a boundary layer and to an effective segregation coefficient k_{eff}. We now consider how the nature and morphology of the solid-liquid interface influences segregation.

8.4.1. Facetted and non-facetted interfaces

If one rapidly decants the melt from crystals grown under conditions which avoid constitutional supercooling (see below) then, in general, one can recognise, with the naked eye, two types of interface. The first is a smoothly curved interface which follows the contour of the isotherm corresponding to thermodynamic equilibrium. The second type is characterised by a large planar area, called a facet, which is a low index plane and which occurs only where the solid surface is convex [19]. The occurrence of facets correlates with a high entropy of fusion (the Jackson α factor) as discussed by Bennema and Gilmer in Chapter 10.

The significance of facets to our present discussion is that the segregation of solute at them is anomalous [19, 20]. For example the segregation coefficient of Te in InSb at dilute solutions is ≈ 0.5 in the non-facetted regions but on the facet is ≈ 4. Fig. 8.17 shows an autoradiograph of a longitudinal section of an InSb crystal grown from a melt by the Czochralski technique

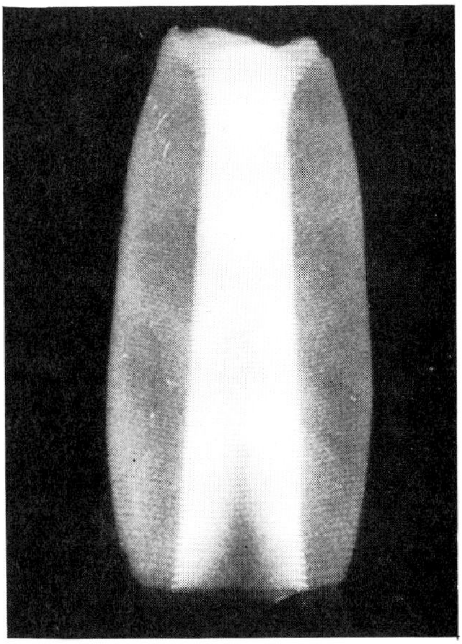

Fig. 8.17. Autoradiograph showing distribution of radio-Te in an InSb crystal.

containing a small amount of radioactive ^{128}Te. The crystal was grown in the ⟨111⟩ direction and there was a large {111} facet in the centre of the solid-liquid interface which produced the central Te-rich core seen in the autoradiograph. Whilst this effect has been studied in detail in a few semiconducting materials only, it is probably a universal phenomenon. Most materials exhibit facets during growth from the melt; in general only the metals, which have very low entropies of fusion, have completely non-facetted interfaces.

8.4.2. The cellular interface morphology

In section 3 we showed that incomplete mixing led to the formation of a boundary layer in the melt adjacent to the solid-liquid interface. For a solute with $k < 1$, the concentration at the interface is enhanced; for $k > 1$ it is depleted. Let us consider how the freezing point (the equilibrium liquidus temperature) varies in this boundary layer. For $k < 1$, the liquidus is downwards sloping (fig. 8.10a) and hence the higher the solute concentration the lower the liquidus temperature. The liquidus temperature in the boundary layer will thus vary as shown in fig. 8.18. The actual temperature at the solid liquid boundary will be at, or slightly below the equilibrium temperature, the amount of undercooling being equal to that required to drive the kinetic

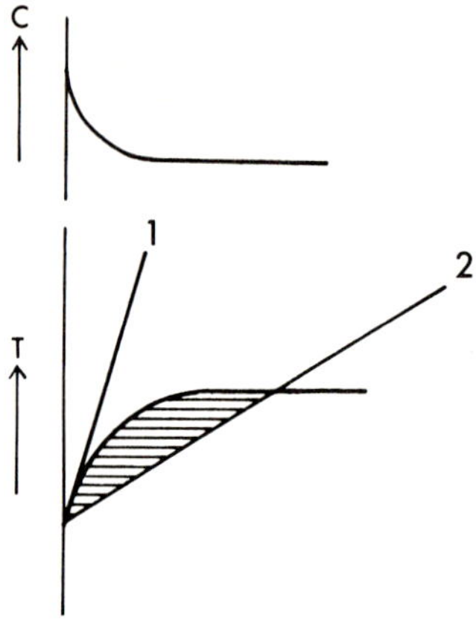

Fig. 8.18. Constitutional supercooling.

processes of crystallization at the imposed rate. In the absence of facets this interface undercooling is, in general, negligible at the usual growth rates. Consider two possible actual temperature distributions in the melt, designated by the lines (1) and (2). Both correspond to a positive temperature gradient. With the steeper gradient the actual temperature in the melt is everywhere greater than the liquidus temperature except actually at the solid-liquid interface. However, with the shallower gradient there exists a region (shaded) for which the actual temperature is below the liquidus temperature i.e. the region is supercooled. The name *constitutional supercooling* is given to this phenomenon [21]. The criterion for the existence of a zone of constitutional supercooling (S) in the melt is

$$\left(\frac{\mathrm{d}S}{\mathrm{d}z}\right)_{z=0} = m\left(\frac{\mathrm{d}C}{\mathrm{d}z}\right)_{z=0} - \left(\frac{\mathrm{d}T}{\mathrm{d}z}\right)_{z=0} > 0, \tag{8.19}$$

where m is the slope of liquidus on the phase diagram, C the solute concentration and T the temperature. The interface lies in the plane $z = 0$.

An analytical expression for the condition $(\mathrm{d}S/\mathrm{d}z)_{z=0} > 0$ is derived by Sekerka. For the present it is sufficient to observe that a large growth rate, a high solute concentration, poor stirring and a large value of $(1-k)$ all tend to produce a large boundary layer and hence pre-dispose to constitutional supercooling. For a given solute concentration in the melt, a slow growth rate, good stirring and a steep temperature gradient avoid it. Constitutional supercooling is undesirable because it produces instability in a planar-solid-liquid interface. A naive explanation of this is as follows: If, by a chance fluctuation, a protuberance forms on the solid surface, it finds itself in a region of supercooling and hence can grow more rapidly. In doing so it will start to segregate solute in the plane of the interface making the liquid adjacent to neighbouring regions of interface more solute-rich and hence with a lower freezing point. The solidification of these regions tends therefore to be suppressed and the protuberance is stabilised. A number of such protuberances may be expected to form a close-packed array separated by distances related to the lateral diffusion distance. The region between these projections will become progressively rich in solute and long solute-rich channels will form in the crystal. This, approximately, is what happens and the structure is referred to as a cellular structure. The above explanation is given merely to give the reader some physical picture; it ignores the effects of the projection on the heat flow. The actual conditions under which an interface becomes unstable must be derived from perturbation theory calculations and this is done in Chapter 15.

The rest of this section is devoted to a description of the morphology of the *cellular structure* in several classes of materials and of the chemical and structural inhomogeneities which it introduces into the crystal. The presence

Fig. 8.19. Corrugations on surface of a doped, ⟨110⟩ Ge crystal. (from Solid State Electronics 3 (1961) 142).

of a cellular structure can be seen by the naked eye on crystals from which the melt has been rapidly decanted. Additionally, small grooves called *corrugations,* are formed where the cellular interface intersects the free surface of the crystal (fig. 8.19).

Etching techniques exist for a number of semiconductor and oxide crystals which enable one to reveal the growth history of a crystal from a study of etched sections cut in the crystal. There are, within most doped crystals, periodic variations in solute concentration caused either by crystal rotation or by non-steady convection motion of the melt. These are the subject of the next section. Etching reveals these variations and hence provides a profile of the shape of the solid-liquid interface at successive cycles. Fig. 8.20 is from the $(1\bar{1}0)$ surface of a Ga doped Ge crystal grown in the [110] direction (which is orthogonal to the section). The growth striae reveal a perturbation in the shape of the interface which is initially sinusoidal but which becomes composed of small {111} facets when the amplitude of the perturbation has grown sufficiently for it to become tangential to the {111} planes. Thus the interface structure of a [110] Ge crystal is composed predominantly of an elongated ridge structure made up of the pairs of {111} planes which lie at

Fig. 8.20. Cellular structure delineated by striae on a $(1\bar{1}0)$ section of a [110] Ga-doped Ge crystal.

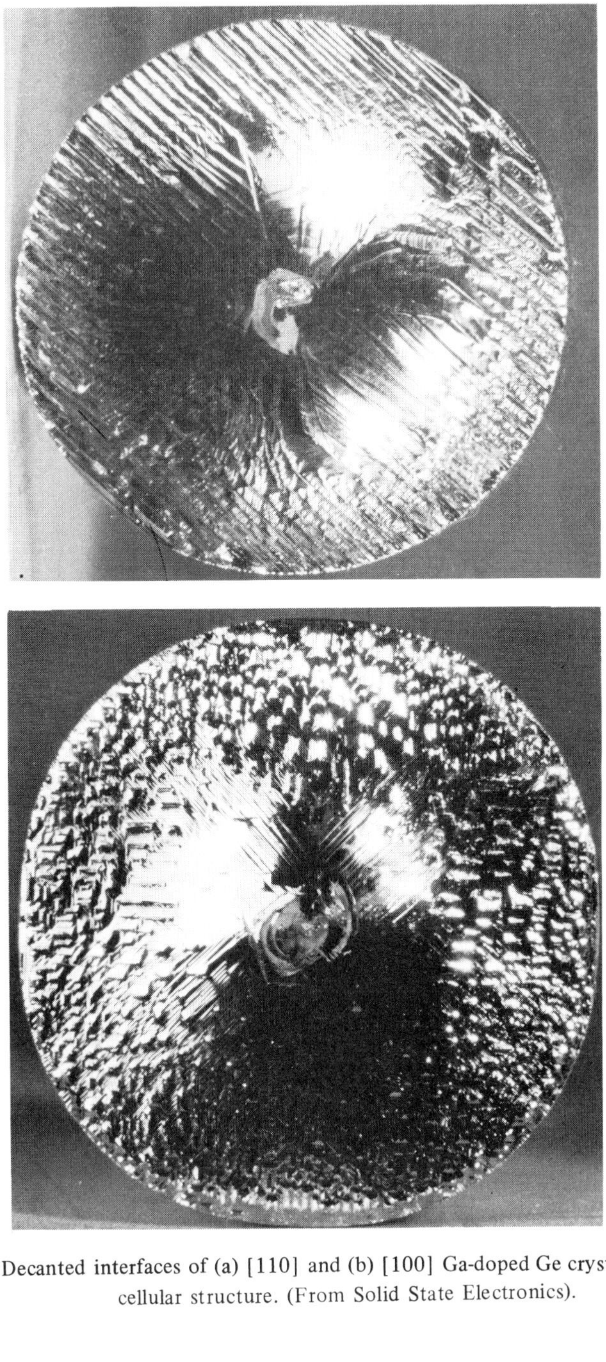

Fig. 8.21. Decanted interfaces of (a) [110] and (b) [100] Ga-doped Ge crystals showing cellular structure. (From Solid State Electronics).

$\approx 35°$ to the (110) surface. Likewise with a [100] axis crystal the interface structure is composed of a set of square-based pyramids formed by the four {111} planes inclined at $\approx 55°$ to the (100) surface. The interfaces of a [110] and [100] Ge crystals are shown in fig. 8.21. A photomicrograph of a section of a (100) surface normal to the growth axis of the crystal shown in fig. 8.21b is shown in fig. 8.22. The solute-rich cell boundaries are clearly seen as are the growth striae and the traces of the intersections of the pairs of {111} planes which make up the pyramidal interface.

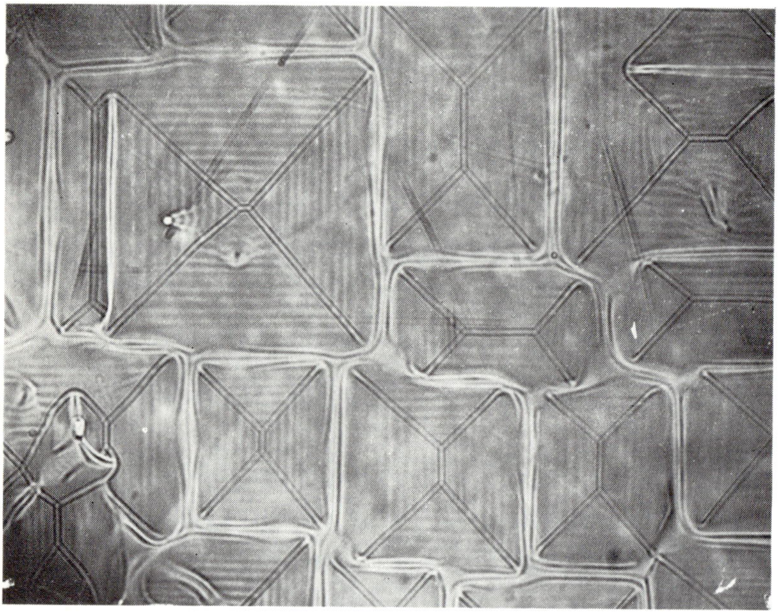

Fig. 8.22. Etched (100) section of magnified portion of crystal shown in fig. 8.21b. (from Solid State Electronics 5 (1962) 395).

As growth proceeds, the cell boundary grooves deepen [22]. If the freezing range of the alloy is large these long liquid channels become unstable and break up into a string of spherical or cylindrical droplets. These droplets dissolve material on the hot side and precipitate it on their colder side i.e. the droplets migrate slowly up the temperature gradient, leaving behind trails of solute-rich material (by the process known as temperature gradient zone melting [2]). Finally, when the crystal is cooled to room temperature, the liquid droplets shrink and ultimately precipitate a second phase when the eutectic composition is reached. If the volume change on solidification is negative, small voids are left in the crystal. Solute trails and second phase precipitates are shown in fig. 8.23 which is of a longitudinal $(1\bar{1}0)$ section of a [110] Ga doped Ge crystal.

The solute trails contain material of different lattice parameter from the crystal matrix and the resultant stress is partially relieved by dislocation generation. The crystalline perfection rapidly degenerates as massive arrays of dislocations form around the solute trails and second phase precipitates. This is illustrated in fig. 8.24 which is a longitudinal $(1\bar{1}0)$ section of a [110] Ga doped Ge crystal showing dislocation etch pits in the vicinity of solute trails.

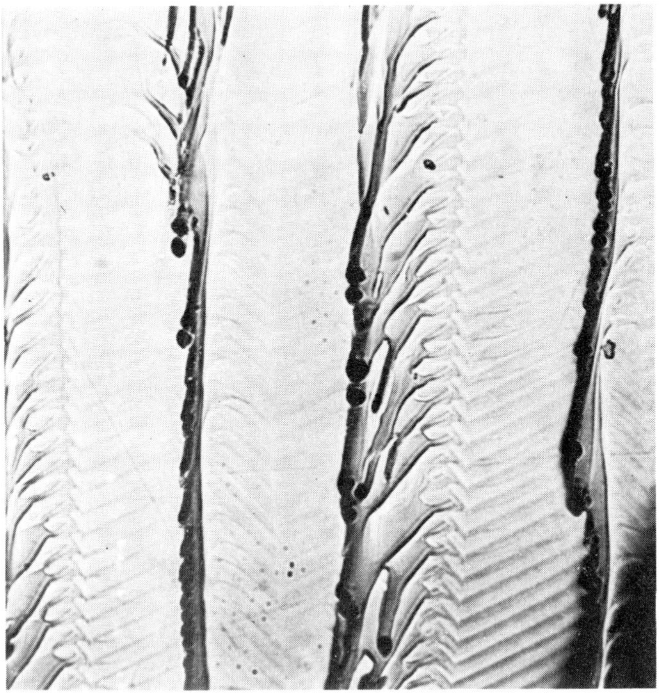

Fig. 8.23. Section as in fig. 8.20 but with solute trails developed.

Whilst the development of the cellular morphology has here been demonstrated with Ge as an example similar behaviour occurs in a wide class of materials. Metals differ in that, because they do not form facets, the cellular morphology is commonly in the form of a close-packed hexagonal array, in which crystallography does not play a dominant part [23].

Melt non-stoichiometry can cause constitutional supercooling during the growth of compounds [24]. Consider, for example the InSb phase diagram

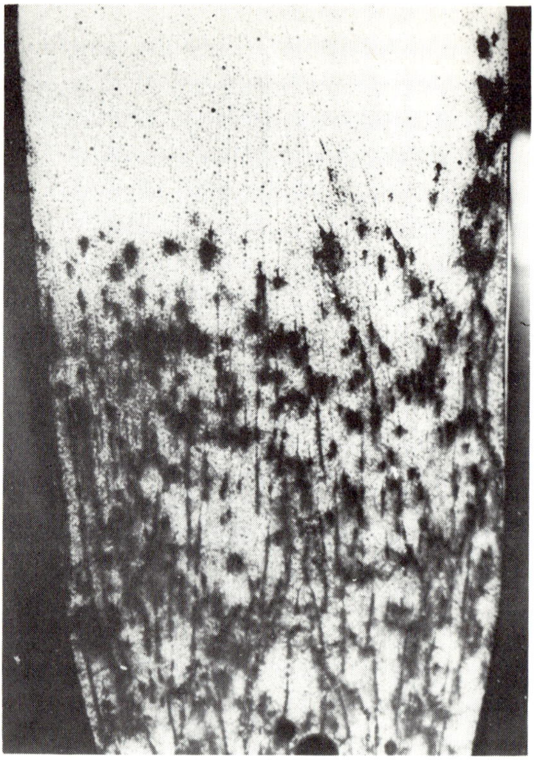

Fig. 8.24. Dislocation etch-pit arrays due to solute trails; $(1\bar{1}0)$ section of a [110] Ga-doped Ge-crystal.

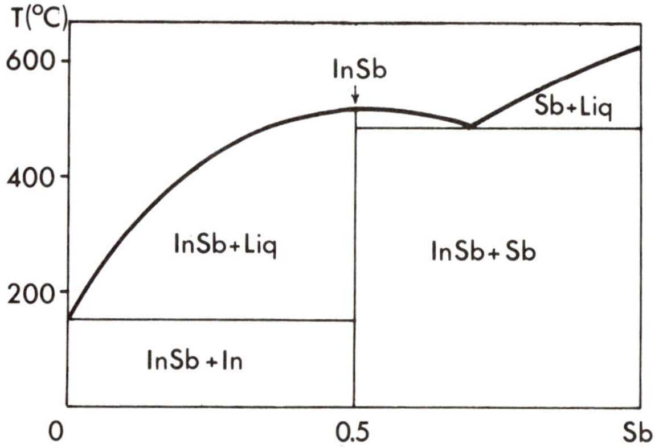

Fig. 8.25. In-Sb Phase diagram.

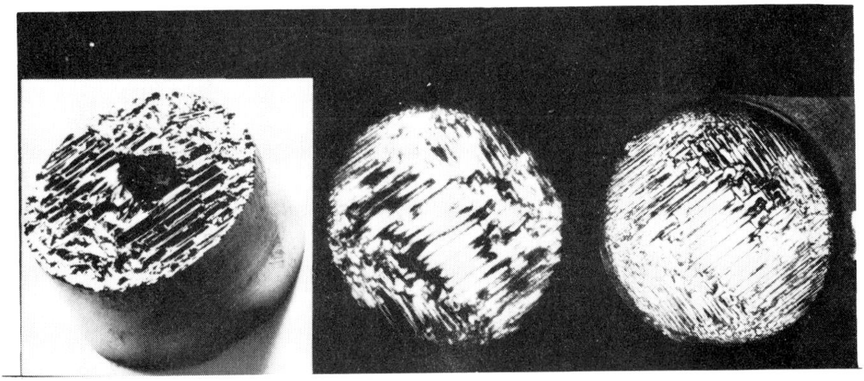

Fig. 8.26. Cellular structure in a [100] InSb crystal (a) decanted interface (b) autoradiograph of (100) section (c) photomicrograph of section (b) after etching (from Solid State Electronics 3 (1961) 142).

shown in fig. 8.25. If one solidifies an In-rich melt, InSb solid is formed and the excess In is rejected into the melt. This excess In lowers the liquidus temperature of the melt in exactly the same way as an impurity and hence can also lead to the formation of a cellular structure. This is illustrated in fig. 8.26. The left hand portion shows the interface of a [100] InSb crystal grown from an In-rich melt; the middle portion is an autoradiograph of a section cut parallel to the macroscopic plane of the interface (the crystal was doped with a trace amount of radioactive Te); the third portion is a photograph of the previous surface after etching. The radioactive Te has been preferentially incorporated on the {111} facets of the cellular interface (the facet effect).

8.5. The nature and effects of convection

We have already seen that convection governs the degree of segregation at the solid-liquid interface and mention has been made that non-steady convection produces fluctuations in the solute concentration called growth striae. We begin this section with a brief outline of the origin of natural convection. The basic phenomena are best illustrated by the *Bénard cell* – a layer of fluid contained between a pair of horizontal surfaces of infinite lateral extent [25]. If the fluid, initially at rest, is heated from below then, provided it has an expansion coefficient $\alpha > 0$, dense cold fluid will be lying on top of hot, less dense, fluid and the free energy of the system would be reduced if the fluid were overturned. To find if the system is dynamically unstable we must perform a perturbation calculation, along similar lines to that of Sekerka for interface instability, to see if there exists a marginal state. When this is done it is found that the fluid will start to convect only when the dimensionless temperature gradient, called the *Rayleigh number*, R exceeds a critical value R_c

$$R = \frac{\alpha g \beta d^4}{\nu a} > R_c .$$

R_c depends on the thermal and mechanical properties of the boundary surfaces and for rigid conducting boundaries $R_c \approx 1707$; a is the thermal diffusivity, ν the kinematic viscosity, d the depth of the cell, β the adverse temperature gradient and g the acceleration due to gravity.

There are two distinct types of convective instability. In linear perturbation theory the time-dependence of a perturbance is taken to be of the form e^{pt}, where p is complex. The marginal state is characterised by $\mathrm{Re}\, p = 0$. The two types of instability are characterised by $\mathrm{Im}\, p = 0$ and $\mathrm{Im}\, p \neq 0$. The former corresponds to a steady, non-periodic motion of the fluid called stationary convection; the latter to a periodic oscillation of the fluid called *overstability*. It can be shown that the onset of convection in a Bénard cell must be of the stationary form [25] and in fact the layer of fluid tessellates into an array of hexagonal cells with fluid rising up the centre and flowing down the sides of each cell.

The other geometry of significance is the vertical slot – that is a layer of fluid confined between parallel vertical surfaces across which is maintained a temperature difference. In this case it can be shown that there is no critical temperature gradient – a steady circulation of the fluid occurs for any finite temperature difference [26]. In fact in any system where there is a horizontal component of temperature gradient convection will occur.

In melt-growth systems this steady convection manifests itself roughly as follows: – In a puller crucible where heating is supplied principally to the side walls of the crucible with some bottom heating a simple flow occurs with melt rising up near the walls, flowing in and descending at the centre. Some radial sub-division can occur giving rise to the familar spoke pattern seen on oxide melts [27]. In horizontal normal-freeze and zone-melting systems a simple circulation with fluid rising at hot surfaces and descending at cold ones occurs. This steady convection is very desirable and produces efficient mixing in the melt. At high Rayleigh number, the rate of heat or solute transport is proportional to R^n where $n \approx 0.5$ to 0.25 depending on the geometry.

There are two basic ways in which non-steady components to the motion can occur. The first is overstability which, in general, can be produced by providing a second, stabilising, body force to the fluid. Examples of such forces are (a) rotation, (b) solute gradients and (c) surface tension gradients. The second way in which non-steady convection can occur is as superimposed motion upon a pre-existing steady convection [28]. At very high Rayleigh number where the number of modes is very large this is properly called turbulence but the transition from steady flow at low R to fully developed turbulence is very complex and typically takes place over several orders of magnitude in R.

We next briefly consider overstability due to rotation and solute gradients in the Bénard geometry. Rotation is characterised by the non-dimensional *Taylor number*

$$T = \frac{4\,\Omega^2}{\nu^2} d^4$$

where Ω is the angular velocity of the fluid. Stability analysis shows that, if the *Prandtl number* $\mathfrak{p} = \nu/a < \mathfrak{p}_c \approx 0.67$ then overstability occurs, with a frequency which depends on T but is typically of the order of the rotation frequency. The application of rotation raises the value of R_c, i.e. it suppresses the onset of convection. In the limit $\mathfrak{p}^2 T \to \infty$, $R_c \propto T^{2/3}$. This phenomenon is of importance with crucible rotation [29], because semiconductor, metal and oxide melts all tend to have low Prandtl number $< \mathfrak{p}_c$.

The presence of a solute gradient which produces a contribution to the density gradient in the fluid which is stabilising (i.e. opposes that due to the temperature gradient) can also produce overstability [30]. The concentration gradients required are very small. As the amplitude of the oscillations builds up it tends to mix the solute thereby destroying the stabilising component of density gradient. Under these conditions the overstable oscillations are replaced by a steady finite-amplitude mode. Clearly the solute gradient ahead of the solid-liquid interface must have a pronouced effect on the natural convection but this has not yet been studied.

Coming now to oscillations superimposed on top of steady convection we must first observe that there is very little known theoretically about these. Such oscillations are very readily demonstrated in low Prandtl number fluids such as the oxide and metal-like liquids commonly encountered in melt-growth [31]. Fig. 8.27 shows regular oscillations obtained from a rectangular volume of liquid gallium heated from the side. These oscillations set in at a critical axial temperature gradient which is a function of the fluid dimensions. The period of the oscillations is proportional to the boat length. As the temperature gradient is increased the amplitude of the oscillations increases,

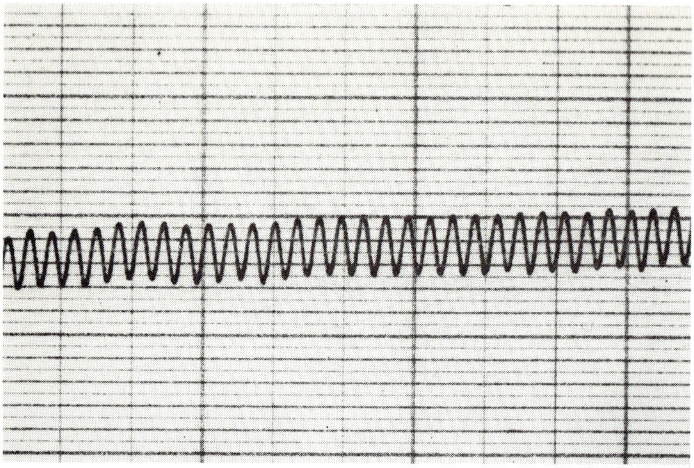

Fig. 8.27. Temperature oscillations in a Ga melt.

harmonics appear and ultimately the signal appears noisy. This latter state is characteristic of real crystal growth melts. The measured amplitudes of the fluctuations in crystal growth melts vary from a few degrees for low melting point materials like InSb to several tens of degrees for high melting point oxides.

In electrically conducting fluids a *magnetic field* can be used to damp out the fluctuations [32, 33]. Fig. 8.28 shows a zone-melting apparatus adapted so as to apply a transverse magnetic field to the molten zone. Fig. 8.29 shows the temperature recorded in the melt (a) with zero magnetic field; (b) with full field of ≈ 3 koersted and (c) with intermediate field (H_c) chosen to give near threshold conditions for oscillation. Zone melting of Te doped InSb under condition (a) resulted in the usual striations: under condition (b) no striations were visible; under condition (c) fine regularly-spaced striations occurred whose spacing was equal to the average amount of crystal grown per cycle. The use of magnetic fields in pulling is less effective because (a) the frequencies observed are lower and require large fields to damp them and (b) it is difficult to generate a large axial field in a puller geometry.

In oxide melts minimisation of temperature gradients and the use of shallow melts (the Rayleigh number is proportional to the fourth power of the depth at constant β) serves to reduce the fluctuations.

We finish with a brief consideration of how the interface responds to thermal fluctuations in the melt. The dependence on frequency is rather

Fig. 8.28. Zone-melting apparatus showing the arrangement of the electromagnet. (from Crystal Growth, ed. by H.S. Peiser (Pergamon Press, New York, 1967)).

complex but a few simple observations can be made [34, 35]. Very high frequencies will have little effect because (a) they are strongly attenuated as they diffuse through the boundary layer to the interface and (b) the resulting striae will be so close together that they will be annealed out by solid-state diffusion. For dilute solute concentrations for which the solute does not measurably lower the freezing point the interface will respond up to frequencies $\omega_t \sim 2a/\delta_t^2$ where δ_t is the thermal boundary layer thickness. For large solute concentrations, attempts to accelerate the interface by lowering the temperature are counteracted by the increased build up of solute which lowers the liquidus temperature. The interface responds much more sluggishly in this case with a cut off frequency $\omega_c \sim 2D/\delta_m^2$ where D is the solute diffusion coefficient and δ_m the mass-flow boundary layer thickness.

Fluctuations decrease the effective segregation and hence reduce the efficiency of the zone-refining process. To demonstrate this, consider the following: – For $k < 1$ increasing the velocity increases the rate of rejection and enhances the interface solute concentration, and hence the solid solute concentration. Consider a velocity of the form $v = v_0 + v_1 \sin \omega t$. During the half cycle for which the velocity is above its mean value v_0 the concentration is also above its mean value. However more crystal is grown during this half

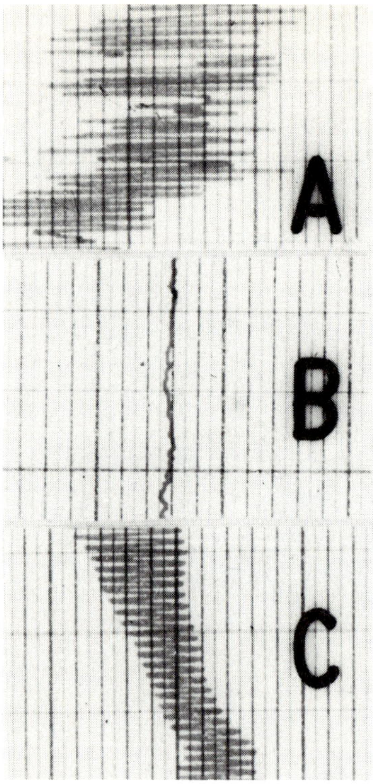

Fig. 8.29. Temperature at a point in the molten zone. A magnetic field $H = 0$; B, $H > H_c$; C, $H = H_c$. (from: *Crystal Growth*, ed. by H.S. Peiser (Pergamon Press, New York, 1967)).

cycle (because the velocity is above the mean) and hence the mean concentration incorporated in the solid is greater than k times the mean concentration in the liquid at the interface. Thus k_{eff} is closer to unity; this can also be verified for a solute with $k > 1$.

The role of convection in crystal growth is currently attracting interest. There would appear to be much to be understood.

Acknowledgement

The author wishes to acknowledge that he is indebted to his colleagues in the Electronic Materials Division of RRE for extensive contributions, arising from many years of collaboration, to the ideas and information presented in this chapter.

References

[1] J.C. Brice, The Growth of Crystals from the Melt (North-Holland, 1965).
[2] W.G. Pfann, Zone Melting (J. Wiley, New York, 1966).
[3] J.J. Gilman, The Art and Science of Growing Crystals (J. Wiley, New York, 1963).
[4] J.R. Carruthers and K. Nassau, J. Appl. Phys. 39 (1968) 5205.
[5] B. Cockayne, J. Crystal Growth 3/4 (1968) 60.
[6] J.B. Mullin, B.W. Straughan and W.S. Brickell, J. Phys. Chem. Solids 26 (1965) 782.
[7] S.J. Bass and P.E. Oliver, J. Crystal Growth 3/4 (1968) 286.
[8] J.B. Mullin, R.J. Heritage, C.H. Holliday and B.W. Straughan, J. Crystal Growth 3/4 (1968) 281.
[9] D.B. Gasson and B. Cockayne, J. Mat. Sci. 5 (1970) 100.
[10] D.B. Gasson, J. Sci. Inst. 42 (1965) 114.
[11] D.A. Jones, R.V. Jones and R.W.H. Stevenson, in: Crystal Growth ed. by H.S. Peiser, (Pergamon Press, 1967) p. 57.
[12] J.A. Burton, R.C. Prim and W.P. Slichter, J. Chem. Phys. 21 (1953) 1987.
[13] J.D. Hunt and K.A. Jackson, Trans. AIME 236 (1966) 843.
[14] J.D. Hunt and D.T.J. Hurle, Trans. AIME 242 (1968) 1043.
[15] K.A. Jackson and J.D. Hunt, Trans. AIME 236 (1966) 1129.
[16] F.R. Mollard and M.C. Flemings, Trans AIME 239 (1967) 1526.
[17] F.S. Galasso, J. Metals 19 (1967) 17.
[18] F.S. Galasso, F.C. Douglas and J.A. Batt, J. Metals 22 (1970) 40.
[19] K.F. Hulme and J.B. Mullin, Phil. Mag. 4 (1959) 1286.
[20] J.A.M. Dikhoff, Solid State Electron. 1 (1960) 202.
[21] J.W. Rutter and B. Chalmers, Canad. J. Phys. 31 (1953) 15.
[22] W. Bardsley, J.S. Boulton and D.T.J. Hurle, Solid State Electron. 5 (1962) 396.
[23] D.T.J. Hurle, Mechanisms of Growth of Metal Single Crystals from the Melt, Prog. Mat. Sci. 10 (1962) 79.
[24] D.T.J. Hurle, O. Jones and J.B. Mullin, Solid State Electron. 3 (1961) 317.
[25] S. Chandresekhar, Hydrodynamic and Hydromagnetic Stability (Oxford University Press, 1961).
[26] G.K. Batchelor, Quart. J. Applied Math. 12 (1954) 209.
[27] B. Cockayne and M.P. Gates, J. Mat. Sci. 2 (1967) 118.
[28] E. Jakeman and D.T.J. Hurle, Reviews of Physics in Technology. 3 (1972) 3.
[29] J.R. Carruthers, J. Electrochem. Soc. 114 (1967) 1077.
[30] T.G.L. Shirtcliffe, J. Fluid Mech. 35 (1969) 677.
[31] D.T.J. Hurle, Phil. Mag. 13 (1966) 305.
[32] H.P. Utech and M.C. Flemings, J. Appl. Phys. 37 (1966) 2021.
[33] H.A. Chedzey and D.T.J. Hurle, Nature 210 (1966) 933.
[34] D.T.J. Hurle, E. Jakeman and E.R. Pike, J. Crystal Growth 3/4 (1968) 633.
[35] D.T.J. Hurle and E.J. Jakeman, J. Crystal Growth 5 (1969) 227.
[36] A.J.R. de Kock, Appl. Phys. Letters 16 (1970) 100.

P. Hartman, ed., Crystal growth: an introduction. © (1973) North-Holland Publ. Co.

INDUSTRIAL MASS CRYSTALLIZATION

M.A. van DAMME – van WEELE

9.1. Introduction

Several chapters in this book are primarily concerned with fundamental theoretical aspects of the various crystallization phenomena. This follows the general trend apparent in world conferences, the literature and periodical publications. However, the majority of people, concerned in their daily life with mass crystallization aspects, are not active in the fundamental theoretical field. Therefore it is appreciated that a place has been given among the other authors to a representative of the experts in industrial mass crystallization.

It is not the intention to give a detailed review of the engineering approach to industrial mass crystallization. This can be studied from the literature to which an introduction is given in the next section. The rest of the chapter has four separate aspects: Firstly, the principles of the evaporation crystallizer are described and the very limited extent to which crystallization theories may be applied is indicated. Secondly, the influence of cost trends on product diversification is briefly indicated using sodium chloride as an example. Thirdly, some problems of "caking" of salt crystals are mentioned. Finally, the development and scope of industrial crystallization is considered and some emphasis is given to financial and general technological and engineering aspects.

It is the author's hope that this chapter will help to arouse the interest of at least some of the people who are active in the more fundamental theoretical aspects of crystallization. This would help to diminish the gap between the rather sophisticated and strongly developed crystallization theory and the vastly expanding, but still predominantly empirically-based mass crystallization practice.

9.2. Introduction to the literature on industrial crystallization

In this introduction five different sources of information will be discussed·

I General engineering series, discussing crystallization aspects

II Monographs on crystallization with emphasis on industrial aspects

III Symposium proceedings

IV General scientific and engineering magazines, regularly containing information on industrial crystallization

V Review publications and coordinating institutes in the industrial crystallization field.

Monographs on crystallization practice began to appear frequently about 1960. These books mostly begin with a broad introduction on general crystallization theory. In this literature review only those chapters will be mentioned, which really discuss the industrial aspects of crystallization. The writers sometimes use different classifications for the various aspects of crystallization. Therefore the classification used in this review does not always include the titles of the chapters and subchapters in the books mentioned.

I General Chemical Engineering Series: The most important volumes containing chapters which provide an introduction to the field are given in ref. [1, 2].

II Monographs on crystallization. These are to be found in ref. [3–7]. Chapters in these monographs which give emphasis to industrial and crystallization aspects are:

Bamforth [3]

Complete (360 pages). Giving: Flowsheets, choice and descriptions of crystallizer equipment, performance data, etc.

Van Hook [4]

1) Data on nucleation and growth of various substances: Chapter 4, pages 154–191.
2) Description Industrial Crystallization Equipment: Pages 209–223.
3) Process descriptions and general information about some industrial products (sugar, salt, heavy chemicals, and fertilizers): Pages 238–268.

Matz [5]

Chapter IV Technik der Kristallisation: Pages 174–363.

1) Thermal aspects: Pages 174–190.
2) Particle size distribution: Pages 190–201.
3) Equipment, choice and description, and performance data: Pages 215–363.

Mullin 6)

Chapter 3. Physical and thermal properties: Pages 44–64.

Chapter 5. Mass transfer: Pages 122–128.
Chapter 7. Methods of industrial crystallization: Pages 158–186.
Chapter 8. Equipment; choice and description: Pages 187–211.
Appendices: Data on solubility, thermal properties.
Strickland–Constable [7]
Chapter VII. Various topics: Industrial bulk crystallization, equipment, choice and description: Pages 294–310.

III There were two recent symposia which included aspects of industrial and engineering practice [8,9].

IV Magazines which contain information, primarily engineering, about industrial crystallization.

German language

1) Fortschritte der Verfahrenstechnik. About 20% of the publications in this journal are focussed on industrial crystallization aspects.

Czechoslovakian language

An increasing number of articles about industrial crystallization, based on an engineering science approach, have recently been published by Nyvlt in:

1) Chemisky Prum
2) Colln. Czech. chem. Commun.

English language

1) Chemical Engineering Progress
2) Industrial Engineering Chemistry
3) Am. Institute of Chemical Engineers Journal
4) Brit. Chem. Eng.
5) Trans. Inst. Chem. Eng.
6) Chem. Eng.

V Review Publications and Coordinating Activities in Industrial Crystallization

1) V.D.I. Zeitschrift
The yearly review on crystallization articles, with about 130–170 references; contains an increasing amount of industrial crystallization titles, from 4% in 1962 to 20% in 1968.
2) Information Services, e.g.
Dechema (Germany)
Chemical Abstracts Condensates (CAC)
3) "Key words" in crystallization
To be published by V.D.I., as a chapter of a general Key word publication, (in German, English, French and Russian).
4) Working Party on Crystallization of the European Federation of Chemical Engineering. This Working Party tries to coordinate and stimulate

activities of national group members working on industrial crystallization.

Those who are interested in industrial crystallization can find their way easily in the abundance of literature, once having absorbed from the quoted literature the systematics and the names of the most prominent writers in the field of industrial crystallization.

9.3. Evaporating Crystallizer

The (continuous) crystallization of sodium chloride will be chosen as an example. Fig. 9.1 shows the very simplified crystallization rules which will be used. The linear growth rate at the chosen temperature is, according to the study by Rumford and Bain [11], more or less diffusion-dependent. The equation $dN/dt = k\sigma^3$ is chosen, very arbitrarily, as the nucleation equation

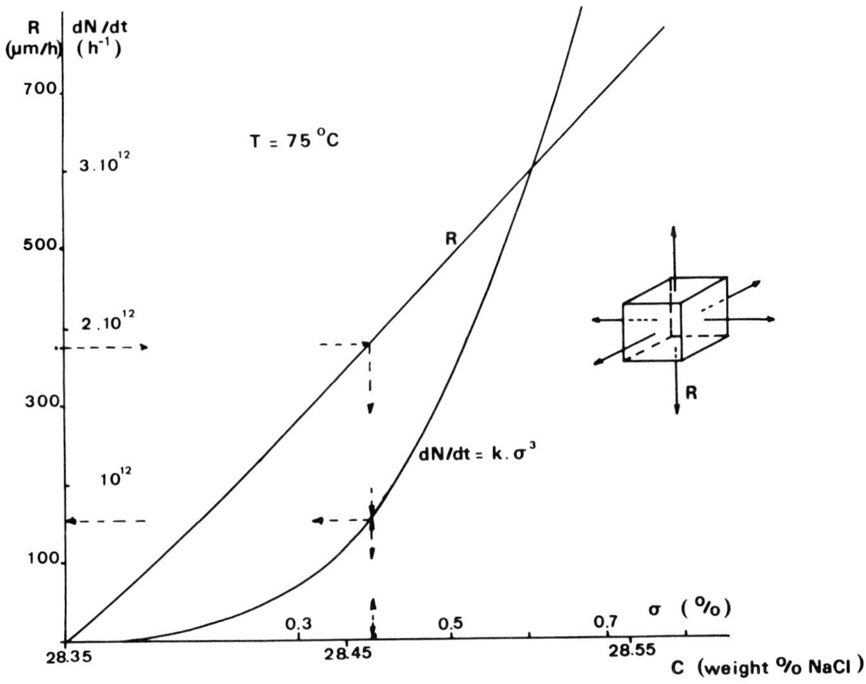

Fig. 9.1 Determining factors in NaCl crystallization.

where σ = relative supersaturation. The figures should give, for a certain residence time of the crystals in an evaporator, the corresponding relation between crystal size, supersaturation and number of nuclei. Growth rate as a function of supersaturation can be, and is, measured using laboratory scale apparatus. Mean crystal size and mean residence time in an industrial crystallizer are already more complex figures. The literature quoted in ch. 9 § 2 gives enough information about complexity and measurement techniques.

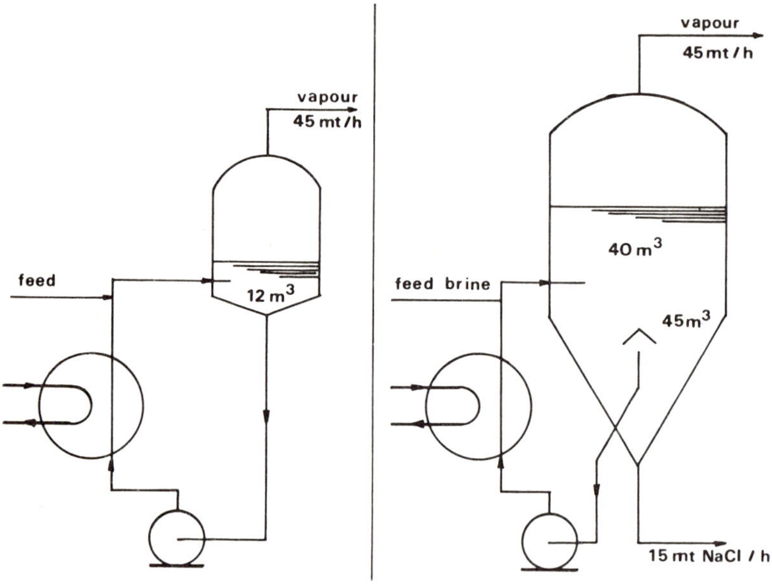

Fig. 9.2 Evaporator (left) and evaporating (NaCl) crystallizer (right).

Fig. 9.2 gives information about the essential parts of an industrial evaporating crystallizer. The "black box" consists of an evaporator, a pump and a heat exchanger. As long as evaporation is the only objective, the surface area depends on the evaporation velocity of the liquid and problems related to mist entrainment. In the second part of the same figure the consequences for the evaporating body are shown when the additional objectives of crystallization and good crystal formation are added to the evaporation task. Volume, as well as surface area of the vessel have increased.

Fig. 9.3 shows what is happening with the salt concentration in a solution "particle", circulating through the crystallizer. The continuous lines are in fact rounded off to the dashed lines – in the heating period because of dis-solution of cocirculating crystals and in the evaporating upper part of the evaporator because crystallization and probably some nucleation takes place.

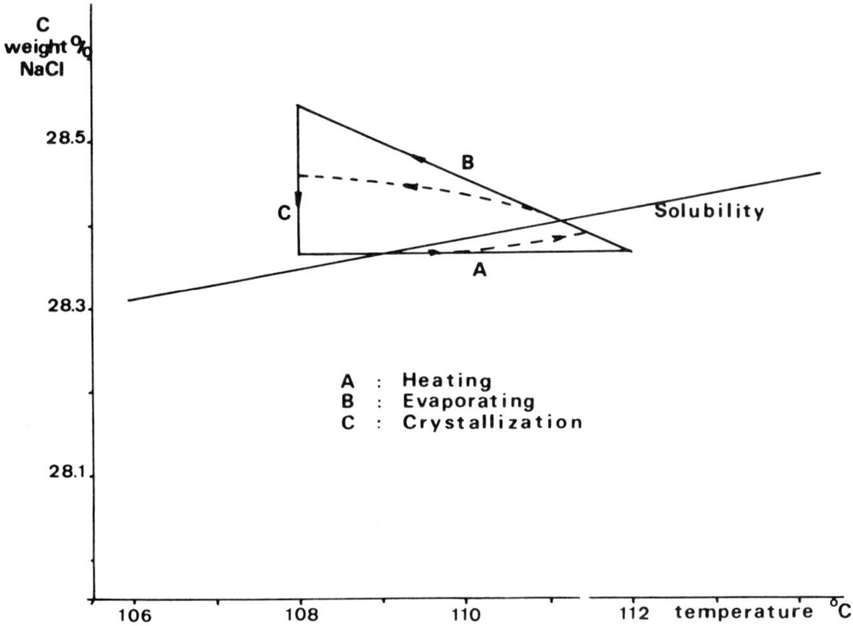

Fig. 9.3 Crystallization cycle in an evaporating crystallizer.

Fig. 9.4 shows the accident that would have taken place if the evaporating vessel had not been enlarged into a crystallizing one. Abundant nucleation would have occurred, destroying the performance of the design with respect to crystal size. Moreover, a tremendous crust formation around the walls of the vessel, especially in the evaporating zone would have occurred, followed by mechanical problems in the pump if large crustations had happened to fall down into the pump suction line. Even if the crystal vessel had been enlarged but the brine inlet had not been lowered in order to bring down the super-saturation by severe mixing of the heated brine coming from the heat ex-

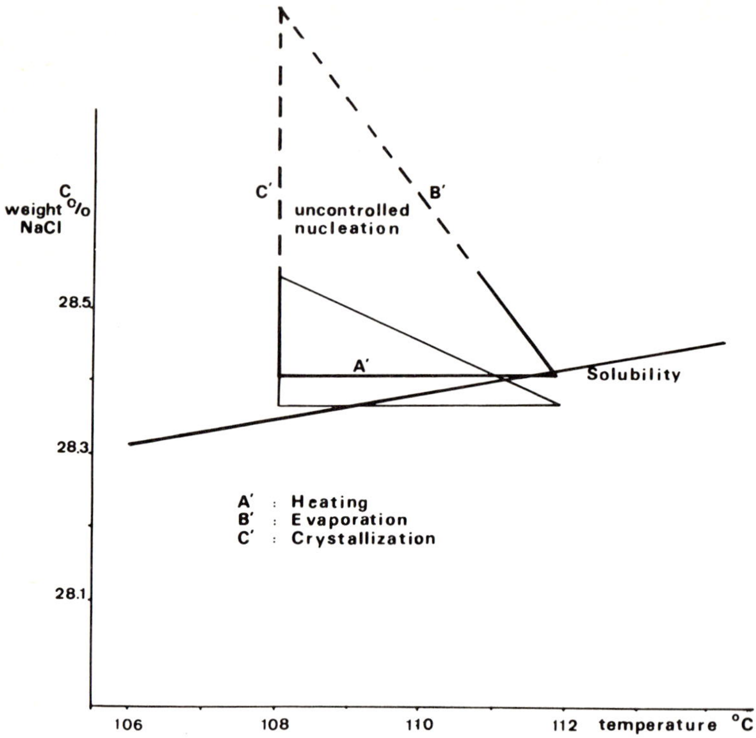

Fig. 9.4 Crystallization cycle in a badly designed evaporating NaCl crystallizer.

changer, with a larger amount of crystal slurry in the vessel, the problems described would still have occurred. Now nothing is more wrong than to conclude that all these failures do not occur (or would not occur) because the chemical engineers can calculate how to design an apparatus, based on the crystallization parameters of the product in question.

Up till now designs have been based predominantly on empirical relations, built up by long experience and small-step changes. Only recently has more information become available about the hydrodynamics that control the mixing conditions, and, through this, the supersaturation conditions. Appropriate information about the influence of a crystal magma on hydrodynamic flow and of crystal interaction with respect to crystal growth phenomena are still lacking. Even the measurement of supersaturation in a working industrial crystallizer is still a problem, especially in the several important areas, which have just been described. Another difficult monitoring problem is that of the magma density. It is the very presence of a crystal slurry that makes all checking of calculations of nucleation velocity in an industrial crystallizer a big problem. Attrition of crystals, either by collision with each other or by pump blades is probably a more important source of "nuclei" than heterogeneous nucleation. The industrial crystallization engineer hardly ever bothers himself about homogeneous nucleation. And finally the "dissolving" power of an evaporating crystallizer for small crystals, either in the heat exchanger or by entrainment in the growing crystals, makes calculation of real nucleation rates impossible. Only approximations, or trends, can be given. This can be important for the chemical engineer in design work and upscaling operations.

9.4. Product Diversification and Cost Trends

Fig. 9.5 gives pictures of various product qualities of evaporated sodium chloride. Granular salt and dendritic salt are made in crystallizers, specially designed for that purpose. The dendritic salt is produced by adding traces of a ferrocyanide salt to the evaporating slurry. This ferrocyanide is a well-known crystallization poison for NaCl. Under very special conditions this results in the formation of dendrites. There is still much work to be done in this

Table 9.1
Important crystallization cost items

	heat exchanger	evaporating vessel	separation and drying	capacity (m.t./year)	relative cost increases
normal evaporated	0	0	0	600 000	0
normal evaporated (forced circulation)	–	–	0	600 000	–
dendritic salt	+	+	+++	< 100 000	++
"granular" salt	0	++++	–	< 100 000	+++

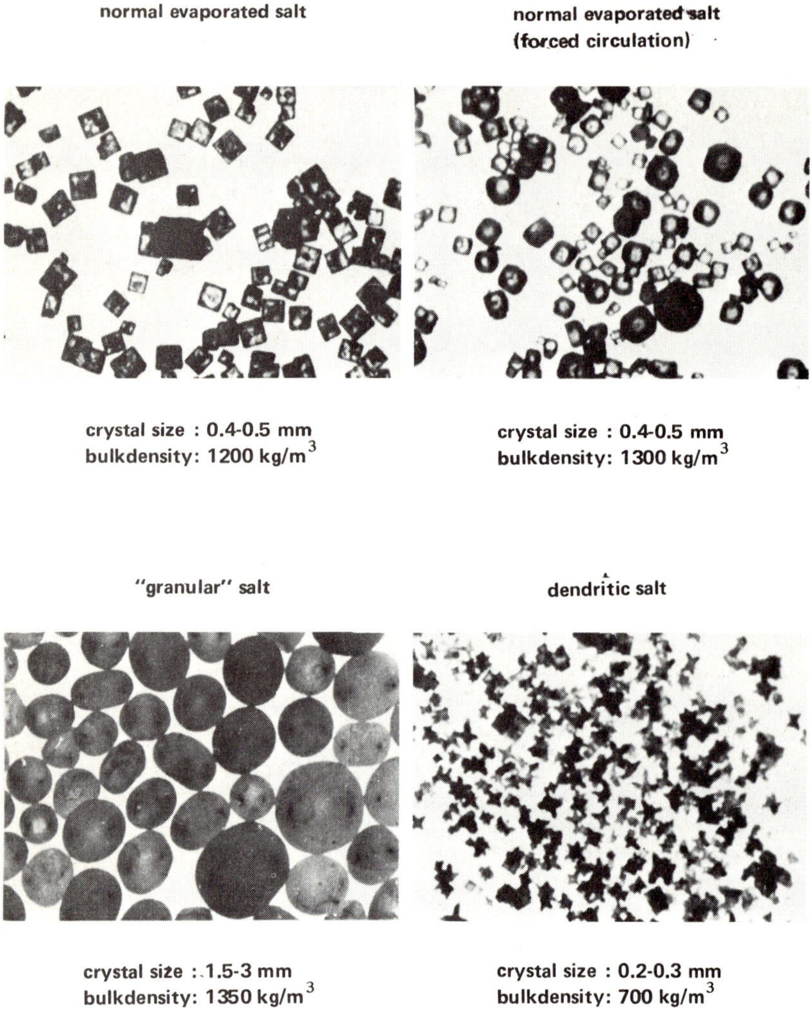

Fig. 9.5 Various types of evaporated salt.

particular, widely explored, field of dendritic sodium chloride production. The salt specifications are needed for special applications in the market. The two types of evaporated salt, the cubic and the more rounded shape are formed in large-type crystallizers. The cubic type mostly in evaporators with internal heat exchangers and natural circulation. The rounded type in evaporators with external heat exchangers with largely forced circulation, developed because of economic reasons with respect to production costs. Table 9.1 gives the influence of the most important cost items. It is the task of the industrial crystallization expert, together with the chemical engineer, to determine how far cost increases, caused by special product wishes, can be lowered through clever research and design work.

9.5. Caking and Anti-Caking

Several scientists in different European countries have obtained, since 1964, doctor's degree based on studies on the influence of ferrocyanide-ions on the crystallization and dissolution behaviour of sodium chloride and potassium chloride. This interesting phenomenon had already been detected in the nineteen-fifties as a result of empirical research on anticaking agents for NaCl. The history of this subject can be traced from the literature references given in ref. [12]. A short publication on the theoretical background to the anti-caking effect appears in ref. [13]. Many crystallization retarding, and sometimes also accelerating additives for different crystalline materials have been studied since then. As far as the author knows none of these combinations is so powerful and so specific in its action as the ferrocyanide ion.

9.6. Development and Scope of Industrial Crystallization

For a person, working in industry, it is interesting to know if the field in which he (or she) has been specialising is likely to become more important in the future. There are of course many ways of defining importance. One of the parameters most easily measured is the predicted output of the industrial activity in the future. In ref. [10] a "guestimate" is given for the growth profile, in the U.S.A., of various branches of industry up till 1982. From this "guestimate" we can extract five important growth areas: chemicals, plastic materials, drugs and medicines, rubber products and man-made fibres. At the same time we see some areas that will lag in growth, among which are foods and beverages, stone, clay and glass, and chemically processed products (e.g. leather).

It should be stimulating for the reader to see that in the areas where crystallization theory is only just beginning (plastic materials and man-made fibres) an excessive growth is foreseen. Also in the production of drugs and medicines crystallization phenomena play an important role, as will be shown later on in the chapter. This area is rather unexplored by professional crystallization experts. The production of chemicals, which from the beginning of crystallization science belonged to the most explored area, will continue to grow vastly. Slow growth is expected for stone, glass, cement, lime, known for their solid or molten state reaction-based crystallization phenomena, which, by and large are too complex for interpretation within the general crystallization theories.

A more detailed picture of the business activity in chemical products is given in table 9.2. This table gives the growth and ranking of inorganic products, organic basic products and organic end-products for American production, measured in thousands of tons/year. It follows from table 9.2 that the well known inorganic bulk products stay in the ranks without too much change and with a rate of growth between 5–10%. In this table an asterisk (*) indicates that the end-product (or an important step in the production process) is performed by crystallization. About 50% of the inorganic products, with a total value of approximately $2,000 million (in the year 1965) appear to be produced with one or more crystallization steps.

From the same table it follows that crystallization in the production of organic basic products is less important. For only one third of the products is a crystallization operation used and the total dollar value (1965) in the U.S.A. for these products amounted to about $ 700 million. However, the growth of these products is generally much larger than in the inorganic section. This can give opportunities, both for theoretically oriented crystallization studies as well as for engineering oriented development.

Finally, table 9.2 shows many examples of plastics and man-made fibres, in the column headed under organic end-products. Here is beginning, as was mentioned before, a relatively new, interesting field of industrial crystallization. One of the fastest growing organic bulk end-products is urea, made by a variety of processes, all of them containing one or more crystallization steps.

Of course the well-known organic product sugar is one of the largest of the organic bulk products with an American production of about 10 million tons a year, being 10% of the world sugar production. It is not shown in table 9.2 because it is not considered a chemical. However, sugar production is one of the most important processes which, as a unit operation, is based on crystallization (see e.g. Van Hook [4], page 238).

Medicines are not shown either in the list of organic end-products. However, ranked by dollar value, some of them would surely appear in the top ranks, as for instance in the case of penicillin (sales of $30 million in 1965 in the U.S.A.). An interesting example of "industrial" production of a medicine is the synthesis of β $^{1-24}$Corticotropin. This is produced in 24 steps, about 12 of them accompanied by crystallization. The world consumption of this product is less than 2 kg a year. The total loss in yield, summed over all the steps, amounts to about 95%. World sales are about $3 million. It is possible that a small increase in yield, e.g. by better understanding of a recrystallization step, could improve the financial results. In that case, what is first needed is that the laboratory approach to the determination of the crystallization parameters be downscaled to a micro-scale: surely an activity to which crystallization engineers are not accustomed. Academic workers in the crystallization field could find here a stimulating area of activity.

Returning to the bulk industrial chemicals we have to add to the list of inorganic products sodium chloride, one of the most studied products, in both the scientific as well as in the engineering field. The world production of NaCl amounts about 235 million tons/year. However, only about 10% of the total production is made by evaporation in industrial crystallizers. The majority of salt is produced by mining or by solar evaporation. This last process of course also requires physical-chemical know-how in order to produce a high quality product. Whereas the product salt is very cheap and because of the competition from rock salt and solar salt, the engineering effort to diminish the costs of NaCl-crystallization by industrial evaporation has been quite considerable. One of the most cited references in crystallization engineering is the work of Rumford and Bain [11]. These workers carried out "fundamental" investigations into the controlled crystallization of sodium chloride both on laboratory and pilot plant scale. This reference is still very valuable for workers who have to assist in the commercial development of a crystallization process.

This chapter has been restricted to chemical products. Of course, in the area of physics and electronics there are also very interesting industrial crystallization processes. Chs. 5–8 give examples of industrial application in this area.

Finally, for some information on the capital costs involved in the crystallization aspects of some chemical products, the reader is referred to the following literature:

Chemical Engineering, June 20, 1966, pages 246–248, is focussed on crystallization equipment and gives some general information about the cost versus capacity, cost versus crystal diameter and cost versus crystallization temperature relations.

Table 9.2
List of chemical products arranged by size of the american production (in 1000 metr. tons)

Organic basic products				Organic end-products				Inorganic products			
	1969	1965	1959		1969	1965	1959		1969	1965	1959
Ethylene	7 030	4 200	2 200	Polyethylene	2 470	1 380	540	Sulfuric acid	25 860	22 310	15 850
Benzene	3 800	2 750	1 130	Urea *	2 460	1 170	570	Ammonia	11 700	7 840	4 070
Propylene	3 500	2 200	1 110	SBR-rubber	1 400	1 030	1 130	Oxygen	11 680	8 540	2 940
Toluene	2 620	1 790	920	PVC and				Caustic soda	8 660	6 120	4 210
Dichloroethane	2 180	1 330	520	copolymers	1 380	830	410	Sulfur	8 640	7 390	6 240
Styrene	2 080	1 300	710	Polystyrene	970	620	280	Chlorine	8 480	5 830	3 860
Formal-				Ethylene glycol	930	820	550	Soda (ash) *	6 300	5 560	2 210
dehyde (37%)	1 900	1 410	790	Nylon fibres	620	430	180	Nitric acid	5 630	4 400	2 770
Methanol	1 880	1 300	800	Polyester fibres	590	170	35	Aluminum			
Ethylbenzene	1 830	1 430	780	Polypropylene	490	140	10	oxide *	5 270	5 040	3 470
Vinylchloride	1 700	910	440	Rayon	490	490	390	Ammonium			
Xylenes	1 470	1 090	800	Phenol-resins	440	420	280	nitrate *	5 030	4 160	2 570
Ethylene oxide	1 450	990	630	Cellulose				Potash *	4 430	4 860	3 630
Butenes	1 390	1 280	1 410	acetate	370	300	240	Phosphoric			
Ethanol	1 000	1 010	720	Cellulose acetate-				acid *	4 430	4 850	4 670
Isopropanol	940	700	510	fibres	300	270	170	Nitrogen	4 070	2 520	490
Acetic				Polyester resins	300	180	80	Normal super-			
anhydride	790	690	500	Polyurethane				phosphates *	4 050	4 950	5 710
Acetic acid	790	610	300	foam	290	180	30	Ammonium			
Phenol *	770	560	310	Alkyd resins	290	290	250	sulphate *	2 340	2 420	1 540
Adipic acid *	750	390	170	Perchloro-				Ammonium			
Acetaldehyde	720	560	450	ethylene	280	190	90	phosphates *	1 860	3 270	650
Acetone	680	510	330	Tel/TML	270	310	220	Hydrochloric			
Cumene	610	300	100	Polybutadiene	264	164	10	acid	1 670	1 230	860

Table 9.2 (cont'd)

Organic basic products	1969	1965	1959
Dimethyl-terephthalate *	590	250	50
Acrylonitrile	520	350	110
Propylene oxide	520	270	130
Acetylene	480	520	320
Tetrachloro-methane	400	270	170
Naphthalene *	380	370	190
Carbon disulfide	360	340	260
Phthalic anhydride *	350	280	160
Dodecyl benzene-sulfonate *	260	210	200
Caprolactam *	150	130	20

Organic end products	1969	1965	1959
Trichloroethylene	260	200	160
Acryl fibres	240	170	60
Methylethyl-ketone	200	140	90
Polymethyl methacrylates	200	110	50
Styrene-buta-diene resins	180	110	100
Polyvinyl-acetate resins	180	140	60
Glycerol	170	160	120
Cellophane	160	180	200
Propylene glycol	160	100	70
Methylene chloride	160	100	50
Coumarone indene resins	150	150	140
Polychloroprene	150	140	130
Freon 12	150	120	70
Methyl-isobutyl-ketone	80	100	60

Inorganic products	1969	1965	1959
Carbon blacks	1 330	1 060	880
Sodium sulphate *	1 320	1 270	980
Sodium tri-polyphosphate *	1 090	830	600
Aluminum sulphate *	1 050	1 000	500
Carbon dioxide	990	970	790
Calcium carbide	760	990	460
Calcium chloride *	730	650	520
Titanium dioxide *	590	520	460
Sodium silicate	590	530	460
Phosphates	570	500	340
Hydrogen	450	310	140
Dicalcium phosphate *	450	230	240
Magnesium chloride *	370	280	120
Aluminum hydroxide *	270	240	180
Potassium sulphate *	220	250	160

Chemical Engineering, June 15, 1970, pages 140–156, gives capital and operation costs for 54 chemical processes, among which are several important crystallization processes like ammonium sulphate, phosphoric acid, acrylonitrile, ammonium nitrate and urea. This article gives more detailed information about which categories are included in capital as well as in operating costs for an industrial product.

Finally Chemical Engineering, March 24, 1969, pages 114–142, gives for the most cost conscious readers an explanation of the way in which the costs of the various capital investments are cumulating, starting from "bare equipment" and using appropriate key factors.

References

[1] D.G. Jordan, in: Interscience library of chemical engineering and processing (Interscience, New York, 1968) Vol. 6, Part 2, pp. 707–788.
[2] C.S. Grove, R.V. Jelinek and H.M. Schoen, in: Advances in chemical engineering (Academic Press, New York, 1962) Vol. 3, ch. 1, pp. 1–60.
[3] A.W. Bamforth, Industrial crystallization (Leonard Hill, London, 1965).
[4] A. van Hook, Crystallization, theory and practice (Reinhold Publ. Co., New York, 1961).
[5] G. Matz, Kristallisation, Grundlagen und Technik (Springer Verlag, Berlin, 1969).
[6] J.W. Mullin, Crystallization (Butterworths, London, 1961).
[7] R.F. Strickland-Constable, Kinetics and mechanism of crystallization (Academic Press, London, 1968).
[8] Crystallization from solutions and melts, ed. by J.A. Palermo and M.A. Larson, Chem. Eng. Progr., Symposium Series No. 95, Vol. 65 (1969) Part I.
[9] Industrial crystallization, Proc. Symp. Inst. Chem. Eng., London, April, 1969.
[10] Chem. Eng. Jan 12th, (1970) 81.
[11] F. Rumford and J. Bain, Trans. Inst. Chem. Eng. (London) 38 (1960) 10.
[12] M.A. van Damme – van Weele, Thesis, Techn. Univ. Twente (Netherlands) (1965).
[13] M.A. van Damme – van Weele, in: Adsorption et croissance cristalline, Coll. C.N.R.S. 152 (1965) 433.

P. Hartman, ed., Crystal growth: an introduction. © (1973) North-Holland Publ. Co.

KINETICS OF CRYSTAL GROWTH

P. BENNEMA and G.H. GILMER

10.1. Analysis of the crystal fluid interface

10.1.1. Presuppositions of the multilayer model of Temkin

Before treating kinetic crystal growth models, we first ought to study the physical properties of the crystal-fluid interface because the whole crystal growth process takes place in the boundary region. We have chosen as the basis for this study the multilayer Temkin model [1] because i) it does not limit the number of layers; it therefore is more general than one or two layer models [2, 3]; ii) it will be shown that upon using a simple thermodynamic derivation analogous to Mutaftschiev's derivation used for his two layer model, the Temkin model in principle is rather easy to understand; iii) it will follow from this derivation that the Temkin model in fact holds for all kinds of fluid-crystal interfaces, where the fluid can be vapour, any solution, melt, etc.

Let us take as the basis for our considerations a tetragonal crystal, where the crystal consists of tetragonal blocks and let us consider the (001) face*. Our crystal is in fact a tetragonal Kossel crystal since we will assume only first neighbour interactions; each tetragonal block has four equivalent "horizontal" "contacts" or bonds and two "vertical" bonds. The whole interface is the region where the tetragonal solid blocks are in contact with the fluid continuum.

The Temkin interface consists of an infinite number of layers with the thickness d_{001}, the interplanar distance of the (001) face, i.e. the "thickness" of the tetragonal blocks in the c direction, where one surface unit of the n'th layer consists of N_{ns} solid blocks and N_{nf} fluid blocks. Since the interface only contains tetragonal blocks we find

* The limitation for the time being to a tetragonal crystal and the (001) face is not essential in the argument; in fact any shape of block and face (hkl) can be studied as will be discussed below.

$$N_{ns} + N_{nf} = N ,$$

where N is the total number of blocks per unit surface. If we define $C_n = N_{ns}/N$ as the fraction of blocks, which are solid in the n'th layer, it then automatically follows that the fraction of fluid blocks is $(1 - C_n)$. In this Temkin model n varies from $-\infty$ to $+\infty$ and we introduce the boundary conditions $C_{-\infty} = 1$ and $C_{+\infty} = 0$, which simply means that upon going from $n = -\infty$ to $n = +\infty$ we travel from the completely solid phase to the completely fluid phase. We assume that solid blocks only can occur on top of solid blocks. Thus upon comparing the $(n+1)^{\text{th}}$ layer with the nth layer, it is supposed that on top of each solid a solid block or a fluid block may occur and that due to the decreasing concentration of solid blocks $C_{n+1} \leqslant C_n$. The assumption, that solid blocks only can occur on top of other solid blocks means that isolated solid blocks in a completely fluid environment are ruled out; the same holds for "overhangs".

We note, that although it may appear that we must consider the fluid phase as consisting of separate fluid blocks, in fact we need only to consider the "fluid space" as a homogeneous fluid continuum, the shape of the interface being defined by the shape of the solid blocks only. We will label the energy of the horizontal bond due to a solid-solid contact of blocks as ϕ_{ss}, the horizontal bond energy between a solid block and the fluid ϕ_{sf} and the energy which is required to separate a surface having the same shape as the side faces of a solid tetragonal block from the fluid continuum as ϕ_{ff}.

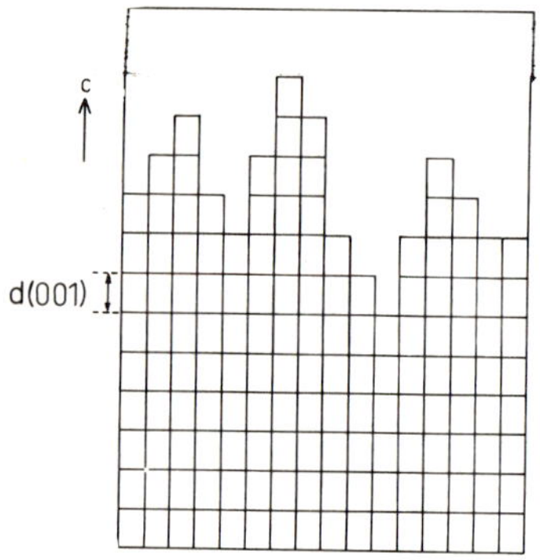

Fig. 10.1. Schematic representation of the crystal-fluid interface. Only solid blocks are shown.

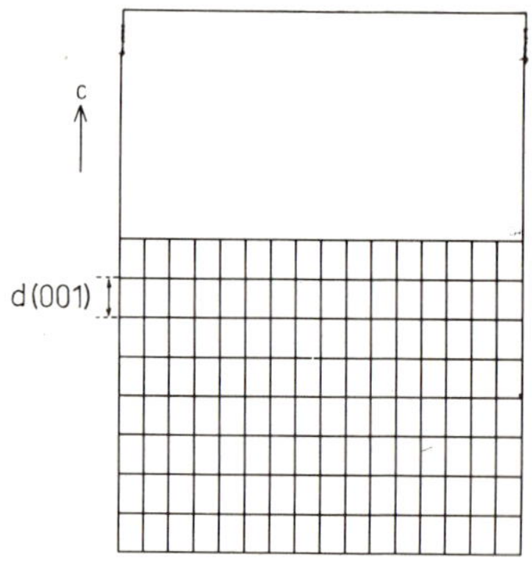

Fig. 10.2. Schematic representation of a completely flat crystal-fluid interface. Only solid blocks are shown.

10.1.2. Derivation of the characteristic Temkin expression

In what follows we will not use Temkin's statistical analysis, but present an alternative derivation analogous to Mutaftschiev's derivation for his two-layer model [2].

For the development of Temkin's formalism we will analyse what happens if a completely flat interface (fig. 10.2) (this is the reference face) is "roughened" (fig. 10.1). For the reference face the following conditions hold: $-\infty < n < 0$, $C_n = 1$, and for $1 < n < +\infty$, $C_n = 0$; this means that we place the reference face between $n = 0$ and $n = 1$.

The change in the Gibbs free energy G upon "roughening" the flat reference plane is given by

$$\Delta G = \Delta G_{\mathrm{f-s}} + \Delta G_{\mathrm{surf}} , \tag{10.1}$$

where $\Delta G_{\mathrm{f-s}}$ is the change in Gibbs free energy due to the replacement of solid blocks by fluid blocks and vice versa (this is a bulk change effect); ΔG_{surf} is the difference in Gibbs free energy due to a change in bond energy of the interface (ΔE_{surf}), and the configuration entropy ($T\Delta S_{\mathrm{surf}}$). ($\Delta G_{\mathrm{surf}}$ is due to a change in the structure of the surface; no volume change is included in ΔG_{surf} since if it does occur it is included in $\Delta G_{\mathrm{f-s}}$). It follows that

$$\Delta G_{\mathrm{f-s}} = \frac{\partial G}{\partial N_{\mathrm{s}}} \Delta N_{\mathrm{s}} + \frac{\partial G}{\partial N_{\mathrm{f}}} \Delta N_{\mathrm{f}} \tag{10.2}$$

and

$$\Delta G_{\mathrm{surf}} = \Delta E_{\mathrm{surf}} - T\,\Delta\, S_{\mathrm{surf}} , \tag{10.3}$$

where ΔN_{s} and ΔN_{f} are the number of solid blocks, which become fluid and the number of solid blocks which are formed in the fluid phase, upon roughening respectively.

10.1.3. Derivation of $\Delta G_{\mathrm{f-s}}$

If we assume for the time being that each solid block and each fluid volume with the same volume as the solid block, does contain one growth unit (i.e. an atom or a molecule), we can write for the derivatives of G with respect to N_{s} and N_{f} the chemical potentials of the same growth units in both phases:

$$\frac{\partial G}{\partial N_{\mathrm{s}}} = \mu_{\mathrm{s}} \text{ and } \frac{\partial G}{\partial N_{\mathrm{f}}} = \mu_{\mathrm{f}} .$$

Upon roughening the flat interface, a fraction of $1 - C_n$ of the n'th layer of the originally solid phase is "filled" with fluid so that for that fraction the change in thermodynamic potential is $\Delta\mu = \mu_f - \mu_s$. In the same way NC_n solid blocks are formed in the n'th layer of the originally fluid "sea" and for those blocks the change in μ is $-\Delta\mu = \mu_s - \mu_f$. It follows from this argument that for the whole interface the change in Gibbs free energy due to the formation and dissolution of solid blocks, is given by

$$\Delta G_{f-s} = \Delta\mu N \left[\sum_{n=-\infty}^{0} (1 - C_n) - \sum_{n=1}^{\infty} C_n \right], \text{ where } \Delta\mu = \mu_f - \mu_s . \qquad (10.4)$$

We note that during the roughening procedure we do not keep the number of solid blocks necessarily constant. (In that special case $\Sigma_{n=-\infty}^{0}(1 - C_n)$ cancels with $\Sigma_{n=1}^{\infty} C_n$).

10.1.4. Derivation of ΔE_{surf}

In order to obtain an expression for ΔE_{surf} we have to carry out a "book-keeping" system based on the "conservation of bonds". Due to our interface model consisting of tetragonal blocks, where bonds are body-body contacts, roughening from the energetic point of view means nothing else than replacing ϕ_{ss} and ϕ_{ff} by ϕ_{sf} bonds. We do not need to consider "vertical" bonds in this bookkeeping system, because upon going from fig. 10.2 to fig. 10.1 it can be seen that independent of the final roughness, the number of vertical ϕ_{sf} bonds remains the same as compared to the reference situation. This is a consequence of the "solid above solid" presupposition. (This is also the reason that we can take tetragonal crystals instead of a cubic crystal with cubic blocks since the vertical bonds need not to be equal to the horizontal bonds). Upon using a block model for the crystal fluid interface it will be shown that in order to form a ϕ_{sf} bond on the average $\frac{1}{2}\phi_{ss}$ and $\frac{1}{2}\phi_{ff}$ bonds must be broken, so that the energy gain upon the formation of a ϕ_{sf} bond is

$$\epsilon = \tfrac{1}{2}(\phi_{ss} + \phi_{ff}) - \phi_{sf} , \qquad (10.5)$$

where we take the absolute values of the bond energies involved. In order to show the validity of (10.5) we will analyse more precisely what happens from an energetic point of view if the flat reference plane is roughened. We will treat one example explicitly for each of the two following cases: (i) a roughening process where the number of solid blocks remains the same; (ii) a roughening process where the number changes.

As an example of (i) let us create a surface vacancy in the flat surface and create an isolated solid block somewhere on the surface. It is easy to see that for this process we loose $4\phi_{ss}$ and $4\phi_{ff}$ bonds and that we gain $8\phi_{sf}$ bonds, which is indeed consistent with the statement (10.5). It can be checked that the same statement results if a solid block in the surface adjacent to a surface vacancy is removed and placed on top of the surface etc.

As an example of (ii) let us take the formation of a solid block on top of a flat part of the surface. Two things happen (a) the chemical potential changes by an amount $-\Delta\mu = \mu_s - \mu_f$ (a bulk change effect); (b) $4\phi_{ff}$ bonds are replaced by $4\phi_{sf}$ bonds, moreover a vertical ϕ_{ff} bond is replaced by a ϕ_{ss} bond. The change in chemical potential resulting from the formation of a solid block is due to (1) a difference in bond energy; (2) a difference in vibrational and rotational entropy; (3) a possible difference in density of fluid and solid phase which gives a volume effect ($p\Delta V$ in ΔG). The bulk change in bond energy corresponds with the change in bond energy if a solid block is formed in a kink. This means that one vertical ϕ_{ss} bond replaces one vertical ϕ_{ff} bond and two lateral ϕ_{ss} bonds replace two lateral ϕ_{ff} bonds. We thus find upon the formation of a solid block on a flat surface $4\phi_{ff} \rightarrow 4\phi_{sf}$ (lateral) and $1\phi_{ff} \rightarrow 1\phi_{ss}$ (vertical). We ought to subtract the bond change included in $\Delta\mu$, which gives $2\phi_{ff} \rightarrow 2\phi_{ss}$ (lateral) and $1\phi_{ff} \rightarrow 1\phi_{ss}$ (vertical). This then is again consistent with (10.5). This result can also be obtained for other roughening processes such as the formation of a solid block adjacent to an isolated solid block on a flat surface etc.

In order to calculate the number of ϕ_{sf} bonds, which are formed by roughening, Temkin uses for every n'th layer a Bragg-Williams approximation, which means that he assumes that a solid block "sees" on the average in his direct environment within the n'th layer the same number of solid and fluid blocks as the average composition of the n'th layer. Preferential clustering is thus ruled out. The chance that a solid-fluid bond is found is therefore proportional to $4C_n(1 - C_n)$ i.e. to the product of the fraction of solid blocks and the fraction occupied by the liquid in the n'th layer. A factor 4 is added since there are four contact possibilities for each block. It thus follows from the given argument that

$$\Delta E_{\text{surf}} = 4\epsilon N \sum_{n=-\infty}^{+\infty} C_n(1 - C_n)\,. \tag{10.6}$$

10.1.5. Derivation of $T\Delta S_{\text{surf}}$

It follows from the Boltzmann formula $S = k \ln w$ (w being the number of configurations), that for the flat reference face $w = 1$ and consequently $S = 0$; thus $T\Delta S_{\text{surf}}$ only results from the roughening procedure.

If we consider the $(n + 1)$'th layer, then we only need to take into consideration the sites above the solid blocks of the n'th layer. Of these NC_n sites, NC_{n+1} are identical solid and $N(C_n - C_{n+1})$ are identical fluid or empty blocks. Hence the number of configurations is given by

$$w_{n+1} = \frac{(NC_n)!}{(NC_{n+1})!\left[N(C_n - C_{n+1})\right]!}. \tag{10.7}$$

It thus follows that ΔS_{surf} for the whole interface is given by

$$\Delta S = k \ln w = k \ln \prod_{n=-\infty}^{\infty} \frac{(NC_n)!}{(NC_{n+1})!\,(NC_n - NC_{n+1})!}. \tag{10.8}$$

We can write for

$$\ln \prod_{-\infty}^{\infty} \frac{(NC_n)!}{(NC_{n+1})!} = \ln \frac{(NC_{-\infty})!}{(NC_{-\infty+1})!} \cdots \frac{(NC_{n-1})!}{(NC_n)!}\,\frac{(NC_n)!}{(NC_{n+1})!} \cdots \frac{(NC_{-1+\infty})!}{(NC_{+\infty})!} =$$

$$= \ln N! \approx N \ln N - N. \tag{10.9}$$

This result follows from the cancellation of all the C_n factors except the first and the last one and from the boundary conditions $C_{-\infty} = 1$ and $C_{+\infty} = 0$; we also used the Stirling formula. We can write for

$$\ln \prod_{-\infty}^{\infty} \left[N(C_n - C_{n+1})!\right]^{-1} =$$

$$= -\sum_{-\infty}^{+\infty} \left[N(C_n - C_{n+1}) \ln N + N(C_n - C_{n+1}) \ln (C_n - C_{n+1}) - N(C_n - C_{n+1})\right] =$$

$$= -N \ln N + N - N \sum_{-\infty}^{\infty} (C_n - C_{n+1}) \ln (C_n - C_{n+1}). \tag{10.10}$$

Here we again used the Stirling formula and the fact that

$$\sum_{-\infty}^{+\infty} (C_n - C_{n+1}) = (C_{-\infty} - C_{+\infty}) = 1,$$

due to the cancellations of all the C_n's except the first and the last one. Upon substituting (10.9) and (10.10) in (10.8) (or adding (10.9) and (10.10)) we find

$$T\Delta S_{\text{surf}} = -kTN \sum_{n=-\infty}^{\infty} (C_n - C_{n+1}) \ln (C_n - C_{n+1}) \,. \tag{10.11}$$

10.1.6. General expression for ΔG

If we now substitute (10.4) for $\Delta G_{\text{s-f}}$, (10.6) for ΔE_{surf}, (10.11) for $T\Delta S_{\text{surf}}$ in (10.1) and divide by NkT we obtain the following expression for ΔG, the change of Gibbs free energy upon roughening a flat reference plane,

$$\frac{\Delta G}{NkT} = \beta \left[\sum_{n=-\infty}^{0} (1 - C_n) - \sum_{n=1}^{\infty} C_n \right] +$$

$$+ \alpha \sum_{n=-\infty}^{\infty} C_n (1 - C_n) + \sum_{n=-\infty}^{\infty} (C_n - C_{n+1}) \ln (C_n - C_{n+1}) \,, \tag{10.12}$$

where

$$\beta = \frac{\Delta\mu}{kT} \,, \tag{10.13}$$

$$\alpha = \frac{4\epsilon}{kT} \,. \tag{10.14}$$

We now have obtained a ΔG expression, which is a function of an infinite number of undetermined C_n's. In order to check whether a stable interface will occur and to calculate the C_n's, we have to check whether ΔG has a minimum for all the C_n's. We therefore must find out whether the condition

$$\frac{\partial \Delta G/NkT}{\partial C_n} = 0 \tag{10.15}$$

holds for all C_n's. If this differentiation is carried out we find

$$-\beta + \alpha - 2C_n\,\alpha + \ln(C_n - C_{n+1}) - \ln(C_{n-1} - C_n) + 1 - 1 = 0$$

or

$$\frac{C_n - C_{n+1}}{C_{n-1} - C_n}\,\mathrm{e}^{-2\alpha C_n} = \mathrm{e}^{-\alpha+\beta}\,. \tag{10.16}$$

The only way to solve (10.16) is by numerical analysis. Upon introducing the new variable

$$z_n = 2\alpha C_n - \alpha + \beta\,, \tag{10.16a}$$

we find for (10.16)

$$\frac{z_n - z_{n+1}}{z_{n-1} - z_n} = \mathrm{e}^{z_n} \tag{10.17}$$

and it follows from the boundary conditions

$$C_{-\infty} = 1;\, C_\infty = 0 \ \text{ that } \ z_{-\infty} = \alpha + \beta;\, z_{+\infty} = -\alpha + \beta\,, \tag{10.18}$$

that (10.17) can be rewritten in the form

$$z_{n+1} = z_n - (z_0 - z_1)\,\exp\Big(-\sum_{k=1}^{n} z_k\Big)\,,$$

$$z_{-(n+1)} = z_{-n} + (z_0 - z_1)\,\exp\Big(-\sum_{k=0}^{n} z_{-k}\Big). \tag{10.19}$$

In order to solve (10.17) or (10.19) Temkin starts with two z_0 and z_1 values, from which all the C_n's can be calculated. From $z_{\pm\infty} = \lim z_n$, using (10.18) β and α can be determined, provided z_0 and z_1 were so chosen, that indeed the boundary conditions hold. We note that the condition (10.15), which gives (10.16), is not a sufficient condition. In order to check whether the minimum is a true minimum and not a saddle point, the second derivatives must also be checked. We refer for this to Temkin's original paper.

10.1.7. Results of the Temkin formalism

Solutions which correspond to a true minimum in the ΔG (C_n's) for $\beta = \Delta\mu/kT = 0$ curve fulfil the following conditions

$$z_0 = -z_1,\ z_{-1} = -z_2 \text{ etc., so that}$$

$$C_0 = 1 - C_1,\ C_{-1} = 1 - C_2 \text{ etc.} \tag{10.20}$$

There is another type of solution for $\beta = 0$, which gives

$$z_0 = 0,\ z_{-1} = -z_1$$

$$C_0 = \tfrac{1}{2},\ C_{-1} = 1 - C_2 \text{ etc.} \tag{10.21}$$

which fulfils (10.15) and (10.16) but which corresponds to a saddle point in the ΔG (C_n's) curve. In fig. 10.3 we see some concentration profiles corre-

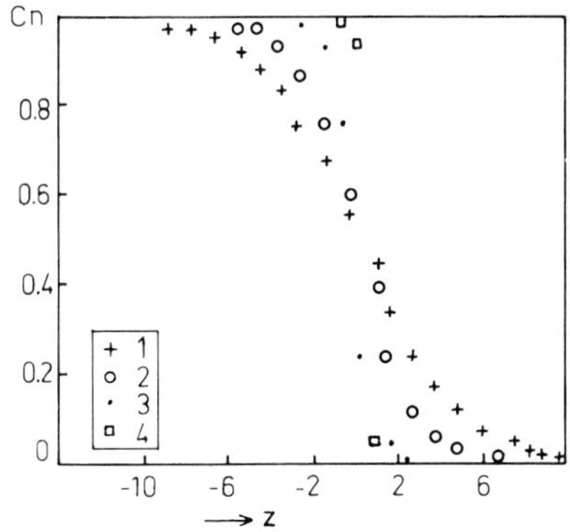

Fig. 10.3. Concentration distribution of "solid" atoms C_n producing a minimum of the free energy of the system for $\beta = 0$ and certain values of parameter $\alpha = 4\epsilon/kT$: 1) $\alpha = 0.446$; 2) 0.769; 3) 1.889; 4) 3.310.

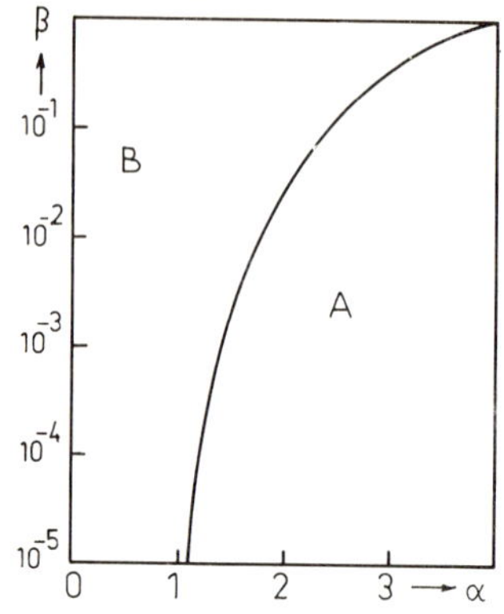

Fig. 10.4. Curve dividing the region of positive values of parameters β and α into subregions (A (stable) and B (unstable)).

sponding to certain $\alpha = 4\epsilon/kT$ values, where C_n is plotted as a function of n. Going from left to right we go from the solid part to the fluid part. It can be seen that the higher α the thinner the interface. For a relatively low value of $\alpha = 3.31$ in fact only two layers, namely a layer with a few ad-atoms on top of an almost flat layer and a few surface vacancies in it are involved.

10.1.8. Layer growth and continuous growth

A very interesting and important result of the multilayer Temkin model is given in fig. 10.4. Here $\ln \beta = \ln(\Delta\mu/kT)$ is plotted against α. We see that the whole field is divided in two parts A and B. If we start at $\alpha = 2$ and then let $\ln \beta$ increase starting with $\log \beta = -5$, we pass through the field A until we arrive at the boundary of A and B. It now follows from the numerical analysis, that in the A field the surface under consideration has a true minimum and it will therefore maintain its (crystallographic) orientation. If, however, $\Delta\mu/kT$ reaches a value in B ($\Delta\mu/kT$ is a measure for the undercooling or supersaturation as will be shown below), there is no stable surface anymore, which means that there will be no minimum in the ΔG (C_n's) curve. Growth can occur (in fact Temkin's model is a static thermodynamic model in which growth is not considered), since there are no thermodynamic barriers (there may be kinetic barriers) and there is no need to maintain the (crystallographic) orientation anymore. It is said that in the region B the crystal will grow without any barriers or continuously. It follows from fig. 10.4, that if α is high enough, also for every high β values the crystal will maintain its "flatness". If on the other hand α is lower than 1.2 it will always grow continuously. If growth occurs in the A field only a layer growth mechanism is possible; that means that growth must occur by a two-dimensional nucleation mechanism or a spiral growth mechanism. This will be the subject of section 2.

We note that the distinction between layer growth and continuous growth also was obtained by Mutaftschiev [2] and Jackson [3] from their two and one layer models respectively. In fact the Temkin model approaches Mutaftschiev's model if α is higher than or equal to four. The expressions of Mutaftschiev and Jackson can be obtained immediately from Temkin's expression (10.12) if, in (10.12), all C_n's are set equal to zero except C_{+1} and C_0, or all are chosen equal to zero except C_0 for the two models respectively.

10.1.9. Generalizations of the Temkin model. Comparison with the Hartman-Perdok theory

First of all it is not necessary to limit ourselves to a tetragonal block crystal model. For each crystallographic face (*hkl*) we want to study, we can choose a suitable block model. Blocks can have any shape, such as identical triclinic blocks (see fig. 10.5a) or identical deformed hexagonal blocks (see fig. 10.5b). We can differentiate between the contacts or bonds 1, 2, 3, 4 in the first case and 1, 2, 3 ... 6 in the other case. The case that opposing bonds as 1 and 3 in fig. 10.5a and 2 and 5 in fig.10.5b are not the same due to an absence of a centre of symmetry can also be included. We can also use block models with second and third neighbours, or models which consist only of centres of gravity, which are mutually connected by bonds. (For Temkin's model second and third neighbour interactions only are allowable within a slice). Instead of the expression for α (10.14) we can write a more general

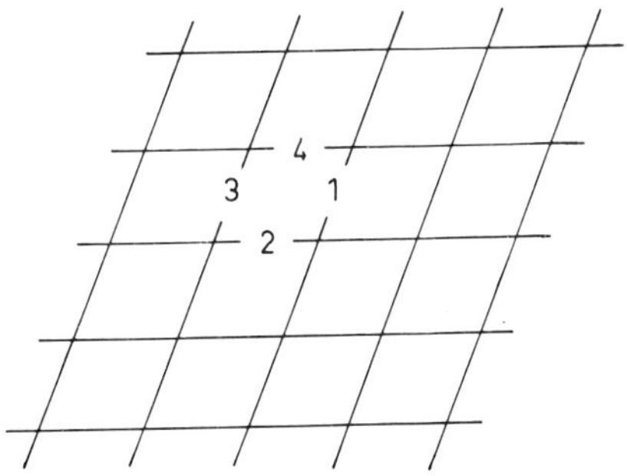

Fig. 10.5*a*

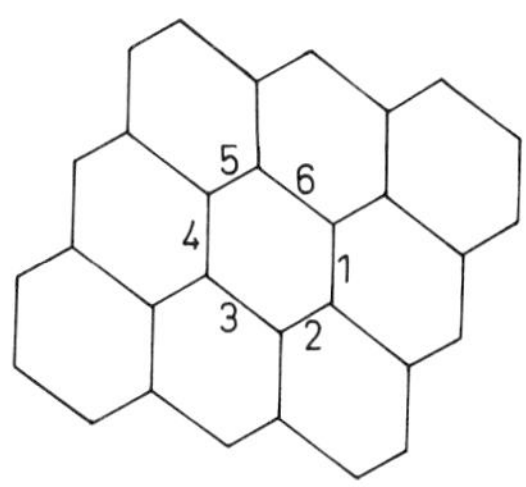

Fig. 10.5*b*

Fig. 10.5. Block crystals with arbitrary block shapes.

expression for α

$$\alpha = \frac{\sum_{i=1}^{m} \{(\frac{1}{2}\phi_{ss,i} + \frac{1}{2}\phi_{ff,i}) - \phi_{sf,i}\}}{kT}, \tag{10.22}$$

because the replacement law of bonds (10.5) holds for each of the contact or bond directions such as 1 to 4 or 1 to 6 of fig. 10.5a and fig. 10.5b respectively, where i is the label for a particular bond direction. We can now define

$$E_{ss}^{slice} = \sum_{i=1}^{m} \tfrac{1}{2}\phi_{ss,i}$$

$$E_{ff}^{slice} = \sum_{i=1}^{m} \tfrac{1}{2}\phi_{ff,i} \tag{10.23}$$

$$E_{sf}^{slice} = \sum_{i=1}^{m} \tfrac{1}{2}\phi_{sf,i},$$

where we deliberately wrote E_{ss}^{slice} since the first expression corresponds indeed with E^{slice} as defined in the Hartman and Perdok theory (see ch. 14). E_{ss}^{slice} is the bond energy per solid block in the layer, with a thickness d_{hkl}. The same holds for the energy per fluid block in a fluid layer and the ϕ_{sf} bonds within a layer. Now the following relations will hold:

$$E_{ss}^{slice} + E_{ss}^{att} = E_{ss},$$

$$E_{ff}^{slice} + E_{ff}^{att} = E_{ff},$$

$$E_{sf}^{slice} + E_{sf}^{att} = E_{sf}.$$

Here E_{ss}^{att}, E_{ff}^{att} and E_{sf}^{att} are the vertical bond energies between adjacent blocks; E_{ss} is the total crystallization energy which is independent of the crystallographic face under consideration if a solid block crystallizes from the

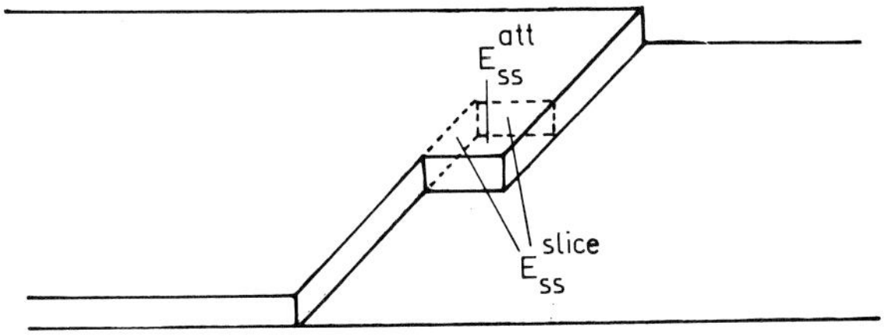

Fig. 10.6. E_{ss}^{att} and E_{ss}^{slice} defined by means of the kink site.

fluid phase. This can be seen from fig. 10.6 because $E_{ss}^{att} + E_{ss}^{slice} = E_{ss}$ (half crystal phase or "Halbkristallage" energy). In an analogous way we can define formally the "crystallization energy" E_{ff} per fluid block and the "crystallization energy" E_{sf} as the crystallization energy of a mixed solid-fluid crystal, all in reference to vacuum. If we now assume that

$$\frac{E_{ss}^{slice}}{E_{ss}} \approx \frac{E_{ff}^{slice}}{E_{ff}} \approx \frac{E_{sf}^{slice}}{E_{sf}}, \tag{10.24}$$

we can write upon substituting (10.23) in (10.24)

$$\alpha = \frac{(E_{ss}^{slice} + E_{ff}^{slice}) - 2\,E_{sf}^{slice}}{kT}$$

$$\approx \frac{E_{ss}^{slice}}{E_{ss}} \left[\frac{(E_{ss} + E_{ff}) - 2\,E_{sf}}{kT}\right] \tag{10.25}$$

where $\xi = E_{ss}^{slice}/E_{ss}$ is the well-known anisotropy factor already introduced by Jackson [3] where he defined ξ as the fraction of bonds within the layer over the total number of bonds. ξ is now defined in terms of the Hartman-Perdok theory. In case of growth from the melt, if we have one particle in a block and if there is complete wetting which means

$$E_{sf} \approx E_{ff}, \tag{10.26}$$

we see that α becomes

$$\alpha = \xi \frac{L}{kT} = \alpha_J \,, \tag{10.27}$$

where α_J is the α factor as originally defined by Jackson [3]. Since in that case

$$E_{ss} - E_{ff} = L \,, \tag{10.28}$$

the heat of fusion, dissolution or evaporation. Thus only if the conditions (10.24) and (10.26) are fulfilled do we find that $\alpha = \alpha_J$. The condition (10.24) seems to be a reasonable condition since the larger d_{hkl} the more contact possibilities and/or the stronger ϕ_{ss}, ϕ_{ff} and ϕ_{sf} bonds. The condition of complete wetting also seems to be a reasonable one, but is by no means general as has been shown recently by Zell and Mutaftschiev [4].

10.1.10. Comparison with the Jackson and Mutaftschiev's models

It follows from Jackson's one layer model (and Mutaftschiev found the same also on using a one layer approximation) that if $\alpha_J > 2$ layer growth will occur and if $\alpha_J < 2$ then continuous growth will occur. This sharp distinction is somewhat questionable because (i) one can see from fig. 10.4 that it does not occur as a result of the ln β versus α plot, which results from Temkin's multilayer model; (ii) an exact calculation of α using (10.27) may be impossible due to the fact that (10.24) and (10.27) may not hold. (In order to determine ξ for different crystallographic faces it is desirable to carry out a PBC analysis (see Hartman-Perdok theory, ch. 14)).

10.1.11. Solid-solution interface

The multilayer Temkin model can be extended to the case where the fluid is a solution which has a different composition from the solid phase, since in our derivation nothing was assumed about the way a solid block is transformed into a fluid block or vice versa. In fact there need not to be a one to one correspondence between the contents of a solid block and a "fluid block" of the same volume. This generalization implies that there will be a difference in density between the homogeneous fluid and solid phases. But as mentioned above in section 4 this is automatically included by definition in the $\Delta\mu$ term.

Also in principle it is not necessary to limit ourselves to solid blocks which consist of only one type of growth unit (i.e. atoms, ions or molecules). In the latter case, however, it is somewhat difficult to define the interaction energies, because we then must differentiate between bonds which are defined as bonds due to interaction of complete blocks mutually and bonds due to mutual interaction of separate growth units. The best way out such problems may be (1) carry out a Hartman-Perdok analysis; (2) on the basis of such an analysis replace the real complex structure by a block structure, for each F face under consideration another block model can be chosen; (3) carry out a Temkin analysis. But even with such a procedure, which may not be unambiguous, it will always be difficult if not impossible to calculate α a priori. It is probable, however, that qualitative relations between two different crystallographic faces as for example

if $E^{\text{slice}}_{\text{ss}\,h_1k_1l_1} > E^{\text{slice}}_{\text{ss}\,h_2k_2l_2}$ then $\alpha_{h_1k_1l_1} > \alpha_{h_2k_2l_2}$ will hold.

10.1.12. Definition of $\beta = \Delta\mu/kT$ for growth from the melt

For growth from the melt, at the melting point, the following condition holds

$$F_{\text{fm}} = E_{\text{fm}} - T_{\text{m}}S_{\text{fm}} = E_{\text{sm}} - T_{\text{m}}S_{\text{sm}} = F_{\text{sm}}$$

where F_{fm} and F_{sm} are the Hemholtz free energies of the fluid (melt) phase and the solid phase respectively, at the melting point T_{m}.

Or $E_{\text{fm}} - E_{\text{sm}} = T_{\text{m}}\,(S_{\text{fm}} - S_{\text{sm}})$

or $\Delta E_{\text{m}} = T_{\text{m}}\,\Delta\,S_{\text{m}}$. (10.29)

If we have a temperature slightly below the melting point we get

$$F_{\text{f}} - F_{\text{s}} = E_{\text{f}} - E_{\text{s}} - T\,(S_{\text{f}} - S_{\text{s}}) \tag{10.30}$$

and if we assume that since we are close to the melting point

$$\Delta E = E_{\text{f}} - E_{\text{s}} \approx \Delta E_{\text{m}} \text{ and } \Delta S = S_{\text{f}} - S_{\text{s}} \approx \Delta S_{\text{m}}\ , \tag{10.31}$$

we can rewrite the expression (10.30)

$$\Delta F \approx \Delta E_{\text{m}} - T\,\Delta\,S_{\text{m}}$$

which gives upon using (10.29)

$$\Delta F \approx \Delta E_{\mathrm{m}} \frac{T_{\mathrm{m}} - T}{T_{\mathrm{m}}} = \Delta E_{\mathrm{m}} \frac{\Delta T}{T_{\mathrm{m}}}.$$

Per growth unit we thus find

$$\beta = \frac{\Delta\mu}{kT} = \frac{\mu_{\mathrm{f}} - \mu_{\mathrm{s}}}{kT} \approx \frac{L}{RT}\frac{\Delta T}{T_{\mathrm{m}}}, \tag{10.32}$$

where L is the molar heat of fusion. Equation (10.32) holds only for low undercoolings.

10.1.13. $\beta = \Delta\mu/kT$ for growth from the vapour and solution

A solution may contain several different kinds of growth units which come together to form the solid blocks. Then the equation (10.2) for the change in free enthalpy as a result of some of the growth units transferring from the solution to the solid phase becomes

$$\Delta G_{\mathrm{f-s}} = \sum_{i=1}^{m} \left[\frac{\partial G}{\partial N_{\mathrm{s}}^{i}} \Delta N_{\mathrm{s}}^{i} + \frac{\partial G}{\partial N_{\mathrm{f}}^{i}} \Delta N_{\mathrm{f}}^{i}\right] \tag{10.33}$$

$$i = 1, 2, \ldots,$$

where N_{s}^{i} is the number of growth units of type i in the solid phase, N_{f}^{i} is the number in the fluid, and m is the number of different types. If one block of the solid phase contains ν_i growth units of the ith type, then

$$\Delta N_{\mathrm{s}}^{i} = \nu_i \Delta N_0 \text{ and } \Delta N_{\mathrm{f}}^{i} = -\nu_i \Delta N_0,$$

where ΔN_0 is the number of solid blocks formed by the transfer. Using these expressions in (10.33) gives

$$\Delta G_{\mathrm{f-s}} = -\Delta N_0 \sum_{i=1}^{m} (\mu_{\mathrm{f}}^{i} - \mu_{\mathrm{s}}^{i}) \nu_i, \tag{10.34}$$

where μ_{s}^{i} is the chemical potential of the ith type growth unit in the fluid, or

$$\Delta G_{\mathrm{f-s}} = -\Delta N_0 \Delta M ,$$

where

$$\Delta M = \sum_{i=1}^{m} (\mu_{\mathrm{f}}^i - \mu_{\mathrm{s}}^i)\,\nu_i$$

plays the same role as $\Delta\mu$ in the one component melt system; it is the change in free energy for the formation of a solid block. In dilute solutions

$$\mu_{\mathrm{f}}^i - \mu_{\mathrm{s}}^i \approx kT \ln(C^i/C_0^i) ,$$

where C^i is the concentration of the ith type in the solution and C_0^i is its equilibrium concentration. Then

$$\frac{\Delta M}{kT} = \ln \prod_{i=1}^{m} (C^i/C_0^i)^{\nu_i} . \tag{10.35}$$

Similarly, for vapour

$$\frac{\Delta M}{kT} = \ln \prod_{i=1}^{m} (p^i/p_0^i)^{\nu_i} . \tag{10.35a}$$

Thus the effective β^{eff} for the Temkin analysis is $\Delta M/kT$.

In the case where the solution has the same stoichiometric composition as the solid it follows that the concentration of components is given by $\nu_1\, C$, $\nu_2\, C$ ... $\nu_i\, C$ ... etc., where C is the concentration in gmol of the stoichiometric "compound"*, so that

$$\Delta M = kT\Big(\sum_{i=1}^{m} \nu_i\Big) \ln (C/C_0) \tag{10.36}$$

* In the case of alum, $KAl(SO_4)_2 . 12H_2O$ dissolves congruently in the aqueous solution and we can define C as the number of "gmol's alum" in the solution.

or

$$\frac{\Delta\mu}{kT} = \frac{\Delta M}{\sum_{i=1}^{m} \nu_i} \cdot \frac{1}{kT} = \ln(C/C_0)\,,$$

where again $\Delta\mu = \mu_f - \mu_s$, which is now the same for all growth units. For the case of low supersaturations we can write

$$\beta = \frac{\Delta\mu}{kT} = \ln\left(1 + \frac{C - C_0}{C_0}\right) = \ln(1+\sigma) \approx \sigma\,, \tag{10.37}$$

where σ is the relative supersaturation.

In the BCF theory we will use σ as defined above. σ Can be replaced by the more general expressions for $\Delta\mu/kT$ or $\Delta M/kT$. It will always be assumed that $\mu_s = \mu_{s0}$ corresponding with equilibrium. It will also then be presupposed that $\mu_s = \mu_{s0}$ at the steps and kinks. The expressions for a generalized $\Delta\mu$ or ΔM are very general and can be used for complex systems occurring in growth from the vapour by transport reactions, crystallization from complex solutions or fluxes etc. ΔM, however, can only be calculated exactly, if the phase diagram of the systems concerned is known together with the actual situation, i.e. the pressure, temperature and composition of the system.

10.1.14. Applications and limitations of the Temkin model

One of the most serious limitations of the Temkin model, together with Mutaftschiev's and Jackson's models is the use of the Bragg-Williams approximation, which rules out preferential clustering. Solid blocks will try to surround themselves by other solid blocks. This will lower the energy and decrease the entropy, so that it is not easy to see offhand what the influence of clustering on ΔG and on the (β,α) plot will be. A way out of this difficulty is to introduce a cluster correction analogous to the corrections in magnetism and which may be called a quasi-chemical or Bethe approximation [5]. Leamy and Jackson [6] reported the results of applying such a correction. These authors found that the concentration profiles did not change very much upon introducing the cluster correction.

A more elegant and, from a physical point of view, more relevant way to overcome the Bragg-Williams approximation, is to simulate the roughening process as described above and as schematized in fig. 10.2 → fig. 10.1 on a fast digital computer. Playing a so-called Monte Carlo game on the surface of a Kossel crystal, one can create and annihilate solid blocks. The chance that solid blocks are created is chosen to be independent of the position, but the chance of annihilation depends on the bond situation. The more bonds the less the chance of being annihilated. Since there is equilibrium on the average as many blocks are created as annihilated. In this way equilibrium surfaces are simply made. Such games were played by Leamy and Jackson [6] and independently by Gilmer and Bennema [7]. As was found by Leamy and Jackson [6] we could confirm that the concentration profiles were about the same as those of Temkin.

Very recently Nason and Tiller [8], in a discussion with Jackson [9], found that for the (001) face of a Kossel crystal the surface free energy γ will become negative below a certain α value, which is an absurd, physically irrelevant result. This conclusion was confirmed by Temkin [10] who calculated that for $\alpha \lesssim 1.79$, $\gamma < 0$. We suspect that this has something to do with the solid above solid approximation. If α becomes very low and the surface very rough, overhangs will become important. This suggests that for low α we need another model which includes overhangs. We note the fact that $\gamma < 0$ if $\alpha_{cr} < 1.79$ only holds for a Kossel crystal.

If

$$\alpha^{att} = \frac{E_{ss}^{att} + E_{ff}^{att} - 2E_{sf}^{att}}{kT} \quad \text{becomes} > \tfrac{1}{2}\alpha^{slice}$$

or $< \frac{1}{2}\alpha^{slice}$, α_{cr} becomes smaller or larger respectively.

We note that the results of Temkin's model and similar block models are the basis for crystal growth kinetic models as recently developed by Jackson [11, 12], Temkin [13, 14, 15] and the Monte Carlo computer simulation growth models of Binsbergen [16, 17] and Gilmer and Bennema [18, 19].

Other important papers on surface structures based on block models dealing also with structures of steps are the study of Chernov and Voronkov [20] and the original work of Burton, Cabrera and Frank [21].

We learned from our computer simulations that the distinction between continuous growth and layer growth is questionable. First of all this distinction is not based on a kinetic model (and this is what growth is all about) and secondly this result is also based on the Bragg-Williams approximation. The Bragg-Williams randomization makes the chance of formation of clusters too low and hence growth too difficult.

Notwithstanding these limitations we consider the Temkin model as one of the best models to start the study of the solid-fluid interface.

10.2. The crystal growth theory of Burton, Cabrera and Frank

10.2.1. Presuppositions of the BCF theory

The BCF theory [21, 22, 23] in essence is developed for crystal surfaces, which are flat or, using the concepts of the Temkin formalism, have a high α, say larger than 3 or 4. That means the interface is thin; it consists of two layers or a reference layer with a low concentration of ad-atoms – or more general growth units – on top of this layer and a low concentration of surface vacancies in it. It has been shown already by BCF that, for a bond energy in the surface $\phi/kT = 4$, or what comes to the same thing $\alpha = 8$, relative supersaturations of 25 to 50% are required to have a measurable growth rate by a two-dimensional nucleation mechanism. It has, however, been shown that close packed crystal surfaces or F (flat) faces can grow at relative supersaturations as low as 10^{-4}.

If steps were available, then growth could occur much more easily, since the bond energy of growth units in the steps is much less and therefore the α of the step is also low. Steps are therefore much rougher and BCF have calculated that for a Kossel crystal with $\phi/kT = 4$ one kink on 5 sites in the closest packed step direction on the average does occur. Now it is an essential presupposition of all the classical crystal growth theories first introduced by Kossel, that growth of F faces does occur by capture of growth units into kinks. Such a growth unit can diffuse over the surface and then leave the surface, or, if it lands on the surface close enough to a step, which automatically has a high concentration of kinks due to thermal roughening, it can enter the step and join the kinks. From there it is very hard to escape since then its bond energy is much higher than on the surface.

The problem which arises is how steps can be formed since for high α values they cannot be formed by thermal fluctuation. For growth to occur

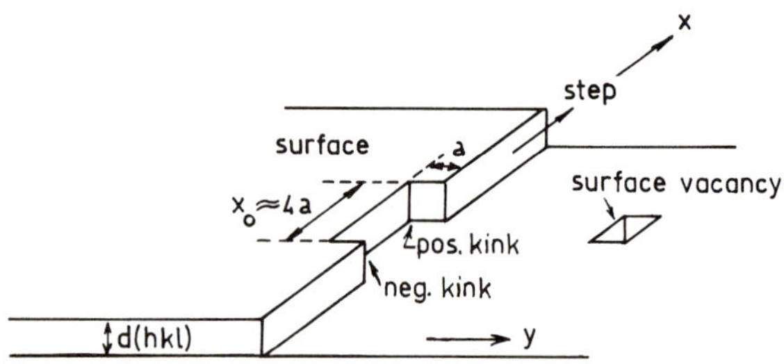

Fig. 10.7. Schematized surface structure with kinks in a step and a surface vacancy.

continuously the availability of a step is not a sufficient condition. Because, if a step were available, the layer would advance over the surface upon imposing a supersaturation on the surface and growth would stop completely once the layer had covered the surface. Therefore, step formation, other than by two-dimensional nucleation, was a very difficult problem to solve.

10.2.2. Frank's growth spiral

To overcome this difficulty Frank [24] "invented" in 1949 the famous growth spiral. Recognizing that crystals are not perfect Frank showed that screw dislocations terminating in the surface will provide the steps required for growth. Such a screw dislocation ending at a fixed point on a crystal surface produces a step, which winds itself into a spiral around that point. Numerous types of such spirals were indeed found (see the books of DeKeyser and Amelinckx [25], Verma [26] and Honigmann [27]). No exact general formulae for the shape of a growth spiral has yet been developed. However, BCF have pointed out that a good approximation, at least for the sequence of equidistant steps not too close to the centre, is an Archimedian spiral given by

$$r = 2r^*\theta , \qquad (10.38)$$

where r and θ are the polar coordinates of the spiral and r^* is the radius of

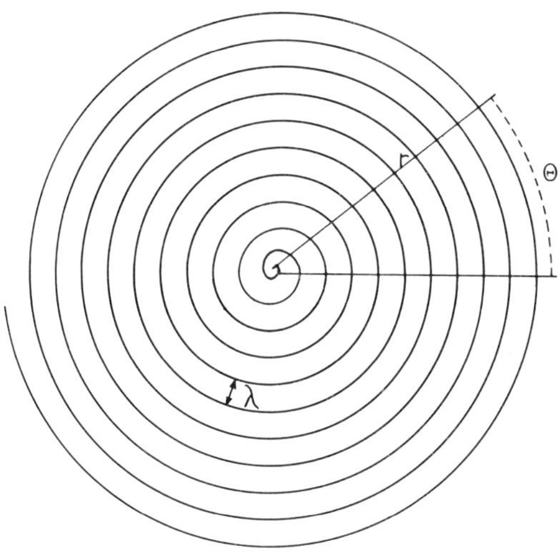

Fig. 10.8. Single Frank spiral.

the critical two-dimensional nucleus which is given by the classical Gibbs-Thomson expression

$$r^* = \frac{\gamma' a}{kT \dfrac{\Delta\mu}{kT}} = \frac{\gamma' a}{kT\sigma}, \tag{10.39}$$

where γ' is the edge free energy of a growth unit in a step, a is the distance between growth units and σ the relative supersaturation. It follows from (10.39) that the distance between steps λ' (see fig. 10.8) is given by

$$\lambda' = r_1 - r_2 = 2r^* \{(\theta + 2\pi) - \theta\} = 4\pi r^* . \tag{10.40}$$

According to Cabrera and Levine [28] a better approximation, which will be used in what follows, is given by

$$\lambda' = 19 r^* . \tag{10.41}$$

We must take account of the fact that several screw dislocations of the same sign, which have an average distance $e < 2r^*$, can co-operate (see fig. 10.9), and in such a case the distance between steps for the co-operating set of spirals is:

$$\lambda' = \frac{19}{s} \frac{\gamma'}{kT} \frac{a}{\sigma}, \tag{10.41a}$$

where s is the number of co-operating spirals, which probably will not be

Fig. 10.9. "Double spiral".

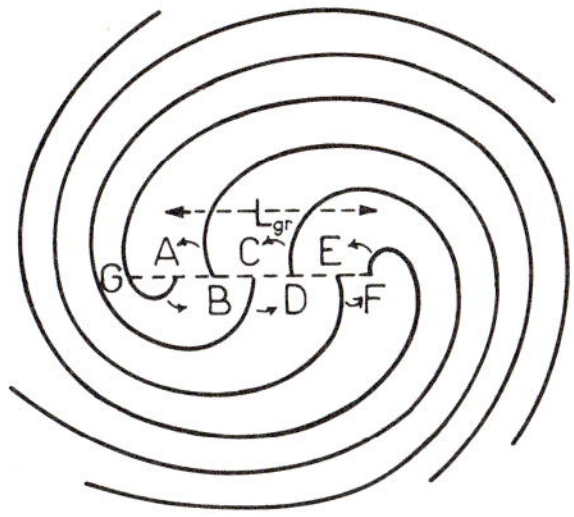

Fig. 10.10. Group of co-operating spirals resulting from a grain boundary.

larger than 5 to 10. s May be a more complex expression. In the case of a set of s' co-operating screw dislocations with the same sign due to a grain boundary, having a length L_{gr}, we find, for the average distance between steps λ,

$$\lambda = \frac{GF}{\frac{1}{2}s'} = \frac{19r^* + L_{gr}}{s'} \text{ and } \frac{1}{s} = \frac{\lambda}{\lambda'} = \frac{19r^* + L_{gr}}{s'\,19r^*}. \text{ Thus}$$

$$s = \frac{s'}{1 + L_{gr}/19r^*} \tag{10.42}$$

From this expression we recognize the definition of s, which is λ' (the distance between steps of a single spiral) divided by the average value of λ (the distance between steps of a set of co-operating spirals at the same supersaturation). s is thus a measure for the "strength" of the step source.

10.2.3. Surface diffusion to steps

Since we now have a continuous source of steps, the whole crystal growth problem is reduced to a diffusion problem i.e. a transport problem of growth units from the bulk of the supersaturated vapour or solution to the surface and from the surface to the kinks of the steps. In this section we will simply assume that the transport of the growth units from the bulk to the surface is not rate determining and focus our attention on the surface diffusion and step integration problem. The problem of coupled transport through the bulk (which will be called volume diffusion) and surface diffusion will be discussed briefly below in section 7.

In what follows we will assume that the first and second Fick laws hold i.e. that the surface diffusion constant is independent of the concentration or supersaturation. This is a reasonable presupposition for high α values, where we have a low concentration of ad-atoms. We will define formally four relaxation times [29a,b,c,] which play a role in the growth process, namely

τ_{desolv} = relaxation time for entering the adsorption layer; therefore a growth unit must overcome a "desolvation barrier";

τ_{deads} = relaxation time for leaving the surface adsorption layer, therefore a growth unit must overcome a deadsorption barrier;

τ_{sdiff} = relaxation time for making a diffusion jump from one equilibrium site on the surface to a neighbouring one on the surface: a growth unit must overcome a surface diffusion barrier;

τ_k = relaxation time for entering the kinks from the surface. We will assume that the ad-atoms are the ones which determine the growth and we will neglect possible contributions to the growth of moving surface vacancies, recombinations of ad-atoms and surface vacancies etc.

It may appear as if our crystal and crystal surface are built up of only one type of growth units. It will, however, hold for such complex compounds as NaCl or $KAl(SO_4)_2 . 12\,H_2O$ (alum) growing from an (aqueous) solution as well since, if we assume local electroneutrality and stoichiometry of the adsorption layer (which can be assumed as being the same as the crystal), the number of growth units, which pass the successive barriers will be the same – also on a very small atomic scale – as in the crystal. Thus for each barrier we can define a common relaxation time, which will be to a large extent determined by the slowest moving component. (In the case of alum it is reasonable to assume that we have as growth units $Al^{3+}(6\,H_2O)$, $K^{+}(6\,H_2O)$ and SO_4'' i.e. Al^{3+} and K^{+} with their hydration shells).

The model may also be valid in cases where the multicomponent solution or vapour has a different stoichiometric composition from the solid, because, in such a case we can also define for each event one relaxation time for all the components in the (stoichiometric) adsorption layer.

We assume also that the relaxation times do not depend on the supersaturations, which may be a reasonable presupposition for low supersaturations or alternatively if we consider only small variations in supersaturation.

Thus, in principle, (provided α of the given interface is high enough) the model to be developed below has a general applicability for all kinds of crystal growth processes (vapour growth, transport growth, solution growth, precipitation, hydrothermal growth, flux growth, electrolytic growth and to a certain extent melt growth), since it is a general phenomenological model. In case of low α, where all the sites (not only ad-atoms and kinks in steps) play their role, we must play Monte Carlo computer games as mentioned in section 10.1.14.

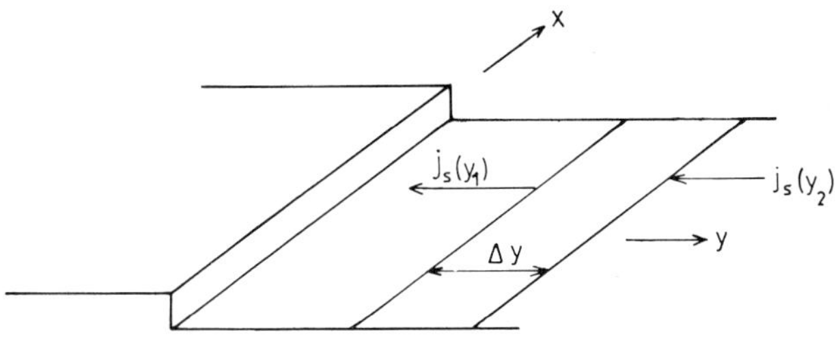

Fig. 10.11. Strip on a surface element between steps.

10.2.4 Differential equation of BCF

The first step in the BCF theory is to develop a differential equation for the potential field or supersaturation field around a straight step. Steps are then considered as non-moving sinks in which growth units disappear, or as non-moving wells in the case of dissolution. The fact that steps are considered as non-moving sinks or wells in the diffusion problem will be justified a posteriori.

Once the expression for the fluxes of growth units arriving at or leaving the steps are found, the velocity of advance of the steps in the growth problem can be calculated. In order to have a mathematically-treatable problem we will consider only systems of straight steps i.e. systems of steps which are far enough from the centre of the growth spiral. If we consider a surface strip of a length of a unit length and a width of Δy, we can distinguish two fluxes:

1) a flux j_s going to the step since a concentration gradient is built up with decreasing concentration towards the step (growth units which "land" closer to the step have a higher chance of being captured).

2) a flux j_v from the bulk to the surface which compensates the flux of growth units to the steps.

Assuming a stationary state, it follows that for a strip the continuity equation holds:

$$\frac{\mathrm{d}j_s(y)}{\mathrm{d}y} - j_v = 0\,. \tag{10.43}$$

For $j_s(y)$ we can write according to the first Fick law

$$j_s = D_s \frac{\mathrm{d}n_s}{\mathrm{d}y} = -D_s n_{s0} \frac{\mathrm{d}\alpha_s}{\mathrm{d}y} = D_s n_{s0} \frac{\mathrm{d}(\alpha - \alpha_s)}{\mathrm{d}y} = D_s n_{s0} \frac{\mathrm{d}(\sigma - \sigma_s)}{\mathrm{d}y} =$$

$$= D_s n_{s0} \frac{\mathrm{d}\psi}{\mathrm{d}y}\,. \tag{10.44}$$

Here n_s is the actual concentration, which in the case of super- or undersaturation is smaller or larger than n_{s0} the equilibrium concentration per cm^2 respectively; D_s is the surface diffusion constant; $\alpha_s'(y) = n_s/n_{s0}$ the supersaturation and $\sigma_s(y) = (n_s - n_{s0})/n_{s0}$ the relative supersaturation; α' is the supersaturation ratio very far from a step and this is then the same as α' for the bulk. In the same way σ is the relative supersaturation very far from a step. We note that $\alpha_s(y)$ and $\sigma_s(y)$ depend on y the distance to the step; $\psi = \sigma - \sigma_s(y)$ is a potential function.

For j_v we can write the expression which is a result of the difference between the flux leaving the surface

$$\frac{n_s(y)}{\tau_{\text{deads}}} = n_{s0}\,\frac{\alpha_s(y)}{\tau_{\text{deads}}}$$

and going into the surface $n_{s0}\,\dfrac{\alpha}{\tau_{\text{deads}}}$

$$j_v = (\alpha - \alpha_s(y))\frac{n_{s0}}{\tau_{\text{deads}}} = \psi\frac{n_{s0}}{\tau_{\text{deads}}}. \tag{10.45}$$

Upon substituting (10.44) and (10.45) in (10.43) it follows that

$$D_s\,\tau_{\text{deads}}\frac{d^2\psi}{dy^2} = \psi \quad \text{or} \quad \lambda_s^2\nabla^2\psi = \psi, \tag{10.46}$$

where $\lambda_s = \sqrt{(D_s\,\tau_{\text{deads}})}$ is the mean displacement.

10.2.5. Solution of the BCF differential equation

A general solution of the differential equation (10.46) is

$$\psi = a\exp\left(\frac{y}{\lambda_s}\right) + b\exp\left(-\frac{y}{\lambda_s}\right). \tag{10.47}$$

In order to solve for a and b we have to introduce boundary conditions which correspond to a certain system of steps.

Following BCF we consider the following four cases:

a) $\lambda_s \gg x_0$ (the mean distance between kinks) and one single step.

In this case steps can be considered as continuous steps since each growth unit can find its kink site. If we take the origin in the step then it follows that the boundary conditions are

$$\text{for} \quad y = \pm\infty; \; \psi = \sigma_{s\infty} - \sigma_{s\infty} = 0;$$

$$y = 0; \; \psi = \sigma_{s\infty} - \sigma_{\text{step}} = \sigma - \sigma_{\text{step}} = \zeta\sigma\,,$$

where we define ζ as

$$\zeta = \frac{\sigma - \sigma_{\text{step}}}{\sigma}. \tag{10.48}$$

ζ is to be considered as a retardation factor. An expression for ζ as a function of τ_k is developed below. It follows from the given boundary conditions that

$$\psi = \zeta\sigma \exp -\frac{y}{\lambda_s}. \tag{10.49}$$

b) $\lambda_s \gg x_0$ and a parallel sequence of equidistant steps. The width of a "terrace" is λ_0 and the origin is chosen in the centre of a "terrace".

Then we have the following boundary conditions* (see fig. 10.12)

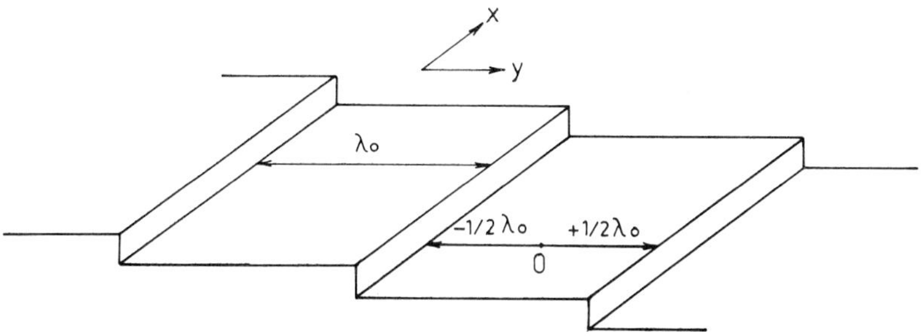

Fig. 10.12. Series of steps.

* Here we have assumed that $\zeta_{\text{high}} = \zeta_{\text{low}} = \zeta$ i.e. that the relation times for entering a step from the upper site is equal to the relaxation time for entering the step from the lower site. Recently it has been shown by Schwoebel that the case τ_k (right) $\neq \tau_k$ (left), has interesting implications for the stability of the step trains. This will be discussed in section 10.4.5.

$$\text{for } y = +\tfrac{1}{2}\lambda_0;\ \psi = \zeta\sigma = a \exp\left(\frac{\lambda_0}{2\lambda_s}\right) + b \exp\left(-\frac{\lambda_0}{2\lambda_s}\right)$$

$$y = -\tfrac{1}{2}\lambda_0;\ \psi = \zeta\sigma = a \exp\left(-\frac{\lambda_0}{2\lambda_s}\right) + b \exp\left(\frac{\lambda_0}{2\lambda_s}\right).$$

This gives

$$\psi = \frac{\zeta\sigma \cosh \dfrac{y}{\lambda_s}}{\cosh \dfrac{\lambda_0}{2\lambda_s}}. \tag{10.50}$$

c) $\lambda_s < x_0$ without diffusion in a step. The solution of the differential equation is more complex, since then for the circular diffusion fields around the single sinks a two-dimensional differential equation must be solved using Bessel functions. This has been done by BCF in their paper and they found for a parallel sequence of steps

$$\psi = \zeta\, c_0\, \sigma \frac{\cosh \dfrac{y}{\lambda_s}}{\cosh \dfrac{\lambda_0}{2\lambda_s}}. \tag{10.50a}$$

The only difference with (10.50) is the factor c_0 which is given by

$$c_0 = \pi \frac{\lambda_s}{x_0} \ln \frac{2\lambda_s}{1.78a}, \tag{10.51}$$

since $\lambda_s < x_0$; then $c_0 < 1$.

d) $\lambda_s \ll x_0$ with diffusion in the step. An expression is derived which is similar to (10.50a) except that c_0 becomes a very complicated expression which is also dependent on $\lambda_0/2\lambda_s$.

10.2.6. Derivation of an expression for the retarding factor ζ for $\lambda_s \gg x_0$

Since ζ is very important for the kinetics of the entry of growth units into steps, a derivation in terms of τ_k for adsorption at a kink site of a step will be given. It follows from $\lambda_s \gg x_0$ that $c_0 = 1$.

The flux into the step from one side is determined by the average number of adsorbed molecules on that side which are within an elementary surface jump distance a of the step. Since n_{step} is the number of adsorbed molecules per unit surface adjacent to the step, an_{step} is the number per unit step length and an_{step}/τ_k is the number per unit step length per time unit going to the step from that side. The net flux j_s to the step is obtained by subtracting from this the number per unit length per unit time leaving the step; but since in equilibrium j_s must be zero we have, with $\sigma_{step} = (n_{step} - n_{s0})/n_{s0}$

$$j_s = \frac{an_{step}}{\tau_k} - \frac{an_{s0}}{\tau_k} = \frac{a}{\tau_k}(n_{step} - n_{s0}) = \frac{an_{s0}}{\tau_k}(1 - \zeta) . \tag{10.52}$$

Here we used (10.48) for ζ. On the other hand

$$j_s = D_s n_{s0} \left.\frac{d\psi}{dy}\right|_{y=\frac{1}{2}\lambda_0} = \left.\frac{D_s n_{s0}\, \zeta\sigma \sinh \dfrac{y}{\lambda_s}}{\lambda_s \cosh \dfrac{\lambda_0}{2\lambda_s}}\right|_{y=\frac{1}{2}\lambda_0} . \tag{10.53}$$

Setting the expression for j_s in (10.52) equal to that in (10.53) we obtain

$$\zeta = \left[1 + \frac{D_s\tau_k}{a\lambda_s} \tanh \frac{\lambda_0}{2\lambda_s}\right]^{-1} . \tag{10.54}$$

10.2.7. Derivation of the velocity of advance of steps; a justification of its neglect

We define M_{hkl} as the mesh area of the F face (hkl), N_m as the number of growth units in a mesh and $f_0 = M_{hkl}/N_m$ as the surface occupied by one growth unit in the crystal (in German $f_0 = m_{\text{Flächenbedarf}}$). To obtain v_∞, the velocity of advance of a straight step, the flux of growth units entering the step, j_s, must be multiplied by f_0 because the area f_0 has to be distributed over the whole unit length being considered. For a parallel sequence of steps (this is the case we need for further development of the theory) it follows from (10.53)

$$v_\infty = 2j_s f_0 = 2\frac{D_s n_{s0} f_0 \sigma}{\lambda_s} \tanh \frac{\lambda_0}{2\lambda_s} . \tag{10.55}$$

(The factor 2 is introduced since growth units enter the step from two sides.) It will be shown that the neglect of the advance velocity of the step in the diffusion problem is justifiable.

Since the mean velocity of a growth unit diffusing on the surface is given by $\lambda_s/\tau_{\text{deads}}$, the step velocity may be neglected if

$$\frac{v_\infty}{\lambda_s/\tau_{\text{deads}}} \ll 1 .$$

After substituting $\lambda_s = \sqrt{(D_s \tau_{\text{deads}})}$ and (10.55) we rewrite the previous condition

$$\frac{v_\infty}{\lambda_s/\tau_{\text{deads}}} = 2\, n_{s0}\, \zeta\, f_0 \sigma \tanh \frac{\lambda_0}{2\lambda_s} \ll 1 .$$

Now $\zeta \leqslant 1$, $\tanh(\lambda_0/2\lambda_s) \leqslant 1$, and also $n_{s0}/f_0 \leqslant n_{s0} M_{hkl}/N_m \leqslant 1$, since the number of growth units in the adsorption layer on a part of surface M_{hkl} of the F face (hkl) must be $\leqslant N_m$, the number of growth units in the crystal corresponding to the same part of the surface.

Therefore the condition holds if $\sigma \ll 1$ (i.e. for low supersaturations). The neglect of the step velocity, which is an essential assumption in the mathematical description of the diffusion problem, is also permissible in dense systems, where $n_{s0} M_{hkl}/N_m \leqslant 1$. It is difficult to give an upper limit below which the quasi-steady state theory of BCF holds. The limit is higher the lower the concentration n_{s0} on the surface and the smaller ζ and $\tanh(\lambda_0/2\lambda_s)$.

10.2.8. Derivation of an expression for the rate of growth of an F face, growing by a spiral growth mechanism

To derive a formula for R, the rate of growth of an F face, for the spiral growth theory, we introduce the following quantities (see fig. 10.13):

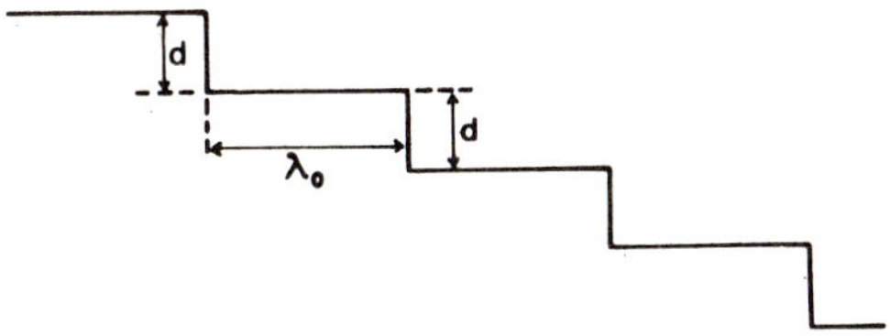

Fig. 10.13. Profile of a step train.

1) the linear density of steps ρ_{st} per unit length on a surface fully covered by a spiral, is given by $\rho_{st} = 1/\lambda_0$;

2) the flux of steps F_{st} advancing in the direction of the Y-axis, is given by $F_{st} = \rho_{st} v_\infty$.

The rate of growth R, perpendicular to a face in unit length per unit time is then given by $R = F_{st} d = v_\infty \rho_{st} d = v_\infty d/\lambda_0$. Using the expression

$$R = v_\infty d/\lambda_0 \tag{10.56}$$

we find, after substituting (10.55) for v_∞, (10.41) for λ_0 (corrected for s) and (10.39) for r^* in (10.56), that

$$R = \frac{\zeta D_s n_{s0} \Omega \sigma^2}{\lambda_s^2 \quad \sigma_1} \tanh \frac{\sigma_1}{\sigma}, \tag{10.57}$$

where $\Omega = M_{hkl}\, d_{hkl}/N_m = f_0 d$ is the mean volume occupied by a growth unit in the crystal and

$$\sigma_1 = \frac{9.5\,\gamma' a}{skT\lambda_s} \text{ and } \frac{\lambda_0}{2\lambda_s} = \frac{\sigma_1}{\sigma}.$$

Conclusion: From the given argument it follows that, for the surface diffusion model, the relation between the rate of growth R and the relative supersaturation σ of an F face, regulated by a spiral growth mechanism, is given by

$$R = C \frac{\sigma^2}{\sigma_1} \tanh \frac{\sigma_1}{\sigma}, \tag{10.57a}$$

in which the two parameters

$$C = \frac{\zeta D_s n_{s0} \Omega}{\lambda_s^2} \tag{10.58}$$

and

$$\sigma_1 = \frac{9.5\,\gamma' a}{skT\lambda_s} \tag{10.59}$$

together with ζ play the essential roles. C determines the absolute value of R while σ_1 and ζ determine the shape of the $R(\sigma)$ curve, as will be shown in the next section.

10.3. Implications of the BCF theory

10.3.1. Mathematical description and physical interpretation of the theoretical $R(\sigma)$ curves

If $\sigma \ll \sigma_1$ we write

$$R = C\frac{\sigma^2}{\sigma}\tanh\frac{\sigma_1}{\sigma_1} = C\frac{\sigma^2}{\sigma_1}\,\frac{\exp\left(2\dfrac{\sigma_1}{\sigma}\right) - 1}{\exp\left(2\dfrac{\sigma_1}{\sigma}\right) + 1} \approx C\frac{\sigma^2}{\sigma_1},$$

representing a parabolic law.
If $\sigma \gg \sigma_1$:

$$R = C\frac{\sigma^2}{\sigma_1}\tanh\frac{\sigma_1}{\sigma_1} = C\frac{\sigma^2}{\sigma_1}\,\frac{\exp\left(2\dfrac{\sigma_1}{\sigma}\right) - 1}{\exp\left(2\dfrac{\sigma_1}{\sigma}\right) + 1} = C\frac{\sigma^2}{\sigma_1}\,\frac{\left(1 + \dfrac{2\sigma_1}{\sigma} + \ldots\right) - 1}{\left(1 + \dfrac{2\sigma_1}{\sigma} + \ldots\right) + 1} \approx C\sigma,$$

a linear relationship.

When the two curves describing each of these relations are plotted they

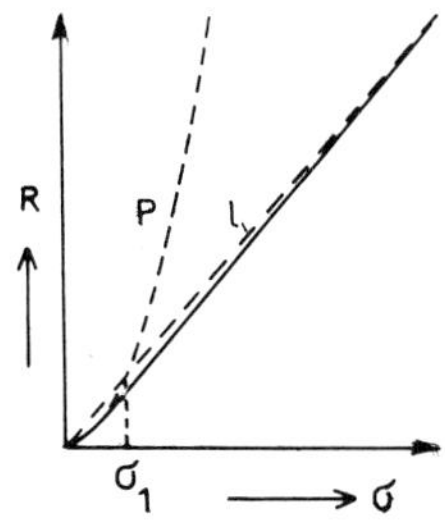

Fig. 10.14. Extrapolated parabolic and linear BCF laws.

intersect in a point which represents σ_1, since, for the point of intersection

$$R = C\frac{\sigma^2}{\sigma_1} = C\sigma, \text{ thus } \sigma = \sigma_1 .$$

As can be seen from fig. 10.14 formula (10.57), which expresses the proper theoretical curve of BCF, gives a very gradual transition from the parabolic to linear law for the approximate range $\frac{1}{2}\sigma_1 \leqslant \sigma \leqslant 5\,\sigma_1$.

If we assume $a < \lambda_s < 10^4 a$, $10^{-1} < \gamma'/kT < 10$ and $1 < s < 10$, we find that σ_1 can have values between $10^{-5} < \sigma_1 < 10^2$.

Fig. 10.15 shows theoretical $R(\sigma)$ curves, with the same arbitrary constant

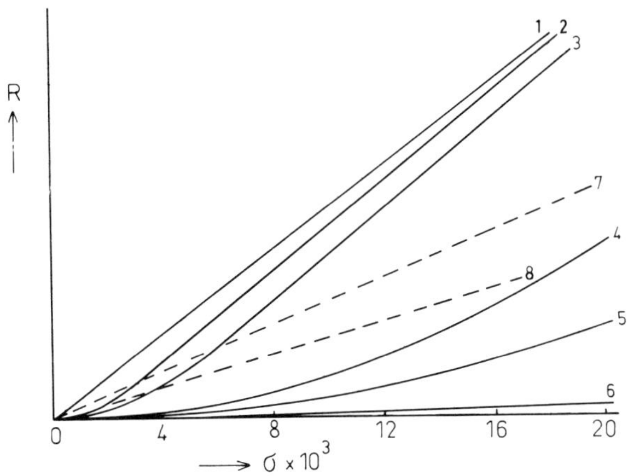

Fig. 10.15. Theoretical BCF curves (1 – 6) with arbitrary C; σ_1 has the following values: 1) 10^{-3} ; 2) 5×10^{-3} ; 3) 10^{-2} ; 4) 5×10^{-2} ; 5) 10^{-1} ; 6) 1. Curves 7 and 8 are two second linear laws with arbitrary C' $(0 < C' < C)$.

C and different values of σ_1. It can be seen that the smaller the values of σ_1, the more nearly linear become the curves.

The following physical interpretation of the form of the $R(\sigma)$ curves can be given. Combining (10.56) and (10.57) gives:

$$R = \frac{v_\infty \bar{d}}{\lambda_0} = C\frac{\sigma^2}{\sigma_1}\tanh\frac{\sigma_1}{\sigma} = C\sigma\left[\frac{\sigma}{\sigma_1}\tanh\frac{\lambda_0}{2\lambda_s}\right].$$

The factor between the brackets can be considered as a correction term, giving a measure of the interaction of diffusion fields around the successive steps.

For low supersaturations $\sigma \ll \sigma_1$ or $\lambda_s \ll \lambda_0$, all growth units arriving in a region of width $2\lambda_s$ around a particular step have a chance of entering that step alone. Since there is no interaction of diffusion fields $\tanh(\sigma_1/\sigma) = 1$. Then it follows from (10.55), that v_∞ (proportional to $\sigma \tanh(\sigma/\sigma_1)$) becomes proportional to σ. The density of steps ρ_{st} ($= 1/\lambda_0 = 1/19r^* = kT\sigma/19\gamma' a$) is also proportional to σ. This gives a parabolic $R(\sigma)$ curve.

On the other hand, for relatively high supersaturations $\sigma \gg \sigma_1$ or $\lambda_s \gg \lambda_0$, the density of steps $1/\lambda_0$ (proportional to σ) will be high. In this case there is a strong interaction of diffusion fields around the steps since any growth unit arriving on the surface has a chance of entering one of several steps. Now the factor $\tanh \sigma_1/\sigma \approx \sigma_1/\sigma$ which makes v_∞ (proportional to $\sigma \tanh \sigma_1/\sigma$) independent of σ. This makes R proportional to σ.

It is evident that the parabolic part will be shifted to lower values of σ and the $R(\sigma)$ curves "linearized" if $\sigma_1 = 9.5\gamma' a/kT\lambda_s$ becomes small. This is possible if γ' is low (for example in the case of adsorption), or λ_s large. We conclude that the value of σ_1 (dependent on γ' and λ_s) is a measure of the interaction of diffusion fields, and so determines the shape of the $R(\sigma)$ curves below.

10.3.2. Occurrence of a second parabolic law; importance of surface diffusion

In the case where the growth units enter the steps directly from the solution into the kinks of the step, we obtain what may be called a second parabolic law. This can be derived immediately as follows. The net number of growth units entering a step per unit length per time unit is given by

$$\frac{l^2 N_0}{\tau_k'}\sigma \qquad (10.60)$$

and the advance velocity of that same step is given by

$$\frac{l^2 N_0 \sigma}{\tau_k'} f_0 \tag{10.60a}$$

Here l is the mean free path of a growth unit in the solution ($l \approx a$), N_0 the concentration of growth units per cm^3 and τ_k' the relaxation time for entering the steps directly. (Note the difference between τ_k and τ_k'. The first one is the relaxation time for entering the step from the surface). Using the expressions (10.41a) for λ_0, (10.56) for R and upon realizing that again $f_0 d = \Omega$, we find

$$R = C' \sigma^2 , \tag{10.61}$$

where

$$C' = \frac{N_0 \, l^2}{\tau_k' a} \frac{s}{19} \frac{kT}{\gamma'} \Omega \approx \frac{N_0 l}{\tau_k'} \frac{s}{19} \frac{kT}{\gamma'} \Omega . \tag{10.62}$$

Let us now assume that there are two ways or two "reaction paths" for the crystallization process namely (i) a direct integration mechanism derived above, (ii) a BCF diffusion mechanism. Let us compare these two mechanisms. Let us assume that the second parabolic law which results from (i) cuts the BCF law which results from (ii) for the value $\sigma = \sigma_2$ (see fig. 10.16). It will be shown that, in most cases, $\sigma_2 \gg \sigma_1$, so that we can take, as discussed in §3.1, the linear BCF law as a basis for comparison. Upon equating (10.61) with (10.57) and substituting (10.62) for C', (10.58) for C and (10.59) for σ_1 and writing for $\lambda_s^2 = D_s \tau_{\text{deads}}$ we find for σ_2, as a result of the equation

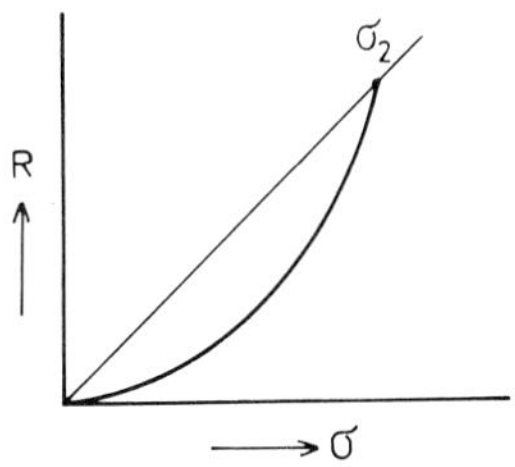

Fig. 10.16. Second parabolic law cutting a linear BCF law.

$C' \sigma_2^2 = C \sigma_2$, the following expression for σ_2,

$$\sigma_2 = \zeta \frac{19}{s} \frac{\gamma'}{kT} \frac{\tau_k'}{\tau_{\text{desolv}}} . \qquad (10.63)$$

We also used the expression $n_{s0} = l N_0 \dfrac{\tau_{\text{deads}}}{\tau_{\text{desolv}}}$ derived below. (See (10.74)). Using (10.59) we find that

$$\frac{\sigma_2}{\sigma_1} = 2\zeta \frac{\tau_k'}{\tau_{\text{desolv}}} \frac{\lambda_s}{a} \qquad (10.64)$$

Since we assume a linear law ζ must be equal to about unity. (This will be discussed below.)

The process of entering the kinks directly from the solution must be, in general, a more complex process than the process of entering the adsorption layer from the solution. Hence, in most cases

$$\tau_k' > \tau_{\text{desolv}} \text{ or } \tau_k' \gg \tau_{\text{desolv}} . \qquad (10.65)$$

We may safely assume that the process of leaving the surface is a much more complex process than making a surface diffusion jump, so that growth units will “stick” on the surface once they enter the adsorption layer. It thus follows that since $\lambda_s = \surd(D_s \tau_{\text{deads}})$ and $D_s = a^2/\tau_{\text{sdiff}}$ from

$$\tau_{\text{deads}} > \tau_{\text{sdiff}} \text{ or } \tau_{\text{deads}} \gg \tau_{\text{sdiff}} \qquad (10.66)$$

that

$$\lambda_s > a \text{ or } \lambda_s \gg a .$$

It can be seen from (10.64), (10.65) and (10.66) that for most crystallization processes

$$\sigma_2 \gg \sigma_1 . \qquad (10.67)$$

A lowest possible value of σ_2 can be estimated by taking the highest possible $s \approx 10$, a lowest possible $\gamma'/kT \approx 10^{-1}$ (for growth from solution), so that it follows from (10.63)

$$\sigma_{2\,\mathrm{min}} \approx 2 \times 10^{-1} \,.$$

We see that in the field, where (R, σ) curves are measured and where the BCF theory applies (say $0 \leqslant |\sigma| \approx 3 \times 10^{-1}$) the curve of the second parabolic law will probably never cut the BCF curve. This argument also holds if we take into consideration that the real linear BCF law probably never will be reached for reasonable relative supersaturations up to a few tens of a percent.

We will now briefly consider the occurrence of a second parabolic law, in which surface diffusion does play some role. If we assume that

$$\frac{D_\mathrm{s}\,\tau_\mathrm{k}}{\lambda_\mathrm{s}\,a} \tanh \frac{\lambda_0}{2\lambda_\mathrm{s}} \gg 1 \,, \tag{10.68}$$

it then follows from (10.54) that again a second parabolic law [29a,b,c] results:

$$R = C'_\mathrm{sdiff}\,\sigma^2 \tag{10.61a}$$

and

$$C'_\mathrm{sdiff} = n_{\mathrm{s}0} \frac{s}{9.5} \frac{kT}{\gamma'} \frac{\Omega}{\tau_\mathrm{k}}. \tag{10.62a}$$

If we write $n_{\mathrm{s}0} = l N_0 \dfrac{\tau_\mathrm{deads}}{\tau_\mathrm{desolv}}$ we then find for C'_sdiff

$$C'_\mathrm{sdiff} = l N_0 \frac{s}{19} \frac{kT}{\gamma'} \frac{\tau_\mathrm{deads}}{\tau_\mathrm{desolv}\,\tau_\mathrm{k}} \Omega \,. \tag{10.62b}$$

Upon comparing (10.62b) with (10.62) we find

$$\frac{C'_\mathrm{sdiff}}{C'} = \frac{1}{2} \frac{\tau_\mathrm{deads}\ \tau_\mathrm{k}'}{\tau_\mathrm{desolv}\ \tau_\mathrm{k}} \,. \tag{10.69}$$

Since in most cases $n_{s0} > l N_0$ because the concentration of growth units in the adsorption layer will be larger than in the solution $\tau_{deads} > \tau_{desolv}$. The process of going directly into the kinks will be more difficult than going from the surface into the kinks hence $\tau_k' > \tau_k$. That means

$$C'_{sdiff} > C' , \tag{10.70}$$

so that a second parabolic law (with surface diffusion) seems to be more probable than a second parabolic law without surface diffusion.

We thus arrive at the important conclusion that in most crystallization processes a mechanism with surface diffusion will operate. A second parabolic law (with surface diffusion) or a parabolic depression due to the fact that the second part in (10.54) is not $\ll 1$, however, may occur.

We note that the surface diffusion itself may be a very complicated process and that the relaxation time for making a surface diffusion jump may be a very long one as compared to the comparable relaxation time for diffusion in the bulk of the solution or the vapour. But as long as $\tau_{deads} > \tau_{sdiff}$ and $\tau_{deads} > \tau_{desolv}$, so that growth units stick on the surface, surface diffusion plays an essential role in the growth process.

The conclusion that surface diffusion is very important in most crystal growth processes also holds if we have other sources of steps than a spiral or if the production of steps is partly disturbed by for example bunching, because the whole argument follows from a comparison of two mechanisms with the same average density of steps, i.e. the same $1/\lambda_0$ where λ_0 is given by (10.41a).

10.3.3. Relation between the BCF theory and the morphological Hartman-Perdok theory

It will be shown that, if we introduce the hypothesis that the higher E_{att} is (as defined in the morphological crystallographic Hartman-Perdok theory), then the higher is λ_s [29] (because the higher E_{att} is, the longer do growth units "stick" on the surface), the empirical rule of the Hartman-Perdok theory [30] (namely that the larger is E_{att} for a certain face (*hkl*), the less is its morphological importance (*MI*) on the growth form), can then be justified. Since in general the morphologically less important faces are the fastest growing ones, we can write, if we compare two different crystallographic F faces

$$E_{att_1} > E_{att_2} \text{ which implies that } MI_1 < MI_2 \text{ and } R_1 > R_2 .$$

With the hypothesis that

$$\lambda_{s_1} > \lambda_{s_2} \text{ if } E_{\text{att}_1} > E_{\text{att}_2}$$

it then follows from (10.59) that if we assume that, for both faces, γ' and s are (about) the same $\sigma_{1,1} < \sigma_{1,2}$. But then it follows, provided we are in the non-linear region of the BCF curves (say $\sigma \lessapprox 5\,\sigma_1$) (see Fig. 10.14), that then indeed $R_1 > R_2$ in agreement with the Hartman-Perdok rule given above.

10.3.4. Occurrence of a second linear law

It will now be shown that a "second linear law" may occur in the case of the combination of spirals as given in Fig. 10.10 [29]. It was shown that

$$s = \frac{s'}{1 + \dfrac{L_{\text{gr}}}{19\,r^*}}\,. \tag{10.42}$$

If $L_{\text{gr}} \ll 19\,r^*$, $s = s'$ and we have the normal case of s cooperating spirals. However, if $L_{\text{gr}} \gg 19\,r^*$ occurring when there are many ends of screw dislocations lying in a line with length L_{gr} then $\lambda_0 = L_{\text{gr}}/s = 2e$, where $2e$ is the mean distance between the dislocations. It follows from the surface diffusion model that an expression for R is given by

$$R = C'\sigma$$

$$\text{with } C' = C\frac{\lambda_s}{e}\tanh\frac{e}{\lambda_s}\,.$$

Now we can distinguish two cases, namely: a) if $\lambda_s/e \ll 1$, then $C' \approx 0$; b) if $\lambda_s/e \gg 1$, then $C' \approx C$. Thus a second linear law $R = C'\sigma$ is obtained with a slope

$$0 \leqslant C' \leqslant C\,.$$

In fig. 10.15 two second linear laws with $0 \leqslant C' \leqslant C$ are given (7 and 8).

It may very well be that for some other reason (bunching of steps, formation of macrosteps, etc.; this will be treated in section 4) the density of steps emitted from the step sources does not depend linearly on the supersatura-

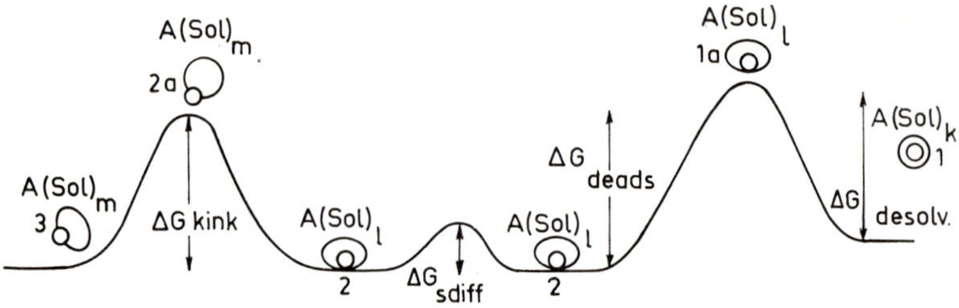

Fig. 10.17. Schematized states of solvated growth units A and potential barriers that must be overcome in order that a growth unit can enter a kink.

tion. In such a case deviations from a parabolic law do occur, whereas, according to the BCF theory, such a law must hold.

10.3.5. Expressions for relaxation times and slopes of the BCF curve

For growth from the vapour or solution we define the following states of a growth unit [29] * (see fig. 10.17):

1) Solvated growth unit in the solution, denoted by $A(Sol)_k$;
1a) Activated partially solvated growth unit for entering the adsorption layer, denoted by $A(Sol)_l$;
2) The same partially dehydrated growth unit in the adsorption layer;
2a) An activated dehydrated growth unit for entering a kink, denoted by $A(Sol)_m$;
3) The same dehydrated growth unit half covered with a solvation shell in a kink, denoted by $A(Sol)_m$.

It follows from the definition given above that $k \geqslant l \geqslant m$.

Expressions for the relative supersaturations are given upon using Eyrings formalism [31a,b]

$$\tau = \nu^{-1} \exp(Q/kT) , \tag{10.71}$$

where Q is ΔG_{desolv}, ΔG_{deads}, ΔG_{sdiff}, ΔG_k respectively for the activation free energies which correspond with τ_{desolv}, τ_{deads}, τ_{sdiff}, and τ_k**. ν^{-1} (ν = the atomic frequency factor) can be taken as $h/kT \approx 10^{-13}$ sec (h = the constant of Planck). If we substitute the expression (10.71) in the

* Even if there is more than one type of growth unit we consider them as one growth unit, since we attribute the same relaxation time to the different growth units.

**We use here Eyrings kinetic formalism; in most cases it is possible to calculate activation (free) energies from empirical data using Arrhenius plots. It does not mean that the simple kinetic picture of Eyring of a single particle jumping over an "activated complex" barrier holds.

expression for λ_s where $\lambda_s = \sqrt{(D_s \tau_{\text{deads}})}$ and $D_s = a^2/\tau_s$ and for ζ (10.54) we find

$$\lambda_s = a \exp\left[(\Delta G_{\text{deads}} - \Delta G_{\text{sdiff}})/2kT\right] \tag{10.72}$$

and

$$\zeta = \left[1 + \exp\left\{\tfrac{1}{2}\left(-\Delta G_{\text{deads}} - \Delta G_{\text{sdiff}} + 2\Delta G_k\right)\right\} \tanh\frac{\sigma_1}{\sigma}\right]^{-1}. \tag{10.73}$$

In order to derive an expression for C (10.58), we assume that the growth units in the adsorption layer are in dynamic equilibrium with the growth units in a neighbouring solution layer with a thickness equal to the mean free path l in the solution.

The number of growth units per cm^2 per second leaving the adsorbed layer must be equal to the flux entering from the adjacent volume or

$$\frac{n_{s0}}{\tau_{\text{deads}}} = l\frac{N_0}{\tau_{\text{desolv}}}, \tag{10.74}$$

so that upon using the expressions for the relaxation times we can write

$$n_{s0} = l\,N_0 \exp\left(\Delta G_{\text{deads}} - \Delta G_{\text{desolv}}\right)/kT\,. \tag{10.75}$$

After substituting (10.75) and (10.72) and $D_s = a^2/\tau_{\text{sdiff}}$ in the expression for C, (10.58), we find

$$C = \nu\,\zeta\,l\,\Omega\,N_0 \exp\left(-\Delta G_{\text{desolv}}/kT\right). \tag{10.76}$$

We note that the expression

$$\nu\,l\,N_0 \exp\left(-\frac{\Delta G_{\text{desolv}}}{kT}\right) = \frac{l\,N_0}{\tau_{\text{desolv}}}$$

is the number of growth units, which hit a surface of 1 cm^2 per second. In the case of growth from the vapour this expression becomes

$$\frac{s_t\,p_0}{(2\pi\,m\,kT)^{\frac{1}{2}}},$$

where s_t is the sticking coefficient $0 \leqslant s_t \leqslant 1$, so that

$$C = \frac{\zeta \, \Omega \, p_0 \, s_t}{(2\pi \, m \, kT)^{\frac{1}{2}}}. \tag{10.77}$$

Here p_0 is the equilibrium pressure, and m is the mass of a growth unit.

10.3.6. Back-stress effect on step sources

Cabrera and Coleman [32] have considered the effect of surface diffusion fields on the spacing λ_0 of the steps produced by a screw dislocation source. They find that the Cabrera and Levine [28] result in eq. (10.41) gives steps which are too closely spaced at high supersaturations; this is due to the assumption that the supersaturation at the center of the spiral step is σ, the supersaturation in the volume, whereas it will actually be somewhat lower due to the flux of adsorbed molecules to neighbouring arms of the spiral.

The surface supersaturation a distance λ_0 from an isolated step is

$$\sigma_{s,\lambda_0} = \sigma \left[1 - \frac{e^{-\lambda_0/\lambda_s}}{1 + \dfrac{D_s \tau_k}{a \lambda_s}} \right], \tag{10.78}$$

according to eq. (10.49) and (10.54). This is approximately the supersaturation "seen" by the central turn of the spiral step and the corresponding r^* is

$$r^* = \frac{\gamma' a}{kT \, \sigma_{s,\lambda_0}}$$

from eq. (10.39) and the distance λ_0 between turns of the spiral is related to this by

$$\lambda_0 = 19 r^* = \frac{19 \gamma' a}{kT \sigma \left[1 - \dfrac{e^{-\lambda_0/\lambda_s}}{1 + \dfrac{D_s \tau_k}{a \lambda_s}} \right]} = \frac{2\sigma_1 \lambda_s}{\sigma} \left[1 - \frac{e^{-\lambda_0/\lambda_s}}{1 + \dfrac{D_s \tau_k}{a \lambda_s}} \right]^{-1} \tag{10.79}$$

(see section 2.2).

When $\lambda_0 \gg \lambda_s$ or $\tau_k \gg a\,\lambda_s/D_s$, the exponential term in the denominator is never large and the Cabrera and Levine equation is accurate. This includes the parabolic region of the growth rate curve, and the second parabolic law where incorporation into the kink sites is rate determining. When $\tau_k < a\,\lambda_s/D_s$ and $\lambda_0 < \lambda_s$, the exponential term is important and in the case of very small λ_0 and τ_k we can expand the exponential to get

$$\lambda_0 = \frac{2\sigma_1\lambda_s^2}{\sigma\,\lambda_0} \quad \text{or} \quad \frac{\lambda_0}{2\lambda_s} = \left(\frac{\sigma_1}{2\sigma}\right)^{\frac{1}{2}} \tag{10.80}$$

This implies that the spacing λ_0 decreases more slowly with increasing σ than predicted by Cabrera and Levine.

Cabrera and Coleman approximated the center of the spiral step by a series of concentric circles and solved the diffusion equations for circular symmetry. This probably gives a better approximation than ours, and they find a relation corresponding to (10.80),

$$\frac{\lambda_0}{2\lambda_s} = \left(\frac{\sigma_1}{\sigma}\right)^{\frac{1}{3}}$$

giving an ever slower decrease in λ_0 with increasing σ.

The implication of the back-stress effect is that the surface of a growing crystal should be quite flat if λ_s is large, since slopes much steeper than $p = a/\lambda_s$ are quite difficult to produce by a screw dislocation source (when $\tau_k < a\lambda_s/D_s$). Impurities which poison kink sites and effectively make $\tau_k \gg a\lambda_s/D_s$ can allow steep slopes to be produced. In many situations the production of whiskers may be related to impurity mechanisms which allow the formation of steep sides around step sources.

The effect of the back-stress effect on the $R(\sigma)$ curve is only to make the transition from the parabolic part to the linear part, which, as can be seen from fig. 10.15, is already very gradual, even more gradual than in the classical BCF curve. The size of the parabolic part is not affected since eq. (10.41a) holds in this region.

10.3.7. Diffusion in the volume

Volume diffusion processes play an important role in the growth of a crystal when large volume concentration gradients appear. The discussion up to this point has neglected any effect due to gradients in the volume, but they will cause the volume diffusion process to be rate-determining if they are so large that concentration differences in the volume are larger than σN_0, the difference between the concentration adjacent to the crystal surface and the equilibrium concentration. This condition is possible in both solution growth and multicomponent vapour growth.

Gilmer, Ghez, and Cabrera [33] have treated the coupled volume and surface diffusion equation which result when the concentration in the volume is allowed to depend upon position. They find that the volume concentration can be approximated by a function depending only on the distance to the crystal surface when the condition is satisfied:

$$\frac{\lambda_s}{l} \ll \frac{\tau_{\text{desolv}}}{\tau_{\text{vdiff}}},$$

where τ_{vdiff} is the average time between diffusion jumps of growth units in the volume. This condition is usually satisfied for multicomponent vapour growth since l is relatively large, but for solution growth it is only true when τ_{desolv} is very large. The physical reason for this condition can be seen quite easily in the special case when $\tau_{\text{vdiff}} = \tau_{\text{desolv}}$ and the condition reduces to $\lambda_s \ll l$. Surface diffusion concentration changes in the y direction occur over regions of about λ_s in length, and since this is smaller than the mean free path in the volume, the volume concentration must be effectively constant over these regions. (An appreciable gradient cannot be maintained over distances less than one mean free path.)

We will follow a procedure equivalent to that used by Bennema [29c] and write

$$N(x,y,z) = N(z) = N_0 \frac{\sigma_v - \sigma}{\delta} z + N_0 (\sigma + 1),$$

where the z axis is perpendicular to the surface, σ_v is the supersaturation at the top of a hypothetical "unstirred layer" of fluid adjacent to the crystal surface, and δ is the thickness of this layer. This function obviously satisfies the steady-state volume diffusion equation

$$\nabla^2 N(x,y,z) = 0$$

and gives the proper supersaturation at $z = \delta$; and at $z = 0$ the supersaturation in the adjacent volume is σ, as in the BCF surface diffusion model.

The flux to the crystal surface is given by

$$j_v = D_v \frac{\partial N}{\partial z} = D_v N_0 \frac{\sigma_v - \sigma}{\delta}$$

and this must be the average number of growth units being incorporated into the growing crystal per unit time per unit area. This is related to the growth rate by

$$R = j_v \Omega .$$

From this we obtain

$$\sigma_v - \sigma = R \frac{\delta}{D_v N_0 \Omega}. \qquad (10.81)$$

The expression in (10.57) for the growth rate due to a supersaturation σ in the adjacent volume can be written

$$R = \frac{\sigma n_{s0} \Omega / \tau_{\mathrm{deads}}}{\dfrac{\lambda_0}{2\lambda_s} \coth \dfrac{\lambda_0}{2\lambda_s} + \dfrac{\tau_k}{\tau_{\mathrm{deads}}} \dfrac{\lambda_0}{2a}}, \qquad (10.82)$$

where we have used the explicit expression for ζ from (10.54), we have replaced σ_1/σ by the exact ratio $\lambda_0/2\lambda_s$, and we have used $\lambda_s^2 = D_s \tau_{\mathrm{deads}}$. Eliminating σ between (10.81) and (10.82) and solving the resulting equation for R gives

$$R = \frac{\sigma_v l N_0 \Omega}{\tau_{\mathrm{desolv}}} \left[\frac{\delta l}{\tau_{\mathrm{desolv}} D_v} + \frac{\lambda_0}{2\lambda_s} \coth \frac{\lambda_0}{2\lambda_s} + \frac{\tau_k}{\tau_{\mathrm{deads}}} \frac{\lambda_0}{2a} \right]^{-1}, \qquad (10.83)$$

where we have used (10.74) and the definition of τ_{desolv} and τ_{deads}. The largest of the three terms in the denominator of (10.83) determines the rate-limiting process: If the first term dominates, then the volume diffusion process is slowest and limits the growth rate; if the second is largest then surface diffusion and/or entry into the adsorbed layer is rate limiting and the equation reduces to a normal BCF relation (10.57) with $\zeta = 1$; if the third term is the largest then the integration of growth units into the kink sites is rate limiting and the expression reduces to (10.61a).

The situation in which the volume diffusion process is rate limiting for some value of λ_0 must be considered in detail since it has not been discussed. For simplicity we will assume that the kink integration term of (10.83) is small compared with the second term, and neglect it. Then the condition that the volume diffusion term dominate for some value of λ_0 (or σ) is

$$\frac{\delta\, l}{\tau_{\text{desolv}} D_{\text{v}}} > 1\,,$$

since the smallest value of $\lambda/2\lambda_s$ coth $\lambda/2\lambda_s$ is 1. Writing $D_{\text{v}} = l^2/\tau_{\text{vdiff}}$, we obtain

$$\frac{\delta}{l}\tau_{\text{vdiff}} > \tau_{\text{desolv}}\,, \tag{10.84}$$

a result we might have expected since τ_{vdiff} determines the rate of transport in the volume and τ_{desolv} determines the rate of entry into the adsorbed layer.

The expression $\lambda_0/2\lambda_s = \sigma_1/\sigma_{\text{v}}$ can be used for low supersaturations and large step spacings provided neither the volume diffusion fields nor the surface diffusion fields of neighbouring steps overlap. Using this and assuming

$$\frac{\delta}{l}\tau_{\text{vdiff}} \gg \tau_{\text{desolv}}\,,$$

we obtain

$$R = \frac{\sigma_{\text{v}} N_0\, \Omega\, l}{\tau_{\text{desolv}}}\left[\frac{\delta\, l}{\tau_{\text{desolv}} D_{\text{v}}} + \frac{\sigma_1}{\sigma_{\text{v}}}\right]^{-1}, \tag{10.85}$$

since coth $\dfrac{\lambda_0}{2\lambda_s} \approx 1$ in the significant part of the curve. This can be rearranged as

$$R = \frac{D_{\text{v}}}{\delta} N_0\, \Omega\, \sigma^2/(\sigma + \sigma^*)\,, \tag{10.86}$$

where $\sigma^* = \sigma_1 l \tau_{\text{desolv}}/\delta\, \tau_{\text{vdiff}}$. Note that $\sigma^* \ll \sigma_1$ from the condition on

τ_{vdiff} and τ_{desolv}. This relation is, strictly speaking, valid only in the parabolic region of the $R(\sigma)$ curve when the diffusion fields of the steps do not overlap, or $\sigma < \sigma^*$. An enhanced back-stress effect will occur at the higher supersaturations since the supersaturation at the center of the spiral step will be reduced due to both the surface diffusion field of neighbouring turns *and* their volume diffusion field. Therefore a slower approach to a true linear $R(\sigma)$ relationship than that predicted by (10.86) is expected.

From (10.86) we note that a strong volume diffusion field can greatly reduce the parabolic region and that the value of R in the linear region is now determined by the volume diffusion process, i.e.,

$$R = \frac{D_{\text{v}}}{\delta} N_0 \, \Omega \, \sigma \quad \text{for } \sigma \gg \sigma^* .$$

10.3.8. Comparison of the BCF theory with experimental data

The BCF theory has been an enormous stimulus, not only to observe crystal surfaces with refined optical and electronmicroscopic techniques as described in the books of Dekeyser and Amelinckx [25] and the work of Sunawaga [34], but also to observe growth patterns as a function of super- or undersaturations. We first mention the work of Bethge et al. [35, 36] who observed evaporation and growth patterns and were able to verify the $\tanh(\lambda_s/\lambda)$ function for the advance velocity of steps [37] in case of evaporation. Also we mention the work of Kaishev and Budevski [38] who carried out fundamental work observing the dependence of spiral growth and nucleation on the $\Delta\mu$ in capillaries (overtension) for electro-crystallization.

Bennema [9a, 39] developed a weighing technique which has recently been automated by Klein Haneveld in collaboration with Van Overbeeke of Groningen University [40, 41, 42]. Bennema [29a, 43] found a linear (R,σ) curve for alum up to σ of about 10^{-2} and for higher σ values "deviating" points situated above the linear law. Recent measurements of Mullin and Garside [44] and Botsaris and Denk [45] using a travelling microscope technique for higher σ values of up to about 10% seem to be in agreement with Bennema's measurements. The slope of Garside and Mullin's curve is, however, four times larger than the slope of the curve of Botsaris and Denk. Two alternative interpretations of the curves of the authors mentioned above seem to be possible: (i) The linear part up till about 1% as measured by Bennema is due to a kind of second linear law (see section 3.4) which, if $\alpha \geqslant 10^{-2}$, changes via transient states (which give more or less non-reproducible results: this may explain the scatter in the deviating points) in a more normal BCF law as measured by Mullin and Garside [44] and Botsaris and Denk [45]. (ii)

Suppose that, for some reason (bunching for example, see the next section), a linear law does occur in the parabolic region of the BCF laws (this cannot be the maximal linear BCF law as shown in fig. 3.1b (see curve 1); in the past we did assume this, not being aware of all the theoretical and experimental evidence against this presupposition). We can then assume, that for values of $\alpha \geqslant 10^{-2}$ a kind of step growth together with two-dimensional nucleation growth may occur. Speculations about such a "compound mechanism" have been made by Bennema, Kern and Simon [46] and Denk and Botsaris [47] and were found as a result of computer simulations (see ref. 19). More work ought to be done in this field. That two-dimensional nucleation plays some role also follows from (R,σ) curves as measured by Troost [48]. An (R,σ) curve resembling a BCF curve has been measured, using the weighing technique, for sodium chlorate by Bennema and a linear (R,σ) curve, using the automated weighing technique, for KCl and NaCl by Klein Haneveld [50], both for growth and dissolution. The observed fact that for dissolution the slope is higher than for growth is a phenomenon which needs further study.

Books dealing with the subject of crystal growth are the books of Van Hook [51], Mullin [52] and Strickland Constable [53].

We note that measurements on $(R,\Delta\mu)$ curves together with a good theoretical interpretation is important for finding optimal growth conditions for the growth of single crystals and to determine optimal growth conditions for industrial crystallization processes.

10.4. The kinematic theory of step motion

10.4.1. Introduction

The ideal theory of crystal growth should provide a function $z = z(x,y,t)$ which gives the shape of a crystal surface during growth. This function will depend on the growth conditions, the initial surface configuration $z=z(x,y,t_0)$ and the defects in the crystal intersecting the initial surface; but with this information one should be able, in principle, to find the rate of growth of each segment of the surface and thereby determine all future configurations.

The BCF theory postulates a uniform distribution of monatomic steps which should appear as a perfectly smooth and flat surface, but in almost every case an examination of the surface of a growing crystal reveals irregularities which are easily visible. Macroscopic steps which are ledges hundreds or thousands of molecular diameters high are frequently visible, and in certain cases the surface breaks up into a number of hills and valleys. Apparently, the surface configuration has an effect on the motion of the crystal surface since growth and dissolution rates should be the same for small departures from equilibrium of equal magnitude for identical surfaces, but in fact dissolution is frequently found to be faster. Clearly it is necessary to determine the cause of surface irregularities and their effect on the growth rate.

10.4.2. Fundamental theory

Frank [54] and Cabrera and Vermilyea [55] have applied the traffic flow theory of Lighthill and Whitham [56] to the motion of steps across the surface of a growing crystal. If we consider the surface $z(x,y,t)$ to be roughly parallel to the (x,y) plane and the rate of growth R to be measured in the z direction, then

$$R = \frac{\partial z}{\partial t}. \tag{10.87}$$

Using the formulation of Cabrera and Coleman [57] we can write the slopes of the surface as measured parallel to the x and y axes as

$$p = -\frac{\partial z}{\partial x}, \text{ and } q = -\frac{\partial z}{\partial y}, \tag{10.88}$$

respectively, and it follows that, if $z\,(x,y,t)$ is an analytic function, then

$$\frac{\partial^2 z}{\partial x \partial t} = \frac{\partial^2 z}{\partial t \partial x},\ \frac{\partial^2 z}{\partial y \partial t} = \frac{\partial^2 z}{\partial t \partial y},\ \frac{\partial^2 z}{\partial x \partial y} = \frac{\partial^2 z}{\partial y \partial x}$$

and therefore

$$\frac{\partial p}{\partial t} + \frac{\partial R}{\partial x} = 0;\ \frac{\partial q}{\partial t} + \frac{\partial R}{\partial y} = 0;\ \text{and } \frac{\partial p}{\partial y} - \frac{\partial q}{\partial x} = 0\,. \tag{10.89}$$

For simplicity we will consider only that part of the surface in which the steps are parallel to the y-axis and therefore $q = 0$, but this does not limit the generality of the results since the coordinate system can be transformed to make any point on the surface satisfy this condition. Equation (10.89) can also be derived from the conservation of mono-molecular steps. When the (x,y) plane is parallel to a singular surface of the crystal, $\rho = p/d$ is the local density of steps* (number per cm) and $J = R/d$ is the flux of steps past a

* Note that $p = 1/\lambda_0$ where λ_0 is the distance between steps of section 10.2.5. Also, d is the height of a monatomic step as before.

point (number passing per second). In this case step conservation implies that

$$\frac{\partial \rho}{\partial t} = \frac{-\partial J}{\partial x} \quad \text{or} \quad \frac{\partial (\frac{p}{d})}{\partial t} = \frac{-\partial (\frac{R}{d})}{\partial x}$$

and this is identical to the first part of (10.89). The previous derivation is more general since it does not require the presence of steps.

In order to obtain useful results from (10.89) we need to make some simplifying assumptions. We will assume that $R(x,y,t)$ depends *only* on the local average slope of the surface p. By the local average slope, we mean the slope of the surface averaged over distances large compared with the spacing between steps, but small compared with the macroscopic dimensions of surface irregularities. Therefore the theory does not apply to microscopic processes occurring between elementary steps. Any effects on the growth rate due to the surface structure in neighbouring regions, changes in conditions with time, or volume diffusion fields are neglected.

With this assumption equation (10.89) becomes

$$\frac{\mathrm{d}R}{\mathrm{d}p}\frac{\partial p}{\partial x} + \frac{\partial p}{\partial t} = 0; \quad \text{or} \quad \frac{\partial p}{\partial t} + c\frac{\partial p}{\partial x} = 0\,, \tag{10.90}$$

where c is given by $c(p) = \mathrm{d}R/\mathrm{d}p$, and it is the x-component of the velocity during growth of points on the surface having a certain slope p. This is easy to see since for p to remain unchanged we must have

$$\Delta p = \frac{\partial p}{\partial x}\Delta x + \frac{\partial p}{\partial t}\Delta t = 0 \quad \text{or} \quad \frac{\Delta x}{\Delta t} = -\frac{\partial p}{\partial t}\Big/\frac{\partial p}{\partial x} = c\,.$$

These regions of constant slope are called "kinematic waves" and, since c depends on p, they will always have the same velocity. Note that the velocity of the kinematic waves is not necessarily the same as the velocity of steps. Since the rate of growth $R = \rho v d$, the step velocity $v = R/p$; and $v \equiv c$ only if R is directly proportional to ρ. The kinematic waves do not necessarily keep the same steps as they move across the surface; the steps will move relative to the kinematic waves unless it happens that $v = c$.

It is instructive to draw successive profiles of the crystal surface as the crystal grows. The trajectory of the kinematic waves is given by lines which have a slope in the (x,z) plane of $\mathrm{d}z/\mathrm{d}x|_p$; z and x are coordinates of points on the crystal surface, and the notation indicates that p is constant. We have

$$\left.\frac{\mathrm{d}z}{\mathrm{d}x}\right|_p = \frac{\partial z}{\partial t}\left.\frac{\mathrm{d}t}{\mathrm{d}x}\right|_p + \frac{\partial z}{\partial x} = \frac{R}{c} - p = \frac{p}{c}(v - c), \tag{10.91}$$

and since $\mathrm{d}z/\mathrm{d}x|_p$ depends only on p, the trajectories are straight lines. In cases where $c > v$, the trajectories will have a negative $\mathrm{d}z/\mathrm{d}x|_p$.

Figure 10.18a shows an $R\,(p)$ curve, Cabrera's type I [39], of the kind expected from growth without impurity effects (see sections 2 and 3). A kinematic wave with inclination p_A will have a velocity $c = \mathrm{d}R/\mathrm{d}p$ given by the slope in the (R,p) plane of the tangent line bc. The step velocity v is R/p, or the slope of the line OA. For any value of p, $v \geqslant c$ and the trajectories will never have a negative $\mathrm{d}z/\mathrm{d}x|_p$. On the other hand, if impurity effects produce an $R\,(p)$ curve similar to that in Fig. 10.18b (type II), a kinematic wave with inclination p_A will have $c > v$ and $\mathrm{d}z/\mathrm{d}x|_p$ is therefore negative. Small areas of the surface having this inclination will tend to disappear from the growth form.

Frank has suggested that an impurity which is absorbed into the growing crystal, but which also reduces the rate of advance of a step, could produce a curve of type II. The amount of impurity encountered by a step, and therefore its rate of advance, will depend on the amount of time that has elapsed since the previous step passed. If the steps are too widely separated, it may be possible that the impurity concentration could build up to such a point that the steps would stop completely, and $R(p)$ would have an initial slope of zero. Immobile impurities of the type discussed by Cabrera and Vermilyea [55] could produce such an effect since it would be very difficult for a step to break through a closely spaced network of impurities unless the supersaturations were extremely high.

Frank also pointed out that an equidistant train of monatomic steps would become unstable under such conditions since, if one step happens to lag

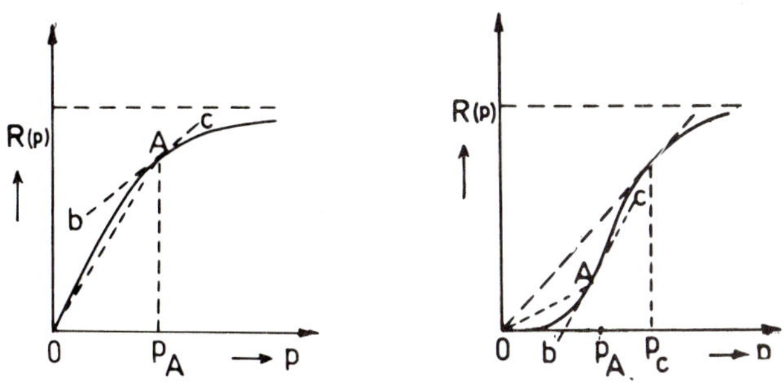

Fig. 10.18. a) Type I curve; b) type II curve.

behind, it encounters a "dirtier" surface and therefore falls even father behind. Since this is a microscopic effect dealing with single steps, it can not be treated with the kinematic theory and we will consider it later when we investigate the stability of a train of steps.

10.4.3. Application to a growth hill

An isolated screw dislocation source of constant strength located at the origin provides a boundary condition which determines the successive shapes of the growth hill which develops. The slope of the surface adjacent to the z-axis must be determined by the source strength R_0; i.e. p_0 must be such that $R(p_0) = R_0$. Using this and the fact that all of the trajectories of kinematic waves with slopes other than p_0 must intersect the origin, we can use the $R(p)$ curve to draw trajectories and estimate the successive hill profiles.

Cabrera [59] has shown that an $R(p)$ curve of type I produces a growth hill as in fig. 10.19 with a smooth transition from the slope p_0 at $x = 0$ to 0 for large x. All slopes from 0 to p_0 appear in the hill. On the other hand, an $R(p)$ curve of type II should produce negative $\mathrm{d}z/\mathrm{d}x|_p$ trajectories for slopes between p_c and 0, and since these trajectories must originate at the origin, the growth hill may not contain slopes in this region. This implies that a source with a strength less than $R(p_c)$ cannot produce a growth hill, and that the growth hills of sources stronger than this value will have a sharp discontinuity in slope at the foot. These same considerations also apply to etch pits, and

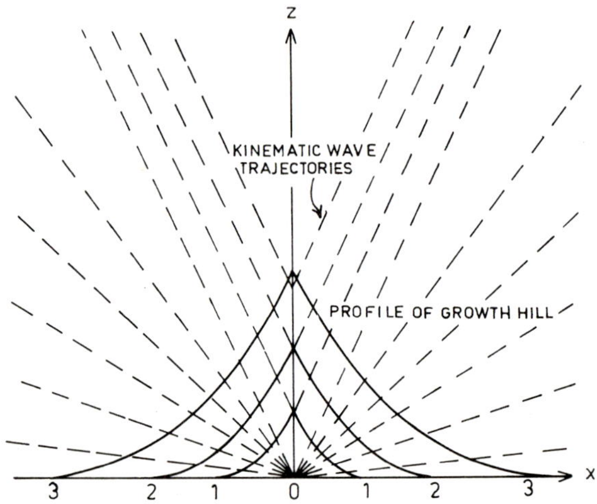

Fig. 10.19. Profile of a growth hill.

Hulett and Young [58] have found results consistent with these predictions for etch pits in copper undergoing anodic dissolution. By varying the amount of a bromide ion impurity and the current density they produced $R(p)$ curves of both types; and the etch pits with a sharp slope discontinuity at the edge were indeed produced by $R(p)$ curves of type II. The trajectories of the kinematic waves in the (x,z) plane were also found to be straight lines.

Straight trajectories are not always found, however, and Ives and Hirth [60] found a distinct curvature in their investigation. They were again examining etch pits, and a strong volume diffusion field could have caused a time dependent change in the conditions.

10.4.4. Formation of shock waves

Fluctuations in the conditions at the step source could produce a local area with a high density of steps. The behaviour of this region can be treated with the kinematic theory by constructing kinematic wave trajectories in the (x,z) plane from representative points on the initial surface profile. The high density region appears as a steeper slope than normal; this is indicated in the initial profile sketched by curve 1 in fig. 10.20. The trajectories for a type I $R(p)$ curve are also indicated in the figure. When the surface has reached the position of curve 2, some of the trajectories have intersected since the high step densities have steeper trajectories than those of the normal surface. The meeting of kinematic waves of different p's implies the presence of a discon-

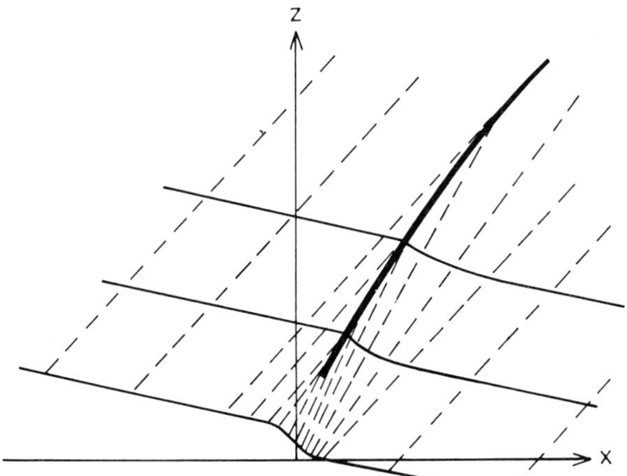

Fig. 10.20. Development of a shock wave.

tinuity in the slope of the crystal surface or a shock wave, also indicated in curve 2. Chernov [61] has shown that the discontinuity will decrease in time provided $d^2R/dp^2 < 0$. This effect is represented in fig. 10.20 since the trajectories of the kinematic waves from the high density region of curve 1 are terminating on the shock wave and disappearing from the profile. Note that the trajectory of the shock wave as indicated by the heavy line does not have a constant slope. Its velocity V is determined by the density of steps on either side since the net flux of steps to the shock wave must be zero. (It is simply a boundary between two regions of different slopes; it is not a pile-up of steps.) Thus

$$\rho_1 (v_1 - V) = \rho_2 (v_2 - V)$$

$$\text{or } V = \frac{R_1 - R_2}{p_1 - p_2}; \qquad (10.92)$$

we have used $R = \rho v d$ and $p = \rho d$, and the subscripts refer to points just to the left and just to the right of the shock wave. For the geometrical construction, it is useful to note that $c_1 = dR/dp|_{p=p_1}$ and $c_2 = dR/dp|_{p=p_2}$ and $V = \Delta R/\Delta p$ is the slope of the chord joining points on the $R(p)$ curve at p_1 and p_2; therefore the shock wave will have a trajectory with a slope (in the (x,z) plane) between the slopes of the kinematic wave trajectories c_1 and c_2 on either side for an $R(p)$ curve of type I.

The kinematic theory does not predict the formation of macro-steps or discontinuities in the height of the surface; in fact it would be inconsistent if it did predict such surface irregularities since it is impossible to define a local average slope p at the edges formed by a macro-step which gives an accurate value for $R(p)$. A shock wave presents the same inconsistency along the line of the density discontinuity, but for small discontinuities the kinematic theory is probably quite accurate in predicting the shock wave trajectory.

The formation of macro-steps must be explained in terms of processes occurring between steps of rather small height or monatomic steps. If two steps happen to meet as a result of encountering an imperfection of the crystal, or due to an instability in the equidistant step spacing, there will be a tendency for them to stick together since steps on surfaces neighbouring singular surfaces have an attractive interaction [57]. The double height step thus formed will move slowly since it requires twice the flux in order to advance at the same rate as a step of unit height. Monatomic steps from the rear may then catch up and join the combined step and form a macro-step.

Mullins and Hirth [62] have considered some of the effects arising from the discrete monatomic steps which do not appear in the classical kinematic theory. Without any of the mechanisms described above they find that in certain cases one step may overtake the one in front and form a double height step. The steps seem to have very little, if any, ability to maintain equidistant spacings if the conditions of growth are perturbed.

10.4.5. Stability of an infinite train of parallel steps

We will examine the stability of equidistant parallel steps against small perturbations in the spacing between adjacent steps. We assume that the velocity of a step is determined only by the distance to the step in front and to the step behind as in the BCF surface diffusion model, but we allow for the possibility that there is a different dependence on the two distances. We therefore write

$$v = f_+(\lambda_+) + f_-(\lambda_-)\,, \tag{10.93}$$

where λ_+ is the distance to the step in front, etc., and in general $f_+(\lambda) \neq f_-(\lambda)$. In the BCF case (eq. 10.55)

$$f_+(\lambda) = f_-(\lambda) = \frac{\zeta D_s n_{s0} f_0 \sigma}{\lambda_s} \tanh \frac{\lambda}{2\lambda_s}\,.$$

A different dependence on λ_+ and λ_- may result from an impurity mechanism of the type proposed by Frank since the distance to the step in *front* determines the amount of impurity encountered. Also, a relaxation time τ_k for entering the step from the front different from that for the back would produce a different dependence on λ_+ and λ_-.

The infinite train of steps is indicated in fig. 10.21. The equidistant points at $y = 0$, $y = \lambda$, $y = 2\lambda$, etc. are the unperturbed positions of the steps, and we

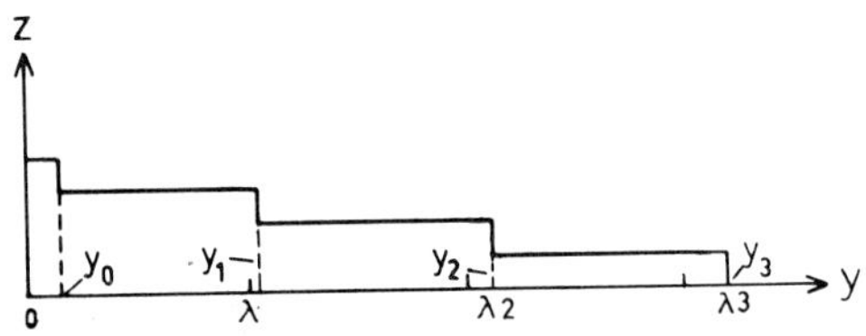

Fig. 10.21. Step train with slight deviations from an equidistant step train.

will examine the growth or decay of small perturbations in the positions of the steps from these equidistant points. If the perturbed position of the n'th step is y_n, we have, from (10.93)

$$v_n = f_+(y_{n+1} - y_n) + f_-(y_n - y_{n-1})$$

or

$$v_n = f_+(u_{n+1} - u_n + \lambda) + f_-(u_n - u_{n-1} + \lambda), \tag{10.94}$$

where $u_n = y_n - n\lambda$ is distance of the n'th step from its unperturbed position. Then $du_n/dt = v_n - v$, where $v = f_+(\lambda) + f_-(\lambda)$ is the average velocity of the steps for very small perturbations. Since $(u_{n+1} - u_n)$ and $(u_n - u_{n-1})$ are assumed to be small compared with λ,

$$\frac{du_n}{dt} \approx f_+'(\lambda)\ [u_{n+1} - u_n] + f_-'(\lambda)\ [u_n - u_{n-1}]$$

or

$$\frac{du_n}{dt} \approx f_+'(\lambda) u_{n+1} - f_-'(\lambda) u_{n-1} + [f_-'(\lambda) - f_+'(\lambda)]\, u_n\,, \tag{10.95}$$

where

$$f_\pm'(\lambda) = \frac{df_\pm}{d\lambda}.$$

These equations are similar to those governing the motion of an infinite chain of masses attached by springs, and we will try a similar solution. Using $u_n(t) = e^{iqn\lambda}\, u_q(t)$, where q is a constant* and $u_q(t)$ is a function to be determined later, we find

$$\frac{du_q}{dt} = [\, f_+'(\lambda)\, e^{iq\lambda} - f_-'(\lambda)\, e^{-iq\lambda} + f_-'(\lambda) - f_+'(\lambda)\,]\ u_q$$

* In the analogous problem of masses attached by springs, $u_n(t)$ is the deviation of the n'th mass from its equilibrium position and q is inversely proportional to the wavelength of the corresponding normal mode of vibration.

or

$$\frac{1}{u_q}\frac{du_q}{dt} = [f'_+(\lambda) - f'_-(\lambda)] \,.\, (\cos q\lambda - 1) +$$

$$+ [f'_+(\lambda) + f'_-(\lambda)] \,.\, \mathrm{i} \sin q\lambda \tag{10.96}$$

and this can be solved by

$$u_q(t) = e^{bt + iwt}, \tag{10.97}$$

where $b = [f'_+(\lambda) - f'_-(\lambda)] \,.\, (\cos q\lambda - 1)$ and $w = [f'_+ + f'_-] \,.\, \sin q\lambda$.

When $f'_+(\lambda) = f'_-(\lambda)$, then $b = 0$ and (10.97) shows that any small perturbation will continue with a constant amplitude. This is the case for a symmetrical dependence of the step velocity on the spacings in front and behind, a disturbance in the spacings will continue unabated. When $f'_+(\lambda) > f'_-(\lambda)$, b is negative since q must be in the range $-\pi/2\lambda < q < \pi/2\lambda$ and $\cos q\lambda < 1$. According to (10.97), the amplitude of the perturbation will decay exponentially to zero. Therefore situations in which the region in front of the step contributes more to the step motion than the region behind will have some stability against perturbations. When $f'_+(\lambda) < f'_-(\lambda)$, the amplitude of the perturbation will grow exponentially; and, if the perturbation growth rate is fast enough, multiple steps should result.

The strength of the stability or instability can be evaluated quantitatively. We will compare the time τ required for the amplitude of a perturbation to grow or decay by a factor of e to the lifetime T of a step. From (10.97)

$$\tau = \frac{1}{b} = \frac{1}{[f'_+(\lambda) - f'_-(\lambda)]\,(\cos q\lambda - 1)}\,. \tag{10.98}$$

Note that perturbations with $|q| \approx \pi/2\lambda$ will grow or decay most rapidly. The lifetime of a step T is determined by the distance L_{st}. It travels from the source to the edge of the crystal or to a point where it is annihilated by meeting a step from another source. Thus

$$T = \frac{L_{st}}{v} = \frac{L_{st}}{f_+(\lambda) + f_-(\lambda)}\,. \tag{10.99}$$

A strong stability or instability is indicated when $\tau/T \ll 1$. This implies, in the case of an instability, that a perturbation will grow to significant size during the lifetime of the steps.

10.4.6. Unequal relaxation times for joining a step

Schwoebel [63] allowed a dependence of the relaxation time for entering the step on the direction of approach and examined the resulting step motion. A molecule approaching the step from the left (see fig. 10.7) may have to pass over a rather large activation energy barrier in order to join the step since a large separation from some of the previous neighbours is required for the move. A molecule approaching from the right has only the activation barrier of a normal surface diffusion move*. This implies that $f_+ > f_-$ for growth since f_+ corresponds to the area to the right of the step; and therefore the step train tends to be stable. In the case of dissolution f_+ corresponds to the area to the left of the step since it will move to the left, and an unstable situation is expected. This result could be expected from qualitative reasoning. If a growing step happens to deviate from its normal position by falling behind, the length of the catchment area in front increases and the area behind decreases. Since the area in front contributes more to the total flux due to the lower activation energy in this direction, this more than compensates for any loss of flux from the rear; and the step increases its velocity and reduces the magnitude of the deviation. However, during dissolution a lagging step has its flux reduced, it slows down further, and the deviation is increased. In this discussion it is assumed that the primary contribution to the flux of matter to the steps comes through adsorbed molecules. In many cases vacancies may dominate and this situation would probably reverse the conclusions above, i.e. the step train would be stable during dissolution and unstable during growth.

We will solve the diffusion equations which determine the fluxes to the step for the general case of unequal relaxation times, and from the expressions for the fluxes we will find the strength of the stability or instability. Consider the diffusion process between two steps, one at $y = 0$ and one at $y = \lambda$. In section 2 the concentration of ad-atoms $n_s(y)$ was shown to be of the form

$$n_s(y) = A\,e^{y/\lambda_s} + B\,e^{-y/\lambda_s} + (\sigma + 1)n_{s0} \tag{10.100}$$

* This statement refers to growth from a single component vapour. Multicomponent vapour growth or solution growth probably have a higher activation energy for entering the step from the right than for surface diffusion due to strongly adsorbed solvent molecules in the step.

(see equation 10.47 and the definition of ψ). It was also shown that the flux of ad-atoms to the step at the origin from the right is given by

$$j_{s,0} = \frac{a}{\tau_{k,0}}(n_s(0) - n_{s0}) = D_s \left.\frac{\partial n_s}{\partial y}\right|_{y=0} ; \tag{10.101}$$

it is directly proportional to the deviation in the concentration of ad-atoms adjacent to the step from the equilibrium concentration, and inversely proportional to the relaxation time for joining the step (see eqs. 10.44 and 10.52). Similarly, the flux to the step at $y = \lambda$ from the left is

$$j_{s,1} = \frac{a}{\tau_{k,1}}(n_s(\lambda) - n_{s0}) = -D_s \left.\frac{\partial n_s}{\partial y}\right|_{y=\lambda} . \tag{10.102}$$

When the expression for $n_s(y)$ in (10.100) is inserted in (10.101) and (10.102) we obtain two algebraic equations for the unknowns A and B:

$$(\frac{\lambda_s}{\Lambda_0} - 1)A + (\frac{\lambda_s}{\Lambda_0} + 1)B = -\frac{\lambda_s}{\Lambda_0}\sigma n_0 , \tag{10.103}$$

$$(\frac{\lambda_s}{\Lambda_1} + 1)\mathrm{e}^{\lambda/\lambda_s} A + (\frac{\lambda_s}{\Lambda_1} - 1)\mathrm{e}^{-\lambda/\lambda_s} B = -\frac{\lambda_s}{\Lambda_1} n_0 , \tag{10.104}$$

where we have defined $\Lambda_1 = D_s \tau_{k,1}/a$ and $\Lambda_0 = D_s \tau_{k,0}/a$ to simplify the expressions. With the expressions for A and B obtained from (10.103) and (10.104), (10.100), (10.101) and (10.102) give

$$j_{s,0} = \frac{D_s \sigma n_0}{\lambda_s} \cdot \frac{\cosh\frac{\lambda}{\lambda_s} - 1 + \frac{\Lambda_1}{\lambda_s}\sinh\frac{\lambda}{\lambda_s}}{\left[(1 + \frac{\Lambda_0\Lambda_1}{\lambda_s^2}\sinh\frac{\lambda}{\lambda_s} + \frac{\Lambda_0 + \Lambda_1}{\lambda_s}\cosh\frac{\lambda}{\lambda_s}\right]} \tag{10.105}$$

and $$j_{s,1} = \frac{D_s \sigma n_0}{\lambda_s} \frac{\cosh\frac{\lambda}{\lambda_s} - 1 + \frac{\Lambda_0}{\lambda_s}\sinh\frac{\lambda}{\lambda_s}}{\left[(1 + \frac{\Lambda_0\Lambda_1}{\lambda_s^2}\sinh\frac{\lambda}{\lambda_s} + \frac{\Lambda_0 + \Lambda_1}{\lambda_s}\cosh\frac{\lambda}{\lambda_s}\right]} . \tag{10.106}$$

Since the contribution of $j_{s,0}$ to the velocity of the step at the origin is $(\Omega/d)\, j_{s,0}$, if the steps are moving to the right we can write $f_+(\lambda) = (\Omega/d)\, j_{s,0}(\lambda)$. Similarly, $f_-(\lambda) = (\Omega/d)\, j_{s,1}(\lambda)$ and these expressions can be used to calculate the strength of the stabilizing influence of the kinetics. The expressions (10.98) for τ and (10.99) for T give the ratio

$$\frac{\tau_M}{T} = \frac{\lambda_s^3}{L_{st}(\Lambda_1^2 - \Lambda_0^2)} \left[\left(1 + \frac{\Lambda_0\Lambda_1}{\lambda_s^2}\right)\sinh\frac{\lambda}{\lambda_s} + \frac{\Lambda_0 + \Lambda_1}{\lambda_s}\cosh\frac{\lambda}{\lambda_s}\right] \cdot$$

$$\cdot \left[2\left(\cosh\frac{\lambda}{\lambda_s} - 1\right) + \frac{\Lambda_0 + \Lambda_1}{\lambda_s}\sinh\frac{\lambda}{\lambda_s}\right], \qquad (10.107)$$

where τ_M is the shortest decay or growth time (for short wavelength perturbations). The ratio τ_M/T is a monotonically increasing function of the spacing between the steps λ and therefore a larger step spacing always implies a weaker stability or instability. In fact, as $\lambda/\lambda_s \to \infty$, the ratio $\tau_M/T \to \infty$ and there is no effect at all. This is to be expected since for large λ/λ_s the diffusion fields of neighbouring steps do not overlap, and the motion of one step is not affected by the motion of its neighbours.

The smallest values of τ_M/T occur, therefore, when λ/λ_s is very small; and then the stability or instability is the strongest. When $\lambda/\lambda_s \ll 1$ the approximations: $\sinh \lambda/\lambda_s \approx \lambda/\lambda_s$ and $\cosh \lambda/\lambda_s \approx 1 + \lambda^2/2\lambda_s^2$ are accurate and from (10.107) we obtain

$$\frac{\tau_M}{T} = \frac{\lambda}{L_{st}}\left[\frac{\Lambda_1 + \Lambda_0}{\Lambda_1 - \Lambda_0} + \frac{\lambda}{\Lambda_1 - \Lambda_0}\left(2 + \frac{\Lambda_1\Lambda_0}{\lambda_s^2}\right)\right]. \qquad (10.108)$$

If we use the definitions of Λ_0 and Λ_1 and we write $\lambda^2 = D_s\tau_{deads}$ and $D_s = a^2/\tau_{sdiff}$, we obtain

$$\frac{\tau_M}{T} = \frac{\lambda}{L_{st}}\left[\frac{\tau_{k,1} + \tau_{k,0}}{\tau_{k,1} - \tau_{k,0}} + \frac{\lambda}{a}\frac{\tau_{sdiff}}{\tau_{k,1} - \tau_{k,0}}\left(2 + \frac{\tau_{k,1}\,\tau_{k,0}}{\tau_{sdiff}\,\tau_{deads}}\right)\right]. \qquad (10.109)$$

A strong stability or instability exists when $\tau_M/T \ll 1$, or if each of the terms on the right hand side of (10.109) is much less than unity. We can then write down three conditions:

$$\tau_{\mathrm{k},1} - \tau_{\mathrm{k},0} \gg \frac{\lambda}{L_{\mathrm{st}}}(\tau_{\mathrm{k},1} + \tau_{\mathrm{k},0}) \qquad \mathrm{A}$$

$$\tau_{\mathrm{k},1} - \tau_{\mathrm{k},0} \gg \frac{2\lambda^2}{L_{\mathrm{st}}a}\tau_{\mathrm{sdiff}} \qquad \mathrm{B} \qquad (10.110)$$

$$\frac{\tau_{\mathrm{k},1}\,\tau_{\mathrm{k},0}}{\tau_{\mathrm{sdiff}}\,\tau_{\mathrm{deads}}} < 2 \qquad \mathrm{C}$$

where condition C assumes that condition B is satisfied. Condition A is to be expected since there must be an appreciable difference in the $\tau_{\mathrm{k},0}$ and $\tau_{\mathrm{k},1}$ to give a difference between the two fluxes. The difference does not need to be large, however, if the step travels a large number of step spacings λ during its lifetime (λ/L_{st} is very small). Condition B is easier to analyse if we replace it with the somewhat more severe condition ($\lambda \to \lambda_{\mathrm{s}}$):

$$\tau_{\mathrm{k},1} - \tau_{\mathrm{k},0} \gg \frac{2a}{L_{\mathrm{st}}}\tau_{\mathrm{deads}} \, .$$

Crystals with a very high density of sources (L_{st} small) and/or a very long τ_{deads} may very well fail to satisfy this condition. This is probably the most stringent of the three conditions for the majority of growth situations. Condition C is related to the second parabolic law of section 2 and arises because of the diminishing surface concentration gradients when it is extremely difficult to enter the kink sites. In this case, $n_{\mathrm{s}}(y) \approx \sigma n_0$, a constant, and there can be very little interaction between the steps due to the weak concentration gradients.

The value of λ_{s} plays an important role in determining the strength of the effect. A large λ_{s} implies more overlap of the diffusion fields for a given supersaturation with a screw dislocation source; and therefore the region where $\lambda < \lambda_{\mathrm{s}}$ and τ_{M}/T reaches its minimum value is more easily reached. On the other hand, the minimum of τ_{M}/T may not be small enough to produce much of an effect if λ_{s} is extremely large, since large λ_{s} implies a long τ_{deads} and the term discussed in condition B becomes large.

The number of sources per unit area of the crystal obviously is important since it determines the lifetime of the steps and the time available for instabilities to build up. Since there should be many more sources for dissolution than for growth, an instability is probably much more important if it occurs during growth than if it occurs during dissolution.

An instability during growth would perhaps explain the experimentally observed inverse correlation between the perfection of the resulting crystals and the rate of growth. A high rate of growth requires a large supersaturation and therefore a small λ for steps produced by a screw dislocation source. This implies a strong instability and therefore the production of a large number of macro-steps. The supersaturation should be held in the range where $\lambda \approx \lambda_s$ to provide the maximum growth rate without macro-steps; the number of macro-steps should increase rapidly as σ rises above this point.

As an example of a situation in which the kinetic instability should be expected, we will consider a nearest neighbour bond model of a crystal evaporating to the vapour. We assume that an ad-atom mechanism is the fastest means of transport, and the (100) face of a simple cubic crystal is evaporating. The bond energy is such that $\phi/kT = 4$, and for surface diffusion we assume a kinetic barrier of $\phi/2$ giving (using eq. 10.71)

$$\tau_{k,0} = \tau_{sdiff} = \nu^{-1}\, e^2 ,$$

where $\tau_{k,0}$ is the relaxation time for joining the step from the lower side. A molecule coming from the high side of the step must stretch its bonds much father in order to join the step, and we will assume an extra activation energy of ϕ/kT over the surface diffusion process, i.e.

$$\tau_{k,1} = \nu^{-1}\, e^6 .$$

An adsorbed molecule should have four nearest neighbours and we take

$$\tau_{deads} = \nu^{-1}\, e^{16} .$$

Then (10.109) becomes

$$\frac{\tau_M}{T} = \frac{\lambda}{L_{st}} + \frac{2a}{L_{st}}\frac{e^{16}}{e^6} = \frac{\lambda}{L_{st}} + \frac{5 \times 10^4 a}{L_{st}} ,$$

where we have approximated λ by λ_s in the second term. Thus, the instability is rather weak; it is significant for distances between sources of $5 \times 10^4 a \approx 2 \times 10^{-3}$ cm; only if the step must travel farther than this is there a danger of one step overtaking another and producing a macro-step by this mechanism.

At high temperature or low bond energies where $\phi/kT = 2$, we have $\tau_M/T = \lambda/L_{st} + 300a/L_{st}$ and the instability could be quite important.

Impurities may have an important effect on the stability or instability of the step train. The mechanism proposed by Frank produces a stronger instability at very low supersaturation than for high supersaturation and should dominate in this case. An impurity which facilitates the growth process, perhaps by providing a path to the kink sites with a lower activation energy, should give stability on growth. Even under extremely pure conditions it is possible that impurity mechanisms dominate in determining the stability or instability.

We note that recently propagation of steps on crystal surfaces also has been simulated on computers by Bertocci [64] and Surek [65].

References

[1] D.E. Temkin, in: Crystallization Processes (Consultants Bureau, New York, 1966) p. 15.

[2] B. Mutaftschiev, in: Adsorption et Croissance Cristalline (Centre Nationale de la Recherche Scientifique, Paris, 1965) p. 231.

[3] K.A. Jackson, Liquid Metals and Solidification (American Society for Metals, Cleveland, 1958).

[4] J. Zell and B. Mutaftschiev, J. Crystal Growth 13/14 (1972) 231.

[5] See among others:
H.J. Eyring, D. Henderson, B.J. Stover and E.M. Eyring, Statistical Mechanics and Dynamics, 303–308.

[6] H.J. Leamy and K.A. Jackson, J. Appl. Phys. 42 (1971) 2121.

[7] G.H. Gilmer and P. Bennema, to be published.

[8] D. Nason and W.A. Tiller, J. Crystal Growth 10 (1971) 117.

[9] K.A. Jackson, J. Crystal Growth 10 (1971) 119.

[10] D.E. Temkin, J. Crystal Growth 10 (1971) 190.

[11] K.A. Jackson, J. Crystal Growth 5 (1969) 13.

[12] K.A. Jackson, Abstract of the International Congress on Crystal Growth (Marseille) (1971) B3–1.

[13] D.E. Temkin, Sov. Phys. Cryst. 14 (1969) 179.

[14] D.E. Temkin, Sov. Phys. Cryst. 14 (1969) 344.

[15] D.E. Temkin, Sov. Phys. Cryst. 15 (1971) 767.

[16] F.L. Binsbergen, Koll. Z. – Z. Polymere 237 (1970) 289.

[17] F.L. Binsbergen, Koll. Z. – Z. Polymere 238 (1970) 389.

[18] G.H. Gilmer and P. Bennema, J. Crystal Growth 13/14 (1972) 148.

[19] G.H. Gilmer and P. Bennema, J. Appl. Phys. 43 (1972) 1347.

[20] V.V. Voronkov and A.A. Chernov, in: Crystal Growth, ed. by H.S. Peiser (Pergamon, Oxford, 1967) p. 593.

[21] W.K. Burton, N. Cabrera and F.C. Frank, Phil. Trans. Roy. Soc. 243 (1951) 299.

[22] A.A. Chernov, Sov. Phys. Uspekhi, Eng. Translation 4 (1961) 129.

[23] R.F. Strickland-Constable, Kinetics and Mechanism of Crystallization (Academic Press, London, 1968).

[24] F.C. Frank, Disc. Farad. Soc. 5 (1949) 48.
[25] W. Dekeyser and S. Amelinckx, Les Dislocations et la Croissance des Cristaux (Masson, Paris, 1955).
[26] A.R. Verma, Crystal Growth and Dislocations (Butterworth, London, 1953).
[27] B. Honigmann, Gleichgewichts und Wachstumformen von Kristallen (Steinkopf Verlag, Darmstadt, 1958).
[28] N. Cabrera and M.M. Levine, Phil. Mag. 1 (1956) 450.
[29] a) P. Bennema, thesis, Delft University of Technology 1965.
b) P. Bennema, J. Crystal Growth 1 (1967) 278.
c) P. Bennema, J. Crystal Growth 5 (1969) 29.
[30] a) P. Hartman and W.G. Perdok, Acta Cryst. 8 (1955) 49, 521, 525.
b) P. Hartman, Z. Krist. 119 (1963) 65.
[31] a) S.C. Glasstone, K.J. Laidler and H. Eyring, The Theory of Rate Processes (McGraw-Hill, London, 1941).
b) H. Eyring and E.M. Eyring, Modern Chemical Kinetics (Reinhold, London, 1963).
[32] N. Cabrera and R.V. Coleman, in: The Art and Science of Growing Crystals (John Wiley, New York, 1963) p. 3.
[33] G.H. Gilmer, R. Ghez and N. Cabrera, J. Crystal Growth 8 (1971) 79.
[34] One out of the numerous publications. H. Komatsu and I. Sunagawa, Min. J. 4 (1964) 203.
[35] H. Bethge, in: Crystal Growth, ed. by H.S. Peiser (Pergamon, Oxford, 1967) p. 623.
[36] K.W. Keller, in: Crystal Growth, ed. by H.S. Peiser (Pergamon, Oxford, 1967) p. 629.
[37] H. Bethge, K.W. Keller and E. Ziegler, J. Crystal Growth 3,4 (1968) 184.
[38] R. Kaishev and E. Budewski, Comp. Phys. 8 (1967) 489.
[39] P. Bennema, Phys. Stat. Sol. 17 (1966) 555.
[40] J. van Overbeeke and P. Bennema, J. Sci. Instr. 43 (1966) 849.
[41] P. Bennema and H.B. Klein Haneveld, J. Crystal Growth 1 (1967) 225.
[42] P. Bennema and H.B. Klein Haneveld, J. Crystal Growth 1 (1967) 232.
[43] P. Bennema, Phys. Stat. Sol. 17 (1966) 563.
[44] J.W. Mullin and J. Garside, Trans Instn. Chem. Engrs. 45 (1967) T285.
[45] G.D. Botsaris and E.G. Denk, I and E.C. Fundamentals 9, 2 (1970) 276.
[46] P. Bennema, R. Kern and B. Simon, Phys. Stat. Sol. 19 (1967) 211.
[47] E.G. Denk and G.D. Botsaris, J. Crystal Growth 6 (1970) 241.
[48] S. Troost, J. Crystal Growth 3,4 (1968) 340.
[49] S. Troost, J. Crystal Growth 13/14 (1972) 449.
[50] H.B. Klein Haneveld, J. Crystal Growth 10 (1971) 111.
[51] A. van Hook, Crystallization Theory and Practice (Reinhold, New York, 1961).
[52] J.W. Mullin, Crystallization (Butterworth, London, 1961).
[53] R.F. Strickland-Constable, Kinetics and Mechanism of Crystallization (Academic Press, London, 1968).
[54] F.C. Frank, in: Growth and Perfection of Crystals (John Wiley, New York, 1958) p. 411.
[55] N. Cabrera and D.A. Vermilyea, in: Growth and Perfection of Crystals (John Wiley, New York, 1958) p. 393.

[56] M.J. Lighthill and G.B. Whitham, Proc. Roy. Soc. 229A (1955) 281.
[57] N. Cabrera and R.V. Coleman, in: The Art and Science of Growing Crystals (John Wiley, New York, 1963) p. 3.
[58] L.D. Hulett and F.W. Young, J. Phys. Chem. Solids 26 (1965) 1287.
[59] N. Cabrera, in: Reactivity of Solids (Elsevier Publ. Comp., Amsterdam, 1961) p. 345.
[60] M.B. Ives and J.P. Hirth, J. Chem. Phys. 33 (1960) 517.
[61] A.A. Chernov, Sov. Phys. Uspekhi 4 (1961) 116.
[62] W.W. Mullins and J.P. Hirth, J. Phys. Chem. Solids 24 (1963) 1391.
[63] R.L. Schwoebel, J. Appl. Phys. 40 (1969) 614.
[64] U. Bertocci, Surface Science 15 (1969) 286.
[65] T. Surek, J. Crystal Growth 13/14 (1972) 19.

P. Hartman, ed., Crystal growth: an introduction. © *(1973) North-Holland Publ. Co.*

EQUILIBRIUM FORMS

† S. TOSCHEV

11.1. Introduction

Crystals display various forms during their growth which is a typical non-equilibrium phenomenon. Nevertheless, the precise derivation of the crystal form at equilibrium conditions has been and still remains a basic problem in the theory of crystal growth. This is not strange considering that apart from its fundamental aspects, the understanding of the equilibrium of the crystals with their environment gives an insight into many morphological and kinetic features of crystal growth and dissolution.

The classical treatment of this problem is based on two approaches which will be briefly outlined in the following. The first approach originating from Gibbs (see e.g. [1]) is purely phenomenological. The second approach developed by Stranski and Kaischew in the nineteen thirties can be termed as atomistic as it considers material exchange across the crystal/ambient phase interface on a microscopic scale.

11.2. Phenomenological treatment

Many years ago Gibbs and Curie defined the equilibrium form as that possessing minimal total surface energy for a given volume. For a given crystal this criterion reads:

$$\Phi = \sum \gamma_i A_i = \text{min.}, \tag{11.1}$$

$$V = \text{const.} \tag{11.1'}$$

Here γ_i and A_i denote the specific surface energy and the surface area of the i-th crystallographic face respectively.

As is known, in contrast to large crystals, finite crystals can be in equilibrium only with supersaturated vapours. The equilibrium of finite crystals is unstable unless very small systems are involved so that minor fluctuations in the linear size could correspondingly affect the chemical potential of the ambient phase. As the mechanism of the equilibrium is immaterial for phenomenological thermodynamics, finite and big crystals can be formally treated in the same way. Let us note that the work of formation of small crystals acting as critical nuclei is minimal for crystals with equilibrium form; that is to say their formation is energetically favoured with respect to crystalline aggregates exhibiting different geometrical forms.

In order to arrive at a more concrete formulation of the equilibrium conditions we shall follow an analysis given by Kaischew [2]. Since the volume of the crystal is

$$V = \tfrac{1}{3} \sum h_i' A_i,$$

where h_i' denotes the length of the normal to the i-th crystallographic face from an arbitrary point P′ inside the crystal, a small variation of the volume can be expressed by

$$\delta V = \tfrac{1}{3} \sum A_i \delta h_i' + \tfrac{1}{3} \sum h_i' \delta A_i.$$

On the other hand it can be directly seen that

$$\delta V = \sum A_i \delta h_i' .$$

Hence, the initial conditions (11.1) and (11.1′) take the form

$$\sum \gamma_i \delta A_i = 0 . \tag{11.2}$$

$$\sum h_i' \delta A_i = 0 . \tag{11.2'}$$

Considering further that the surface areas are not independent as for a closed polyhedron

$$\sum \mathbf{a}_i A_i = 0$$

$\mathbf{a}_i$ being the unit normal vector with respect to the corresponding face, we have to take into account the additional condition

$$\sum \mathbf{a}_i \delta A_i = 0. \tag{11.2''}$$

To solve the variational problem defined by (11.2), (11.2′) and (11.2″) we multiply (11.2′) by the indefinite scalar $-\lambda$ and (11.2″) by the indefinite vector

$-\mathbf{b}$ and sum with (11.2). We get

$$\sum [\gamma_i - h_i'\lambda - (\mathbf{a}_i, \mathbf{b})]\,\delta A_i = 0$$
$$\gamma_i = \lambda[h_i' + (\mathbf{a}_i, \mathbf{b}/\lambda)].$$

If we put the origin of the vector $-\mathbf{b}/\lambda$ at the initially chosen point P′ we realize at once that the sum in the brackets defines a new distance h_i to be measured from a new point P coinciding exactly with the end of the vector $-\mathbf{b}/\lambda$ (fig. 11.1).

In this way it follows for each crystallographic face

$$\gamma_i = \lambda h_i, \tag{11.3}$$

where λ depends solely on the absolute size of the crystal. An equivalent formulation reads

$$\gamma_1 : \gamma_2 : \gamma_3 : \ldots = h_1 : h_2 : h_3 : \ldots \,. \tag{11.4}$$

Relationships (11.3) and (11.4) express the well known Wulff rule (Gibbs-Wulff theorem) for the equilibrium form. According to this rule the equilibrium form can be constructed by the following geometrical procedure: From an arbitrary point vectors are drawn normal to possible crystallographic planes. Lengths proportional to the corresponding γ values are plotted on

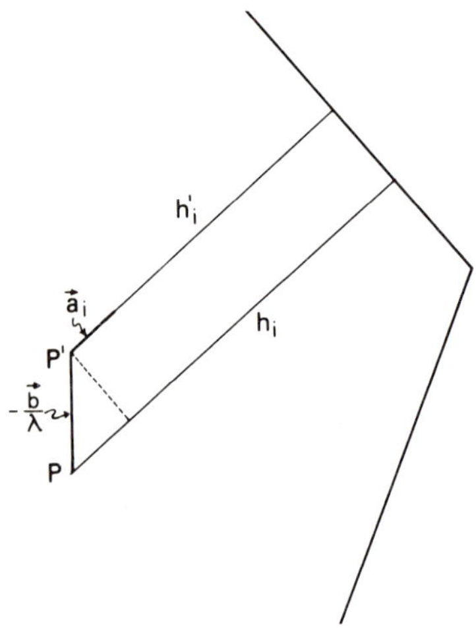

Fig. 11.1. Diagram illustrating the derivation of the Gibbs-Wulff theorem.

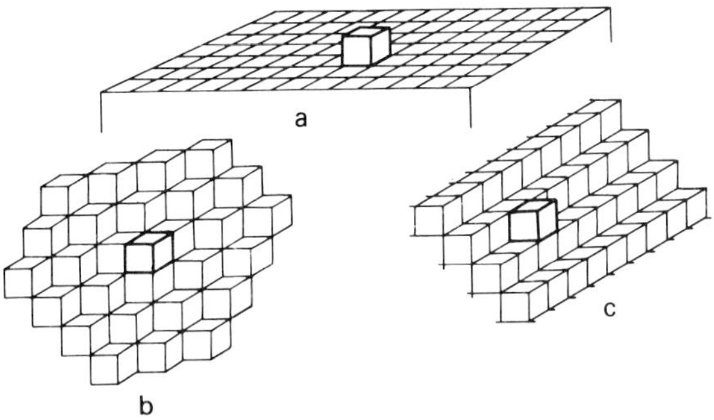

Fig. 11.2. On the evaluation of the specific surface energies of {100} (a), {111} (b) and {110} (c) faces of a Kossel crystal (after [4]).

each vector obtaining thus a family of similar polyhedra. The polyhedron with the same volume as that of the given crystal represents its equilibrium form. In the case of big crystals however, the volume cannot be predicted at a given temperature. A major difficulty in this construction is that the specific surface energies of solids, not to mention those of separate crystallographic faces, are normally not available as experimental values. Nevertheless, simple model considerations can sometimes be very helpful.

Let us consider for instance a homopolar crystal with simple cubic lattice. What is its equilibrium form assuming attractive forces only between nearest atoms in the lattice? The γ values of the {100}, {111} and {110} faces (fig. 11.2a, b and c) of such a crystal, frequently called a Kossel crystal, can be evaluated as follows [3]

$$\gamma_{100} = \frac{\phi}{2d^2} \tag{11.5}$$

$$\gamma_{111} = \frac{\phi\sqrt{3}}{2d^2} \tag{11.5'}$$

$$\gamma_{110} = \frac{\phi}{d^2\sqrt{2}} \tag{11.5''}$$

Here ϕ is the work needed to disrupt the bonds between two neighbouring particles, their centres being spaced at a distance d apart. Applying the Wulff rule we obtain

$$\frac{h_{111}}{h_{100}} = \sqrt{3} \quad \text{and} \quad \frac{h_{110}}{h_{100}} = \sqrt{2},$$

i.e. the octahedral and rhombododecahedral faces obviously do not belong to the equilibrium form because they have only common corners and edges respectively with the cube. Consequently, the equilibrium form is a cube. Its absolute size can be calculated if the supersaturation is known.

It must be noted that the above given formulation of the equilibrium criterion can be appropriately modified to account for the case when one or more crystallographic faces are in contact with a foreign body. As was shown by Kaischew [1] in this instance

$$h_1 : h_2 : h_3^{(s)} : .... = \gamma_1 : \gamma_2 : \gamma_3 - \beta : ...,$$

where h_1, h_2 ..., γ_1, γ_2 ..., retain their usual meaning, $h_3^{(s)}$ being the distance from the Wulff point to the face which contacts the substrate and β – the specific adhesion energy across the crystal/substrate interface for the corresponding plane. It follows that $h_i^{(s)} < h_i$, i.e. in the presence of a foreign substrate the i-th distance is shortened as compared to its original value within a homogeneous phase.

The Gibbs-Wulff theorem holds also for two-dimensional crystals formed on top of three-dimensional crystals. The equilibrium condition is then

$$\sum \kappa_i l_i = \text{min.}, \tag{11.6}$$

at constant surface area. Here κ_i and l_i are the specific edge energy and the length of the normal drawn from a point inside the polygon to the i-th side respectively. It can be shown that (11.6) is satisfied by the analogous relation

$$\kappa_i / l_i = \text{const.}$$

11.3. Atomistic approach *

The mechanism by which the equilibration of a crystal with its surrounding is maintained can be understood when the elementary evaporation and condensation processes on the crystal surface are considered. Kossel and Stranski [5] were the first to realize the importance of such considerations by introducing the energy required for a single atom on the crystal surface to be transferred into the ambient phase. The work of separation, depending on

* For a detailed and comprehensive treatise see the book of Honigmann [4]. References of Kossel's works which are somewhat difficult of access can also be found there.

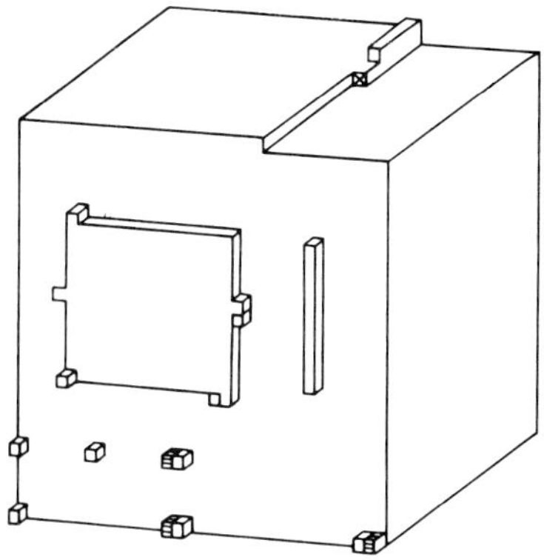

Fig. 11.3. Various positions of surface atoms in a crystal according to Stranski [5].

the location of the atoms on the crystal surface, is a measure of the probability of evaporation. Of all possible positions depicted in fig. 11.3 the so called half crystal position (marked by a cross on the figure) is the most significant one. It is easily seen that an atom located in this position is twice as weakly bonded as in the bulk of the crystal. Considering a sufficiently big crystal, ignoring thus the role of edges and corners, we have to admit that the half crystal position is the repeatable step during growth or evaporation of the top layer. As a matter of fact, the condensation or evaporation of atoms at this position affects in no way the energetics and the geometry of the crystal surface. Therefore, the work needed to separate an atom from the half crystal position, to be denoted by $\varphi_{1/2}$ is the energetic characteristic of the crystal/vapour equilibrium. Atoms which are more weakly bonded than $\varphi_{1/2}$ tend to evaporate and should not be considered as possible constituents of completely built up crystallographic planes. Starting from such premises Stranski and Kaischew were able, in 1931, to predict the equilibrium form of large homopolar crystals imposing the condition that all atoms for which $\varphi < \varphi_{1/2}$ should not be present on the crystal surface. Such an analysis is possible only if the elementary binding energies are amenable to quantitative evaluation. In the case of homopolar bonds φ can be simply calculated by introducing the energy required to disrupt the bonds between the chosen atom and its neighbours, their number being fixed for a given structure type. Taking again the simple cubic crystal we find

$$\varphi_{\frac{1}{2}} = 3\phi_1$$

or

$$\varphi_{\frac{1}{2}} = 3\phi_1 + 6\phi_2$$

or

$$\varphi_{\frac{1}{2}} = 3\phi_1 + 6\phi_2 + 4\phi_3 \text{ , etc. ,}$$

depending on whether the attractive forces between first nearest, second nearest, or third nearest neighbours have been accounted for. Similarly for the corner atom which occupies the site with lowest binding energy within the surface layer the following relations hold

$$\varphi_c = 3\phi_1 \text{ ,}$$
$$\varphi_c = 3\phi_1 + 3\phi_2 \text{ ,}$$
$$\varphi_c = 3\phi_1 + 3\phi_2 + \phi_3 \text{ , etc.}$$

Hence, the cubic form can be retained only if interaction between first nearest atoms is operative in the lattice. When second nearest atoms are involved, $\varphi_c < \varphi_{1/2}$ and some atoms will leave the crystal thus revealing new crystallographic faces. To sum up, provided that the crystal structure as well as the distance at which interatomic forces are still effective are known, the equilibrium form of big crystals can be easily derived. Table 11.1 shows the results obtained by Stranski and Kaischew [3] for homopolar crystals. The Gibbs-

Table 11.1
Equilibrium forms of homopolar Crystals
after Stranski and Kaischew

Lattice type	Number of neighbours around a central atom	1st nearest atoms	1st and 2nd nearest atoms	1st, 2nd and 3rd nearest atoms
simple cubic	3/6/4	100	100, 110, 111	100, 110 111, 211
face centered cubic	6/3/12	111, 100	111, 100, 110	111, 100 110, 311 210, 531
body centered cubic	4/3/6	110	110, 100	110, 211 100, 111
hexagonal close packed structure	6/3/1		0001, $10\bar{1}0$ $10\bar{1}1$	0001, $10\bar{1}0$ $10\bar{1}1$, $10\bar{1}2$ $11\bar{2}0$

Wulff theorem yields the same results if the γ values are calculated on the basis of the above outlined model.

Let us now examine the equilibrium of finite crystals. From an atomistic point of view the fact that finite crystals can be equilibrated only with supersaturated vapours is explained by the increased role of the corners and edges of the crystal in the material exchange between the two phases. Also, on such a crystal there is not a single position which can be viewed as representative during the successive completion of the top layer. The equilibrium of a small crystal is therefore described by the mean work of separation $\overline{\varphi}$ which is the average (per atom) of all separation energies from all sites revealed by the successive evaporation of the upper layer. This very important concept introduced by Stranski and Kaischew in 1934 [6–8] is the corner stone of the classical atomistic theory of crystal growth. That it is the mean work of separation which is responsible for the equilibrium of small crystals can be seen from the following simple argument [9]. The probability of evaporation of a single atom from the crystal surface can be expressed by

$$w_i\uparrow \;=\; K_i\uparrow \exp\left(-\varphi_i/kT\right)$$

whereas the probability of condensation depends primarily on the vapour pressure, i.e.

$$w_i\downarrow \;=\; K_i\downarrow p$$

Here $K_i\uparrow$ and $K_i\downarrow$ are quantities which are characteristic for the chosen site. The attainment of the equilibrium state will mean that the numbers of atoms leaving and joining the crystal per unit time are equal so that no macroscopic movement of crystal planes occurs. To give a kinetic interpretation of the equilibrium a feasible mechanism must be devised which need not necessarily be the real one, the sole requirement being that the treatment is self-consistent. Let us suppose that the top layer evaporates as a whole, disintegrating at the same time into its constituents and that the reverse process occurs by simultaneous condensation of the same number of atoms. Considering the elementary acts of condensation and evaporation as independent events, the probabilities that the described processes will take place are:

$$W\uparrow \;=\; \exp\left(-\sum_1^{n^2} \varphi_i/kT\right)\prod_1^{n^2} K_i\uparrow$$

and

$$W\downarrow \;=\; p^{n^2}\prod_1^{n^2} K_i\downarrow\,,$$

where n^2 is the number of the atoms involved. Since at equilibrium

$$W\uparrow = W\downarrow,$$

the following expression for the vapour pressure is obtained

$$p = \frac{1}{\tilde{\nu}} \exp(-\bar{\varphi}/kT)\,. \tag{11.7}$$

Here

$$\bar{\varphi} = \frac{1}{n^2} \sum \varphi_i$$

is the mean work of separation and $\nu_n = [\sqrt[n^2]{\Pi_1^{n^2} K_i\uparrow/K_i\downarrow}\,]^{-1}$ is the mean geometric of the average vibrational volumes of the atoms forming the top layer of the crystal [10]. Considering that for a large crystal

$$p_{\text{eq}} = \frac{1}{\tilde{\nu}_0} \exp(-\varphi_{\frac{1}{2}}/kT) \tag{11.7'}$$

equations (11.7) and (11.7′) yield

$$kT \ln \frac{p}{p_{\text{eq}}} = \varphi_{\frac{1}{2}} - \bar{\varphi} \tag{11.8}$$

The last relation is nothing other than the Thomson-Gibbs equation for the vapour pressure of small crystals in terms of an atomistic treatment of the equilibrium [6–8]. As $\varphi_{1/2}$ is a constant for a given crystal, the vapour pressure defines the value of $\bar{\varphi}$ and vice versa.

To illustrate the method based on the mean work of separation and its further implications for the equilibrium form, let us again take advantage of the very simple Kossel crystal as a convenient model. Evaporating the top layer as indicated in fig. 11.4*

$$\bar{\varphi} = \frac{1}{n^2}[(n-1)^3\, 3\phi + 2(n-1) 2\phi + \phi] = \varphi_{\frac{1}{2}} - \frac{2\phi}{n}. \tag{11.9}$$

* It must be explicitly noted that the value of $\bar{\varphi}$ is invariant with respect to the mode of removal of the atoms from the top layer. The example given illustrates a convenient procedure. Taking into account that the total number of bonds to be disrupted per atom for the given model is $(1/n^2)[2n(n-1)+n^2]$ expression (11.9) follows again.

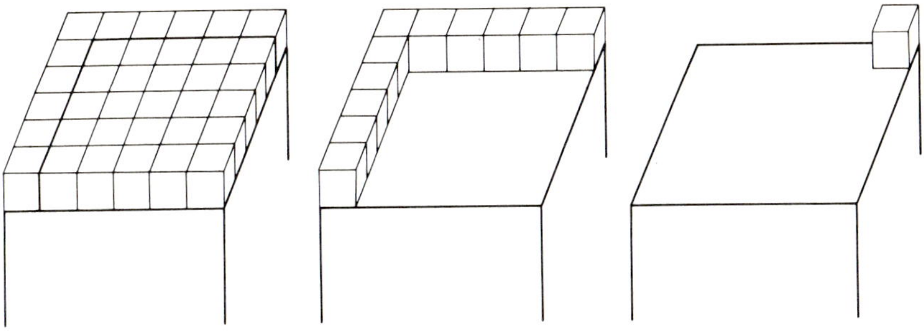

Fig. 11.4. On the evaluation of the mean separation energy according to Stranski and Kaischew [6–8].

Substituting (11.9) in (11.8) we obtain

$$kT \ln \frac{p}{p_{eq}} = \frac{2\phi}{n} \tag{11.10}$$

Using relation (11.5) and writing $L^{(3)} = nd$ and $v_0 = d^3$ the last expression turns in the well known equation

$$kT \ln \frac{p}{p_{eq}} = \frac{4\gamma v_0}{L^{(3)}} \,. \tag{11.11}$$

In a similar way the equilibrium of a quadratic cluster formed on top of cube is defined by the mean work of separation of the atoms building the side of the two-dimensional crystal. In this case

$$\bar{\varphi} = \varphi_{\frac{1}{2}} - \frac{\phi}{n'} \tag{11.9$'$}$$

and

$$kT \ln \frac{p}{p_{eq}} = \frac{\phi}{n'} \tag{11.10$'$}$$

Considering that $\kappa = \phi/2d$ and noting $A_0 = d^2$, $L^{(2)} = n'd$ equation (11.10′) transforms into the well known equation defining the equilibrium of two-dimensional critical nuclei

$$kT\ln\frac{p}{p_{eq}} = \frac{2\kappa A_0}{L^{(2)}} \ . \qquad (11.12)$$

The meaning of the symbols used so far is as follows: $L^{(3)}$ – linear size of the cubic crystal, $L^{(2)}$ – linear size of the quadratic crystal, ν_0 – molecular volume, A_0 – surface area corresponding to one atom, k – Boltzmann constant, T – absolute temperature. Comparing (11.10) and (11.10′) we see that under equilibrium conditions the linear size of the three-dimensional crystal is twice as big as the linear size of the two-dimensional crystal, i.e. $n = 2n'$. More important is the fact that at equilibrium the $\bar{\varphi}$ values pertaining to peripheral rows of two-dimensional crystals and to complete surface layers of three-dimensional crystals are equal for all crystallographic faces belonging to the equilibrium form.

We are now able to formulate the method outlined by Stranski and Kaischew for the determination of the equilibrium form of finite crystals. For this purpose starting from a simple crystallographic form all atoms with lower binding energies than $\bar{\varphi}$ are removed from the crystal surface. In this way all crystallographic planes belonging to the equilibrium form will appear on the crystal. The second stage involves an appropriate variation of the surface areas of the planes revealed until the mean separation energies calculated for each plane become equal to

$$\bar{\varphi} = \varphi_{\frac{1}{2}} - kT\ln\frac{p}{p_{eq}} \ .$$

As an example we shall consider a hypothetical quadratic crystal assuming interaction forces between first nearest and second nearest atoms (fig. 11.5a,b). The binding energy at the half crystal position is $\varphi_{\frac{1}{2}} = 2\phi_1 + 2\phi_2$.

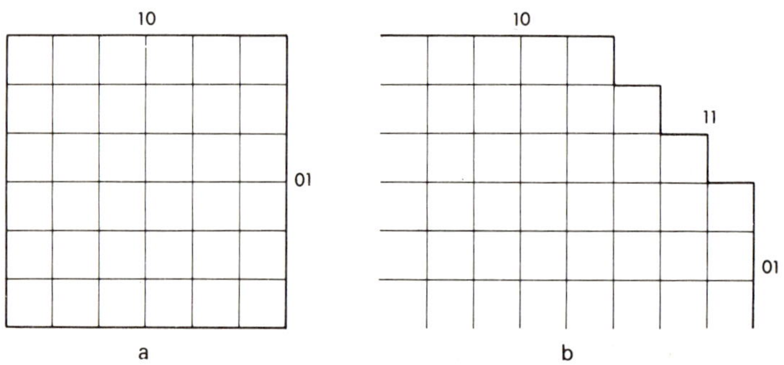

Fig. 11.5. Two-dimensional crystal model illustrating the effect of the supersaturation on the equilibrium form.

Considering first a square (fig. 11.5a) we see that this form will be stable if the separation energy of the corner atom $\varphi_c = 2\phi_1 + \phi_2$ is not less than $\bar{\varphi} = \varphi_{1/2} - kT \ln p/p_{eq}$. Hence, the criterion is $p/p_{eq} \geqslant \exp(\phi_2/kT)$.
At higher supersaturations the quadratic form will remain unchanged, a reduction in size being the only result. At lower supersaturations, $\varphi_c < \bar{\varphi}$ and the corner atoms will leave the crystal revealing a new "plane". The equilibrium form now should satisfy the condition

$$\bar{\varphi}_{10} = \bar{\varphi}_{11} \equiv \bar{\varphi} \,.$$

Since according to fig. 11.5b

$$\bar{\varphi}_{10} = \frac{1}{n_{10}}[(n_{10} - 1)(2\phi_1 + 2\phi_2) + (\phi_1 + \phi_2)] = \underbrace{2\phi_1 + 2\phi_2}_{\varphi_{\frac{1}{2}}} - \frac{\phi_1}{n_{10}}$$

$$\bar{\varphi}_{11} = \frac{1}{n_{11}}[(n_{11} - 1)(2\phi_1 + 2\phi_2) + (2\phi_1 + \phi_2)] = \underbrace{2\phi + 2\phi_2}_{\varphi_{\frac{1}{2}}} - \frac{\phi_2}{n_{11}}$$

we find

$$\frac{n_{10}}{n_{11}} = \frac{\phi_1}{\phi_2} > 1 \,.$$

In this way the equilibrium form including the absolute dimensions of the crystal is completely determined for each value of the supersaturation ratio. The idealized example given above illustrates an important consequence of the theory, that at higher supersaturations the equilibrium form simplifies. As is seen, the model considerations make it possible to predict the critical supersaturations causing changes in the equilibrium form. Another aspect of this problem is connected with the evaluation of the interatomic forces in the structure by measuring the threshold supersaturations responsible for the appearance and disappearance of certain crystallographic surfaces.

To conclude this section* we note that, ignoring entropy contributions which can be entirely suppressed only at the absolute zero temperature, the binding energy of a single atom on the crystal surface equals (with opposite

* Further applications of the method of mean separation energies to various problems of crystal growth and nucleation can be found for instance in [11–13].

sign) the chemical potential at the chosen site. The mean work of separation corresponds to the mean chemical potential $\bar{\mu}_i$ (per atom) of the corresponding crystallographic face. At equilibrium $\bar{\mu}_i = \bar{\mu} \equiv \mu$, μ being the chemical potential of supersaturated vapours at the same temperature. Equation (11.8) can then be rewritten to give

$$kT \ln \frac{p}{p_{\mathrm{eq}}} = \bar{\mu} - \mu_0 = \varphi_{\frac{1}{2}} - \bar{\varphi} ,$$

where μ_0 refers to the macroscopic two-phase equilibrium.

11.4. On the correlation between equilibrium form and growth form

The unstable character of the equilibrium of finite crystals with their environment is the main obstacle for the direct observation of equilibrium forms. As far as large crystals are concerned, their equilibrium with saturated vapours is in a way indifferent in the sense that the thermodynamic driving force arising from variations in the $\bar{\varphi}$ values of separate crystallographic planes is much too low to cause any spontaneous transformations leading to the attainment of a form with prescribed central distances. Considering also that a sufficiently big crystal is the product of prolonged growth only those planes with lowest rate of advancement will be preserved on the crystal obtained, i.e. some of the faces belonging to the equilibrium form may have disappeared at an intermediate stage. Conversely, the initial stages of growth can yield valuable information about the equilibrium form since all faces pertaining to the latter should appear on the crystal. As a matter of fact, in the absence of screw dislocations these surfaces grow by the formation and propagation of two-dimensional nuclei which is a thermally activated process. Naturally, the absolute dimensions of the planes observed on a macrocrystal cannot be predicted by the Wulff theorem. In other words, a qualitative comparison between theory and experiment is possible in this way. Nevertheless, the order in which equilibrium form faces disappear on the growth form upon increasing the supersaturation is the same as if the equilibrium form itself were involved. The same is true for the corresponding supersaturation thresholds. The latter can serve as a source of information about the interatomic forces in the crystal structure. The validity of the theory is confirmed by a large number of experiments [4, 15–20].

References

[1] M. Volmer, *Kinetik der Phasenbildung* (Th. Steinkopf Verlag, Dresden-Leipzig, 1939).
[2] R. Kaischew, Bull. l'Acad. Bulg. Sci. (phys.) 2 (1951) 191.
[3] I.N. Stranski, R. Kaischew, Z. Kristallogr. 78 (1951) 25.
[4] B. Honigmann, *Gleichgewichts- und Wachstumsformen von Kristallen* (Steinkopf Verlag, Darmstadt, 1958).
[5] I.N. Stranski, Z. physik. Chem. 136 A (1928) 259.
[6] I.N. Stranski and R. Kaischew, Z. physik. Chem. 26B (1934) 100.
[7] R. Kaischew and I.N. Stranski, Z. physik. Chem. 26B (1934) 114.
[8] I.N. Stranski and R. Kaischew, Z. physik. Chem. 26B (1934) 312.
[9] R. Kaischew, Jahrb. Univ. Sofia, physik. math. Fak. 38 (2) (1940/1941) 207.
[10] R. Kaischew, Z. physik. Chem. 48B (1940) 82.
[11] I.N. Stranski, Bull. Soc. Minér. Crist. Franç. 79 (1956) 360.
[12] R. Kaischew, Fortschr. Miner. 38 (1960) 7.
[13] R. Lacmann, Erg. exakt. Naturwiss. 44 (1968) 1.
[14] I.N. Stranski and R. Kaischew, Ann. Phys. 23 (1935) 330.
[15] R. Kaischew, L. Keremidtschiew and I.N. Stranski, Z. Metallkunde 34 (1942) 201.
[16] R. Kaischew, Jahrb. Univ. Sofia, physik. math. Fak. 43 (1946/1947) 99.
[17] R. Kaischew, E. Budevski and J. Malinowski, C.R. Acad. Bulg. Sci. 2 (1949) 29.
[18] I.N. Stranski and B. Honigmann, Z. physik. Chem. 194A (1950) 180.
[19] R. Kaischew and D. Nenow, C.R. Acad. Bulg. Sci. 12 (1959) 199.
[20] R. Kaischew and Chr. Nanev, Phys. Stat. Sol. 10 (1965) 779.

P. Hartman, ed., Crystal growth: an introduction. © *(1973) North-Holland Publ. Co.*

GENERALIZED HERRING TREATMENT OF THE EQUILIBRIUM FORM

P. BENNEMA

12.1. Introduction

In ch. 11 Toschev has given a survey of the classical equilibrium form concept and Hartman (see ch. 13) has shown how Bienfait and Kern used this concept to explain the observed morphology of very small crystals in small droplets. These authors have implicitly assumed that the equilibrium form is a polygonal form limited by well defined crystallographic closest-packed or F faces. By exploring the generalized Wulff plot, i.e. a plot in which the surface free energy γ is plotted as a function of all possible orientations, Herring [1,2] summarized in a few theorems the properties of such a plot. On the basis of these theorems it is possible (i) to analyze the importance of all kinds of faces on the equilibrium form; thus also all kinds of crystallographic and non-crystallographic faces (with and without well defined Miller indices respectively), (ii) to study equilibrium forms, which are partly or completely limited by curved areas (thus not only polygonal equilibrium forms).

In what follows we will only give a brief survey of Herring's theorems and some of their implications. For a further, more complete, survey, analysis and application of these concepts we refer to the excellent survey papers of Herring [2], Mullins [3] and Cabrera [4]. We also refer to the book of Strickland-Constable [5].

12.2. Properties of a generalized Wulff plot

In fig. 12.1a a generalized Wulff plot (which will be called simply a Wulff plot in what follows) is presented in two dimensions and in fig. 12.1b in three dimensions. In both cases the equilibrium form is drawn inside the Wulff plot.

It follows from the Wulff plots of fig. 12.1a and fig. 12.1b that, due to the occurrence of sharp cusps, the equilibrium form has a "classical" polygonal

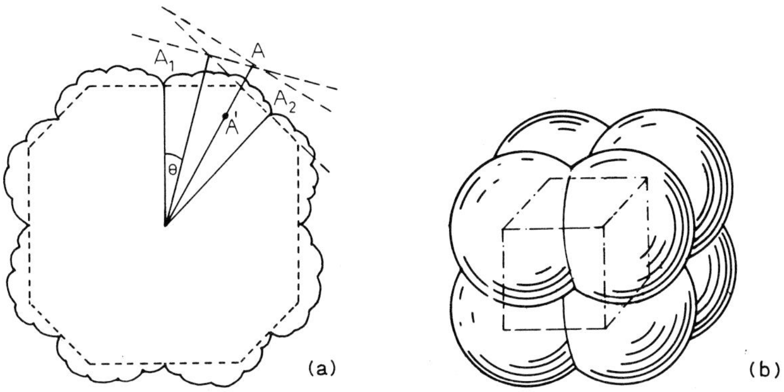

Fig. 12.1. a) A $\gamma(\theta)$ plot in two dimensions; b) a $\gamma(\theta,\phi)$ plot in three dimensions. (From [1]).

shape. The crystallographic face perpendicular to, for example, the normal OA in fig. 12.1a, simply does not occur on the equilibrium form because the γ value for this direction, as given by OA, is too high to let this face truncate a corner from the equilibrium form. The only thing that can be said, according to Cabrera [4], is that the area of A and its direct environment is metastable in reference to the somewhat more remote environment of A, because of a weak cusp.

If, for this direction, the length had been OA′ (see fig. 12.1a) the face normal to OA′ would have been a part of the equilibrium form. In order to obtain the equilibrium form from the Wulff plot we used a generalization of the well-known property derived by Toschev in ch. 11, §2.

$$\gamma_i/h_i = \text{const}\,, \tag{12.1}$$

where i represents any direction (θ,ϕ) (θ and ϕ being the polar coordinates). Having made, from the Wulff plot, an equilibrium form using eq. (12.1), Herring then formulates the following generalized Wulff theorem:

Theorem I

Draw a plane through the end of each radius vector and perpendicular to the radius vector. Then, according to Wulff, the body formed by all points reachable from the origin without crossing any of these planes will be geometrically similar to the equilibrium shape.

This theorem can be verified upon considering fig. 12.1a. For a proof of this theorem we refer to Herring [1, 2].

12.3. Theorem concerning the thermodynamic stability of a hill and valley structure

In order to investigate under which circumstances it is possible to decrease the free energy of an initially flat surface by rearranging the surface into hills and valleys, consider again fig. 12.1a. Consider any of the directions OA_1, OA_2 and OA_3 (the third dimension is not drawn in fig. 12.1a) having positive projections on OA which correspond to the three equilibrium faces A_1, A_2 and A_3 (perpendicular to OA_1, OA_2 and OA_3, respectively).

We will assume that these three faces form the sides of the hill and valley structure. The surface free energy γ of such a hill and valley structure is given by

$$\gamma = \gamma_1 f_{A_1} + \gamma_2 f_{A_2} + \gamma_3 f_{A_3} , \tag{12.2}$$

where γ_1, γ_2 and γ_3 correspond to the surface free energies of the faces A_1, A_2 and A_3 respectively. f_{A_1}, f_{A_2} and f_{A_3} are the areas of surfaces of these three faces in the hill and valley structure per unit projected area in the plane of the macroscopic surface.

We note that f_{A_1}, f_{A_2} and f_{A_3} vary for each direction (θ, ϕ) enclosed by the three equilibrium surfaces. Herring now introduces the three unit vectors $\boldsymbol{e}_1^*$, $\boldsymbol{e}_2^*$, $\boldsymbol{e}_3^*$ which are erected at the origin of the Wulff plot perpendicular to the three equilibrium faces A_1, A_2 and A_3 respectively. Then the unit vector $\boldsymbol{e}^*$ in the direction OA is defined as

$$\boldsymbol{e}^* = f_{A_1}\boldsymbol{e}_1^* + f_{A_2}\boldsymbol{e}_2^* + f_{A_3}\boldsymbol{e}_3^* . \tag{12.3}$$

Now a new set of "reciprocal" unit vectors $\boldsymbol{e}_1$, $\boldsymbol{e}_2$ and $\boldsymbol{e}_3$ is introduced as follows: $\boldsymbol{e}_1$ is the unit vector perpendicular to the plane $(\boldsymbol{e}_2^*, \boldsymbol{e}_3^*)$, $\boldsymbol{e}_2$ the unit vector perpendicular to $(\boldsymbol{e}_1^*, \boldsymbol{e}_3^*)$ etc. So $\boldsymbol{e}_1^* \cdot \boldsymbol{e}_1 = 1$, $\boldsymbol{e}_1^* \cdot \boldsymbol{e}_2 = 0$ or in general

$$\boldsymbol{e}_i^* \cdot \boldsymbol{e}_j = \delta_{ij} , \tag{12.4}$$

where $\delta_{ij} = 1$ if $i = j$ (i and j are 1, 2, 3 respectively) and $\delta_{ij} = 0$ if $i \neq j$.

(Eq. (12.4) follows immediately from the definition of $\boldsymbol{e}_i^*$ in reference to $\boldsymbol{e}_j$. If $i \neq j$, $\boldsymbol{e}_i$ is perpendicular to $\boldsymbol{e}_j$ and hence $\delta_{ij} = 0$. If $i = j$, they are not perpendicular and hence $\delta_{ij} = 1$.)

Upon using (12.3) and (12.4), (12.2) can be rewritten as follows:

$$\gamma = (\gamma_1 e_1 + \gamma_2 e_2 + \gamma_3 e_3) \cdot (f_{A_1} e_1^* + f_{A_2} e_2^* + f_{A_3} e_3^*)$$

or

$$\gamma = c \cdot e^* , \tag{12.5}$$

where e^* is given by (12.3) and c is defined as

$$c = \gamma_1 e_1 + \gamma_2 e_2 + \gamma_3 e_3 . \tag{12.6}$$

In order to visualize the significance of the vector c we first note that e_1, e_2 and e_3 are parallel to the edges where the faces A_2 and A_3, A_3 and A_1 and A_2 and A_1 cut each other respectively (see fig. 12.2). This follows from definitions of e_i. For example e_1 is perpendicular to the plane (e_2^*, e_3^*) of which e_2^* and e_3^* are in turn perpendicular to A_2 and A_3 respectively, it then follows that e_1 is parallel to the edge of A_2 and A_3. This being so, it is suggested that c represents the vector OC (see fig. 12.2), which is the vector joining the origin and the corner C of the equilibrium form where the faces A_1, A_2 and A_3 meet each other. This also can be demonstrated in this way: it follows from (12.5) that

$$c \cdot e_1^* = \gamma_1, \quad c \cdot e_2^* = \gamma_2, \quad c \cdot e_3^* = \gamma_3 , \tag{12.7}$$

which means that the projection of c on the unit vector e_1^* is γ_1, etc. in agreement with our interpretation of c given above. We note that c corresponds to the maximum possible surface free energy γ_{max} of the Wulff plot. Since e^* corresponds with the unit vector of a particular direction we can rewrite (12.5) by

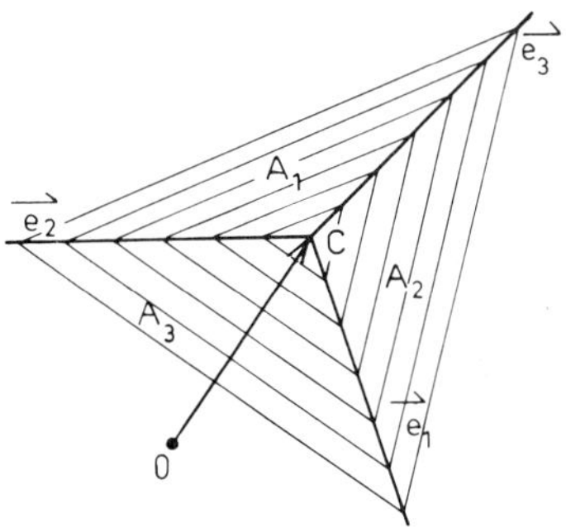

Fig. 12.2. Faces A_1, A_2 and A_3 and their corresponding edges.

$$\gamma = \boldsymbol{c} \cdot \boldsymbol{e}^* = \gamma_{max} \cos \Theta, \dagger \tag{12.8}$$

where Θ is the angle between $\boldsymbol{c}$ (OC) and the particular direction OA under consideration. (Θ is the angle between γ and γ_{max}; it is different from θ the spherical coordinate.) This means that upon using OC as the diameter of a sphere the length of OA can be obtained from the sphere (see fig. 12.3) which then gives γ of the hill and valley structure for any direction OA.

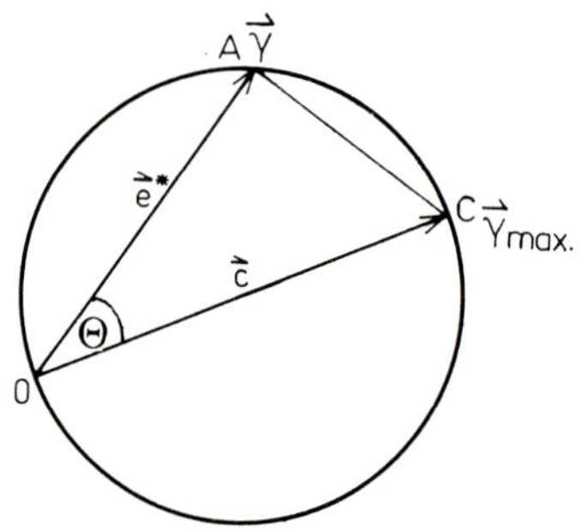

Fig. 12.3. Representation of $\gamma = \gamma_{max} \cos \Theta$.

From these considerations we see immediately that if the projection of OC on a direction OA (see fig. 12.1a) is smaller than the γ value of the Wulff plot, the hill and valley structure is more stable. In the reverse case the hill and valley structure is less stable and the face corresponding to OA will occur on the equilibrium form.

We can summarize these considerations in theorem II as formulated by Herring:

Theorem II

If a given macroscopic surface of a crystal does not coincide in orientation with some portion of the boundary of the equilibrium shape, there will always exist a hill and valley structure which has a lower free energy than a flat surface, while if the given surface does occur on the equilibrium shape, no hill and valley structure can be more stable.

† γ or γ_{max} do not have a vectorial character. By taking $\boldsymbol{e}^* \cdot \gamma$ where $\boldsymbol{e}^*$ is some unit vector corresponding to a particular direction, we can give γ a vectorial character. In that case in what follows we will write $\boldsymbol{\gamma}$.

The implications of this result depend very much on the nature of the equilibrium shape. If the equilibrium shape is a polyhedron then a crystal surface chosen with a random macroscopic orientation will almost always prefer to have a hill and valley structure. At the other extreme, if the equilibrium shape has no sharp corners at all, then surfaces of every orientation will prefer to remain smooth.

12.4. Broken-bond models and the equilibrium form

So far the description of the equilibrium form has been completely macroscopic and phenomenological. No attempt was made to correlate interatomic forces with surface free energies and the Wulff plot. In order to do so we will consider simple crystal structures with only first neighbour interactions. Let us assume that we have three F faces which, for simplicity, will be given the indices (100), (010), (001), which are a part of the equilibrium form. If we assume that the surface free energy is only due to energy which is required to break first neighbour bonds and that no surface deformation takes place upon the formation of a surface, we can write for the surface free energy γ_{hkl} for a particular face (hkl) (h, k and l need not to be small integers but can be any integer),

$$\gamma_{hkl} \cdot M_{hkl} = h\gamma_{100} \cdot M_{100} + k\gamma_{010} \cdot M_{010} + l\gamma_{001} \cdot M_{001} \,. \tag{12.9}$$

Here M_{hkl} is the mesh area of the face (hkl), γ_{hkl} the surface free energy of the face (hkl). We now write

$$\gamma_{100}M_{100} = E \,, \quad \gamma_{010}M_{010} = F \quad \text{and} \quad \gamma_{001}M_{001} = G \,, \tag{12.10}$$

where E, F and G represent the number of broken bonds per mesh area M_{100}, M_{010} and M_{001} respectively, which are formed if a crystal is split according to these faces. It follows from these definitions that the number of bonds which need to be broken per surface mesh area M_{hkl} is given by (12.9). We now use the concepts and definitions of the reciprocal crystallographic lattice. Then analogous to (12.4) to the unit vectors of the elementary Bravais cell $\boldsymbol{a}, \boldsymbol{b}, \boldsymbol{c}$, are added the reciprocal vectors, $\boldsymbol{a}^*, \boldsymbol{b}^*, \boldsymbol{c}^*$ so that

$$\boldsymbol{a} \cdot \boldsymbol{a}^* = V = \boldsymbol{a} \cdot (\boldsymbol{b} \times \boldsymbol{c}) \text{ cycl.} \tag{12.11}$$

and $\boldsymbol{a} \cdot \boldsymbol{b}^* = 0$ cycl.

V is the volume of the elementary Bravais cell. Substituting (12.10) and (12.11) in (12.9), we then find

$$\gamma_{hkl} \cdot M_{hkl} = (Ea + Fb + Gc) \cdot (ha^* + kb^* + lc^*)/V \,. \tag{12.12}$$

Upon realizing that

$$a^* = V/a = M_{100} \quad \text{etc.} \tag{12.13}$$

and that

$$M_{hkl} = ha^* + kb^* + lc^* \,, \tag{12.14}$$

we then find from (12.12)

$$\gamma_{hkl} = \gamma_{max} \cos \Theta, \tag{12.15}$$

where γ_{max} is defined as

$$\gamma_{max} = \frac{Ea + Fb + Gc}{V} \tag{12.16}$$

and Θ is the angle between the vectors $Ea + Fb + Gc$ and $ha^* + kb^* + lc^*$. Here γ_{max} has a fixed position, while (12.14) varies for each orientation. We note that γ_{max} is the vector joining the origin and the corner of the equilibrium form, because it can be rewritten in the following way

$$\gamma_{max} = \frac{E}{(b \times c)} + \frac{F}{(c \times a)} + \frac{G}{(a \times b)} \,,$$

which means that γ_{max} is composed of the energies per surface unit area of the three reference planes. γ_{max} is therefore the same as $\boldsymbol{c}$ given by (12.6), except that Herring's definition is general and phenomenological. In this connection we want to stress that in these broken-bond models the entropy term $T\Delta S$ (as defined for example in Temkin's model, see ch. 10, §1.5 by Bennema and Gilmer) is completely neglected; in fact these considerations strictly speaking hold only for undeformed surfaces in reference to the crystal at $T = 0°$K.

It follows from (12.15) that we have now a recipe to construct Wulff plots on the basis of broken-bond models. Upon constructing a sphere with a diam-

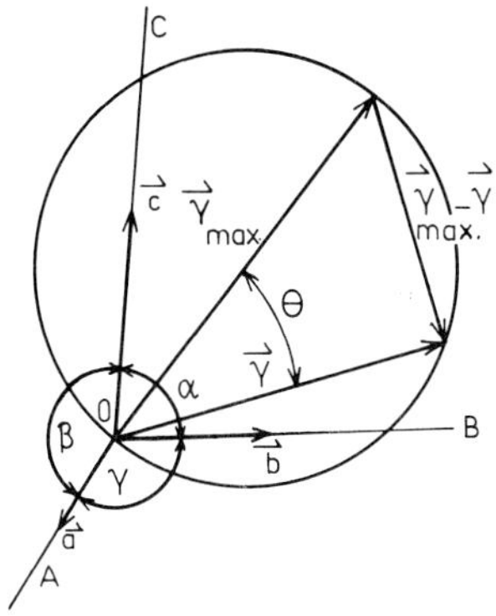

Fig. 12.4. A schematized (two-dimensional) construction of γ. OA, OB, OC are three crystallographic axes which are parallel to $\boldsymbol{e}_1$, $\boldsymbol{e}_2$ and $\boldsymbol{e}_3$ of fig. 12.2. α, β and γ are the angles between pairs of axes. For each direction the vector $\boldsymbol{\gamma} = \boldsymbol{\gamma}_{\max} \cos\Theta$ where Θ is the angle between $\boldsymbol{\gamma}_{\max}$ and $\boldsymbol{\gamma}$.

eter of $\gamma_{\max}$, any other $\boldsymbol{\gamma}$ can be constructed within the corner of say OABC where OA, OB and OC are three crystallographic axes (see fig. 12.4). Extensive work has been carried out over the last ten years on the basis of the derivations given above by Wolff and collaborators [6–8]. Numerous plots were constructed for all kind of homopolar crystal structures as diamond, ZnS, GaAs, etc. Special emphasis was laid on non centro-symmetric structures. In fig. 12.5a,b two examples of Wulff plots as resulting from special structures are presented.

It is interesting to note that the whole Wulff plot is composed from spheres. These spheres cut each other in sharp steep cusps, which correspond to flat faces occurring on the equilibrium form and steep sharp valleys. These valleys are perpendicular to edges of the equilibrium forms and these are PBC's (see ch. 14, §2.3). Also the normals connecting O and the valleys are the same as a set of S faces, all containing that one PBC. We note that the result of these broken-bond models are quite analogous to the considerations concerning the stability of the hill and valley structures of §3. In fact we can consider all regions of the crystal surface whose normals lie within the spacial angle defined by the normals to the three faces of the equilibrium form as regions of K faces (see ch. 14, §1.5) which have a kind of hill and valley structure since K faces are composed of fractions of F faces. In the same way S faces corresponding to the valleys are composed of fractions of two F faces. We can summarize this argument in the following theorem:

Theorem III

Wulff plots can be constructed from broken-bond models. It then turns out that those Wulff plots are composed from spheres which meet each other in steep cusps and valleys. Cusps correspond with F faces occurring on the equilibrium form and valleys to S faces all having the same PBC. The spherical areas correspond to K faces.

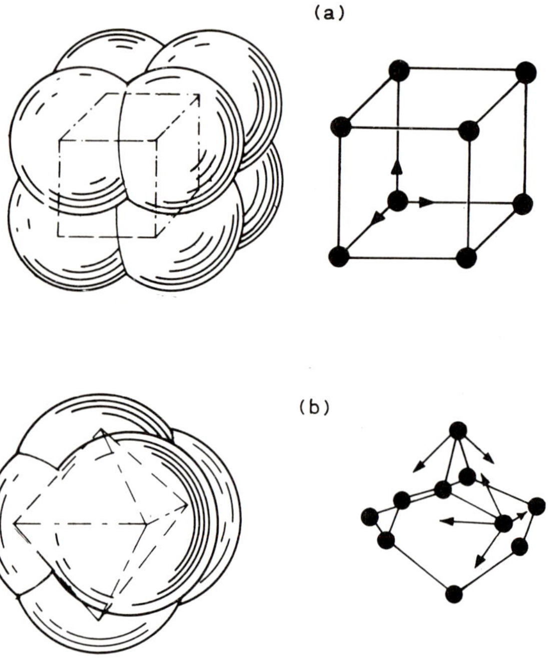

Fig. 12.5. a) Wulff plot of a crystal of simple cubic structure. Only nearest neighbour interaction is taken into account. Each cusp corresponds to an equilibrium face $\{001\}$; each valley connecting the cusps correspondingly represents an edge. This edge is $\langle 100 \rangle$ and is the common boundary to two adjacent $\{001\}$ faces. The valleys are "complete", i.e. they can be traced 360° from a cusp around the plot along one zone. A complete valley corresponds to a straight array of atoms. The equilibrium form $\{001\}$ is inscribed. b) Wulff plot for a crystal with diamond structure. Only nearest neighbour interaction is taken into account. The cusps correspond to $\{111\}$ faces and the valleys connecting them correspond to $\langle 011 \rangle$ edges which are common to two neighbouring $\{111\}$ faces. The valleys are "incomplete", i.e. they cannot be traced 360° around the plot. An incomplete valley corresponds to a zig-zag array of atoms. The equilibrium form $\{111\}$ is inscribed. (From [6]).

12.5. Implications of deviations of the idealized plots resulting from broken-bond models

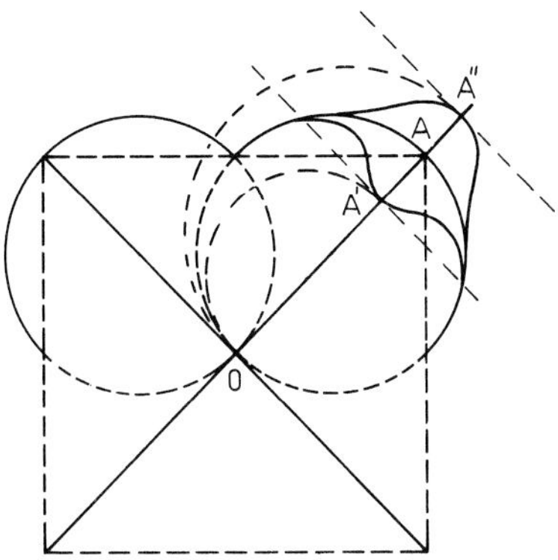

Fig. 12.6. OA′ corresponds to case (i) and OA″ to (ii). The directions OA′, OA and OA″ are for practical reasons chosen as particular directions, but the argument holds for any direction.

Let us consider an idealized Wulff plot resulting from a broken-bond model. A real Wulff plot can have two types of deviations (i) in a certain area γ will be less than the ideal Wulff plot, (ii) it will be greater. In fig. 12.6 OA′ corresponds with (i) and OA″ with (ii). It is easy to see from fig. 12.6 that the face perpendicular to OA′ will be a part of the equilibrium form and in case (ii) this will not be so. Let us draw for the cases (i) and (ii) a circle going through O and tangent to the Wulff plot in A′ and A″ respectively. It can be seen from fig. 12.6 that in case (i) the Wulff plot is situated outside the circle while in case (ii) it is situated inside the circle. In the case of the ideal broken-bond model the Wulff plot and the circle coincide. We thus arrive at the following theorem:

Theorem IV

If OA is a radius vector of the γ-plot in whose neighbourhood the derivatives of γ are continuous, then a necessary and sufficient condition for the direction OA to occur among the normals of the equilibrium shape is that the γ-plot nowhere passes inside the sphere drawn through the origin and tangent to the γ-plot at A. When the γ-plot is outside this sphere at all points except A, all orientations in the neighbourhood of OA will occur among the normals of the equilibrium form, i.e. the latter will be continuously curved in the *region where its normal has the direction OA. Combination of this result with that of the preceding §3 leads to the further conclusion: A continuously curved surface with normals in the neighbourhood of OA will be stable with respect to formation of a hill and valley structure if, and only if, the γ-plot nowhere passes inside the sphere through the origin and tangent at A.*

For a more detailed treatment of this theorem we refer to Herring [1, 2] and Mullins [3].

12.6. Exploration of properties of cusps and valleys

It is clear from our previous discussions that in order for a boundary of the equilibrium form to consist in part of a finite area it is necessary that A be a sharp cusp of the γ-plot. This also can be checked in the same way as in section 5. Let us take again as a reference situation the ideal broken-bond model given by plot 1 in fig. 12.7. Two deviations are possible (see fig. 12.7), (i) the cusp is sharper than the ideal Wulff plot (see curve 2), (ii) the cusp is less sharp, it is "blunted off" (see curve 3). In case (i) the only face occurring on the equilibrium form is the face perpendicular to OA, while in case (ii) faces corresponding to the environment of A additionally occur on the equilibrium form. This can be seen immediately, but it also is a special case of the theorem IV of the previous section. In case (ii) the Wulff plot lies outside the "tangent circle" for the environment of A, which means that those faces do

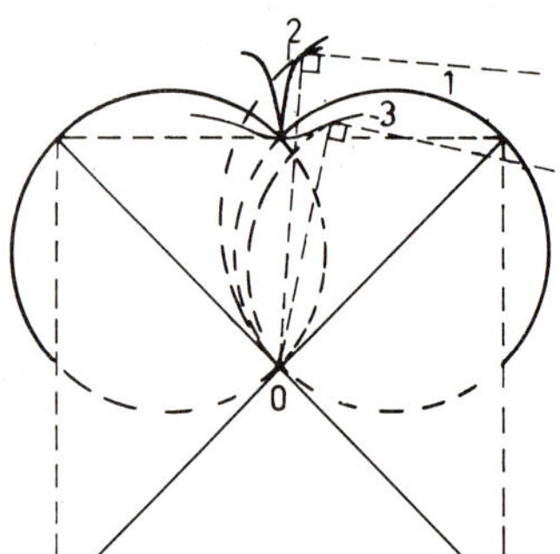

Fig. 12.7. Two possible deviations of the "ideal broken-bond Wulff plot" are represented for a cusp and its environment by curves 2 and 3.

occur on the equilibrium form. In case (i), however, the Wulff plot lies inside the tangent circle or coincides with it and hence the faces occurring in the environment of A do not occur on the equilibrium form *. At the cusp the γ-plot is completely outside the tangent circle. This then leads to the following theorem.

Theorem V

In order that the surface of the equilibrium form have a flat portion of finite extent with an orientation normal to a given vector OA of the γ-plot, it is necessary and sufficient that the γ-plot have a pointed cusp at A and simultaneously that there be some sphere through A and the origin which lies entirely inside the γ-plot. In the same way we find that, in order that the surface have a finite cylindrical portion with surface normals corresponding to a certain plane sector of the γ-plot, it is necessary and sufficient that the γ-plot have a knife-edge cusp for all directions in this sector, and simultaneously that for each point A of the cusp in this sector it be possible to draw a sphere through the origin, tangent to the knife-edge cusp at A, and lying entirely inside the γ-plot.

See further Herring [1, 2] and Mullins [3].

12.7. On the relation between step models and cusps in the Wulff plot

Following Mullins [3] let us take an F face as a reference plane. We want to check the behaviour of γ of stepped faces, which make a (slight) angle θ with this plane.

We will take again as an example a (two-dimensional) Kossel crystal. It can be seen from the particular case shown in fig. 12.8a where $\theta = \text{tg}^{-1}\, 2/5$ that any surface (with rational indices) can be built up by an appropriate periodic sequence of spacings between monosteps. The work W required to create the surface, together with its equivalent mating surface shown in fig. 12.8b is

$$W(\theta) = \left(\frac{1}{a} + \frac{1}{a}\,\text{tg}\,\theta\right)\phi\,, \tag{12.17}$$

* It is possible to have a plot in the environment of the cusp for which the Wulff plot lies outside the tangent circle for case (ii) (curve 2). But such a tangent circle will cut the γ-plot somewhere, in conflict with the criterion for the occurrence of the equilibrium form.

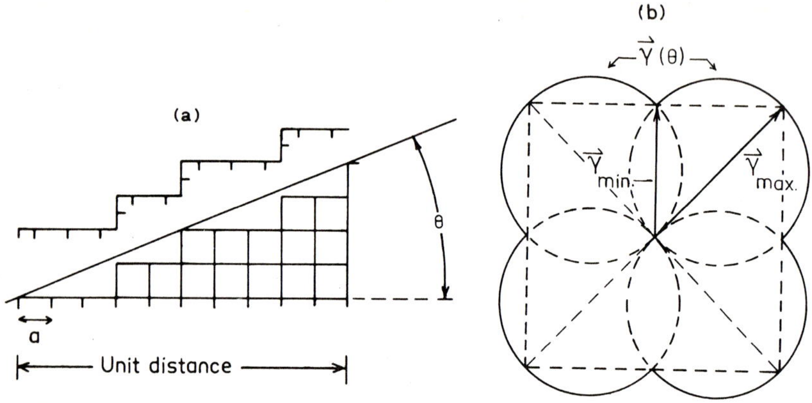

Fig. 12.8. a) Stepped surface of a Kossel crystal with $\theta = \text{tg}^{-1}\ 2/5$; b) γ-plot resulting from a step model. (From [3]).

where ϕ is the bond energy between the blocks of the Kossel crystal. This is because the inclined surface has a surplus of free bonds, over and above the number of $1/a$ of the flat surface ($\theta = 0$), which is equal to the number of steps present (since the number of vertical free bonds is independent of θ).

To calculate γ, we must divide the work $W(\theta)$ by the total area $2/\cos\theta$, of new surface created which gives using (12.17)

$$\gamma(\theta) = \frac{W(\theta)}{2/\cos\theta} = \frac{\phi}{2a}(\cos\theta + \sin\theta) = \frac{\phi}{\sqrt{2}a}\cos\left(\theta - \frac{\pi}{4}\right), \tag{12.18}$$

where θ, the argument of γ, is interpreted as the angle between the normal to the inclined surface and the positive y axis, perpendicular to the (01) base. Equation (12.17) without the restriction on θ, describes a circle passing through the origin and having a diameter of length $\phi/(\sqrt{2}a)$ extending from the origin along a line inclined 45° to the positive y axis (part of this circle appears as the solid curve in the upper left-hand quadrant). It is clear from symmetry, however, that the preceeding argument applies to $-\theta$ and to the other three (10) type surfaces, thereby restricting the validity of equation (12.17) to $0 < \theta < \pi/2$ and yielding for the γ-plot the outside bonding curve, shown as solid in fig. 12.8b, of the four circles comprising the original one and the other three demanded by fourfold symmetry.

The features of most general validity emerging from the γ-plot for our particular crystal model are the anisotropy or dependence of γ on crystallographic orientation, and the appearance of cusps ((10) orientations) at which the derivative $d\gamma/d\theta$ is discontinuous. In fact this step model, which again leads to the "raspberry" type Wulff plots can be considered both as a special case of a broken-bond model or as a hill and valley model. This result is therefore not surprising.

The special, very simple, two-dimensional case we have discussed above is an illustration of a general theorem which can be formulated as follows:

Theorem VI

Wulff plots with spherical areas with steep sharp cusps and steep sharp valleys can be constructed on the basis of step models with only first neighbour interactions.

For a more extensive treatment of this theorem we refer to Mullins [3] and especially Cabrera [4]. According to Cabrera sharp cusps become less sharp if steps attract each other. Due to those interactions the $\gamma(\theta)$ surface will consist of a continuously curved surface as represented in fig. 12.7. For lower (Temkin) α factors (see ch. 10, § 1.6, especially eqs. (10.14) and (10.5)) which may be due to higher temperatures and/or adsorption, sharp cusps may be blunted off and the equilibrium form is limited by less extended flat areas. We note that in our recent Monte Carlo computer games [9] we indeed found interaction of steps at α values of 3. In such a case the cusp must become less sharp than the ideal Wulff plot.

12.8. Frank's and Meijering's reciprocal (γ^{-1}) plot

Frank [8] and independently Meijering [11] discovered that, instead of plotting $\gamma(\theta, \phi)$ as a Wulff plot, there was a considerable advantage in plotting the reciprocal γ^{-1} as a function of (θ, ϕ) which we will call Frank–Meijering plot. We will briefly mention the relation between the Wulff plot and the F–M plot.

Let us take again an ideal broken-bond Wulff plot (see fig. 12.9a) and compare this with a F–M plot (see fig. 12.9b). (For simplicity we take as an example a two-dimensional figure.) In the direction OM we will now have a maximum and in the direction OA a minimum. If we connect A and M in the F–M plot by a straight line we see immediately that for a given direction OA (specified by θ), the angle between γ_{max} and OA_1 in the Wulff plot and γ_{max} and OA_1 in the F–M plot is the same since

$$\cos\theta = \frac{\gamma}{\gamma_{max}}\ (\text{in W plot}) = \frac{\gamma_{max}^{-1}}{\gamma^{-1}}\ (\text{in F–M plot})\ ,$$

which means that in the F–M plot indeed

$$OA_1 = \gamma^{-1}\ . \tag{12.19}$$

This means that the arc MA_1A of the Wulff plot corresponds with the line MA_1A of the F–M plot.

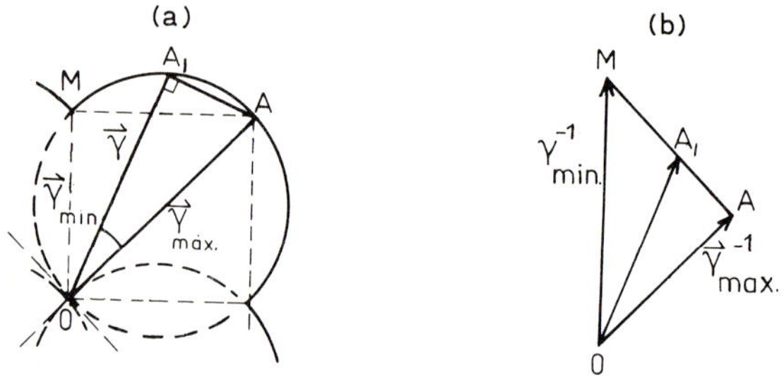

Fig. 12.9. a) Part of an ideal Wulff plot; b) the corresponding F–M plot.

It is easy to see that upon generalizing (12.19) as schematized in fig. 12.9b the spherical areas from the Wulff plot correspond to flat faces in the F–M plot, the steep valleys of the Wulff plot correspond to the edges of the F–M plot and that the cusps of the Wulff plot correspond to the corners of the polygon. As an example we have plotted again the Wulff plot for a simple cubic crystal of which the equilibrium form is a cube (see fig. 12.10a). In fig. 12.10b we have plotted the corresponding F–M plot which is an octahedron. We can summarize these result in the following theorem:

Theorem VII

Flat areas of an F–M plot correspond with spherical areas in a Wulff plot; edges in an F–M plot correspond with valleys in a Wulff plot; corners of an F–M plot correspond with cusps in a Wulff plot and consequently corners in an F–M plot correspond with faces on the equilibrium form; edges in an F–M plot are perpendicular to edges in the equilibrium form; flat areas in an F–M plot correspond to K faces not occurring on the equilibrium form. When cusps in the Wulff plot are blunted off, corners of the F–M plot are rounded off; when the Wulff plot shows a deviation from the ideal broken-bond shapes, the F–M plot will show corresponding "opposite" deviations.

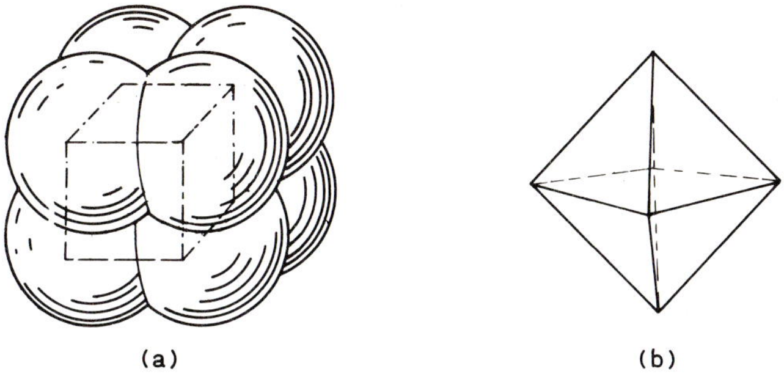

Fig. 12.10. a) Ideal Wulff plot of a Kossel crystal with inscribed equilibrium form, which is a cube; b) the corresponding F–M plot which is now an octahedron.

For a more extensive treatment of the "duality" between equilibrium forms and the F–M plots we refer to the excellent article of Frank [10] and the paper of Meijering [11]. We note that Herring's theorems discussed above are applied to interpret (i) the equilibrium form and faceting phenomena of small metal needle tips in field ion microscopes, (ii) small metal crystals and wires, (iii) etch pits, (iv) formation of macrosteps. Also they can be used to interpret equilibrium forms simulated on a computer. (Work in progress.)

References

[1] C. Herring, Phys. Rev. 82 (1951) 87.
[2] C. Herring, in: Structure and Properties of Solid Surfaces, ed. by R. Gomer and C.S. Smith (Chicago Press, 1953) Chapter I.
[3] W.W. Mullins, in: Metal Surfaces, ed. by the American Society for Metals (Ohio, 1963) Chapter II.
[4] N. Cabrera, Surface Science 2 (1964) 320.
[5] R.F. Strickland-Constable, Kinetics and Mechanism of Crystallization (Academic Press, 1968) Chaper III.
[6] G.A. Wolff and J.G. Gualtieri, Amer. Miner. 47 (1962) 562.
[7] G.A. Wolff, Z. Phys. Chem. "Neue Folge" 31 (1962) 1.
[8] G.A. Wolff, in: Intermetallic Compounds, ed. by J.H. Westbrook (John Wiley and Sons, 1965) Chapter 5.
[9] G.H. Gilmer and P. Bennema, J. Appl. Phys. 43 (1972) 1347.
[10] F.C. Frank, in: Metal Surfaces, ed. by the American Society for Metals (Ohio, 1963) Chapter I.
[11] J.L. Meijering, Acta Met. 11 (1963) 847.

P. Hartman, ed., Crystal growth: an introduction. © *(1973) North-Holland Publ. Co.*

THE EQUILIBRIUM FORM IN A PHASE OF SMALL DIMENSIONS *

P. HARTMAN

13.1.1. The problem

The thermodynamic nucleation theory leads to the concept of a critical nucleus, which occurs when the increase of the free enthalpy reaches a maximum. This nucleus is in labile equilibrium with the surrounding medium and has the shape of the equilibrium form. Growth forms that result from such a critical nucleus are most likely to differ from the equilibrium form. Moreover, as the rate of change towards the equilibrium form in a saturated vapour or solution is extremely slow, no equilibrium forms are observed in practice.

However, when the volume of a closed crystal – solution or crystal – vapour system is small, a stable equilibrium exists besides the labile one. This crystal has a much larger size and its form corresponds to the Wulff theorem.

13.1.2. Thermodynamic considerations

We shall restrict the following considerations to the case of a crystal growing in a drop of solution in equilibrium with its vapour at a given temperature and pressure. The system crystal – solution – vapour is closed.

The supersaturated solution, initially free of a crystal, contains n_0 particles of the solute. The supersaturation is:

$$\beta_0 = a_0/a^{c,1} = (\gamma_0 n_0)/\gamma^{c,1} n_\infty \,, \tag{13.1}$$

where a_0 is the activity and γ_0 the activity coefficient of the n_0 particles, $a^{c,1}$ is the activity and $\gamma^{c,1}$ the activity coefficient of the n_∞ particles in a

* This contribution is largely based on and follows closely the publications of Bienfait and Kern [1] and of Bienfait [2].

saturated solution of infinite dimensions.

The crystallization of one crystal in this drop of solution brings about a change in the free enthalpy

$$dG = dG_v + dG_s , \tag{13.2}$$

where the indices v and s refer to the volume and the surface term, respectively.

$$dG_v = \mu^c dn^c + \mu^l dn^l , \tag{13.3}$$

with μ the chemical potential of the solute in the crystal (c) or in the liquid (l), and n the number of solute particles in the crystal or in the liquid. Now because

$$n^c + n^l = n_0 , \tag{13.4}$$

$$dG_v = (\mu^c - \mu^l)\, dn^c . \tag{13.5}$$

The surface term is:

$$dG_s = \sigma^l ds^l + \sum_j \sigma_j^c ds_j^c , \tag{13.6}$$

where σ is the specific surface free enthalpy of the surface s, and the summation over j extends to all crystal faces. The first term concerns the surface of the drop which is thought to be spherical with a radius r. Then:

$$ds = (2/r)\, dv \quad \text{and} \quad dv = dn^c(V^c - \overline{V}^l)/N ,$$

where V^c is the molar volume of the crystal, $\overline{V}^l$ is the partial molar volume of the solute in solution and N is Avogadro's number.

Further we put for the second term concerning the crystal-liquid interface:

$$s_j^c = c_j(n^c)^{2/3} , \tag{13.7}$$

where c_j is a geometrical factor. This relation holds for a form that remains homothetic to itself during growth from 0 to n^c. We now find:

$$dG = (\mu^c - \mu^l)\, dn^c + \frac{2\sigma^l}{r}\left(\frac{V^c}{N} - \frac{\overline{V}^l}{N}\right) dn^c + \frac{2}{3}(n^c)^{-1/3} \sum_j \sigma_j^c c_j\, dn^c . \tag{13.8}$$

The first term can be evaluated as follows:

$$\mu^1 = kT\ln a^1 = kT\ln \gamma^1 n^1 = kT\ln \gamma^1(n_0 - n^c)\,.$$

μ^c is equal to μ^1 at equilibrium, so

$$\mu^c = \mu^{c,1} = kT\ln a^{c,1} = kT\ln \gamma^{c,1} n_\infty\,.$$

We further put:

$$A = 2\sigma^1(V^c - \bar{V}^1)/(rN)\,;$$

now (13.8) becomes:

$$\mathrm{d}G = -kT\ln\frac{n_0 - n^c}{n_\infty}\,\mathrm{d}n^c - kT\ln\frac{\gamma^1}{\gamma^{c,1}}\,\mathrm{d}n^c + A\,\mathrm{d}n^c + \frac{2}{3}(n^c)^{-1/3}\sum_j \sigma_j^c c_j\,\mathrm{d}n^c. \tag{13.8a}$$

For a crystal consisting of n^c particles the total change in G is obtained by integrating eq. (13.8a):

$$\Delta G(n^c) = \int_0^{n^c} (\mathrm{d}G)\,\mathrm{d}n^c\,. \tag{13.9}$$

When $n^c \ll n_0$, r is large and the activity coefficients almost equal we find the expression of Volmer [3]:

$$\Delta G(n^c) = -n^c kT\ln\beta_0 + (n^c)^{2/3}\sum_j \sigma_j^c c_j\,. \tag{13.9a}$$

13.1.3. The equilibria

Depending on its parameters the function $\Delta G(n^c)$ has two, one or no extrema. Consider the case (fig. 13.1) where there is a labile maximum at n^* and a stable minimum at n_E.

When r is not too small ($> 10^{-6}$ cm), in all cases $n^* \ll n_0$. Neglecting the

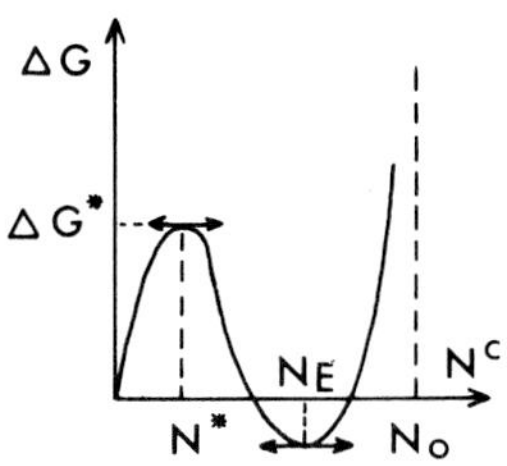

Fig. 13.1. Variation of Gibbs free energy when a crystal grows in a phase of small dimensions. Apart from the usual labile maximum there is now a stable minimum (after Bienfait and Kern [1]).

differences in activity coefficient, we obtain from (13.8a):

$$n^* = \frac{8}{27}\left[(\sum_j \sigma_j^c c_j)/(kT\ln\beta_0 - A)\right]^3 . \tag{13.10}$$

If we put

$$A = kT\ln\beta_c$$

equation (13.10) shows that in a solution of finite size no nucleation occurs unless a critical supersaturation β_c is exceeded.

At the stable minimum we have the approximation:

$$n_E \approx n_0 - n_\infty ,$$

and from (13.8a) we obtain:

$$n_E = n_0 - n_\infty \frac{\gamma^{c,1}}{\gamma^1} \exp\left[\{A + \tfrac{2}{3}(n_0 - n_\infty)^{-1/3} \sum_j \sigma_j^c c_j\}/kT\right] . \tag{13.11}$$

In this case there exists a residual supersaturation

$$\beta_r = (n_0 - n_E)/n_\infty . \tag{13.12}$$

For NaCl in water at 20°C, Bienfait [2] calculated the following values for β_c and β_r:

r(cm)	10^{-6}	10^{-5}	10^{-4}	10^{-3}
$\beta_c - 1$	6×10^{-2}	6×10^{-3}	6×10^{-4}	6×10^{-5}
$\beta_r - 1$	8×10^{-2}	8×10^{-3}	8×10^{-4}	8×10^{-5}

13.1.4. The equilibrium forms

The form of the crystal in equilibrium at n^* or at n_E is considered as a juxtaposition of pyramids with height h_j (central distance) and basal surface s_j.

The volume v^c is:

$$v^c = (\sum_j h_j s_j)/3 ,$$

and

$$dv^c = (\sum_j s_j \, dh_j)/3 + (\sum_j h_j \, ds_j)/3 .$$

On the other hand, neglecting higher order terms:

$$dv^c = \sum_j s_j \, dh_j ,$$

hence

$$2dv^c = \sum_j h_j \, ds_j .$$

With $dn^c = (N/V^c)\, dv^c$ we obtain from (13.8a) and (13.7):

$$dG = \frac{N}{2V^c} \left\{ A - kT \ln \frac{(n_0 - n^c)\gamma^1}{n_\infty \gamma^{c,1}} \right\} \sum_j h_j \, ds_j + \sum_j \sigma_j^s \, ds_j . \tag{13.13}$$

The equilibrium condition implies that for each face j the partial derivative $\partial G/\partial s_j$ must be zero, from which we obtain:

$$\frac{\sigma_j^c}{h_j} = \frac{NkT}{2V^c} \left(\ln \frac{(n_0 - n^c)\gamma^1}{n_\infty \gamma^{c,1}} - \frac{A}{kT} \right) . \tag{13.14}$$

This equation is valid only for $n^c = n^*$ and $n^c = n_E$. In both cases σ_j^c/h_j is a constant. Hence the equilibrium forms at n^* and at n_E differ only in size.

13.1.5. Kinetical considerations

Although the foregoing considerations have shown that a stable equilibrium form exists, the question remains whether a crystal with a form that deviates from the equilibrium form will change its form so that the equilibrium form is attained within a suitable period of time. This kinetic problem was solved by Bienfait and Kern by the application of the theory of fluctuations as given by Landau and Lifshitz [4].

It is found then that n^c and σ_j show fluctuations at constant temperature and pressure. If these fluctuations are supposed to have a Gauss distribution around the equilibrium values, the mean square deviations are:

$$\overline{(\delta n^c)^2} = kT\left(\frac{\partial n^c}{\partial \mu}\right)_{\sigma_j}, \tag{13.15}$$

$$\overline{(\delta \sigma_j)^2} = kT\left(\frac{\partial \sigma_j}{\partial s_j}\right)_{n^c}. \tag{13.16}$$

From equations (13.8) and (13.8a):

$$\Delta\mu = -kT\ln\left[\frac{(n_0 - n^c)\gamma^1}{n_\infty \gamma^{c,1}}\right]$$

and

$$\partial n^c/\partial\mu = (n_0 - n^c)/kT \approx n_\infty/kT.$$

The relative root mean square deviation is then:

$$\left(\overline{(\delta n^c)^2}\right)^{1/2} / n^c = \{n_\infty^{1/2}(\beta_0 - 1)\}^{-1}. \tag{13.17}$$

The fluctuations are largest, and therefore the equilibrium is attained most rapidly when:

(a) the drop is small and the solubility is low (n_∞);

(b) the dimension of the crystal is small with respect to the dimension of the drop (β_0).

The fluctuations in σ_j can be correlated with those of h_j using the condition that σ_j/h_j = constant. Then:

$$\delta\left(\frac{\sigma_j}{h_j}\right)=\frac{\delta\sigma_j}{h_j}-\sigma_j\frac{\delta h_j}{h_j^2}=0\,,$$

so

$$\overline{(\delta\sigma_j)^2}=(\sigma_j/h_j)^2\,\overline{(\delta h_j)^2}\,.$$

The relative root mean square deviation of h_j becomes:

$$\frac{\left(\overline{(\delta h_j)^2}\right)^{1/2}}{h_j}=\left[\frac{kT}{\sigma_j^2}\left(\frac{\partial\sigma_j}{\partial s_j}\right)_{n^c}\right]^{1/2}. \tag{13.18}$$

For a cube $\partial\sigma/\partial s$ is explicitly $\sigma/2s=\sigma/(48h^2)$ which gives:

$$\left(\overline{(\delta h_j)^2}\right)^{1/2}/\,h_j=[kT/(48\sigma h^2)]^{1/2}\,. \tag{13.18a}$$

From (13.18a) it is concluded that the equilibrium is attained most rapidly when:

(c) the crystal is small (h);

(d) the surface energy is small (σ).

For a cube of NaCl with an edge of 10^{-2} mm in a drop of 0.1 mm^3, Bienfait and Kern [1] calculated, using the theory of absolute reaction rates, that the fluctuations amount on a K face to about 10 layers per second, and on an F face to only 10^{-4} layers per second.

13.2.1. Experimental observations

The method is based on experiments made by Kliya [5] and based in turn on observations by Lemmlein [6]. Bienfait and Kern [1] use the following method (see fig. 13.2). On an object glass A for microscopic examination a ring B is glued that has a diameter of 1 cm and a height of a few tenths of a millimeter. A saturated solution is dispersed by means of a vaporisator into small droplets within the ring. Each drop evaporates a little and when the first crystals can be seen the system is hermetically closed by a cover glass C. In order to obtain spherical drops the surface of A is made hydrophobic by a very thin layer of silicone oil.

Experiments have been carried out with NH_4Cl. At first dendrites appear

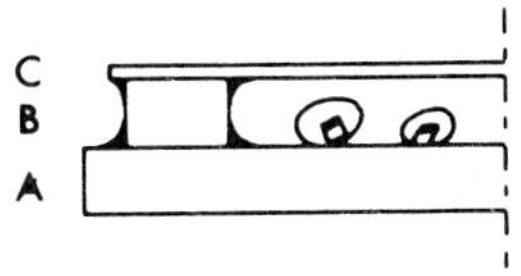

Fig. 13.2. Experimental set-up for obtaining crystals in a phase of small dimensions. A is an object glass, B a metal ring and C a cover glass (after Bienfait and Kern [1]).

with their axes parallel to ⟨100⟩. These are transformed within a few hours into isometric rounded crystals. After several days faces {110} and edges ⟨111⟩ are developed, in accordance with the theoretical prediction that {110} is the equilibrium form.

With NaCl the following elucidating experiment was done. A saturated NaCl solution containing 8 g/l of $CdCl_2$ was used. The NaCl crystals formed in the supersaturated droplets had the octahedral habit {111}. After a certain time {100} faces appear. Their surface area increases while the area of the {111} faces decreases, although they remain planar and do not show any roughening. The final form is {100}. This shows that the habit modification caused by the addition of $CdCl_2$ is essentially a kinetic effect.

However, with other impurities as $HgCl_2$, glycine and $K_4Fe(CN)_6$ no evolution from the growth form ({110} in the two former cases and dendrites in the latter case) towards the equilibrium form was observed. It may be that the impurity is strongly adsorbed, with the result that the frequency of the fluctuations is considerably reduced. The effect of $K_4Fe(CN)_6$ on the dissolution rate of NaCl is well known [7].

13.2.2. Other experiments

It was shown by Bienfait and Kern [1] that a stable equilibrium form is also obtainable in the case of an isolated small solid-vapour system. Experiments in this range are those of Honigmann [8] on hexamethylene tetramine. The growth form {110} of the cubic crystals showed after one or a few days also the forms {100} and {112}.

Extremely small metal crystals in equilibrium with their vapours show the equilibrium form. This has been observed on crystals of Au, Ag, Cu, Ni, α Fe and γ Fe of about 1 μm by Sundquist [9]. Also field emitter crystals with a clean surface show equilibrium forms in the field ion microscope (see Drechsler and Müller [10]). In addition to planar surfaces the equilibrium forms of metals show rounded parts.

References

[1] M. Bienfait and R. Kern, Bull. Soc. franç. Minér. Crist. 87 (1964) 604.
[2] M. Bienfait, Thesis Nancy 1966.
[3] M. Volmer, *Kinetik der Phasenbildung* (Steinkopff, Leipzig, 1939).
[4] L.D. Landau and E.M. Lifshitz, *Statistical Physics* 5 (Pergamon Press, London, 1959).
[5] M.O. Kliya, Dokl. Akad. Nauk SSSR 100 (1955) 259.
[6] G. Lemmlein, Dokl. Akad. Nauk SSSR 98 (1954) 973.
[7] M.A. van Damme-van Weele, Coll. Internat. CNRS 152 (1965) 433.
[8] B. Honigmann and I.N. Stranski, Z. Elektrochem. 56 (1952) 338.
[9] B.E. Sundquist, Acta Metallurg. 12 (1964) 67.
[10] M. Drechsler and A. Müller, J. Crystal Growth 3/4 (1968) 518.

P. Hartman, ed., *Crystal growth: an introduction.* © (1973) North-Holland Publ. Co.

STRUCTURE AND MORPHOLOGY

P. HARTMAN

14.1. Relations between crystal structure and crystal morphology

14.1.1. *Crystallographic symbols*

(hkl) – crystal face or lattice plane.
$\{hkl\}$ – crystal form, comprising all faces equivalent by symmetry to (hkl).
$[uvw]$ – crystal edge or zone axis or lattice row.
$\langle uvw \rangle$ – all edges, zones axes or lattice rows equivalent by symmetry to $[uvw]$.
$[x,y,z]$ – fractional coordinates of an atom in a unit cell.
$\mathrm{A}(i,uvw]$ – an atom $\mathrm{A}(i)$ that underwent a lattice translation $[uvw]$.
d_{hkl} – spacing between two consecutive lattice planes.

Frequent use will be made of the following relations (cf. ref. [1, 2]):

$$d_{nh,nk,nl} = (d_{hkl})/n\,. \tag{14.1}$$

A crystal face (hkl) lies in a zone $[uvw]$ when:

$$hu + kv + lw = 0. \tag{14.2}$$

14.1.2. *Definition of terms*

Crystal habit is the overall shape of a crystal and is most conveniently described in qualitative terms as plate-like, tabular, equant or isometric, cubic, octahedral, prismatic, pyramidal, acicular, needle-like and so on. The term "*combination of forms*" is used to sum up all forms displayed on the crystal (cf. figs. 14.1 and 14.2).

The morphological features of a crop of crystals found in nature or prepared in the laboratory can be described quantitatively by some statistical parameters [3].

The *persistence P* of a form $\{hkl\}$ denotes the percentage of different

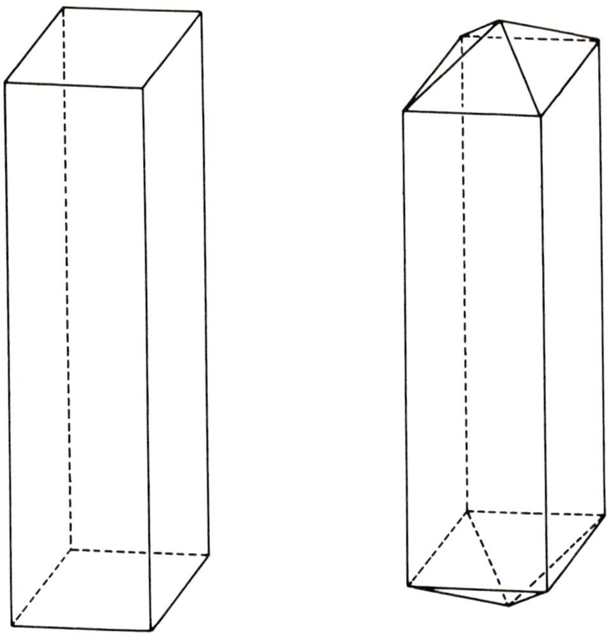

Fig. 14.1. Two crystals with the same habit but different combination of forms.

combinations of forms in which the form has been observed. A certain crystal form is reckoned to have been observed whenever at least one face of the form has been observed.

For minerals also the parameter F (*Fundort persistenz*) is used, denoting the percentage of deposits where the crystal form has been observed.

14.1.3. *Morphology and interplanar spacings*

Law of Bravais–Friedel. Observations revealed a correlation between the importance of a crystal form $\{hkl\}$ (frequency of occurrence, relative area) and the interplanar spacing d_{hkl}, the importance decreasing with d_{hkl}. This is known as the law of Bravais–Friedel [4]). Centering of the unit cell halves certain distances d_{hkl}.

For the various types of Bravais lattices the halving occurs for the following faces:

I-lattice, when $h + k + l$ is odd;

F-lattice, when h, k and l are not all even or odd;

C-lattice, when $h + k$ is odd;

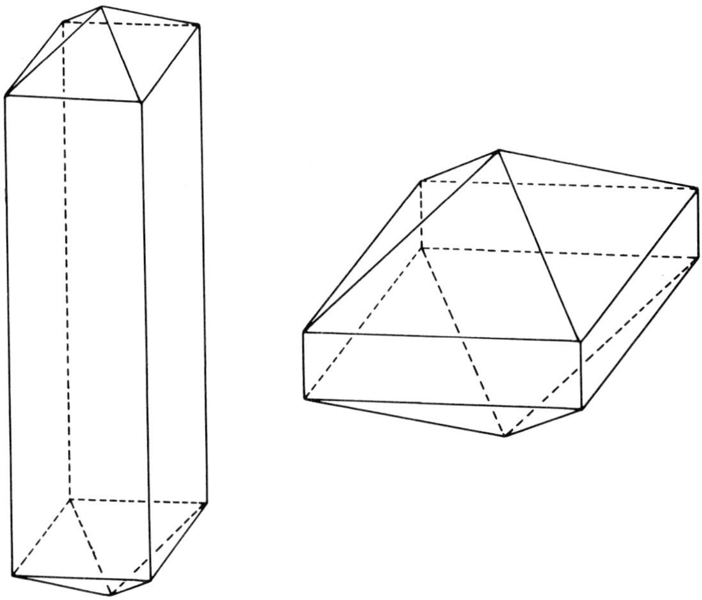

Fig. 14.2. Two crystals with the same combination of forms, but different habits.

B-lattice, when $h + l$ is odd;
A-lattice, when $k + l$ is odd;

For the cubic lattices the list of decreasing d_{hkl} values runs:

P-lattice: {100}, {110}, {111}, {200}, {210}, ...
I-lattice: {110}, {200}, {211}, {220}, {310}, ...
F-lattice: {111}, {200}, {220}, {311}, {222}, ...

The doubling of the indices follows from equation (14.1). This means that crystals with a P-lattice should show the cube as the most important form, followed by {110} and {111}. Crystals with an I-lattice should show {110} and {200} as the most important forms.

Examples: *urotropine* or hexamethylene tetramine usually crystallizes from the vapour in rhombic dodecahedra. The space group is I $\bar{4}$ 3 m. Another example is *tungsten*, space group I m 3 m.

Crystals having a face centered lattice should show the octahedron {111} as the most prominent form, followed by the cube. Example: f.c.c. metals like Ag and Au.

Law of Donnay–Harker. Some observations violate the Bravais–Friedel law. For example, pyrite, FeS_2, has the space group P a 3 and should be ex-

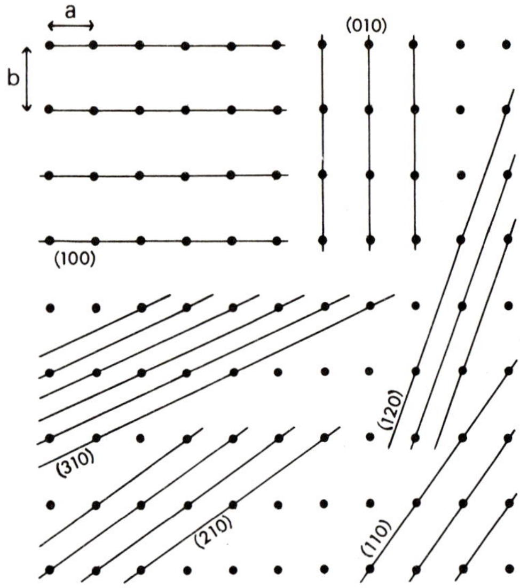

a

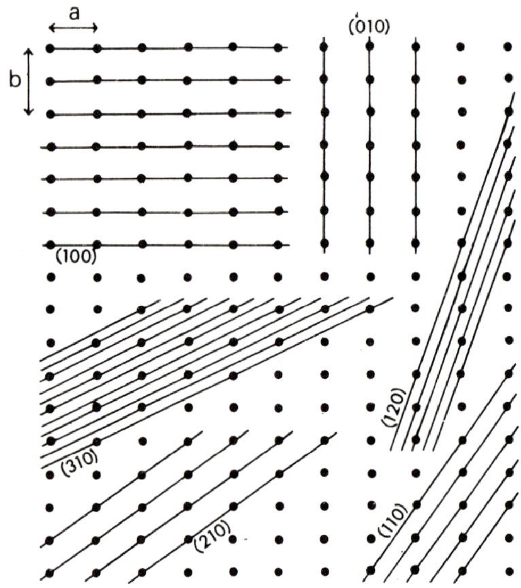

b

Fig. 14.3. a) An orthorhombic lattice with lattice planes ($hk0$); b) the action of a glide plane a parallel to (001): the repeat distance d_{hk0} of the surface energy is halved when h is odd.

pected to show {100}, {110} and {111} as most prominent faces. In fact it has {111}, {100} and {210} as the most important crystal forms, {110} being rather rare. The violation is due to the effect of the glide planes and screw axes as was shown first by Niggli [5], then by Donnay and Harker [6]. A screw axis 2_1 halves the interplanar distance of a face perpendicular to it, because the same surface configuration is repeated after a period $\frac{1}{2}d_{hkl}$. In the same way the interplanar distance of a face perpendicular to a glide plane may be halved (fig. 14.3). For example, a glide plane *a* parallel to (001) halves the interplanar spacings of faces (*hk*0) when *h* is odd. A glide plane *b* (001) halves d_{hk0} when *k* is odd and a glide plane *n* (*d*) parallel to (001) halves (quarters) d_{hk0} when $(h + k)$ is odd (not a fourfold). Note that all these conditions of halving, trisecting, quartering are the same that apply to systematic extinctions of diffracted X-rays.

For *pyrite*, P a 3, the list of decreasing d_{hkl} taking into account the extra halvings now runs:

{111}, {200}, {210}, {211}, {220}, ...

in accordance with observations. Another example is *yttrium iron garnet*, space group I a 3 d, which gives:

{211}, {220}, etc.

Extension of the Donnay–Harker law by Donnay and Donnay. The law of Donnay and Harker is sometimes violated. In many cases this is due to pseudosymmetric features [7] in the crystal structure. This means that the structure shows pseudoperiods or sub-periods in one, two or three directions, leading to extra halving, trisecting, etc. of d_{hkl}. An example is *dibiphenylene ethylene* [8] (fig. 14.4), which crystallizes in the space group P c a n and has a unit cell with dimensions a = 17.22 Å, b = 36.9 Å and c = 8.23 Å. According to the Donnay–Harker law the list of forms with decreasing importance is:

{020}, {110}, {130}, {200}, {111}, ...

The crystals are needles parallel to the *c*-axis which are bounded by {130} dominant and {100} small. The crystal structure may be described as consisting of disc-like molecules piled one upon the other in the direction of the *c*-axis. These piles are arranged at lattice points of centered pseudo cells with a *b*-axis equal to one third of the structural period (fig. 14.4). This leads to [10] a pseudoperiod d_{060} of the face (010) and a pseudoperiod of d_{330} of

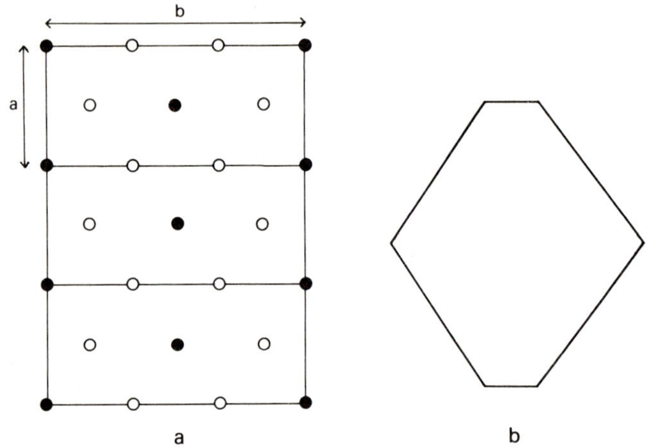

Fig. 14.4. a) Three unit cells of dibiphenylene ethylene. Closed circles represent a pile of crystallographically equivalent molecules seen end on. Open circles are another set of piles of molecules. b) Cross section of crystal bounded by {130} and {100}.

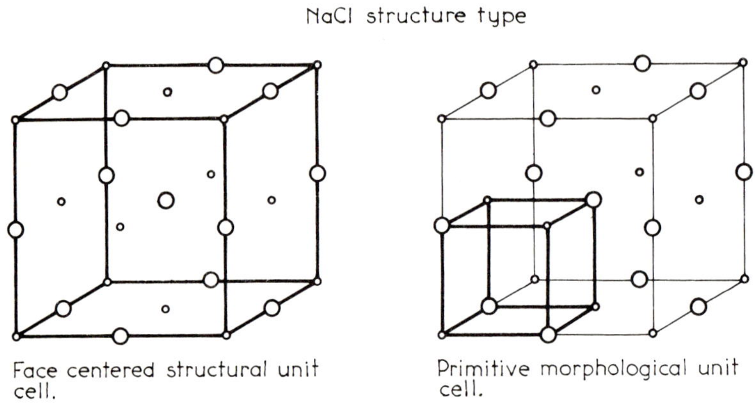

Fig. 14.5. Structural unit cell and morphological pseudo unit cell of NaCl.

(110), while d_{130} and d_{200} are not affected. Hence the list of decreasing d_{hkl} values, taking into account the pseudoperiods is:

{130}, {200}, ...

in excellent agreement with the observed morphology.

Another example is the NaCl structure. According to the space group F m 3 m the octahedron {111} should be the most prominent face. Actually it is the cube, which is due to the fact that the structure can be considered as a three-dimensional array of point charges. As in the growth process no difference is made between a positive and a negative ion, the pseudo unit cell has an edge equal to half the structural edge (fig. 14.5), This pseudo unit cell is primitive so the cube should be the most important form, in accordance with observation.

14.1.4. *The physical basis of the geometrical laws*

As the correspondence between importance of a crystal face and its d_{hkl} value is generally rather good, there must be a physical basis. This is to be sought [11] in the fact that d_{hkl} represents the period with which the surface energy is repeated (or nearly so in the case of pseudoperiods). In the case of layer growth d_{hkl} is the thickness of the elementary growth layer.

This layer is bounded by a surface with a structure that ensures a minimum surface energy for the face considered. A large d_{hkl} involves a small mesh area of the corresponding lattice plane (*hkl*) and, because the interaction energy of atoms and molecules in general decreases with increasing distance, the interaction energy between a growth layer and the underlying crystal is small for large d_{hkl} values. This question will be dealt with later on in section 2.

14.1.5. *Morphology and bonding in the crystal structure*

Consider a two-dimensional crystal (fig. 14.6). The deposition of a growth unit at A in a kink is more probable than deposition at C, when A is attached to the crystal with at least one strong bond A–B more than C. In that case A will not re-evaporate so easily as C. Hence the crystal is bounded by edges that are parallel to directions in which there is an uninterrupted chain of strong bonds between the building units. Such a chain is called a *Periodic Bond Chain (PBC)* [8].

Now consider a three-dimensional crystal as an array of PBC's. In fig. 14.7 the PBC at the step at A has a greater probability of being deposited than PBC's at C, when the PBC's parallel to the considered face are bonded by strong bonds A–B. But this means that there must be at least one more PBC parallel to the face. Hence, layer growth is possible when the layer contains two or more PBC's. If the PBC's are not bonded (fig. 14.8), layer growth does not occur because the probability of deposition of another PBC is everywhere the same. The growth rate is faster than in the foregoing case, because deposition of new PBC's occurs simultaneously.

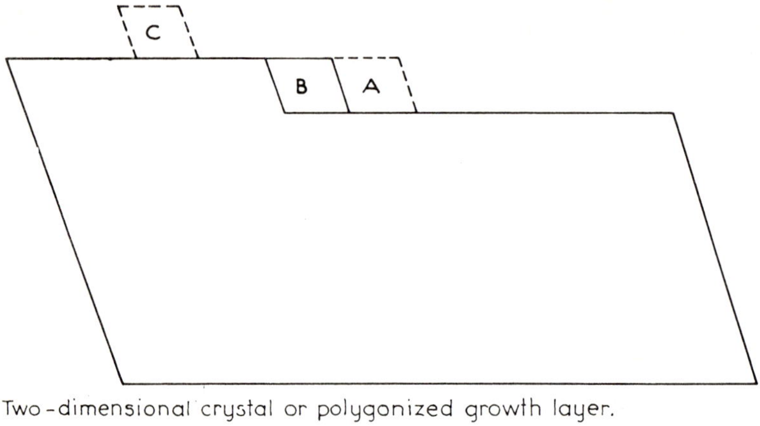

Fig. 14.6. Probability of attachment of building units at A is larger than at C when there exists a strong bond between A and B.

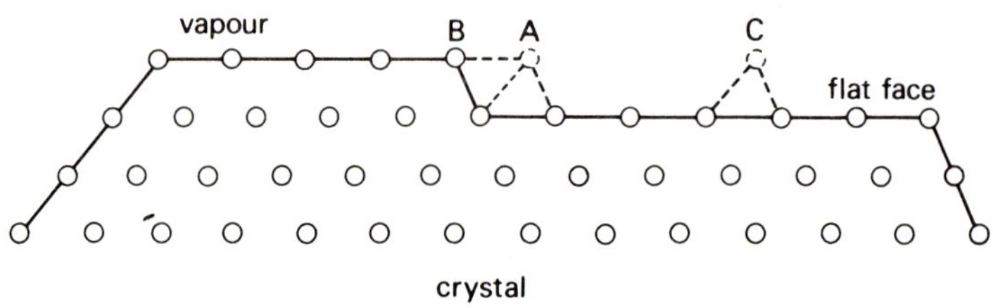

Fig. 14.7. Each circle represents a PBC seen end on. Layer growth occurs when strong bonds between PBC's A and B exist.

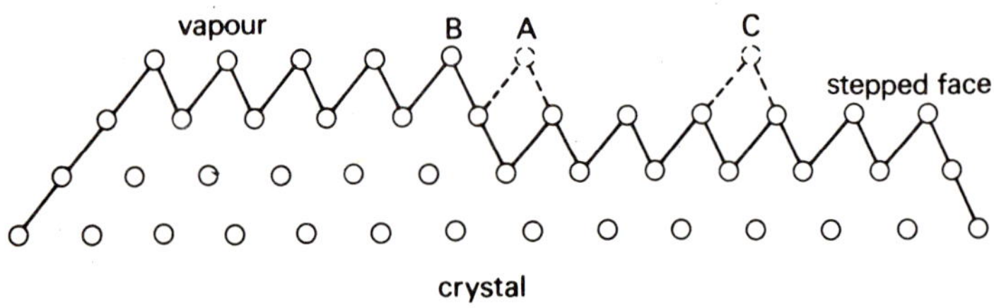

Fig. 14.8. Cf. fig. 14.7. No layer growth occurs when strong bonds between A and B do not exist. The probability of attachment is the same at A and C.

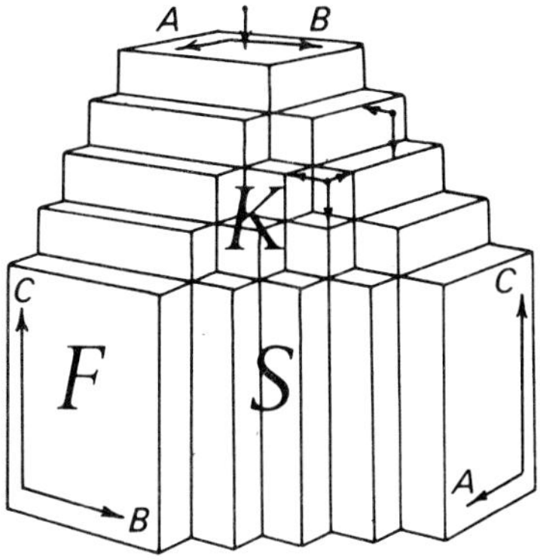

Fig. 14.9. Hypothetical crystal with three PBC's: A // [100], B // [010] and C // [001]. Flat faces are: (100), (010) and (001). Stepped faces are: (110), (101) and (011). Kinked face is (111).

Using this concept three categories of faces can be distinguished [9–11] (fig. 14.9):

F faces or flat faces, containing two or more PBC's in a layer d_{hkl}. Such a layer is called a slice.

S faces or stepped faces, containing only one PBC in a layer d_{hkl}.

K faces or kinked faces, containing no PBC in a layer d_{hkl}.

The three categories have three different growth mechanisms. F faces grow according to a layer mechanism and therefore are slowly growing and hence important faces. S faces grow according to one-dimensional nucleation. K faces need no nucleation at all and hence grow fast and are not normally found on crystals.

The *strong bonds* that form a PBC are bonds in the first coordination sphere of an atom, ion or molecule. They are formed *during the crystallization process*. From the periodic structure of crystals it follows that all PBC's parallel to one and the same direction are identical or symmetrically equivalent and must have the same composition as the crystal. With ionic crystals a slice must have no dipole moment across the face, in order to avoid high surface energies.

Similarly a PBC should have no dipole moment perpendicular to its direction.

Example: *f.c.c. metal* Ag. Strong bonds occur between nearest neighbours only. Because all atoms are crystallographically identical, the period of a PBC consists of only one strong bond. Hence the PBC's are in the directions $\langle\frac{1}{2}\frac{1}{2}0\rangle$.

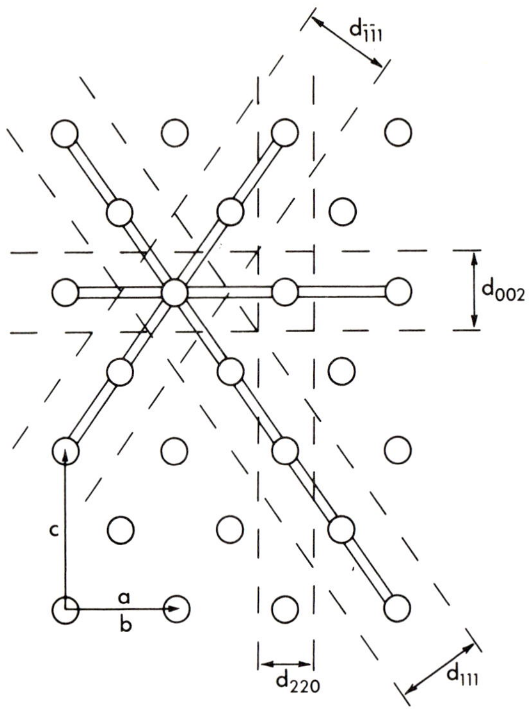

Fig. 14.10. Projection of a cubic face centered structure (silver) parallel to $[1\bar{1}0]$. Circles represent PBC's seen end on. Slices of three F faces are shown: (111), (002) and $(\bar{1}\bar{1}1)$. Face (220) is S face because no strong bond exists between adjacent PBC's.

These directions determine the F forms {111} and {001}. Fig. 14.10 shows a projection along $[1\bar{1}0]$. Each circle represents a pile of Ag atoms seen end on and forming a PBC. In the slice d_{111} neighbouring PBC's are bonded, hence it is an F face. The same applies to the slice d_{002}. In the slice d_{220} neighbouring PBC's are not bonded, so (110) is an S face. Quite generally $\{hhl\}$ are S forms ($h \neq 0$; $h \neq l$) and $\{hkl\}$ with $h \neq k \neq l$ are K forms. Crystals of Ag are thus expected to be bounded by {111} and {001} faces.

14.1.6. *Development of a zone parallel to a PBC*

In the normal case some faces in a zone are F faces and all other faces are S faces. There are two cases in which a K face may occur in such a zone [12]: (1) d_{hkl} is so small due to extra halving, trisecting etc., that no PBC occurs in

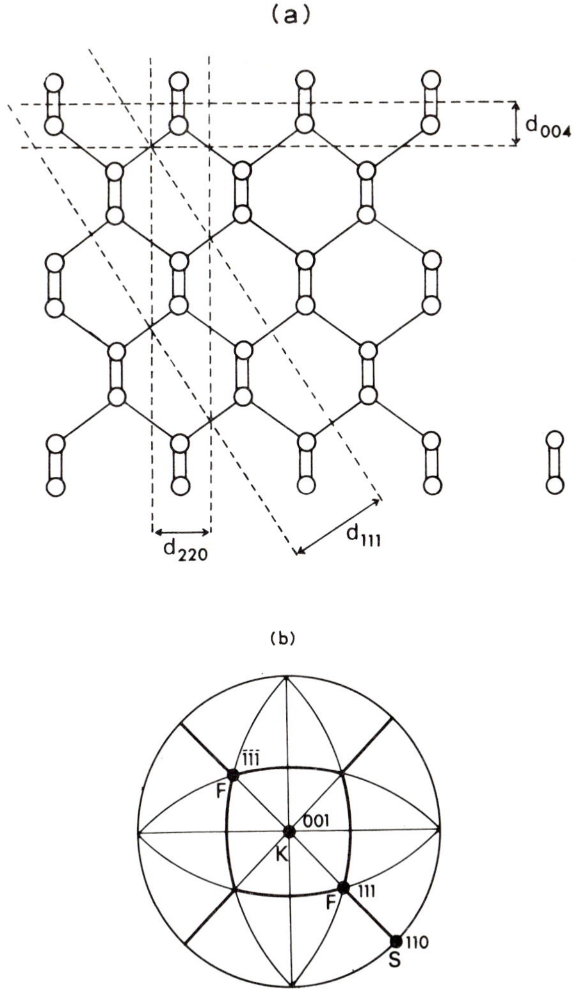

Fig. 14.11. Projection of the diamond structure along [$\bar{1}10$]. Each circle represents a carbon atom. Adjacent PBC's are bonded in slice d_{111}, not in d_{220}. In layer d_{004} no PBC occurs. Stereogram shows part of zones where S faces occur in heavy lines.

the layer. This face is then a K face and all faces between $F_1 - K - F_2$ are also K faces because their surface structure will always contain part of the surface structure of the K face.

Example: *diamond*. There is one type of strong bond between carbon atoms at [0, 0, 0] and at $[\frac{1}{4}, \frac{1}{4}, \frac{1}{4}]$. This bond, together with the bond between atoms at $[\frac{1}{4}, \frac{1}{4}, \frac{1}{4}]$ and $[\frac{1}{2}, \frac{1}{2}, 0]$, forms the period of a PBC $[\frac{1}{2}\frac{1}{2}0]$. Thus there are zigzag PBC's in the directions ⟨110⟩. According to the projection along $[1\bar{1}0]$ (fig. 14.11) (111) is an F face, as is $(\bar{1}\bar{1}1)$. The space group of diamond is F d 3 m which implies the halving of d_{hhl} when $(h+l)$ is uneven and quartering for (001). It is seen from fig. 14.11 that neighbouring PBC's in d_{220} are not bonded. Hence (110) is an S face. The layer d_{004} is too thin to contain a PBC, so (001) is a K face. Therefore all faces between $(\bar{1}\bar{1}1)$ – (004) – (111) are to be considered as K faces.

(2) An S face between two F faces occurs in two symmetrically equivalent zones. In this case the surface will be divided into domains. In one kind of domain it is an S face parallel to one zone, in the other kind it is an S face in the other zone. Where two domains meet, kinks are formed (even at 0°K), as pictured in fig. 14.12.

Example: *sphalerite*, ZnS. Strong bonds occur between Zn and S in tetrahedral coordination, just as in the diamond structure. Therefore the PBC's are the same: ⟨110⟩. The space group F $\bar{4}$ 3 m gives for (*hhl*): halving when $(h + l)$ is odd. Now a layer d_{002} contains PBC's $[\frac{1}{2}\,\frac{1}{2}\,0]$ which are not bonded laterally, so (001) is an S face. But the same applies to the PBC's $[\frac{1}{2}\frac{\bar{1}}{2}0]$. Hence (001) is an S face in two symmetrically equivalent zones and in fact has to be considered as a K face.

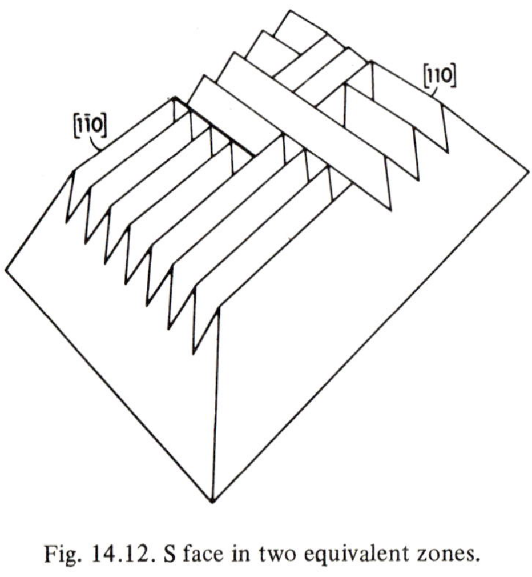

Fig. 14.12. S face in two equivalent zones.

14.2. Energy considerations

14.2.1. *Attachment energies*

The equilibrium form of a crystal is found by means of the Wulff plot. Similarly the growth form is obtained by plotting the growth rates of the various faces along their normals. Current crystal growth theories do not always indicate one particular step in the growth process that is rate-determining.

When two-dimensional nucleation occurs, the formation of the nucleus is the rate-determining step. The energy associated with this process is the energy per molecule released when a slice d_{hkl} is formed (we neglect the influence of the edge energies). This energy, E_{sl}, is related to the crystal energy E_{cr} by:

$$E_{\mathrm{cr}} = E_{\mathrm{sl}} + E_{\mathrm{att}} \,, \tag{14.3}$$

where E_{cr} is defined as the energy per molecule released when the crystal is formed from the crystallizing particles and where E_{att}, the attachment energy, is defined as the energy per molecule released when a new slice is attached to the crystal face. E_{cr} is therefore the half crystal energy (see ch. 10, § 1.9 and ch. 11, § 3), which equals the evaporation energy, when crystallization takes place from the vapour.

We now put the hypothesis that the attachment energy E_{att} is a measure for the growth rate in general and for F faces is proportional to the growth rate. A justification is hard to give (see ch. 10, § 1.9 and § 3.3) but the habits deduced from this simple hypothesis are not in conflict with observations. Compared with the Donnay–Harker law the geometrical value $1/d_{hkl}$ is replaced by the energy value E_{att}.

For crystals with *covalent* and *Van der Waals* bonding a broken-bond model gives satisfactory results.

Example: *naphthalene* structure type [13]. The unit cell contains two molecules per cell in the space group $P2_1/a$. The center of molecule M(1) is at [0, 0, 0], that of molecule M(2) at $[\frac{1}{2}, \frac{1}{2}, 0]$. For naphthalene $a = 8.218$ Å, $b = 5.990$ Å, $c = 8.640$ Å and $\beta = 122°55'$. Each molecule is surrounded by twelve other molecules in a distorted cubic closest packing, involving four different Van der Waals bonds, named d, e, f and g (cf. fig. 14.13).

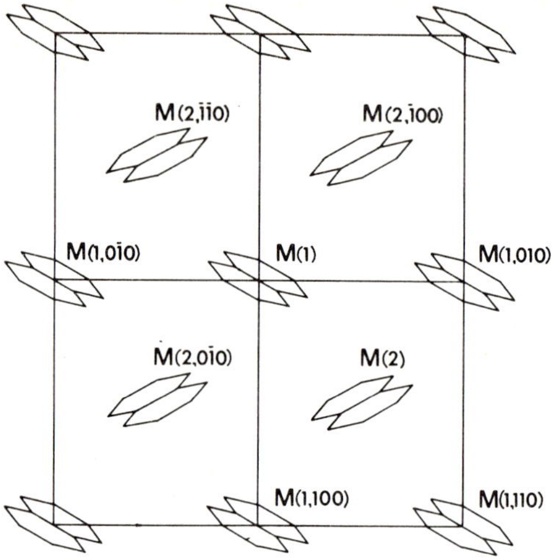

Fig. 14.13. Projection along the *c*-axis of the naphthalene crystal structure. Four unit cells are shown.

Bonds in the naphthalene crystal structure

d	M(1) – M(2)	*e*	M(1) – M(1,010)	*g*	M(1) – M(2,001)
d	M(1) – M(2,0$\bar{1}$0)	*e*	M(1) – M(1,0$\bar{1}$0)	*g*	M(1) – M(2,0$\bar{1}\bar{1}$)
d	M(1) – M(2,$\bar{1}$00)	*f*	M(1) – M(1,001)	*g*	M(1) – M(2,$\bar{1}$0$\bar{1}$)
d	M(1) – M(2,$\bar{1}\bar{1}$0)	*f*	M(1) – M(1,00$\bar{1}$)	*g*	M(1) – M(2,$\bar{1}\bar{1}\bar{1}$)
d	M(2) – M(1)	*e*	M(2) – M(2,010)	*g*	M(2) – M(1,00$\bar{1}$)
d	M(2) – M(1,010)	*e*	M(2) – M(2,0$\bar{1}$0)	*g*	M(2) – M(1,01$\bar{1}$)
d	M(2) – M(1,100)	*f*	M(2) – M(2,001)	*g*	M(2) – M(1,101)
d	M(2) – M(1,110)	*f*	M(2) – M(2,00$\bar{1}$)	*g*	M(2) – M(1,111)

The PBC's in the structure are:

[010]	M(1) – M(1,010) and M(2) – M(2,010)
[001]	M(1) – M(1,001) and M(2) – M(2,001)
[110]	M(1) – M(2) – M(1,110)
[112]	M(1) – M(2,001) – M(1,112) .

Fig. 14.14a shows a projection along the *b*-axis. F faces are (001), (100) and (20$\bar{1}$). The other figures give projections along the other PBC directions.

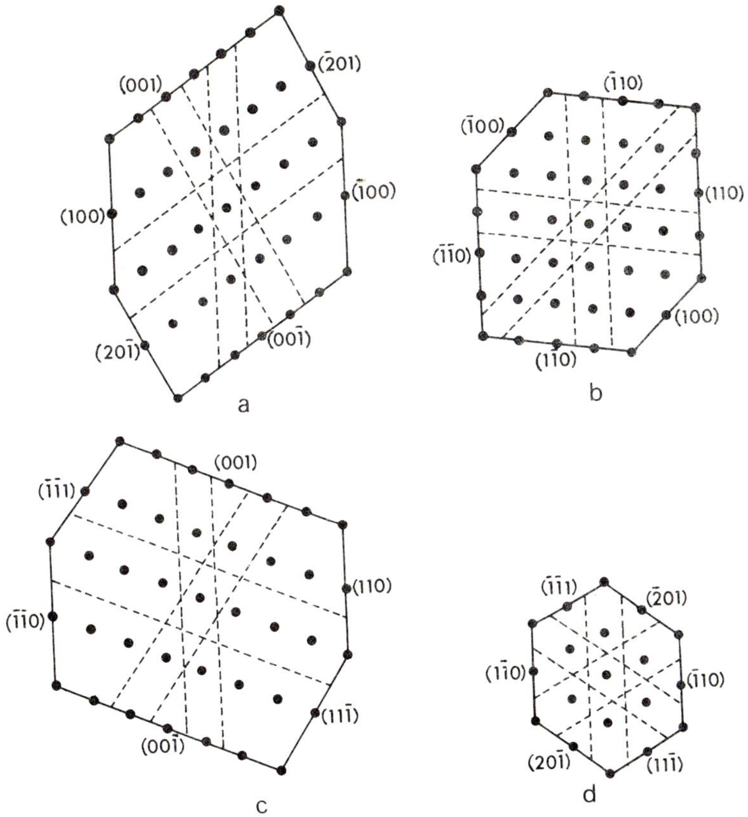

Fig. 14.14. Projections of the naphthalene crystal structure along different PBC's. Each dot represents a PBC seen end on. Slices of F faces are shown. Projections are along: a) [010], b) [001], c) [1$\bar{1}$0] and d) [112].

These determine two other F faces namely (110) and (11$\bar{1}$). The attachment energies are:

(001): $f + 2g$
(110): $d + e + g$
(20$\bar{1}$): $2d + f$
(11$\bar{1}$): $d + e + f + g$
(200): $2d + 2g$.

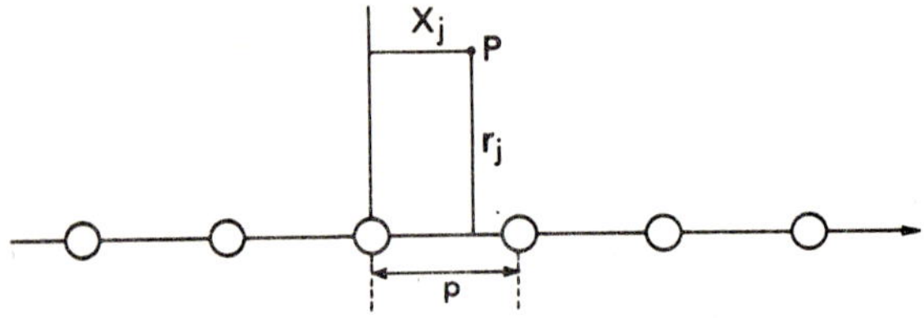

Fig. 14.15. Coordinates of $P(x,r)$ with respect to row of ions.

The strongest bonds are presumably d and e, while f and g will be weaker. This gives a tabular habit $\{001\}$, with smaller faces $\{110\}$ and $\{20\bar{1}\}$, while $\{11\bar{1}\}$ and $\{200\}$ faces are least important among the F faces. Observed morphology: plates $\{001\}$, bounded by $\{110\}$ and $\{20\bar{1}\}$ with sometimes $\{11\bar{1}\}$; not observed: $\{100\}$.

For *ionic* crystals a point charge model can be used. The Madelung [14,15] method of calculating the electrostatic potential is best suited for this purpose. Consider a row of $2N$ equally spaced point charges, $z_j e$ with a period p, extending along the x axis. The potential in a point P at a distance r_j from the row (fig. 14.15) and coordinate x_j is for $r_j \neq 0$:

$$V_j(r,x) = \frac{2z_j e}{p}\left[\sum_{l=1}^{\infty} \pi i \mathrm{H}_0^{(1)}\left(\frac{2\pi i l r_j}{p}\right) \cos\left(\frac{2\pi l x_j}{p}\right) + \ln\frac{2p}{r_j} + \ln N\right]. \qquad (14.4)$$

For $r_j = 0$ and $x_j \neq 0$:

$$V_j(0,x) = -\frac{z_j e}{p}\left[\psi\left(\frac{x_j}{p}\right) + \psi\left(1-\frac{x_j}{p}\right) - \frac{p}{x_j} - \frac{p}{p-x_j}\right] + \frac{2z_j e}{p}\ln N. \qquad (14.5)$$

For $r_j = 0$ and $x_j = 0$:

$$V_j(0,0) = \frac{2z_j e}{p}\,[C + \ln N]\,, \qquad (14.6)$$

where $i\,\mathrm{H}_0^{(1)}\,(2\pi i l r_j/p)$ is the Hankel function of zero order with argument $(2\pi l r_j/p)$ as tabulated in Jahnke, Emde and Lösch, Funktionentafeln, 1960 [16],

$$\psi(y) = \frac{\mathrm{d}\ln y!}{\mathrm{d}y}$$

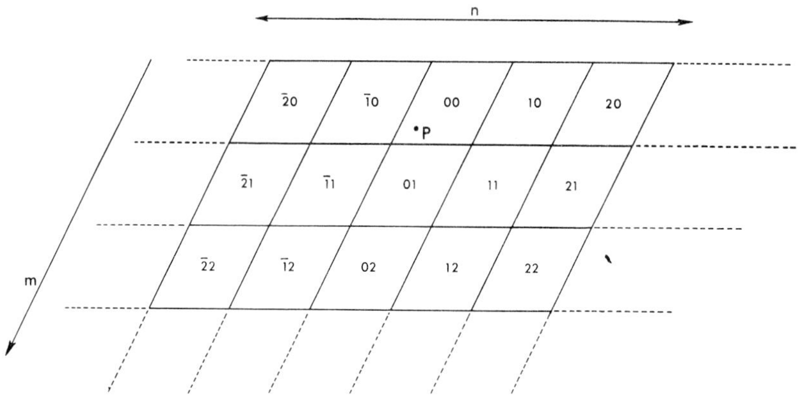

Fig. 14.16. Scheme for calculating the electrostatic potential in P. The row of ions in fig. 14.15 is now perpendicular to plane of drawing.

and $C = -\psi(0) = 0{,}577216$ ... (Euler's constant).

Consider now a crystal structure divided into a set of parallel ionic chains (n, m) where m is the number of a slice and n the number of an ionic chain within a slice (fig. 14.16). The potential at point P with respect to the chain $(0, 0)$ is:

$$V_{00} = \Sigma_j V_j \,, \tag{14.7}$$

where the summation extends over all ions per period p in the chain. When the ionic chain is neutral $\Sigma_j (2z_j e \ln N)/p = 0$. The potential at P with respect to all chains in the zero slice is:

$$V_0 = \sum_{n=-\infty}^{+\infty} V_{n0} \tag{14.8}$$

and similarly for any slice m:

$$V_m = \sum_{n=-\infty}^{+\infty} V_{nm} \,. \tag{14.9}$$

The potential due to the whole crystal is then:

$$V_{cr} = \sum_{n=-\infty}^{+\infty} \sum_{m=-\infty}^{+\infty} V_{nm} \, . \tag{14.10}$$

The energy is obtained by calculating the potential for each different ion in the unit cell:

$$E = \Sigma_k \, z_k e V \, . \tag{14.11}$$

The attachment energy is:

$$E_{att} = \sum_{m=1}^{\infty} \sum_{n=-\infty}^{+\infty} \sum_{k} z_k e V_{nmk} \tag{14.12}$$

where V_{nmk} is the potential at the site of the k'th ion with respect to the chain (n, m). A good convergence of the sums in (14.8) and (14.9) is obtained when terms $+n$ and $-n$ are put together [17].

Example: *rutile*, TiO_2, space group P 4_2/mnm, two molecules of TiO_2 in a unit cell with dimensions $a = 4.594$ Å and $c = 2.959$ Å. F forms are {110} and {101}. Fig. 14.17 shows a projection along the PBC [$\bar{1}$11] which is seen at the intersection of the slices d_{110} and $d_{0\bar{1}1}$. The energy of the PBC is calculated [18] to be: (ions Ti^{4+} and O^{2-}): $E_{00} = -15.918 \, e^2$ Å^{-1} per molecule (1 e^2 Å^{-1} per mole equals 331.83 kcal/mole, which is equivalent to 1.43990 eV).

For (110) the energy per molecule in the zeroth slice amounts to -18.338 e^2 Å^{-1} while the attachment energy is $-0.641 \, e^2$ Å^{-1}. For (0$\bar{1}$1) these values are $-17.712 \, e^2$ Å^{-1} and $-0.954 \, e^2$ Å^{-1} respectively.

If the growth rates of the faces are proportional to their attachment energies the growth form in fig. 14.18 results [18]. An improvement may be obtained by taking into consideration Born repulsion and Van der Waals attraction energies as well as partial covalency of the Ti–O bonds.

14.2.2. *Zonal development*

In ionic structures the interaction energy of two parallel ionic chains is dependent on their relative position in space [19]. When their distance is large with respect to their period p (in practice $r > 2p$) the term with the Hankel function in (14.4) can be neglected. For neutral ionic chains the logarithmic term remains and (14.4) changes into:

$$V_j(r, x) = -(z_j e/p) \ln (r_{jk}^2 / 4p^2) \, , \tag{14.13}$$

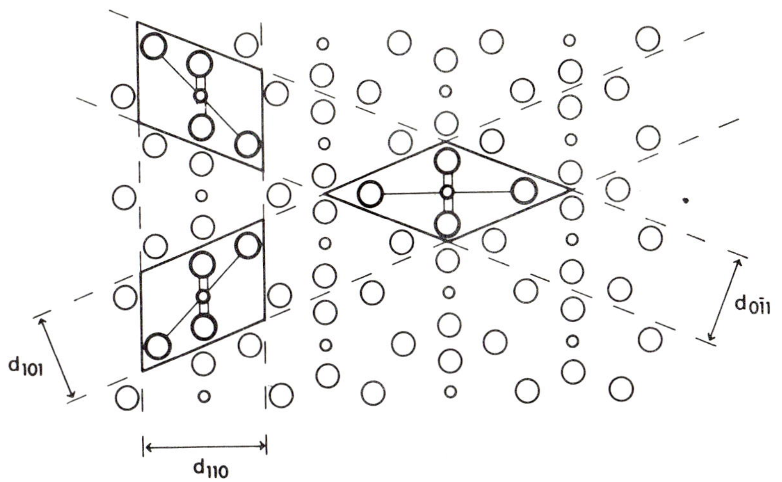

Fig. 14.17. Projection of the rutile structure along [$\bar{1}$11]. Small circles: 2 superposed Ti atoms. Large circles: O atoms.

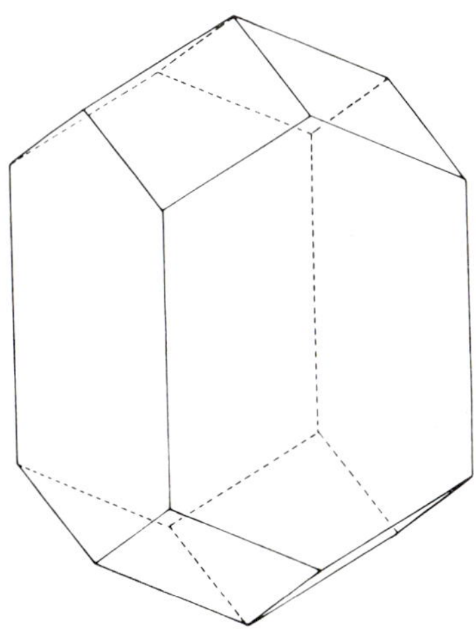

Fig. 14.18. Growth form of rutile.

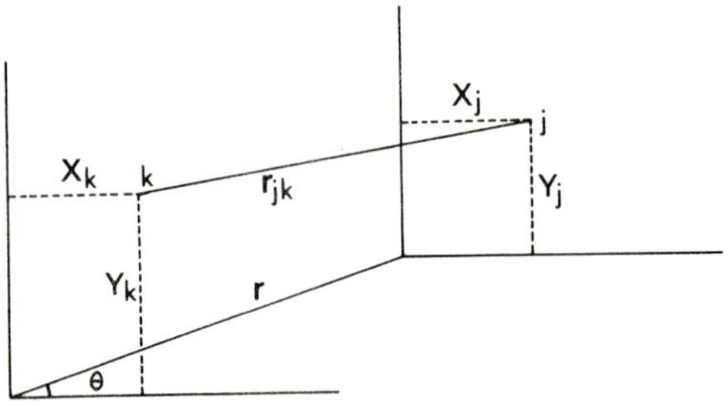

Fig. 14.19. Coordinates used for calculating interaction energy of two ionic chains perpendicular to plane of drawing.

where r_{jk} is the distance between ions j and k projected on a plane perpendicular to the chain (fig. 14.19).

The interaction energy of ionic chains is:

$$E_{jk} = -z_j z_k (e^2/p) \ln (r_{jk}^2/4p^2) . \qquad (14.14)$$

With:

$$r_{jk}^2 = r^2 + 2r(X_j - X_k)\cos\theta + 2r(Y_j - Y_k)\sin\theta$$
$$+ (X_j - X_k)^2 + (Y_j - Y_k)^2 \qquad (14.15)$$

the logarithmic term in (14.14) can be expanded into a series of powers of r.

When both ionic chains are identical and neutral and have no dipole moment perpendicular to their direction, that is:

$$\Sigma z_j = \Sigma z_k = 0 , \quad \Sigma z_j X_j = \Sigma z_j Y_j = \Sigma z_k X_k = \Sigma z_k Y_k = 0 ,$$

(14.14) can be written as:

$$E_{jk} = K_4 \sin 4(\theta - \theta_o)/r_{jk}^2 , \qquad (14.16)$$

where $K_4 = 3\, e^2\, Q_4/p \cos 4\theta_o = -3\, e^2\, P_4/p \sin 4\theta_o$

$$\tan 4\theta_o = -P_4/Q_4$$

$$P_4 = (\Sigma z_j X_j^2 - \Sigma z_j Y_j^2)^2 - 4(\Sigma z_j X_j Y_j)^2$$

$$Q_4 = 4\Sigma z_j X_j Y_j (\Sigma z_j X_j^2 - \Sigma z_j Y_j^2) .$$

The angular dependence of the interaction energy causes differences in attachment energies, and therefore in growth rates, of S faces parallel to one and the same PBC.

Using equation (14.16) it has been found that faces with relatively high indices are parallel to a direction in which the interaction energy E_{jk} is near to zero.

Example: in *quartz* the face $(51\bar{6}1)$ includes with $(10\bar{1}0)$ an angle of $12°1'$ while E_{jk} is zero at a point on the zone $[01.\bar{1}]$ at $10°41'$ from $(10\bar{1}0)$.

14.2.3. *The Wolff method*

In cases where the broken-bond model is applicable another method exists, devised by Wolff and his collaborators [20, 21], which gives the F faces and PBC's from the γ-plot where γ is the specific surface energy. Details of the method are given by Bennema in ch. 12, §4. It should be mentioned here that the method is applicable for covalent and Van der Waals bonding and interesting results have been obtained for polar crystals (e.g. those with the wurtzite crystal structure, see [48, 49].

14.2.4. *Polar crystals*

When the surface energy is defined as the work necessary to split a crystal isothermally and reversibly into two half crystals along the surfaces (hkl) and $(\bar{h}\bar{k}\bar{l})$, then these surfaces will have the same surface energies. When (hkl) and $(\bar{h}\bar{k}\bar{l})$ are not equivalent their surface energies and growth rates differ which causes a polar habit. The difference in surface energy is caused by a rearrangement in the structure of the surface either by displacement of the atoms with respect to their positions in the interior of the crystal or by a change in the electron distribution around the outer atoms.

Theory of Gatos. For covalent crystals Gatos and collaborators [22, 23] explained the different etching behavior of polar surfaces by the interaction of the "dangling bonds" caused by non-bonded electrons.

Example: AB structures of the sphalerite or wurtzite type with A an element of columns I, II or III of the periodic table, B an element of columns VII, VI or V, respectively. For a GaAs, Ga has an empty orbital in the tetrahedral sp^3 bonding at the surface, while As has an unshared filled fourth orbital (fig. 14.20). In the etching processes used, oxidizing (electrophilic) agents are involved, so As atoms are the more reactive ones. Etch pits were observed on the Ga surface, not on the As surface.

Theory of Monier. For ionic crystals Monier and his collaborators [24–26]

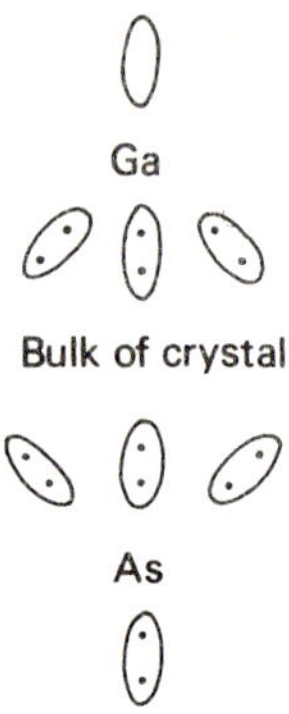

Fig. 14.20. Dangling bonds on As side $(\bar{1}\bar{1}\bar{1})$ of GaAs.

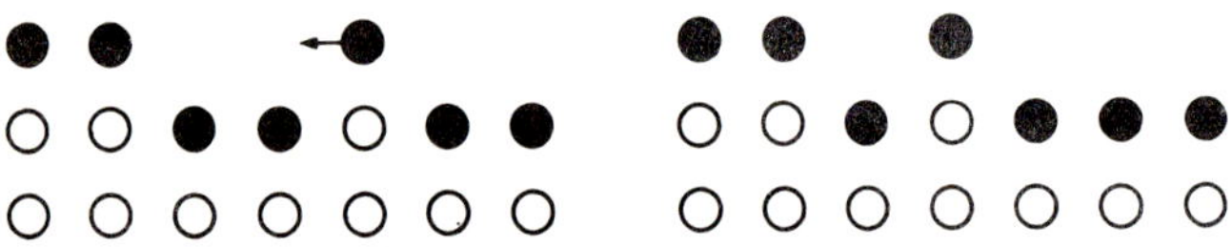

Fig. 14.21. Polarisation changes during surface diffusion. Closed circles: polarised atoms. Open circles: unpolarised atoms.

considered the polarisation of the ions in the outermost layers. With respect to the undeformed state, the polarisation will lower the surface energy. In the growth process it is mainly the surface diffusion that is affected, because the activation energy increases by the polarisation energy necessary to change the polarisation state of the underlying atoms (fig. 14.21). Cadoret [27] showed that when the BCF theory is applicable or when two-dimensional nucleation occurs, the slower growing face has in its outermost layer the most polarisable ions.

Example: *tetraethylammonium iodide* $(C_2H_5)_4NI$ (Cadoret), space group $I\bar{4}$ with $a = 8.87$ Å and $c = 6.95$ Å with 2 molecules per unit cell (fig. 14.22). Every ion I^- is tetrahedrally surrounded by four $(C_2H_5)_4N^+$ ions. This leads to PBC's ⟨100⟩ and ⟨111⟩, defining the F forms {011} and $\{0\bar{1}\bar{1}\}$, both tetragonal disphenoids. At low supersaturation the combination of forms was: $\{0\bar{1}\bar{1}\}$ dominant, {011} very small. Determination of the absolute configuration [27] showed that $(0\bar{1}\bar{1})$ has I^- ions in the outermost layer and these are the most polarisable ions.

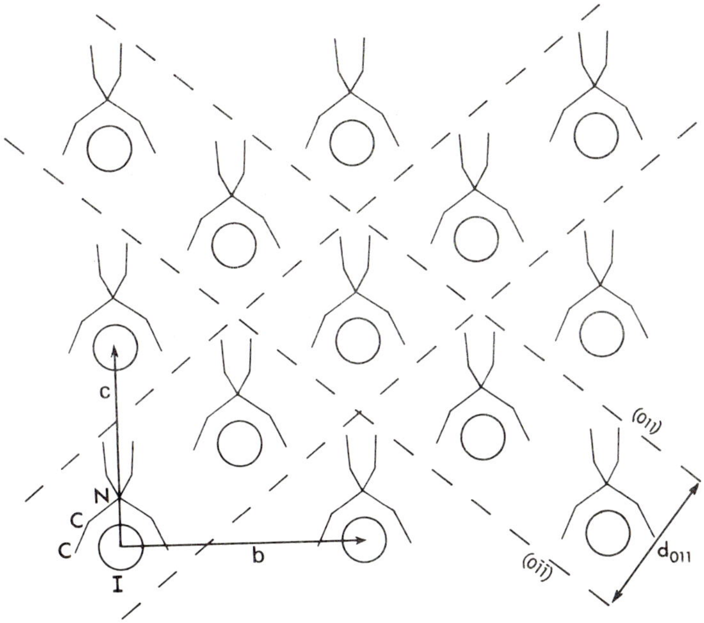

Fig. 14.22. Projection of the $(C_2H_5)_4NI$ crystal structure on (100).

14.3. Environmental influences on crystal morphology

14.3.1. *Environmental symmetry*

According to the principle of Curie the symmetry of a crystal is the symmetry common to crystal and environment. This statement is based on the assumption that equivalent faces have the same growth rate when the environment has spherical symmetry [28].

Example: a cube (m3m) growing on the bottom of a beaker (environmental symmetry ∞ mm) develops into a flattened cube with geometrical symmetry 4 mm.

14.3.2. *Supersaturation*

Kern and his collaborators [29, 30] showed in a number of experiments that ionic crystals may change their habit when the supersaturation exceeds a certain critical value.

Examples: NaCl crystallizes as cubes {100} at low supersaturation, but as octahedra {111} at high supersaturation. Similarly CsCl changes habit from {110} to {100} at high supersaturation.

The structural characteristics of the faces produced at high supersaturation is that they have alternating layers of positive and negative ions. A face like (111) of NaCl has a very high electrostatic potential at its surface when the outermost layer contains ions of one kind only. A rearrangement in which the outermost layer contains $\frac{1}{4}$ of the ions, the next layer $\frac{1}{2}$ and the third layer $\frac{3}{4}$ of all sites, gives a lower potential and a stable surface. The surface structure is that of a K face. The electrostatic field strength normal to this surface is particularly high so that dipole molecules are strongly adsorbed. At high supersaturation, and therefore at high growth rates the removal of the adsorbed water dipoles may become the rate determining step. As the field strength of the cube face is much lower, the adsorption here is much weaker. Hence it is possible that the growth of {111} is slowed down even below $1/\sqrt{3}$ times growth rate of the cube.

14.3.3. *Impurities*

Although strictly speaking the solvent itself has to be considered as an impurity, the term is restricted to compounds other than solute and solvent. The effect is manyfold.

(1) Kink poisons. Adsorption is very strong and takes place at kinks. This reduces rates of growth and of dissolution enormously.

Examples: A well known effect is that of $Fe(CN)_6^{4-}$ ions on NaCl [31], which has an industrial application (see ch. 9, §5). This impurity causes a dendritic growth of the NaCl cubes.

Another example [32, 33] is the effect of Fe^{3+} ions on LiF, probably through the formation of the complex ion FeF_6^{3-}.

(2) Adsorption on the surface of an F face [30]. The growth rate is slowed down due to the necessity of removal of the impurity.

(3) Adsorption at the ledge of a growth layer on an F face [30]. This adsorption will decrease the ledge energy γ' ("surface" energy of the ledge per unit of length). As this ledge energy enters into the expression for the growth rate of a layer, it appears that this kind of adsorption increases the growth rate (cf. ch. 10, §2.2 and §2.8).

Example: *sodiumtriphosphate hexahydrate*, $Na_5P_3O_{10}.6H_2O$ [34]. The compound is triclinic. It usually shows the F forms {001}, {100} and {010}. The influence of DOBS (sodium-p-dodecylbenzenesulphonate) is shown in fig. 14.23 for a supersaturation equal to 0.80. The sharp drop to zero growth rate of $(0\bar{1}0)$ is caused by the appearance of the face $(0\bar{1}1)$ that has an extremely slow growth rate. The phenomena can be understood by assuming

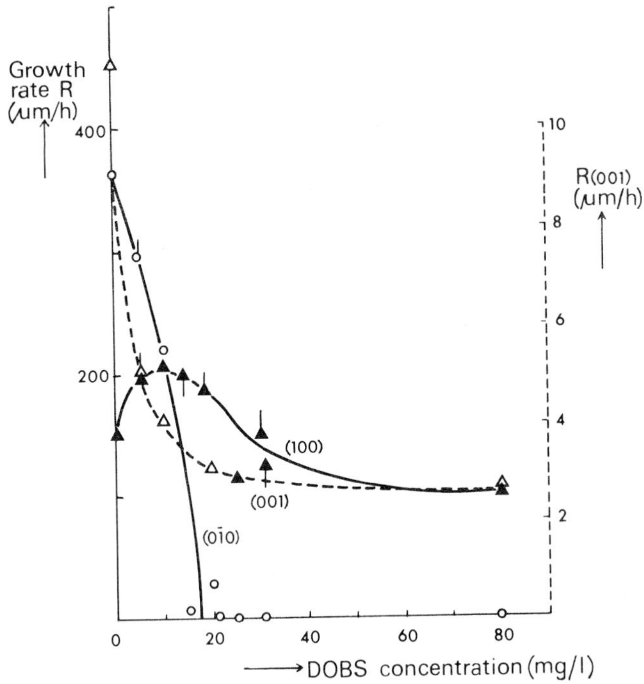

Fig. 14.23. Effect of DOBS concentration at a supersaturation of 80% on growth rates of three faces of sodium triphosphate hexahydrate.

preferential adsorption of DOBS on the (0$\bar{1}$1) face. At the same time the ledges [011] of the growth layers on (100) have a strong adsorption, so the growth rate of this face is enhanced. Higher concentrations of DOBS adsorption on the other faces, (001) and (100), decrease again their growth rates.

(4) Adsorption on S and K faces [35]. Sometimes new faces appear that are brilliant and grow according to a layer mechanism. Formally this means that the adsorption layer should provide the missing PBC's, which leads to the hypothesis [36, 38] that the adsorption layer has a kind of structure that has an epitactic relationship with the substrate (the crystal face).

Example: $Hg(CN)_2$, space group I $\bar{4}$ 2 d with $a = 9.693$ Å and $c = 8.896$ Å shows the F forms {200}, {101}, {112} and {$\bar{1}\bar{1}\bar{2}$}, when crystallized from vapour or from various solvents. From a solution in methyl alcohol above 23°C the crystals show the K form {$\bar{2}\bar{1}\bar{1}$} which often predominates over the F forms (fig. 14.24), as was found by Ledésert and Monier [39]. Below 23°C

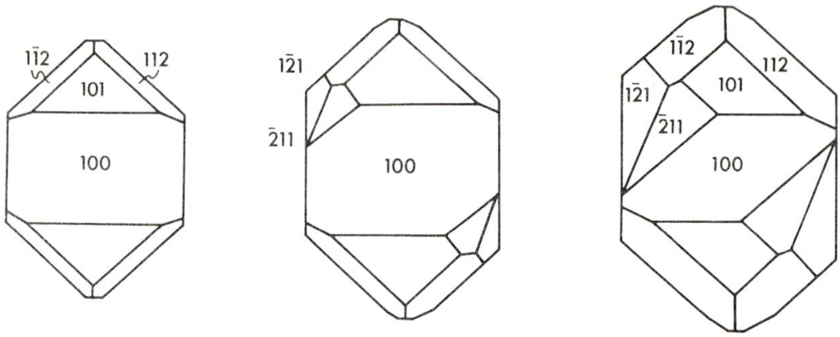

Fig. 14.24. Growth of $Hg(CN)_2$ crystal in CH_3OH. At left the original crystal bounded by F faces only. In next figures appearance and development of K form $\{\bar{2}\bar{1}\bar{1}\}$.

the solvate $Hg(CN)_2.CH_3OH$ was formed which has an orthorhombic crystal structure (Frey [40]) with space group P n m a, $a = 20.192$ Å, $b = 4.140$ Å and $c = 7.572$ Å. The solvate has a plate-like habit {100}, which is an F face, and this face has epitactic relationships with $\{\overline{2}\overline{1}\overline{1}\}$ of $Hg(CN)_2$:

solvate	$Hg(CN)_2$	misfit
[001] = 7.572 Å	$\frac{1}{2}[11\bar{1}]$ = 8.171	− 7.3 %
3 [010] = 12.420 Å	[011] = 13.156	− 5.6 %
angle = 90°	angle = 93°3′	

14.3.4. *Combined effect of supersaturation and impurities*

Kern [29] showed that, in order to obtain a habit modification, a critical supersaturation has to be exceeded, depending on the concentration of the impurity in the solution. In a plot of supersaturation against impurity concentration, lines can be drawn that separate fields with different crystal habits (fig. 14.25). Such a plot is called a *morphodrome* [41]. In all examples in the foregoing section the effect of supersaturation has been shown to exist.

14.3.5. *Composition of disordered phase*

When the crystal consists of atoms of different species or when it contains more than one kind of cation or anion, the composition of the disordered phase (vapour, solution or melt) may affect the growth rates of different crystal forms in different ways. This results in habit changes.

Example: Chase and Osmer [42] found that yttrium gallium garnet, $Y_3Ga_5O_{12}$ crystallizes from a $PbO{-}PbF_2$ flux rich in Y_2O_3 with {211}

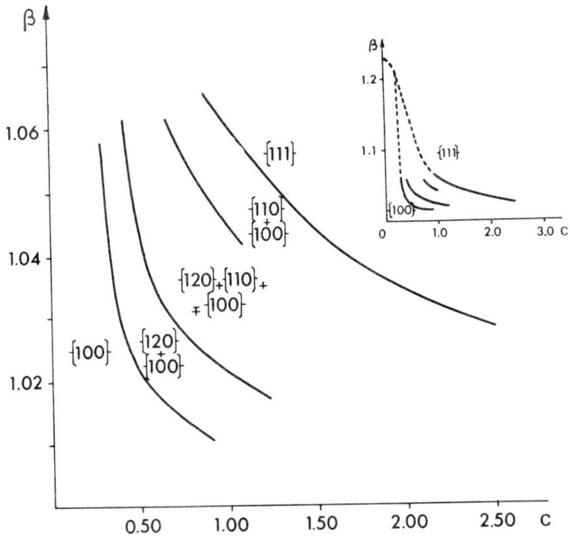

Fig. 14.25. Morphodrome of NaCl with $K_3Fe(CN)_6$.

faces exclusively. From a flux rich in Ga_2O_3 small faces {110} developed on the crystals. On the other hand $Y_3Al_5O_{12}$ crystals from a flux rich in Al_2O_3 show {110} only, while small {211} faces are developed when the flux is rich in Y_2O_3. One might conclude that Y_2O_3 favours the development of {211}, but this is at variance with the observation of Gendelev [43] who found {110} dominating on $Y_3Fe_5O_{12}$ when the melt is rich in Y_2O_3, whereas {211} is prominent on crystals from a Fe_2O_3 enriched melt. If crystallization is considered as a consecutive series of formations of chemical bonds, the kinetics may well be determined by the ion present in smaller concentrations. The same processes apply to the uptake in the crystal structure of isomorphous impurities in different growth sectors [44]. The distribution coefficients k (see ch. 8, §3) may vary for different faces. When the impurity is associated with a colour, only some growth sectors show a colour while others do not (as for instance in hour-glass structures).

14.3.6. *Negative crystals*

When a hole is drilled in a crystal and the material is dissolved away by a continuous stream of undersaturated solution, the hole will be bounded by planar crystal faces. The habit and the combination of forms are similar to those of the growth form. The hole or a gaseous or fluid inclusion is often referred to as a negative crystal. Impurities may cause a habit modification.

Example: Glycine changes the habit of NaCl from {100} to {110} both on the growth form and on negative crystals [30].

14.3.7. *Polar crystals*

The morphological polarity of crystals may be affected by any of the external factors that can change the habit of the growth form.

Examples: Ewing and Greene [45] found that the growth rates of {111} and $\{\bar{1}\bar{1}\bar{1}\}$ faces of GaAs (outermost layers respectively Ga or As) are dependent on the Ga/As ratio in the vapour phase. The $(\bar{1}\bar{1}\bar{1})$ As face has the slower growth rate when the Ga/As ratio lies between 0.2 and 2.5 at 780°C. At Ga/As ratios above 2.5 the (111) Ga face becomes the slower growing face.

Cadoret [46] observed an inversion of the growth rates of the disphenoids $\{0\bar{1}\bar{1}\}$ and {011} of $(C_2H_5)_4NI$ by increasing the supersaturation.

14.4. Morphology deduced from crystal structure

The application of the PBC method [10] is usually done in several stages.

First, a table is made of strong bonds for each of the atoms in the crystal structure. The choice of the bonds sometimes may cause difficulties, but in most cases the first coordination sphere is well established (see also ch. 14, §1.5).

Second, projections of the crystal structure along various short lattice translations [*uvw*] onto a plane perpendicular to the lattice row are made. For faces (*hkl*) parallel to the lattice row [*uvw*] equation (14.2) holds.

Angles between faces in the zone [*uvw*] can be calculated [2] directly from the dimensions of the reciprocal lattice or from the dimensions of the real lattice using a stereographic projection and spherical trigonometry.

Interplanar distances are calculated from lattice dimensions or determined by making an X-ray diffractogram.

A projection can be made without the aid of a computer by observing that the distance of an atom A with coordinates $[x, y, z]$ to the lattice plane (*hkl*) going through the origin is equal to: $(hx + ky + lz)\, d_{hkl}$, where h, k and l have no common factor. When two planes $(h_1k_1l_1)$ and $(h_2k_2l_2)$ are suitably chosen the atomic coordinates in the projection are readily found.

Third, either *PBC's* are sought from the table of bonds or *slices* of possible F faces are drawn, at the intersection of which a PBC common to both slices is found.

For ionic structures slices should have no dipole moment perpendicular to them. In practice this often means that in the center of the slice a symmetry element is found (a symmetry plane, a symmetry axis or inversion axis, a symmetry center). For the thickness of the slices the space group extinctions and eventual fortuitous extinctions must be taken into account.

Following these lines the finite set of PBC's and the classification into F, S and K faces may be obtained.

14.4.1. *Hexagonal closest packing.*

Space group P 6_3/mmc. Extinction: $(hh\bar{2}hl)$ present with $l = 2$. There are 2 atoms per unit cell. If $a = 1$ then $c = \sqrt{(8/3)}$. Atomic positions (origin at center of symmetry):

atom	x	y	z
A(1)	1/3	2/3	1/4
A(2)	2/3	1/3	3/4

Bonds:		
A(1) – A(1,100)		A(1) – A(2)
– A(1,110)		– A(2,010)
– A(1,010)		– A(2,$\bar{1}$00)
– A(1,$\bar{1}$00)		– A(2,00$\bar{1}$)
– A(1,$\bar{1}\bar{1}$0)		– A(2,01$\bar{1}$)
– A(1,0$\bar{1}$0)		– A(2,$\bar{1}$0$\bar{1}$)

Projection along [100] (fig. 14.26). In this direction there exists one bond per period, namely A(1) – A(1,100). Therefore the directions ⟨100⟩ are PBC's ([100], [110] and [010]). In the unit cell there is a second PBC [100], namely A(2) – A(2,100). In fig. 14.26 each circle represents a PBC [100] seen end on. They are bonded laterally in the slices d_{0002}, $d_{01\bar{1}1}$ and $d_{01\bar{1}0}$. Hence these faces are F faces. Each atom is surrounded by twelve others. Taking all bonds between nearest neighbours equal, and denoting the energy by a, we find that $E_{cr} = 6\,a$. In the slice d_{0002} each atom is surrounded by 6 others, so $E_{sl} = 3\,a$, from which $E_{att} = 3\,a$. For a slice of $(01\bar{1}0)$ the atom A (1) is surrounded by 6 others and so is A (2). Hence $E_{sl} = 3\,a$ per atom and again $E_{att} = 3\,a$. This value is the mean for A (1) and A (2). Separately, the attachment energy of A (1,010) to the slice $d_{01\bar{1}0}$ drawn in fig. 14.26 is $2\,a$, while for A (2,010) it is $4\,a$. Finally in a slice of $(01\bar{1}1)$ each atom is surrounded by 5 others, so $E_{sl} = 2.5\,a$ from which $E_{att} = 3.5\,a$, which is actually the mean of $4\,a$ for A (1) and $3\,a$ for A (2).

According to the attachment energies, {0001} and {10$\bar{1}$0} are the most

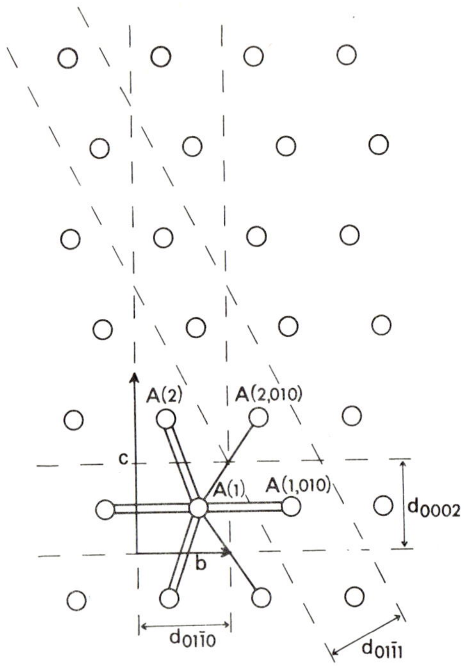

Fig. 14.26. Projection of h.c.p. structure along [100].

important F faces, followed by $\{10\bar{1}1\}$. Note that these faces also appear first in the derivation of the equilibrium form (ch. 11, §3).

Projection along [001] (fig. 14.27). A PBC in this direction is formed by the bonds A (1) – A (2) – A (1,001). These PBC's are bonded to neighbouring chains in the slices $d_{10\bar{1}0}$ and $d_{01\bar{1}0}$, hence $\{10\bar{1}0\}$ is an F form. The face $(1\bar{1}00)$ does not contain the same PBC, but instead a slice contains another PBC that is symmetrically equivalent to the first one, namely either A (1) – A (2,010) – A (1,001) or A (1) – A $(2,\bar{1}00)$ – A (1,001). These latter PBC's are also present in slices of $(10\bar{1}0)$ and $(01\bar{1}0)$ respectively. Other prisms $\{hki0\}$ are S forms.

Projection along [011] (fig. 14.28). The PBC [011] consists of the following bonds: A (1) – A (2,010) – A(1,011). Note that the latter bond is given in the table above as A (1) – A $(2,00\bar{1})$. From the figure it is seen that a PBC is bonded to neighbouring ones in slices $d_{01\bar{1}\bar{1}}$, $d_{1\bar{1}01}$ and $d_{10\bar{1}0}$, so again $\{10\bar{1}1\}$ and $\{10\bar{1}0\}$ are found as F forms.

When all symmetrically equivalent zones ⟨100⟩, ⟨001⟩ and ⟨011⟩ are drawn,

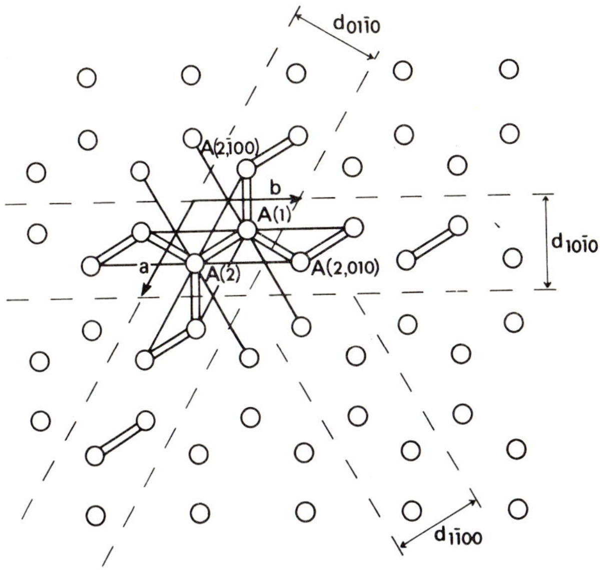

Fig. 14.27. Projection of h.c.p. structure along [001].

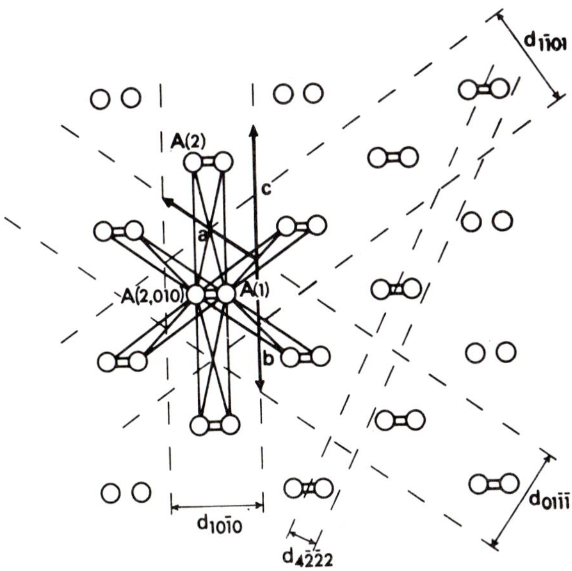

Fig. 14.28. Projection of h.c.p. structure along [011].

they intersect in the forms {0001}, {1011}, {1010} and {1121}. According to the extinctions the interplanar distance of (1121) is halved. In fig. 14.28 a layer $d_{4\bar{2}\bar{2}2}$ is indicated lying in the equivalent zones [011] and $[\bar{1}\bar{1}1]$. The face $(2\bar{1}\bar{1}1)$ is to be considered as an S face in two equivalent zones, and therefore is rather a K face.

14.4.2. TiO_2, *anatase modification*

Unit cell: $a = 3.785$ Å. $c = 9.514$ Å. $Z = 4$.

Space group I 4_1/amd. Extinctions: (hkl) present with $(h+k+l) = 2n$

$(hk0)$	,,	,,	$h = 2n, k = 2n$
$(0kl)$	,,	,,	$k + l = 2n$
(hhl)	,,	,,	$2h + l = 4n$

Atomic positions (origin at point with symmetry $\bar{4}$m2):

atom	x	y	z	atom	x	y	z
Ti(1)	0	0	0	O(1)	0	0	0.2064
Ti(2)	0	1/2	1/4	O(2)	0	0	−0.2064
				O(3)	0	1/2	0.4564
				O(4)	0	1/2	0.0436

Bonds:		
Ti(1) – O(1)		Ti(2) – O(3)
– O(2)		– O(4)
– O(4)		– O$(2,\frac{1}{2}\frac{1}{2}\frac{1}{2})$
– O$(4,0\bar{1}0)$		– O$(2,\frac{\bar{1}}{2}\frac{1}{2}\frac{1}{2})$
– O$(3,\frac{1}{2}\frac{\bar{1}}{2}\frac{\bar{1}}{2})$		– O(1)
– O$(3,\frac{\bar{1}}{2}\frac{\bar{1}}{2}\frac{\bar{1}}{2})$		– O(1,010)

Projection along [100] (fig. 14.29). Each titanium atom is surrounded by six oxygen atoms in a distorted octahedron. A bond chain Ti (1) – O $(3,\frac{1}{2}\frac{\bar{1}}{2}\frac{\bar{1}}{2})$ – Ti (1,100) is easily found. In the unit cell there are four such bond chains through the four titanium atoms. The composition of such a chain is not stoichiometric and we have to add one more oxygen atom. This could be any of the three oxygen atoms around Ti (1) that appear "alone" in the figure. Let us take O (4). This PBC, consisting of Ti (1), O $(3,\frac{1}{2}\frac{\bar{1}}{2}\frac{\bar{1}}{2})$ and O (4) has an electrostatic dipole moment perpendicular to its direction, which preferably it should not have (see ch. 14, § 1.5). By combining this PBC with the equivalent one containing Ti (2), O $(2,\frac{1}{2}\frac{1}{2}\frac{1}{2})$ and O (1) we obtain a configuration without such a dipole moment. This configuration of two PBC's is repeated in the slice $d_{0\bar{1}1}$. Neighbouring PBC's are bonded, so {011} is an F form.

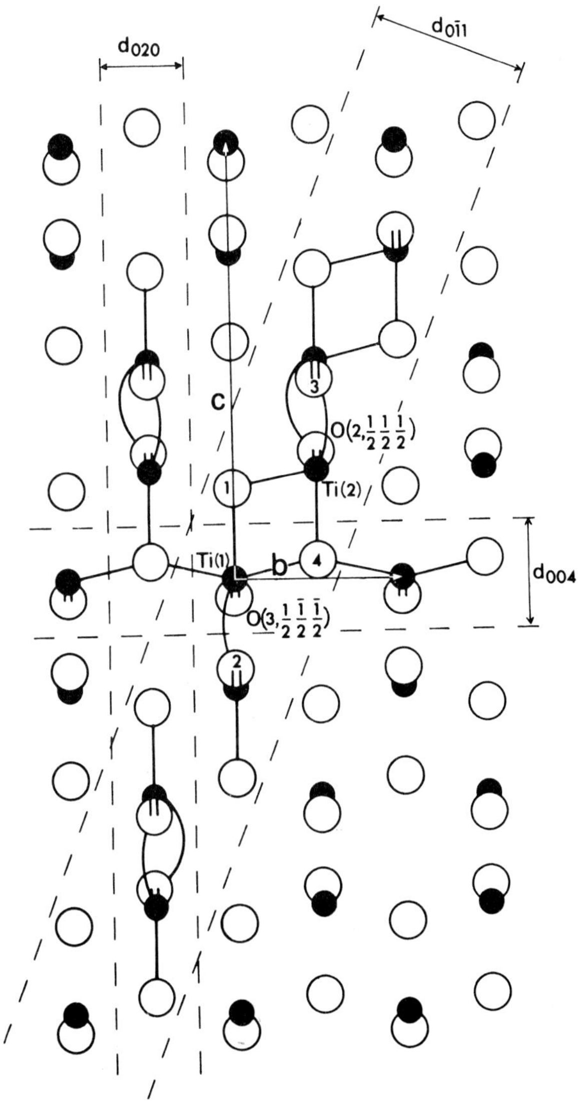

Fig. 14.29. Projection of the anatase structure along [100]. Small circles represent titanium atoms, large circles oxygen atoms.

If, instead of taking O (4) as the oxygen atom to be added to the bond chain of composition TiO, we would have taken O (1), the same configuration of bond chains and the same slice would have been found. However, in this case also another combination without dipole moment is possible, namely with the chain containing $\mathrm{Ti}\,(2,\frac{1}{2}\frac{\bar{1}}{2}\frac{\bar{1}}{2})$. Such a combination is shown in the lower left half of fig. 14.29. It is seen that it lies also in the same slice as the other combination. This second combination is seen to occur in a slice d_{020}. There are no strong bonds between neighbouring chains, so {010} is an S face.

The PBC's are also bonded in the direction of the b axis to their neighbours, suggesting that (001) is an F face. Because of the extinctions the thickness of a slice is d_{004} and even in this slice neighbouring PBC's are bonded, and the slice has no electrostatic dipole moment. So effectively {001} is an F form.

All other {0*kl*} forms are of the S type.

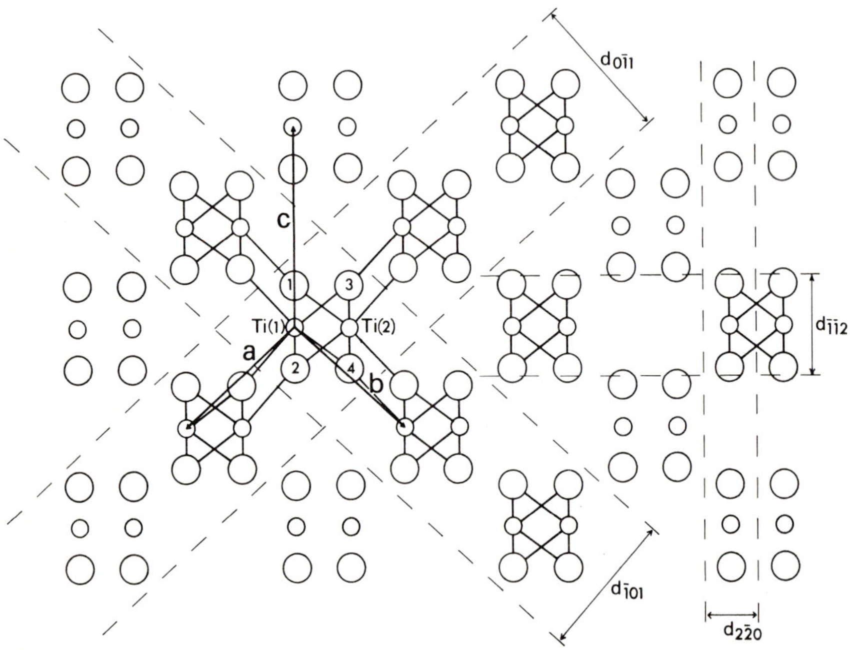

Fig. 14.30. Projection of the anatase structure along [111]. Small circles: Ti; large circles: O.

Projection along [111] (fig. 14.30). The fact that $(0\bar{1}1)$ was found to be an F face implies that there must be other PBC's parallel to this face. The next shortest lattice translation is $[\frac{1}{2}\frac{1}{2}\frac{1}{2}]$ and indeed a PBC can be found of the form:

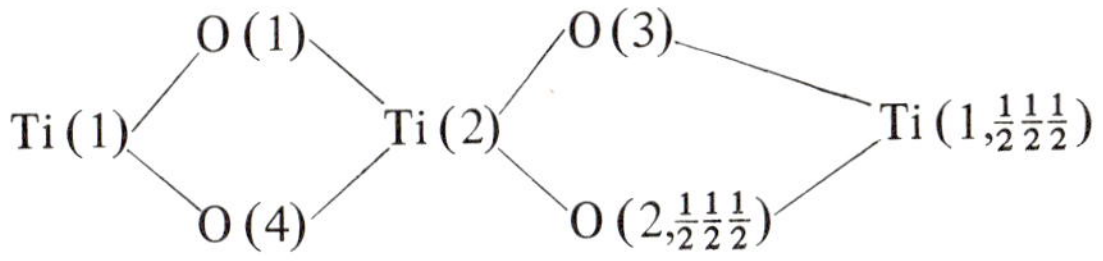

This PBC, which is easily recognized in fig. 14.30, is bonded to its neighbours in slices of $(0\bar{1}1)$ and $(\bar{1}01)$. This is not so in the layer $d_{\bar{1}\bar{1}2}$, so {112} is an S face. Only half a PBC occurs in the layer $d_{2\bar{2}0}$, so {110} is a K face. No other PBC's occur in this structure, so {001} and {011} are the F forms. This is in agreement with the persistences calculated by Parker [47]:

{011} P = 81,2 ; {001} P = 50,0 ; {112} P = 48,4 ; {017} P = 36,5

{011} F = 98,5 ; {001} F = 65,0 ; {112} F = 27,5 ; {017} F = 27,5 .

Among the {0*kl*} forms {017} has the highest persistence. This is an anomaly, because the S face in the zone [100] with the highest d_{hkl} value is (013). It can be shown (see ch. 14, § 2.2) that the interaction energy of neighbouring PBC's in the layer d_{017} is almost zero, which is characteristic for high index faces with high persistencies [19].

References

[1] F.C. Phillips, *An introduction to crystallography* (Longmans, London, 1971).
[2] P. Terpstra and L.W. Codd, *Crystallometry* (Longmans, London, 1961).
[3] P. Niggli, *Lehrbuch der Mineralogie und Kristalchemie* (Borntraeger, Berlin, 1941) Vol. I.
[4] G. Friedel, Bull. Soc. franç. Minér. 30 (1907) 326.
[5] P. Niggli, *Geometrische Kristallographie des Diskontinuums* (Borntraeger, Leipzig, 1919).
[6] J.D.H. Donnay and D. Harker, Amer. Miner. 22 (1937) 446.
[7] J.D.H. Donnay and G. Donnay, Compt. Rend. Acad. Sci. Paris 252 (1961) 908.
[8] C.P. Fenimore, Acta Cryst. 1 (1948) 295.
[9] P. Hartman and W.G. Perdok, Acta Cryst. 8 (1955) 49, 521.
[10] P. Hartman, Z. Krist. 119 (1963) 65.
[11] P. Hartman, in: *Growth of crystals*, ed. by N.N. Sheftal' (Consultants Bureau, New York, 1969) Vol. 7, p. 3.
[12] P. Hartman, Z. Krist. 121 (1965) 78.
[13] P. Hartman, in: *Physics and chemistry of the organic solid state*, ed. by D. Fox, M.M. Labes and A. Weissberger (Interscience, New York, 1963–1965) Vol. I, p. 369, Vol. II, p. 873.

[14] E. Madelung, Phys. Z. 19 (1918) 524.
[15] W. Kleber, N. Jahrb. Miner. Geol. Pal. Beil. Bd. A 75 (1939) 72.
[16] E. Jahnke and F. Emde, *Tables of higher functions* (Teubner, Leipzig, 1960).
[17] P. Hartman, Acta Cryst. 9 (1956) 569.
[18] C.F. Woensdregt and R.O. Felius, to be published.
[19] P. Hartman, Acta Cryst. 11 (1958) 365.
[20] G.A. Wolff, Z. phys. Chem. 31 (1962) 1.
[21] G.A. Wolff and J.G. Gualtieri, Amer. Miner. 47 (1962) 562.
[22] H.C. Gatos and M.C. Lavine, J. electrochem. Soc. 107 (1960) 427.
[23] H.C. Gatos and M.C. Lavine, J. Phys. Chem. Solids 14 (1960) 169.
[24] R. Kern and J.C. Monier, Bull. Soc. franç. Minér. Crist. 78 (1955) 585.
[25] R. Kern and J.C. Monier, Bull. Soc. franç. Minér. Crist. 79 (1956) 129.
[26] J.C. Monier and R. Kern, Bull. Soc. franç. Minér. Crist. 79 (1956) 495.
[27] R. Cadoret and J.C. Monier, J. Crystal Growth 1 (1967) 59.
[28] I.I. Shafranovskiy, *Lektsiy po kristallomorfologiy* (Moskva, 1968).
[29] R. Kern, Bull. Soc. franç. Minér. Crist. 76 (1953) 325, 365; 78 (1955) 461, 497.
[30] R. Kern, in: *Growth of Crystals*, ed. by N.N. Sheftal' (Consultants Bureau, New York, 1969) Vol. 8, p. 3.
[31] M.A. van Damme – van Weele, in: *Adsorption et Croissance cristalline*, Coll. C.N.R.S. 152 (1965) 433.
[32] J.J. Gilman, W.G. Johnston and G.W. Sears, J. appl. Phys. 29 (1958) 747.
[33] G.W. Sears, J. chem. Phys. 33 (1960) 1068.
[34] S. Troost, J. Crystal Growth 3/4 (1968) 340.
[35] I.N. Stranski, Bull. Soc. franç. Minér. Crist. 79 (1956) 359.
[36] G.A. Wolff, Z. phys. Chem. 31 (1962) 1.
[37] P. Hartman and R. Kern, Compt. Rend. Acad. Sci. Paris 258 (1964) 4591.
[38] P. Hartman, in: *Adsorption et croissance cristalline*, Coll. C.N.R.S. 152 (1965) 477.
[39] M. Ledésert and J.C. Monier, in: *Adsorption et croissance cristalline*, Coll. C.N.R.S. 152 (1965) 537.
[40] M. Frey, Thesis Caen, 1969.
[41] M. Bienfait, R. Boistelle and R. Kern, Compt. Rend. Acad. Sci. Paris 256 (1963) 2189.
[42] A.B. Chase and J.A. Osmer, J. Crystal Growth 5 (1969) 239.
[43] S.Sh. Gendelev, Kristallografiya 8 (1963) 431; Soviet Phys. Cryst. 8 (1963) 335.
[44] V.Ya. Khaimov-Mal'kov, Z.I. Zhmurova, Kh.S. Bagdasarov and E.M. Akulyonok, Kristallografiya 7 (1962) 437; Soviet Phys. Cryst. 7 (1962) 345.
[45] R.E. Ewing and P.E. Greene, J. electrochem. Soc. 111 (1964) 1266.
[46] R. Cadoret and J.C. Monier, in: *Adsorption et croissance cristalline*, Coll. C.N.R.S. 152 (1965) 559.
[47] R.L. Parker, Z. Krist. 59 (1923) 1.
[48] G.A. Wolff, in: *Intermetallic compounds*, ed. by J.H. Westbrook (Wiley, New York, 1967) p. 85.
[49] G.A. Wolff, M. Weinstein and B.N. Das, J. Crystal Growth 13/14 (1972) 186.

P. Hartman, ed., Crystal growth: an introduction. © (1973) North-Holland Publ. Co.

MORPHOLOGICAL STABILITY

R.F. SEKERKA

15.1. Introduction

The subject of morphological stability deals with the stability or instability of the shape of the interface which separates a crystal from a second phase from which it is crystallizing or melting. This second phase may be a solid, a liquid or a gas depending upon the crystal growth process under consideration. Similar instability phenomena have been observed for each second phase although, of course, the details and the growth conditions are generally quite different. For the sake of concreteness, we shall confine our discussion in the first two sections to the case of growth from the melt. In particular, we shall be considering the morphological stability of a solid-liquid interface during the solidification of a pure substance or an alloy. In sections 3 and 4, however, we shall also consider the case of an isothermal diffusion-controlled transformation and the case of a melting interface in a liquid-solid transformation.

Let us first be a little more explicit about what we mean by morphological stability or instability. We refer to fig. 15.1 which depicts a crystal (Region I) growing from a melt (Region II) surrounded by some external boundary. This figure is only intended to be schematic. In fact, the external boundary might intersect the crystal itself; it is only a boundary where certain boundary conditions are prescribed.

It certainly would not be unexpected if the solid-liquid interface had a shape that was influenced by the overall boundary conditions. For instance, if the boundary in fig. 15.1 were isothermal, we might expect the solid-liquid interface to have an indentation somewhat similar to that of the external boundary; we would not call this a situation of morphological instability. On the other hand, if the solid-liquid interface were to appear as in fig. 15.2, where there is no apparent correlation between the interface shape and the external boundary conditions, we would anticipate a situation of morphological instability.

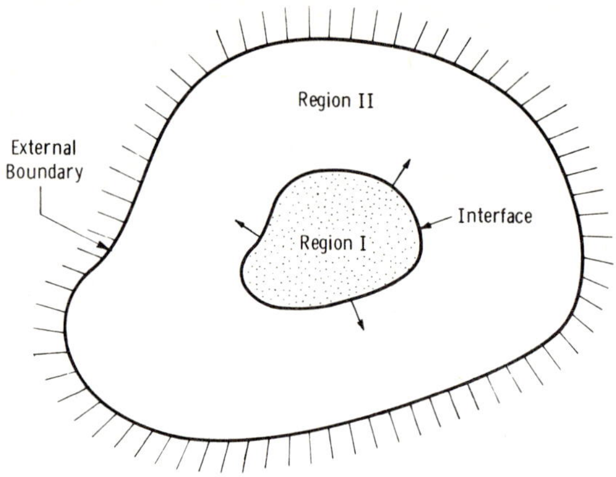

Fig. 15.1. Region I, containing phase I, is growing at the expense of region II containing phase II. The two regions are separated by a smooth interface whose shape reflects that of the external geometry.

We emphasize that the scale of the undulations in an unstable solid-liquid interface is typically 10–100μm. These growth features therefore have no direct connection to smoothness or roughness on an atomic scale such as considered in the atomistic theory of crystal growth. In fact, the resulting

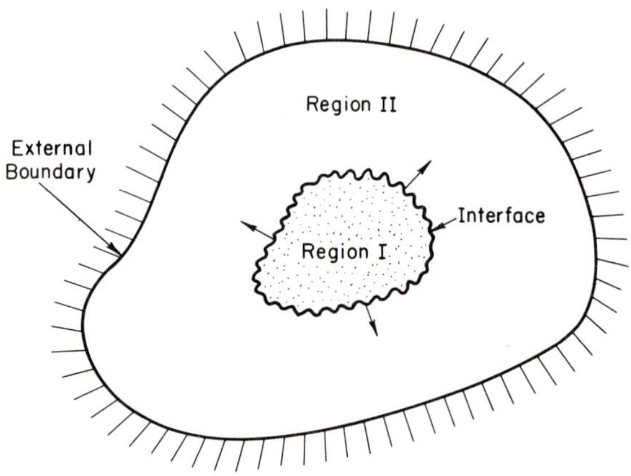

Fig. 15.2. Region I, containing phase I, is growing at the expense of region II containing phase II. The two regions are separated by a corrugated interface whose shape is not just a reflection of the external geometry.

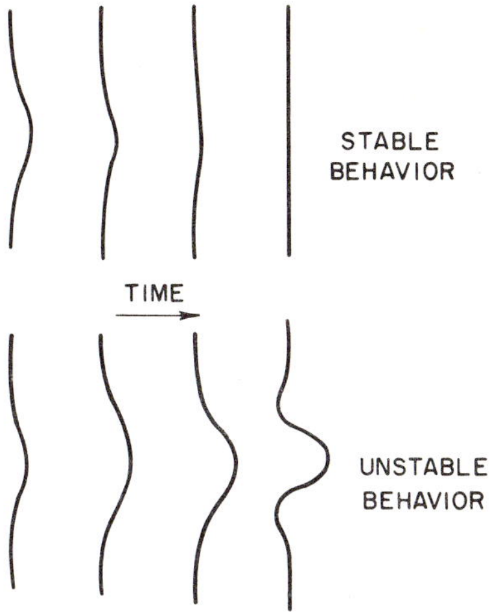

Fig. 15.3. Successive positions of the solid-liquid interface for stable and unstable behavior (after [36]).

growth features are often directly observable under a light microscope for transparent materials. For opaque materials, the presence of morphological instability can often be detected in terms of variations in structure or composition.

It should also be apparent from figs. 15.1 and 15.2 that we are considering growth conditions and materials such that our crystals do not grow with facets. These conditions will generally be met, according to the theory of Jackson [1], for materials whose entropies of fusion, ΔS, satisfy the relation $(\xi \Delta S/R) < 2$ where ξ is a crystallographic factor that is less than one and R is the gas constant. This condition is generally met for most metals, for "a" axis growth in ice, and for a number of transparent organic materials. Interface instability phenomena can also be present during the growth of crystals with facets; however, their analysis is somewhat different from that considered here.

Another way to view the phenomenon of morphological instability, and one which is the key to a theoretical analysis of the problem, is to consider the differences in dynamical behavior that lead to the results depicted in figs. 15.1 and 15.2. In any real physical situation, imperfections of some type are always present. These may be actual crystalline imperfections, thermal fluctuations, inhomogeneities in composition etc. At various stages during the growth process, these imperfections will cause small perturbations in the shape of the solid-liquid interface such as depicted by the curves at the left in fig. 15.3. Under stable conditions (top of fig. 15.3) the perturbation will decay with time and the interface will return to a smooth shape (fig. 15.1). For unstable conditions (bottom of fig. 15.3) the perturbations will be enhanced in amplitude; this will cause a still greater disturbance, lateral spreading, and eventually a distorted and often corrugated interface (fig. 15.2).

In the next section, we shall briefly discuss the phenomenological theory of growth from the melt for pure and impure substances. We shall then discuss solutions for a few simple shapes, ignoring the possibility of morphological instability; these solutions will form the basis for ch. 15 §§ 3–5 where such shapes will be perturbed and tested for instability.

15.2. General phenomenological background

We seek to describe a solidification process wherein a liquid is transformed into a solid along some moving boundary, the solid-liquid interface. In order for this to occur, the latent heat of fusion, which is evolved everywhere along the solid-liquid interface at a rate proportional to the local freezing velocity, must be transported into the adjacent solid and liquid which, in turn, might possibly increase in temperature or simply serve to transport the heat to some external boundary. In addition, if we are dealing with alloys instead of pure materials, a solute may be rejected or, alternatively, preferentially incorporated by the freezing crystal. This, in effect, makes the solid-liquid interface a source or sink of solute which must be carried away or supplied by transport in the adjacent phases.

Several different modes of heat and solute transport could be considered here. Except for substances with high melting points, the dominant modes of transport will be conduction (diffusion) and convection. Except for a few remarks to be made in ch. 15 §4, we shall only consider cases where convection can be ignored; convection is considered in ch. 8 §5. We also note that typical diffusion coefficients in the liquid are $\sim 10^3$ larger than those in the solid. Therefore, whatever solute freezes into the solid phase will, effectively, be "frozen in place" and we can neglect diffusion in the solid.

In the liquid phase, we shall have an equation for the conduction of heat,

$$\nabla^2 T_L = \frac{1}{a_L} \frac{\partial T_L}{\partial t} \quad \text{(in the liquid)} \tag{15.1}$$

where a_L is the thermal diffusivity of the liquid (assumed constant) and an equation for the diffusion of solute,

$$\nabla^2 C = \frac{1}{D}\frac{\partial C}{\partial t} \quad \text{(in the liquid)}, \tag{15.2}$$

where C is the concentration of solute in mass per unit volume and D is the interdiffusion coefficient (assumed constant). If there is more than one solute of importance, an equation similar to eq. (15.2) would be needed for each. In the solid phase, where we shall neglect diffusion as discussed above, we shall need only an equation for conduction of heat

$$\nabla^2 T_S = \frac{1}{a_S}\frac{\partial T_S}{\partial t} \quad \text{(in the solid)}. \tag{15.3}$$

Eq. (15.1–3) will have to be solved subject to certain boundary conditions. The most important and complex of these conditions must be met at the solid-liquid interface; they are:

(i) The conservation of (heat) energy at the interface:

$$Lv \cdot \hat{n} = (-K_L \nabla T_L + K_S \nabla T_S) \cdot \hat{n} \quad \text{(at the interface)}. \tag{15.4}$$

Here L is the latent heat of fusion per unit volume *, K_L and K_S are the thermal conductivities of the liquid and solid, respectively, v is the local growth velocity, $\hat{n}$ is a unit normal to the solid-liquid interface and points into the liquid and $\cdot$ denotes the scalar product.

(ii) The conservation of solute at the interface:

$$(C - C_S)v \cdot \hat{n} = -D\nabla C \cdot \hat{n} \quad \text{(at the interface)}. \tag{15.5}$$

Here C_S is the concentration in the freezing solid at the interface. Eq. (15.5) looks similar to eq. (15.4) except that there is no counterpart for the term $K_S \nabla T_S$ since we are ignoring diffusion in the solid.

(iii) Continuity of temperature at the interface:

$$T_L = T_S \quad \text{(at the interface)}. \tag{15.6}$$

(iv) Relationship of temperature and composition at the interface to thermodynamics and (or) kinetics.

In the most general case, the composition and temperature at the interface will depart slightly from those values corresponding to thermodynamic equili-

* Density differences of the phases are ignored.

brium in order to provide a driving force for the local kinetic processes which must take place at the interface. For situations in which these departures from equilibrium are very large, one has the limiting case of "interfacial control"; this could be the situation, for instance, for stable faceted growth. For the opposite case, where these departures from equilibrium are negligibly small, one has a situation of "transport control" or, in our case, "diffusion control". This is the case of interest for our initial discussions of morphological stability; later we shall mention some cases where there is some degree of interfacial control.

It is therefore sufficient to assume that local equilibrium prevails at the solid-liquid interface and hence that

$$C_S = kC \text{ (at the interface),} \tag{15.7}$$

where k is a constant that may be evaluated from an appropriate phase diagram. For dilute binary alloys, we can have either case "a" or "b" of fig. 15.4 where, for simplicity, we assume the solidus and liquidus lines to be straight. Case "a" pertains to the rejection of solute on freezing and corresponds to $k < 1$; case "b" pertains to enhanced incorporation of solute on freezing and thus to $k > 1$. The parameter m is the slope of the liquidus line for which we adopt the sign convention $m(1 - k) < 0$. Thus, if T_i is the temperature of the interface, then we have

$$T_i = T_M + mC \text{ (at a planar interface)} \tag{15.8}$$

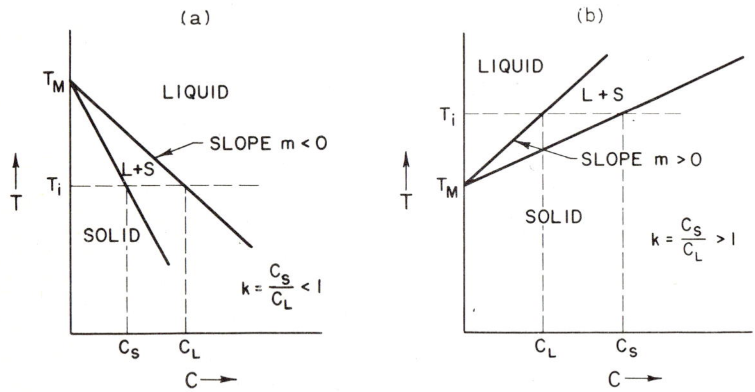

Fig. 15.4. Portions of idealized phase diagrams for dilute binary alloys. (a) $k < 1$, (b) $k > 1$.

which "couples" the solute and concentration fields at the interface. We note that eq. (15.8) only holds if the solid-liquid interface is planar. Otherwise, eq. (15.8) must be corrected for curvature because of the Gibbs–Thomson effect; the appropriate equation is [2]

$$T_i = T_M + mC - T_M \Gamma \left(\frac{1}{R_1} + \frac{1}{R_2}\right), \tag{15.9}$$

where R_1 and R_2 are the principal radii of curvature of the interface (positive for a convex crystal) and $\Gamma = \gamma/L$ where γ is the solid-liquid surface free energy per unit area (surface tension).

Finally, we must specify some external boundary conditions for the problem. These are often given in terms of a specification of the concentration and temperature far from the solid-liquid interface. If the external boundary conditions are more complicated or are applied at finite distances from the interface, the result is usually a more complicated (or impossible) mathematical problem.

A few examples of solved problems will now be discussed.

(i) *Unidirectional unconstrained solidification of a pure substance:*

We refer to the situation depicted in fig. 15.5. The entire liquid as well as the solid-liquid interface are at the melting temperature T_M while the plane $z = 0$ is held at temperature T_0. Freezing takes place via heat conduction in the solid where the temperature can be described * by the expression [3]

$$T_S = T_0 + (T_M - T_0)\,\frac{\operatorname{erf}(\xi)}{\operatorname{erf}(K)}\,. \tag{15.10}$$

Here, $\xi = (z^2/4a_S t)^{1/2}$ and the solid-liquid interface is located at the position

$$z_o = K(4a_S t)^{1/2}\,, \tag{15.11}$$

where K satisfies the equation

$$\tfrac{1}{2}\sqrt{\pi}\, K e^{K^2} \operatorname{erf}(K) = \tfrac{1}{2}(T_M - T_0)\left(\frac{K_L}{a_L L}\right). \tag{15.12}$$

* $\operatorname{erf}(\xi) \equiv (2/\sqrt{\pi})\int_0^{\xi} \exp(-u^2)\,du$ denotes the error function with argument ξ. $\operatorname{erf}(0) = 0$, $\operatorname{erf}(\infty) = 1$ and $\operatorname{erf}(\xi)$ is an odd monotone increasing function of ξ. Tables may be found in the appendix of ref. [3].

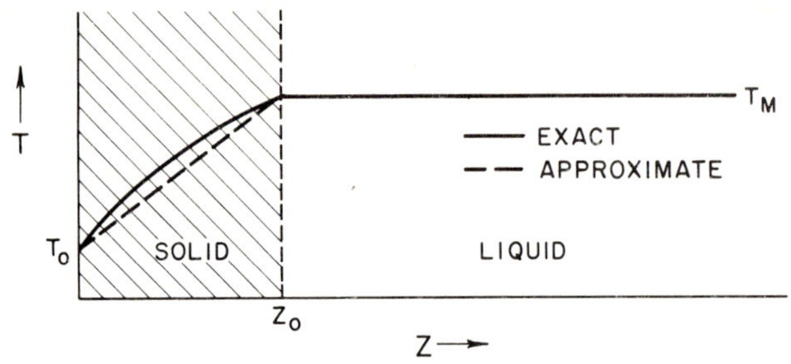

Fig. 15.5. Temperature profile for unconstrained unidirectional solidification of a pure substance from an isothermal liquid at the melting point. The dashed line pertains to an approximate solution.

The quantity on the right-hand side of eq. (15.12) is the "driving force" for solidification; it is a "normalized undercooling", being one half the quotient of $T_M - T_0$ and the quantity $La_L/K_L = L/C_v$ where C_v is the specific heat per unit volume. For tin, for instance, $L/C_v = 232°C$ so the quantity

$$\epsilon \equiv \tfrac{1}{2}(T_M - T_0)\left(\frac{K_L}{a_L L}\right) = \tfrac{1}{2}\,\frac{(T_M - T_0)}{(L/C_v)} \tag{15.13}$$

is often quite small. For such a case, an approximate solution to eq. (15.10) is

$$K \approx \epsilon^{1/2}\,. \tag{15.14}$$

K is often called the "*parabolic rate constant*" because of eq. (15.11).

While eq. (15.10) is an exact solution, the fact that ϵ is small can lead to a very useful approximate solution to the problem. In fact, if we expand erf(ξ) and erf(K) to first order in ξ and K, eq. (15.10) can be rewritten approximately as

$$T_S = T_0 + (T_M - T_0)\,\frac{z/\sqrt{4a_S t}}{K} = T_0 + (T_M - T_0)\,\frac{z}{z_0}\,. \tag{15.15}$$

Eq. (15.15) corresponds to the "straight line" approximate solution indicated by the dashed line in fig. 15.5. It is also immediately recognized to be a solution to the equation

$$\frac{\mathrm{d}^2 T_\mathrm{S}}{\mathrm{d}z^2} = 0\,, \tag{15.16}$$

which is just the appropriate one dimensional version of eq. (15.3) with the term $\partial T_\mathrm{S}/a_\mathrm{S}\partial t$ omitted. This is therefore a "quasi-steady-state" approximation; it is "steady-state" because of eq. (15.16) but only "quasi-steady-state" because z_0 changes slowly with time (see eq. (15.15)).

(ii) *Unidirectional solidification of a dilute binary at constrained velocity V.* This situation of "*constrained growth*" corresponds to the case where we have a very long sample which is caused to freeze unidirectionally at constant speed. This could be accomplished, for instance, by a travelling furnace. It is a very important case because it approximates the situation in many practical cases of crystal growth.

To handle this case approximately, it is useful to transform all equations into a reference frame which moves along with the solid-liquid interface. If we take the motion to be at uniform speed V in the z direction and assume that a steady state will prevail in the *moving* coordinate system, then eqs. (15.1)–(15.3) are transformed into

$$\frac{\mathrm{d}^2 T_\mathrm{L}}{\mathrm{d}z^2} + \frac{V}{a_\mathrm{L}}\frac{\mathrm{d}T_\mathrm{L}}{\mathrm{d}z} = 0\,, \tag{15.17a}$$

$$\frac{\mathrm{d}^2 C}{\mathrm{d}z^2} + \frac{V}{D}\frac{\mathrm{d}C}{\mathrm{d}z} = 0\,, \tag{15.17b}$$

and

$$\frac{\mathrm{d}^2 T_\mathrm{S}}{\mathrm{d}z^2} + \frac{V}{a_\mathrm{S}}\frac{\mathrm{d}T_\mathrm{S}}{\mathrm{d}z} = 0\,. \tag{15.17c}$$

For typical cases which are generally of interest in crystal growth, $D/V \approx 50\mu\mathrm{m}$ whereas $a_\mathrm{L}/V \approx a_\mathrm{S}/V \approx 50$ cm. As we shall be interested in changes of T_L, T_S and C which take place over distances $\approx 50\mu\mathrm{m}$, eqs. (15.17) can be further simplified to

$$\frac{d^2 T_L}{dz^2} = 0 \,, \tag{15.18a}$$

$$\frac{d^2 C}{dz^2} + \frac{V}{D}\frac{dC}{dz} = 0 \,, \tag{15.18b}$$

and

$$\frac{d^2 T_S}{dz^2} = 0 \,. \tag{15.18c}$$

We are now ready to write down a solution to eqs. (15.18) in our moving coordinate system where we take the solid-liquid interface to lie in the plane $z = 0$. The result, for a set of external boundary conditions to be discussed momentarily, is

$$T_L = T_0 + G_L z \,, \tag{15.19a}$$

$$C = C_\infty + C_\infty \left(\frac{1-k}{k}\right) \exp\left[-\frac{Vz}{D}\right] , \tag{15.19b}$$

and

$$T_S = T_0 + G_S z \,. \tag{15.19c}$$

The external boundary conditions are that $C \to C_\infty$ as $z \to \infty$ (far into the liquid) and $(dT_L/dz)|_{z=0} = G_L$. The composition in the liquid at the interface is evidently

$$C_0 = \frac{C_\infty}{k} , \tag{15.20}$$

whereas the temperature at the interface is (see eq. (15.8))

$$T_0 = T_M + mC_0 \,. \tag{15.21}$$

We also note that G_S is not an independent parameter once G_L and V are specified; via eq. (15.4) one obtains

$$V = \frac{1}{L}(K_S G_S - K_L G_L) \,. \tag{15.22}$$

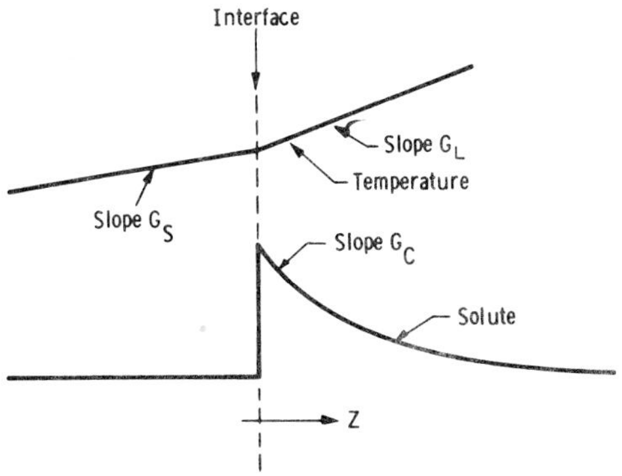

Fig. 15.6. Temperature and solute profiles ($k < 1$) as a function of distance z from the solid-liquid interface for constrained growth of a dilute binary alloy at velocity V.

We note for future reference that the concentration gradient at the interface is

$$G_c \equiv \left.\frac{dC}{dz}\right|_{z=0} = -C_\infty \frac{V}{D} \frac{1-k}{k}. \tag{15.23}$$

A sketch of the solutions represented by eqs. (15.19) is shown in fig. 15.6 for the case $k < 1$. We note especially the solute rich boundary layer that drops off significantly in a distance $\approx D/V$.

(iii) *Unconstrained growth of a cylinder of a pure substance from a supercooled melt.*

We can easily describe the growth of a cylinder of a pure substance if we make the assumption that the growth is sufficiently slow that we can use the "quasi-steady-state" approximation. Then the temperature at radius r in the liquid is *

$$T_L = T_B + (T_i - T_B) \frac{\ln(r/R_B)}{\ln(R/R_B)}, \tag{15.24}$$

where $r = \sqrt{(x^2 + y^2)}$, $r = R$ is the equation of the solid-liquid interface, R_B is the radius of a large coaxial cylinder on which the temperature is T_B, and T_i

* In cylindrical coordinates, the appropriate "quasi-steady-state" equation for radial symmetry is $\nabla T_L = (1/r)(\partial/\partial r)[r(\partial T_L/\partial r)] = 0$. Hence $r(\partial T_L/\partial r) = (\text{constant})_1$ and $T_L = (\text{constant})_1 \ln r + (\text{constant})_2$ by integration.

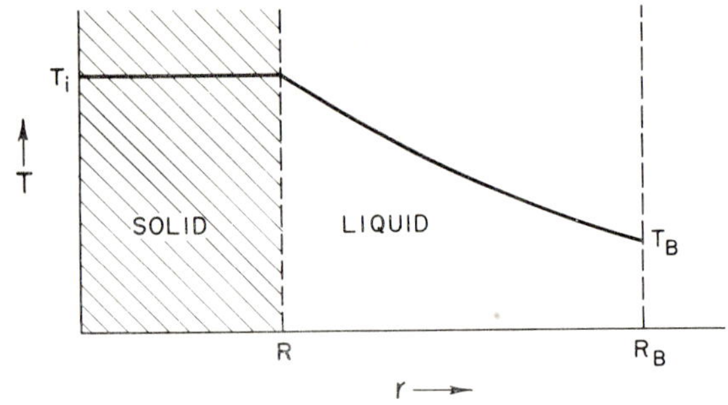

Fig. 15.7. Temperature profile for the growth of a circular cylinder of a pure substance.

is the temperature at the solid-liquid interface. In this approximation, the solution diverges as $R_B \to \infty$ so a "finite cut off" must be retained. Fig. 15.7 is a sketch of the temperature profile.

The freezing temperature at the surface of a cylinder is dependent upon the curvature via eq. (15.9) which, for a cylinder, yields

$$T_i = T_M \left(1 - \frac{\Gamma}{R}\right). \tag{15.25}$$

If $(T_M - T_B) \gg T_M \Gamma/R$ we can neglect the dependence of T_i on R in eq. (15.24); otherwise this term must be retained. Since

$$\left.\frac{dT_L}{dr}\right|_{r=R} = (T_i - T_B)\,\frac{1}{R\ln(R/R_B)} \tag{15.26}$$

we see that the "curvature correction", i.e. the Γ/R term of eq. (15.25), will decrease the magnitude of the temperature gradient and thus decrease the growth rate.

(iv) *Unconstrained growth of a sphere of a pure substance in a supercooled melt.*

For a sphere growing slowly from a supercooled melt at temperature T_B, the temperature in the liquid is given approximately by the equation *

* In spherical coordinates, the appropriate "quasi-steady-state" equation for radial symmetry is $\nabla^2 T = (1/r^2)(\partial/\partial r)[r^2(\partial T_L/\partial r)] = 0$. Hence $r^2(\partial T_L/\partial r) = (\text{constant})_1$ and $T_L = -(\text{constant})_1 r^{-1} + (\text{constant})_2$.

$$T_L = T_B + (T_i' - T_B)\,\frac{R}{r}\,, \tag{15.27}$$

where

$$T_i' = T_M\left(1 - \frac{2\Gamma}{R}\right) \quad \text{and} \quad r = \sqrt{(x^2 + y^2 + z^2)}\,. \tag{15.28}$$

No "finite cut-off" radius appears in eq. (15.27) as was needed in eq. (15.24). The additional factor of 2 in eq. (15.28), as compared with eq. (15.25), stems from the fact that there are 2 finite principal radii of curvature for a sphere whereas there is only one for a cylinder.

From eq. (15.27), we readily obtain

$$\frac{dT_L}{dR} = -\frac{(T_i' - T_B)}{R}\,, \tag{15.29}$$

which, when combined with eq. (15.4) (note $v = dR/dt$) gives

$$R\,\frac{dR}{dt} = \frac{K_L}{L}\,(T_M - T_B), \tag{15.30}$$

to the approximation that T_i' is independent of R. Eq. (15.30) integrates to

$$R^2 = 2\,\frac{K_L}{L}\,(T_M - T_B)\,t + \text{const.} \tag{15.31}$$

which is essentially the same parabolic rate law as expressed by eq. (15.11). If T_i' depends significantly on R, then the growth rate is slower and has a more complicated dependence on time.

15.3. A detailed stability analysis

In this section we shall present a fairly detailed analysis of the stability of the solid-liquid interface for one of the cases discussed in ch. 15 §2. In particular, we shall consider the case of unidirectional solidification of a dilute binary alloy at constrained velocity V. We shall first give an elementary analysis of morphological stability by the application of thermodynamic arguments; this will lead to the concept of constitutional supercooling and an approximate assessment of the growth conditions where stability or instability can be expected. Then we shall indicate how the problem may be treated more quantitatively with the aid of perturbation theory; this latter approach will also lead to a description of the early stages of growth under conditions where instability is present.

The application of thermodynamic arguments to analyze the morphological stability [4, 5] can easily be made with the aid of the considerations discussed in ch. 15 §1. We refer again to fig. 15.3 and to the argument that stable or unstable behavior will occur according as a perturbation will decay or grow with time. If this perturbation is imagined to be a protuberance into the liquid, then we would be led to inquire if conditions in the liquid would favor melting or further freezing of the protuberance. If melting would tend to occur, we would anticipate stability; otherwise instability would be expected.

Let us suppose that the protuberance disturbs the situation very little so that eqs. (15.19) are still valid. Then the thermodynamic freezing temperature at each position in the liquid can be obtained by substitution of eq. (15.19b) into eq. (15.8). To remind ourselves that we are not confining ourselves to the solid-liquid interface, we shall call the resulting quantity T^* instead of T_i. Thus

$$T^* = T_M + m\left[C_\infty + C_\infty\left(\frac{1-k}{k}\right)\exp(-Vz/D)\right]$$

or

$$T^* = T_0 - mC_\infty\,\frac{(1-k)}{k}\,[1-\exp(-Vz/D)] \tag{15.32}$$

where T_0 is given by eq. (15.21). We wish to compare this with the actual temperature in the liquid given by eq. (15.19a), namely $T_L = T_0 + G_L z$. This is done graphically in fig. 15.8 which corresponds to both $k < 1$ and $k > 1$ because of our sign convention $m(1-k) < 0$. For case "a", there is a zone in which T^* lies above T_L; in this zone the liquid is supercooled. Because of the role of solute in producing this supercooling, it has been termed *constitutional supercooling*. For case "b", on the other hand, there is no zone of constitutional supercooling near the interface. By the arguments advanced above, we would expect the presence of constitutional supercooling to correspond to morphological instability and its absence to correspond to morphological stability.

Comparison of the initial slopes in fig. (15.8) with the aid of eq. (15.32) shows that constitutional supercooling is present whenever

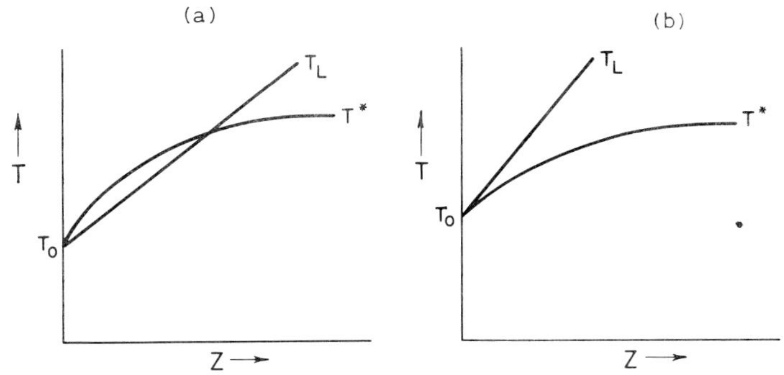

Fig. 15.8. Sketches of the actual temperature and the equilibrium temperature as a function of distance z from the solid-liquid interface for the constrained growth of a dilute binary alloy at velocity V.

$$\left.\frac{\mathrm{d}T^*}{\mathrm{d}z}\right|_{z=0} > G_L \tag{15.33}$$

or

$$-mC_\infty \frac{(1-k)}{k}\frac{V}{D} > G_L\,. \tag{15.34}$$

With the aid of eq. (15.23), this may also be written as

$$mG_c > G_L\,. \tag{15.35}$$

If we replace the inequality sign in eq. (15.34) by an equality sign, we can graphically represent the demarcation between the presence and absence of constitutional supercooling. With G_L/V and C_∞ as independent variables, the appropriate equation is

$$\frac{G_L}{V} = (-m)\frac{(1-k)}{Dk}C_\infty \tag{15.36}$$

and a straight line, as depicted in fig. 15.9, is the appropriate demarcation.

Eqs. (15.33)–(15.36) and fig. 15.9 can actually lead to a semi-quantitative understanding of morphological stability. They focus attention on the fact that a large positive G_L provides a stabilizing influence which is less effective

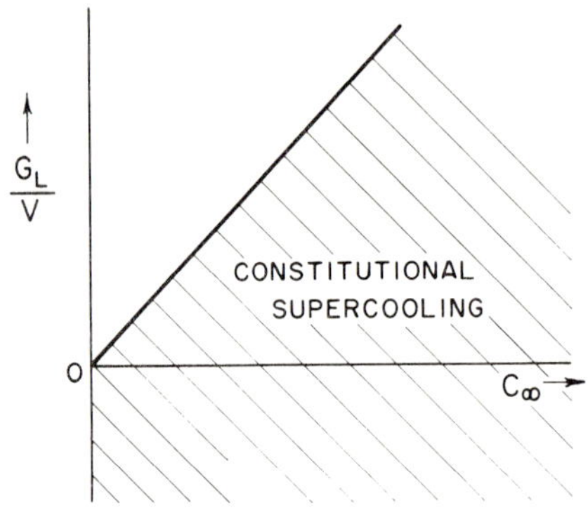

Fig. 15.9. Diagram to illustrate the growth conditions under which constitutional supercooling is present.

the higher the velocity V. The destabilizing influence is provided by the impurity concentration C_∞. We note that the above considerations will always predict instability if $G_L < 0$; for $C_\infty = 0$ this would be supercooling in the ordinary sense.

We must now ask why the principle of constitutional supercooling should not be used to provide a rigorous test of morphological stability. The answer is that the principle of constitutional supercooling is based on the application of thermodynamic arguments to the liquid adjacent to the solid-liquid interface; further, this liquid is regarded as a closed thermodynamic system. In reality, however, the situation is one in which *kinetic* processes play an important role, heat and solute are being transported throughout the system and their transport may even be influenced by the perturbation itself.

We therefore turn to consider the theory of morphological stability which provides a self-consistent kinetic analysis of the instability phenomenon [6]. This theory makes use of the following steps:

(i) Obtain an analytical solution to the unperturbed (smooth interface) problem.

(ii) Imagine that there occurs an arbitrary infinitesimal perturbation in the shape of the interface.

(iii) Obtain, via classical perturbation theory, an approximate solution for the case of the perturbed interface.

(iv) Examine the solution for the perturbed case to see if the perturbations will grow or decay with time. In this manner one can establish a stability criterion – i.e. growth conditions under which the interface will be stable or unstable.

(v) For growth conditions where the interface is unstable, deduce the manner in which the interface shape will change with time.

We shall now indicate how steps (i)–(v) above can be carried out.

Step (i) has already been discussed in ch. 15 §2; for further developments we shall utilize the case of "unidirectional solidification of a dilute binary alloy at constrained velocity V" which led to eqs. (15.19)–(15.23).

To carry out step (ii) above, we must find a way to represent an arbitrary infinitesimal perturbation in the shape of the interface. Since the original unperturbed interface lies in the plane $z = 0$, we can measure a distance along the interface by means of coordinates x and y such that the x, y and z axes form a Cartesian coordinate system. Then the perturbed interface can be represented by an equation of the form

$$z = \Phi(x, y, t) \tag{15.37}$$

Actually eq. (15.37) is much too general to permit further progress. The problem is only tractable for the case when Φ is sufficiently small that all equations can be linearized. In particular, Φ must be small compared to some distance characteristic of the lateral extent of the perturbation, i.e. its extent in the x and y directions. Furthermore, so long as the crystal can be considered to be isotropic in the x and y directions, we will lose no generality if we take Φ to depend only on x and t.

For mathematical reasons it is more convenient to work with the Fourier transform of $\Phi(x, t)$ rather than with $\Phi(x, t)$ itself. In other words, we can solve the problem for a perturbed interface of the form

$$z = \delta \sin(\omega x)\,, \tag{15.38}$$

where δ depends on t and ω is a spatial frequency which corresponds to a wavelength $\lambda = 2\pi/\omega$. An arbitrary perturbation* can be built up by a superposition of sinusoidal waves such as represented by eq. (15.38); these will be considered to have small amplitudes, compared to their lateral extents, if $\delta/\lambda \ll 1$. In such a case, the perturbed growth velocity in the z direction is just

* Actually, terms in $\cos(\omega x)$ would also be needed in the most general case. However, eq. (15.38) can be shown to suffice unless fluid convection effects are important. A general treatment can be found in reference [7].

$$v = V + \dot{\delta} \sin(\omega x) \tag{15.39}$$

where V is the constant unperturbed growth speed and $\dot{\delta} = \mathrm{d}\delta/\mathrm{d}t$. So long as $\delta/\lambda \ll 1$, there is essentially no component of v in the x direction.

The equations which must be solved to accomplish step (iii) above are obtained by transformation of eqs. (15.1)–(15.3) into a coordinate system which is moving along with the unperturbed interface. This leads to equations similar to eqs. (15.18) except that there are now terms containing $\mathrm{d}/\mathrm{d}x$; specifically

$$\frac{\partial^2 T_\mathrm{L}}{\partial x^2} + \frac{\partial^2 T_\mathrm{L}}{\partial z^2} = 0\,, \tag{15.40a}$$

$$\frac{\partial^2 C}{\partial x^2} + \frac{\partial^2 C}{\partial z^2} + \frac{V}{D}\frac{\partial c}{\partial z} = 0\,, \tag{15.40b}$$

and

$$\frac{\partial^2 T_\mathrm{S}}{\partial x^2} + \frac{\partial^2 T_\mathrm{S}}{\partial z^2} = 0\,. \tag{15.40c}$$

Strictly speaking, the zeros on the right hand sides of eqs. (15.40) should be replaced by

$$\frac{1}{a_\mathrm{L}}\frac{\partial T_\mathrm{L}}{\partial t}\,,\ \frac{1}{D}\frac{\partial c}{\partial t} \quad \text{and} \quad \frac{1}{a_\mathrm{S}}\frac{\partial T_\mathrm{L}}{\partial t}\,,$$

respectively; however, these terms can be shown [7, 8] to lead to uninteresting transients in many cases and so have been omitted.

For an interface of the form represented by eq. (15.38), we can write an approximate solution to eqs. (15.40) by the addition of a few more terms to the unperturbed solutions given by eqs. (15.19). Thus we take

$$T_\mathrm{L} = T_0 + G_\mathrm{L} z + A_\mathrm{L} \mathrm{e}^{-\omega z} \sin(\omega x)\,, \tag{15.41a}$$

$$C = C_\infty + C_\infty \left(\frac{1-k}{k}\right) \exp(-Vz/D) + B\mathrm{e}^{-\omega^* z} \sin(\omega x)\,, \tag{15.41b}$$

and

$$T_S = T_0 + G_S z + A_S e^{+\omega z} \sin(\omega x)\,, \tag{15.41c}$$

where A_L, B and A_S are constants and

$$\omega^* = \frac{V}{2D} + \left[\left(\frac{V}{2D}\right)^2 + \omega^2\right]^{1/2}. \tag{15.42}$$

The "perturbed parts" of eqs. (15.41) are chosen to die out for distances far away from the interface, i.e. $z \to +\infty$ in the liquid and $z \to -\infty$ in the solid.

All that really remains is to substitute eqs. (15.41), (15.39) and (15.38) into the four boundary conditions given by eqs. (15.4), (15.5), (15.6) and (15.8) and solve for the four unknowns A_L, B, A_S and $\dot{\delta}/\delta$. This process involves quite a bit of algebra which is not suitable for presentation here. We only remark that it is important to separate each equation into "unperturbed parts" and "perturbed parts" where the "perturbed parts" are linear in $\sin(\omega x)$; the "unperturbed parts" and "perturbed parts" can be then equated to each other.

The central result of the calculation just outlined is the following expression for $\dot{\delta}/\delta$:

$$\frac{\dot{\delta}}{\delta} = \frac{V\omega\{-2T_M\Gamma\omega^2[\omega^* - pV/D] - (\mathcal{G}' + \mathcal{G})[\omega^* - pV/D] + 2mG_c[\omega^* - V/D]\}}{(\mathcal{G}' - \mathcal{G})[\omega^* - pV/D] + 2\omega m G_c}$$

$$\equiv f(\omega)\,. \tag{15.43}$$

In eq. (15.43), we have employed the conductivity-weighted temperature gradients

$$\mathcal{G} \equiv \frac{K_L G_L}{\bar{K}} \tag{15.44a}$$

and

$$\mathcal{G}' \equiv \frac{K_S G_S}{\bar{K}} \tag{15.44b}$$

where $\bar{K} = \frac{1}{2}(K_S + K_L)$ is the average thermal conductivity of the system. The quantity $p = 1 - k$. Integration of eq. (15.43) gives

$$\delta = \delta_0 e^{f(\omega)t}\,, \tag{15.45}$$

where δ_0 is the value of δ at $t = 0$.

Analysis of eq. (15.45) shows that Fourier components whose corresponding $f(\omega)$ is positive will increase exponentially with time while those whose corresponding $f(\omega)$ is negative will decay exponentially with time. In fig. 15.10, $f(\omega)$ is sketched for several possible cases. In case 3, all Fourier components will decay with time; this case corresponds to morphological stability. In case 1 there is a range $\omega_0 < \omega < \omega_{00}$ where $f(\omega)$ is positive; the corresponding Fourier components will increase exponentially with time and this will constitute morphological instability. Components with $\omega < \omega_0$ (large λ) are stabilized by the lack of long range lateral diffusion whereas components with $\omega > \omega_{00}$ (short λ) are stabilized because of the effect of surface tension. Case 2 represents the demarcation between cases 1 and 3; a detailed analysis leads to the following stability criterion:

$$\tfrac{1}{2}(\mathcal{G}' + \mathcal{G}) > mG_c\mathcal{S}(A, k) \tag{15.46}$$

where $\mathcal{S}(A, k)$ is a function displayed graphically in fig. 15.11. $\mathcal{S}$ depends on the parameter

$$A = \frac{k^2}{(1-k)} \frac{\Gamma V}{D} \frac{T_M}{(-mC_\infty)} \tag{15.47}$$

and the distribution coefficient k. It is usually true that $A \ll 1$ in which case

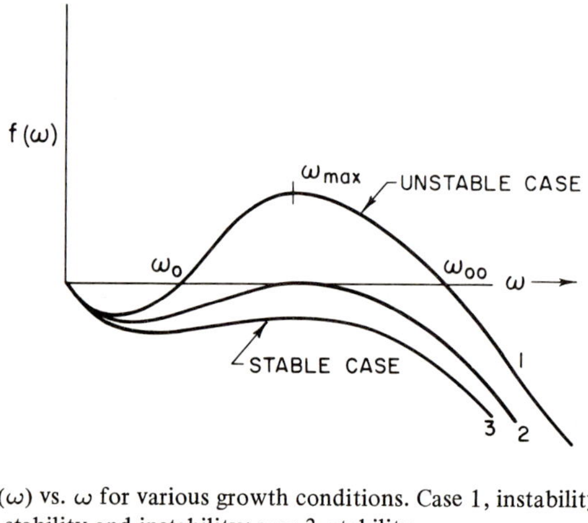

Fig. 15.10. $f(\omega)$ vs. ω for various growth conditions. Case 1, instability; case 2, demarcation between stability and instability; case 3, stability.

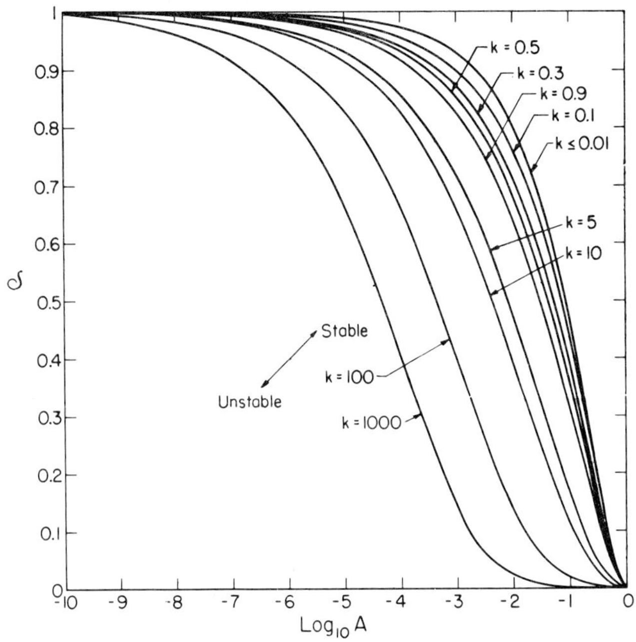

Fig. 15.11. Plot of the stability function $\mathcal{S}$ versus $\log_{10}A$ for several values of k. The difference between $\mathcal{S}$ and 1 represents the tendency of the solid-liquid surface tension to stabilize the interface shape (after [9]).

$\mathcal{S}$ is of the order of 0.9. Comparison of eq. (15.46) with eq. (15.35) (the former inequality is for stability while the latter is for instability) reveals a striking resemblance. Eq. (15.46) shows that the relevant parameter for comparison with mG_c is not G_L but, instead,

$$\tfrac{1}{2}(\mathcal{G}' + \mathcal{G}) = \frac{K_L G_L + K_S G_S}{K_S + K_L}, \tag{15.48}$$

which is a conductivity-weighted-average temperature gradient. Thus the temperature gradient in the solid provides a stabilization that is roughly equivalent to that in the liquid. We note that eq. (15.46) can be rewritten in the form [10]

$$\frac{G_L}{V} + \frac{L}{2K_L} < \frac{(-mC_\infty)}{D} \frac{1-k}{k} \frac{K_S + K_L}{2K_L} \mathcal{S} , \tag{15.49}$$

which should be compared with eq. (15.34).

Under conditions where instability is encountered, it is evident from eq. (15.45) and fig. 15.10 that the Fourier component corresponding to ω_{max} will grow at the fastest rate. We would anticipate, therefore, that the interface might eventually evolve into a sinusoid with wavelength $\lambda_{max} = 2\pi/\omega_{max}$. Actually, however, the approximations of the theory might lead to serious errors before this occurs; in particular, the amplitudes of the perturbations will become large in comparison to their wavelengths.

We shall close this section by briefly mentioning what happens when an actual perturbation grows with time. In that case, the time-dependent interface shape can be approximately characterized by the Green's function

$$G_+(x;t) = e^{t/\tau} \cos\left(\frac{2\pi x}{\lambda_{max}}\right) \frac{\exp(-x^2/4\mathcal{D}t)}{\sqrt{(4\pi\mathcal{D}t)}} \tag{15.50}$$

where τ is a characteristic time and $\mathcal{D}$ is an effective diffusivity which describes the lateral spreading of the perturbation. A graph of $G_+(x;t)/G_+(0;t)$ is shown in fig. 15.12. For a metal with $k \ll 1$ and $V = 10^{-3}$ cm sec^{-1}, typical values are $\tau = 0.33$ sec, $\lambda_{max} = 6.9\mu$m and $\mathcal{D} = 5.5 \times 10^{-8}$ cm^2 sec^{-1}. For more detail the reader is referred to refs. [10] and [11].

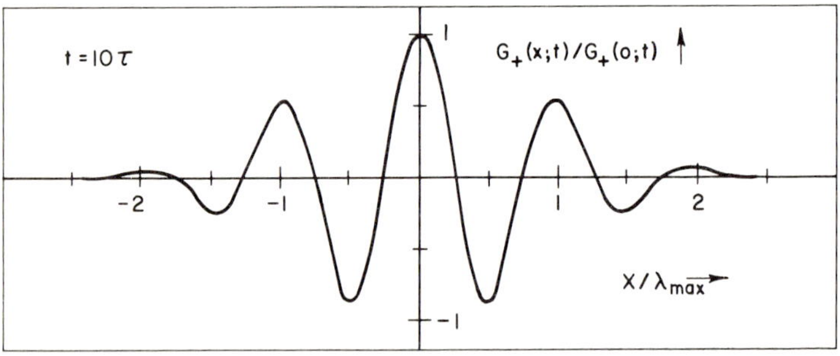

Fig. 15.12. Graph of the normalized Green's function $G_+(x;t)/G_+(x;0)$. For sufficiently long times, $G_+(x;t)$ is proportional to the actual interface shape that results from a small perturbation initiated at $x = 0$ at time $t = 0$.

15.4. Survey of stability analyses

In ch. 15 §§2 and 3 we discussed the general theory of morphological stability in a rather detailed fashion for the case of unidirectional solidification of a dilute binary alloy. In the present section, we shall give a less detailed survey of other work on morphological stability.

(i) *Other geometrical shapes*

We shall first consider the application of morphological stability theory to solid-liquid interfaces with geometrical shapes other than a plane. The other shapes for which detailed analyses have been successful are the sphere [12] and the circular cylinder [13]. Indeed, these analyses were successful because the plane, the sphere and the circular cylinder are all shapes with constant surface curvature, this curvature being 0 for a plane, $2/R$ for a sphere of radius R and $1/R$ for a circular cylinder of radius R. Thus the "curvature correction" in eq. (15.9) does not introduce severe mathematical complications and an exact solution of the unperturbed problem is possible within the quasi-steady state approximation.

For a *sphere* of a pure substance which is solidifying from a supercooled melt, the surface perturbations can be described in terms of spherical harmonics $Y_{lm}(\theta, \varphi)$, which are solutions of the Laplace Equation that depend upon the polar angle θ and the azimuthal angle φ of a spherical coordinate system. Thus a perturbed interface can be represented by an equation of the form

$$r = R + \delta_{lm} Y_{lm}(\theta, \varphi) . \tag{15.51}$$

Eq. (15.51) represents a slightly distorted sphere if δ_{lm} is small. An arbitrary perturbation can be represented by superposition of the Y_{lm} in a manner analogous to Fourier analysis.

For the sphere one can proceed in a manner similar to that described in ch. 15 §3. The main result is [12]

$$\frac{\dot{\delta}_l}{\delta_l} = \frac{(l-1)K_\mathrm{L}}{RL} \left\{ |G_\mathrm{L}| - \frac{T_\mathrm{M}\Gamma}{R^2} \left[(l+1)(l+2) + l(l+2) \frac{K_\mathrm{S}}{K_\mathrm{L}} \right] \right\} , \tag{15.52}$$

where we have omitted the subscript m on δ because m does not appear on the right hand side of eq. (15.52). $|G_\mathrm{L}|$ is the magnitude of the (negative) temperature gradient in the liquid at the interface and is given by the expression (compare eqs. (15.28) and (15.29))

$$|G_L| = \frac{T_M - T_\infty}{R} - \frac{2T_M\Gamma}{R^2} . \tag{15.53}$$

Evidently the term $|G_L|$ in eq. (15.52) is destabilizing while that containing l is stabilizing; this stabilizing term is a result of surface tension effects and therefore vanishes when $\Gamma = 0$. For any given l there will exist a value of R such that the lth harmonic will be unstable; this will occur whenever

$$\frac{T_M - T_\infty}{T_M} > \frac{\Gamma}{R}\left[(l+1)(l+2) + 2 + l(l+2)\frac{K_S}{K_L}\right] . \tag{15.54}$$

For $l = 2$, $\Gamma = 10^{-8}$ cm, $(T_M - T_\infty)/T_M \approx 0.1$ and $K_S/K_L = 2$, eq. (15.54) is satisfied for $R > 300$Å. For larger R, the sphere becomes unstable with respect to harmonics with larger l.

The behavior just discussed appears to be somewhat different from that encountered for a plane but, in reality, it is not. The radius R only enters the stability criterion because of its influence on the magnitude of G_L and the "curvature effect". Indeed, we can pass to a continuum theory by considering the case $l \gg 1$ and introducing * the variable $\omega = 2\pi/\lambda \equiv l/R$. Then eq. (15.52) takes the form

$$\frac{\dot\delta}{\delta} = \frac{\omega K_L}{L}\left\{|G_L| - T_M\Gamma\omega^2\left(1 + \frac{K_S}{K_L}\right)\right\} = \frac{\omega\bar{K}}{L}\{|\mathcal{G}| - 2T_M\Gamma\omega^2\} \tag{15.55}$$

which compares favorably with eq. (15.43) for the special case that $G_S = G_c = 0$.

For a circular *cylinder*, the appropriate perturbed shape is [13]

$$r = R + \delta_\Phi \cos(k\Phi) + \delta_z \cos(k\Phi)\cos(\omega z) . \tag{15.56}$$

The term $\delta_\Phi \cos(k\Phi)$ gives rise to so-called "Φ *perturbations*" (see fig. (15.13)) while the term $\delta_z \cos(k\Phi)\cos(\omega z)$ represents "*z perturbations*" (see fig. (15.14) for the case $k = 0$). A cylinder perturbed by a superposition of both types of perturbations is depicted in fig. 15.15. The corresponding expressions for $\dot\delta_\Phi/\delta_\Phi$ and $\dot\delta_z/\delta_z$ are different in detail from those discussed previously; however, the same general considerations are still applicable. We shall

* The justification for the introduction of the variable $\omega = l/R$ is that Y_{lm} has l nodal planes.

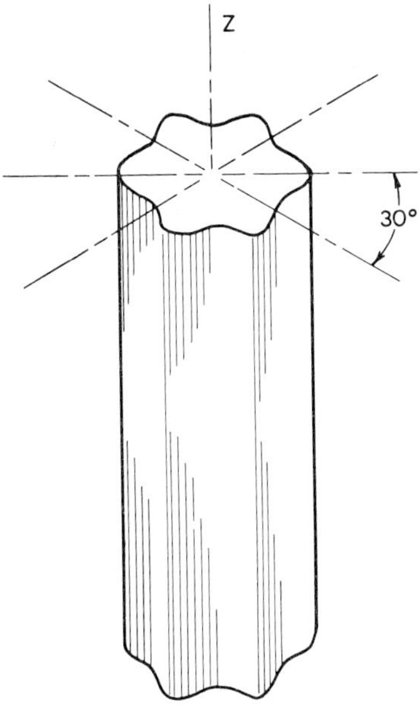

Fig. 15.13. An infinitely long circular cylinder with Φ perturbations ($k = 6$).

return to this case in ch. 15 §5 when we discuss experimental aspects of the subject.

The only other interface shape for which a detailed stability analysis has been attempted is a *paraboloid of revolution*. This is an important problem insofar as it relates to the side branching of a dendrite whose primary stalk can be approximated by a "needle crystal" with the approximate shape of a paraboloid of revolution. Unfortunately, no exact solution to the "unperturbed problem" is possible because of the complication of the "curvature correction" in eq. (15.9). Therefore it is not clear if the perturbation analysis supplies a correction to the unperturbed problem or really accounts for changes due to the perturbations themselves. The problem is mathematically quite complicated and the interested reader is referred directly to Kotler's paper for details [14].

(ii) "*Time-dependent*" *treatments*

In obtaining eqs. (15.40) from eqs. (15.1)–(15.3), the terms $\partial T_L/a_L\partial t$, $\partial c/D\partial t$ and $\partial T_S/a_S\partial t$ were omitted. This approximation is not essential and

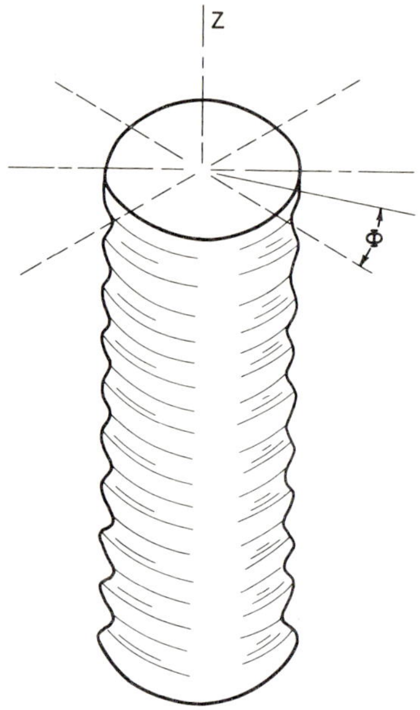

Fig. 15.14. An infinitely long circular cylinder with z perturbations ($k = 0$ in eq. 15.51).

has, in fact, been eliminated in more precise treatments [7, 8, 15]. It turns out that the terms $\partial T_L/a_L \partial t$ and $\partial T_S/a_S \partial t$ give rise to thermal transients and are much less important than the term $\partial c/D\partial t$ which gives rise to transients in the solute concentration field; this is true because $a_L \sim a_S \gg D$ for most systems.

"Time-dependent" treatments generally make use of the Laplace Transform or some equivalent device to handle this more complicated problem. Their results indicate a dependence of the amplitude of a Fourier component on time which is not simply exponential as portrayed by eq. (15.45). However, aside from a few uninteresting transients, the results agree with the simpler "quasi-time-independent" treatments. The main importance of time-dependent treatments has probably therefore been their assistance in the clarification of the general aspects and approximations of the theory. For a detailed numerical comparison in a specific case, the reader is referred to ref. [16].

(iii) *Interface kinetics.*

Up to now we have assumed a situation of diffusion control and have therefore employed eqs. (15.7) and (15.9) at the solid-liquid interface. This

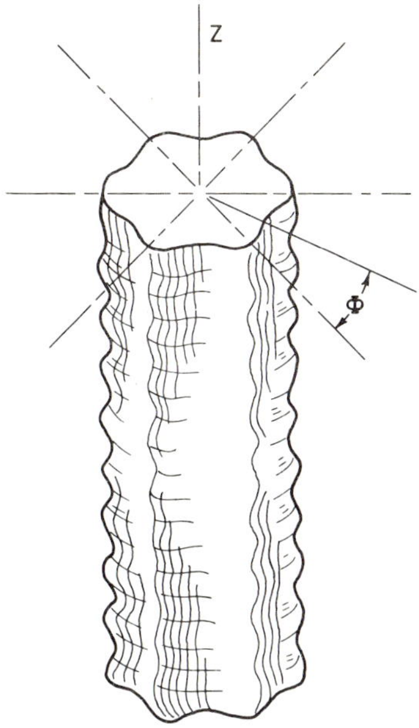

Fig. 15.15. An infinitely long circular cylinder with a superposition of z perturbations and Φ perturbations.

approximation is not essential and has, in fact, been alleviated by several workers [15, 17–23]. The essential modification is to replace eq. (15.9) by an equation of the form

$$T_{\mathrm{i}} = T_{\mathrm{M}} + mC - T_{\mathrm{M}}\Gamma\left(\frac{1}{R_1} + \frac{1}{R_2}\right) - g(v) \qquad (15.57)$$

where $g(v)$ represents a departure from thermodynamic equilibrium because of the growth velocity v. $g(v)$ can have many forms (this general problem will be discussed in ch. 10 by Bennema and Gilmer); however, in any case, $g(0) = 0$. Generally speaking, $g(v)$ gives rise to a reduction in the driving force available for solidification.

Two effects of $g(v)$ are important insofar as considerations of morphological stability are concerned. These can be explained formally for the simple case when g depends only on the magnitude of v. We take $v = V + \dot{\Phi}$ and associate V with the unperturbed problem and $\dot{\Phi}$ with the growth or decay of a perturbation. Then

$$g(v) \approx g(V) + \frac{\mathrm{d}g}{\mathrm{d}v}\dot{\Phi} \equiv \Delta T_0 + \frac{\dot{\Phi}}{\mu}, \tag{15.58}$$

where μ is an appropriate kinetic coefficient. The term ΔT_0 in eq. (15.58) is independent of $\dot{\Phi}$ and gives rise to a "static shift" in the unperturbed temperature of the solid-liquid interface. The term $\dot{\Phi}/\mu$, on the other hand, gives rise to a "dynamic drag" on the growth of a perturbation.

As a specific example of the modifications brought about by interface kinetics, we shall consider how eq. (15.55) would be modified when interfacial kinetics are taken into consideration. The result is [20]

$$\frac{\dot{\delta}}{\delta} = \frac{\omega K_{\mathrm{L}}}{L} \frac{\left\{|G_{\mathrm{L}}'| - T_{\mathrm{M}}\Gamma\omega^2\left(1 + \dfrac{K_{\mathrm{S}}}{K_{\mathrm{L}}}\right)\right\}}{\left\{1 + \dfrac{\omega(K_{\mathrm{S}} + K_{\mathrm{L}})}{L\mu}\right\}} \tag{15.59}$$

where

$$|G_{\mathrm{L}}'| = |G_{\mathrm{L}}| - \frac{\Delta T_0}{R}. \tag{15.60}$$

The reduction of the effective gradient, according to eq. (15.60), is the "static effect" of interface kinetics; the reduction of $\dot{\delta}/\delta$ by the denominator of eq. (15.59) represents the "dynamic effect".

The effect of interface kinetics will therefore be to favor morphological stability. The exact magnitude of this effect will, however, depend upon the values of ΔT_0 and μ which are not well known for many materials. A detailed study by Coriell and Parker [20], however, has shown that considerable stabilization can be achieved for a substance with slow kinetics like salol (an increase in the critical radius for instability by a factor of about 3000) but that very little stabilization is achieved for a substance with fast kinetics like tin.

(iv) *Effects of surface diffusion during isothermal growth*

So far we have only considered binary alloys, where both heat flow and solute diffusion are important, and pure substances where only heat flow is important. However, we can also consider the case of an isothermal system where only solute transport is important, e.g. solution growth or vapor growth. This problem has been treated by Coriell and Parker [24] who consider both volume diffusion (diffusion coefficient D) and surface diffusion (diffusion coefficient D_{s}) as transport mechanisms.

In particular, they consider Φ perturbations (see eq. (15.56)) of an infinitely long circular cylindrical precipitate (concentration C) which is growing from a medium whose bulk concentration is c_∞. The equilibrium concentration in the medium adjacent to the precipitate is taken to be c_0 and c_R denotes the quantity $c_0 + c_0\Gamma/R$ (where Γ is the capillarity constant for the correction of equilibrium composition for interface curvature) [13]. For a cylinder of unperturbed radius R it is found that [24]

$$\frac{\dot{\delta}_\Phi}{\delta_\Phi} = \frac{D(k-1)}{R^2(C-c_R)} \left\{ G_c - \frac{c_0\Gamma}{R} k(k+1) - \frac{(C-c_R)\Gamma\Omega^{1/3}}{R^2} \frac{D_s}{D} k^2(k+1) \right\} \tag{15.61}$$

where G_c is the concentration gradient at the precipitate-matrix interface [13] and Ω is the atomic volume of the precipitate.

The first term in the brackets of eq. (15.61), namely G_c, is destabilizing and results from the point effect of diffusion. The term $(-c_0\Gamma/R)k(k+1)$ is stabilizing and represents the stabilizing effect of capillarity. Both of these terms have direct analogs in the solidification case considered heretofore (see eq. (15.52)). The remaining term in eq. (15.61), however, contains D_s and represents the additional stabilizing effect of surface diffusion. Surface diffusion is therefore a very important stabilizing effect whenever

$$\frac{C}{c_0} \frac{\Omega^{1/3}}{R} \frac{D_s}{D} \gg 1 . \tag{15.62}$$

In eq. (15.62), the factor C/c_0 stems from the difference in driving forces for surface and volume diffusion while the factor $\Omega^{1/3}/R$ stems from the difference in diffusion paths for these two processes. Under conditions where eq. (15.62) is valid, surface diffusion has the effect of smoothing out the concentration differences that stem from volume diffusion and lead to instability.

(v) *Convection effects*

Let us now return to the case of unidirectional solidification of a dilute binary alloy and see what effects convection might have on morphological stability.

Hurle has employed the boundary layer analysis of Burton et al. to study morphological stability in the presence of convection by both the application of the constitutional supercooling principle [25] and the application of stability theory [26]. According to that analysis, convection is supposed to keep the concentration equal to c_∞ for $z \geqslant \Delta$ where Δ is the thickness of the "solute boundary layer". Thus eq. (15.19b) is replaced by the equation

$$C = \begin{cases} k_{\text{eff}} \left[c_\infty + c_\infty \left(\frac{1-k}{k} \right) \exp \left(-\frac{Vz}{D} \right) \right], 0 \leqslant z \leqslant \Delta . \\ c_\infty, z \geqslant \Delta . \end{cases} \tag{15.63}$$

Here,

$$k_{\text{eff}} \equiv \frac{k}{k + (1-k) \exp(-V\Delta/D)} \tag{15.64}$$

is the "*effective distribution coefficient*" (cf. eq. 8.17). The concentration that freezes out in the solid is

$$C_S = k_{\text{eff}} c_\infty = kC(z = 0) . \tag{15.65}$$

If we inquire as to the effect of convection on morphological stability, we should be careful to ask the right questions. Just asking "Does convection favor morphological stability?" is not enough – we must specify what is to be held constant in the comparison. If we focus on the constitutional supercooling principle for an initial comparison, eq. (15.63) leads to the expression

$$G_c \equiv \left(\frac{\partial c}{\partial z} \right)_{z=0} = -k_{\text{eff}} \frac{V}{D} c_\infty \left(\frac{1-k}{k} \right) = -\frac{V}{D} C_S \left(\frac{1-k}{k} \right) . \tag{15.66}$$

If we now consider the constitutional supercooling principle (eq. (15.33)) for a given G_L, eq. (15.66) reveals that convection has no effect on stability if a crystal of a given C_S is grown but has a very definite effect (via the factor k_{eff}) if c_∞ is regarded as fixed. By inspection of eq. (15.64), we see that $k_{\text{eff}} < 1$ if $k < 1$ and $k_{\text{eff}} > 1$ if $k > 1$; therefore when c_∞ is regarded as fixed, convection produces less constitutional supercooling for $k < 1$ and more for $k > 1$.

When the same boundary layer model is employed along with a perturbation analysis of morphological stability, it is found that G_c, given by eq. (15.66) above, may be inserted directly in eq. (15.46) provided that $\mathcal{S}$ is appropriately changed. Hurle [26] finds, that $\mathcal{S} \approx 1$ for weak convection while $\mathcal{S} \approx 1 - k\Delta V/D$ for strong convection (provided that k is not too large). More interesting, perhaps, are his results

$$\lambda_0 = \frac{2\pi}{\omega_0} \approx (G_L V)^{-1} \tag{15.67}$$

for strong convection while $\lambda_0 \approx (G_L V)^{-\nu}$ where $\nu \approx \frac{1}{3}$ for weak convection. It is generally believed that the cell size during cellular solidification obeys an equation of the form of eq. (15.67); these results are therefore quite suggestive.

Delves [27, 28] has attempted to refine the theory of morphological stability when convection is important to account for flow along the solid-liquid interface within the solute boundary layer. A typical flow pattern, for a sinusoidally perturbed interface, is depicted in fig. 15.16. The corresponding analysis is sufficiently complicated that no general results will be presented here. However, it is noteworthy that Delves [27, 28] finds a behavior that can be characterized by an equation of the form

$$\Phi(x, t) = \delta_0 e^{t/\tau'} \sin\left[\frac{2\pi}{\lambda'}(x - v_p t)\right], \tag{15.68}$$

where v_p is a phase velocity. For no convection, the corresponding result would be

$$\Phi(x, t) = \delta_0 e^{t/\tau} \sin\left(\frac{2\pi x}{\lambda}\right), \tag{15.69}$$

which would result from eq. (15.45). The phase velocity in eq. (15.68) is caused by, and leads to motion in the direction of the unperturbed flow velocity; it is, however, many orders of magnitude smaller.

(vi) *Eutectic-like crystals*

Up to now we have only considered the morphological stability of an interface that bounds a single-phase crystal. It is also of some interest, however, to consider crystallization processes in which there is coupled growth of two phases, e.g. in the solidification of eutectics. Eutectic solidification would

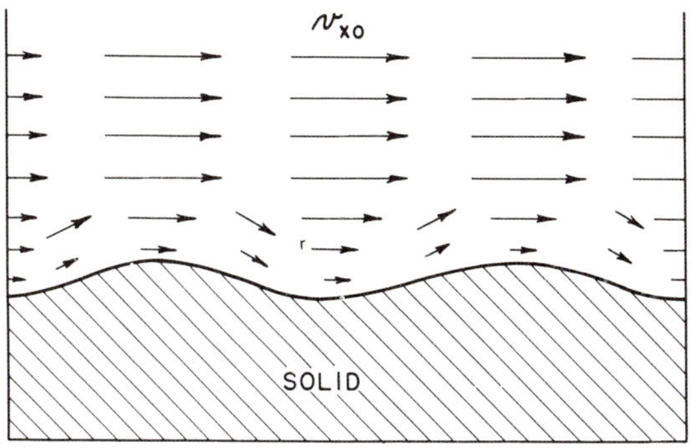

Fig. 15.16. Fluid flow in the boundary layer near a sinusoidally perturbed interface.

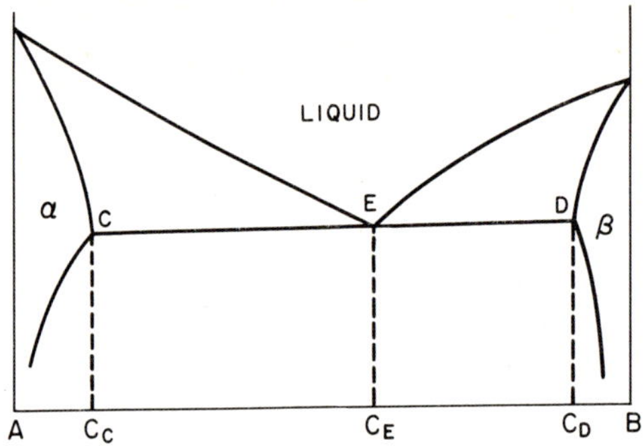

Fig. 15.17. Simple eutectic phase diagram.

refer strictly to the solidification of liquid of composition C_E in fig. 15.17. However, with sufficiently large temperature gradients in the liquid it has been found that coupled "eutectic-like" growth can be obtained for a range of overall compositions roughly between C_C and C_D in fig. 15.17. This has been demonstrated in the Pb–Sn system, for instance, by Mollard and Flemings [29, 30].

The question naturally arises as to the conditions under which coupled "eutectic-like" growth will be stable with respect to the formation of pro-eutectic dendrites and "true eutectic" of composition C_E. This is currently an active research topic where there is little universal agreement. Nevertheless, several treatments are worthy of mention.

About the simplest possible treatment of the problem was that of Mollard and Flemings [29, 30] who simply made an effective single-phase problem out of the actual problem without becoming deeply involved with any detailed theory of eutectic solidification. They assumed that the composition in the liquid at the solid-liquid interface was the eutectic composition C_E and that * c_∞ was both the average composition of the eutectic-like solid and of the liquid far from the solid-liquid interface. Then for purely diffusive transport in the liquid, they obtained an equation analogous to eq. (15.19b), namely

$$C = c_\infty + (C_E - c_\infty) \exp\left[-\frac{Vz}{D}\right]. \tag{15.70}$$

From eq. (15.70) the gradient of composition at the solid-liquid interface may be computed and an equation analogous to eq. (15.34), namely

$$-m(C_E - c_\infty)\frac{V}{D} > G_L, \tag{15.71}$$

* Mollard and Flemings actually used the symbol c_0 instead of c_∞.

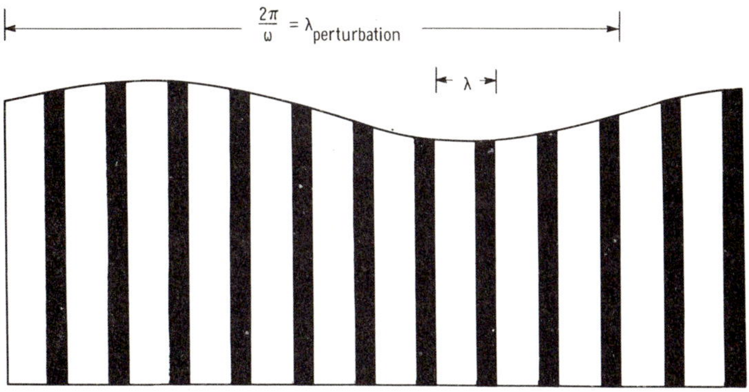

Fig. 15.18. Solid-liquid interface of a lamellar eutectic perturbed with a sinusoid whose wavelength is much larger than the lamellar spacing (after [36]).

is generated to predict the presence of constitutional supercooling. In eq. (15.71), m is the slope of the appropriate liquidus near the point E; consistent with our previous convention on the sign of m, we now have $-m(C_E - c_\infty) \geqslant 0$.

Eq. (15.66) was, in fact, found to be in fair agreement with experiments in the Pb–Sn system [30] and in the Ni–Cr system [31] although detailed quantitative agreement was not obtained. Nevertheless, it is important to recognize that we are dealing here with an instability phenomenon that is closely related to that encountered in the growth of single phase crystals.

Several workers [32, 33] have tried to give a more refined analysis of the morphological stability of lamellar eutectics by combination of the theory of Jackson and Hunt [34] on eutectic solidification with a perturbation analysis of the type we have been discussing. Such theories have only been tractable in the limiting case where the wavelength of the perturbation can be considered to be much larger than the lamellar spacing; this situation is depicted schematically in fig. 15.18. Under such conditions, one can average over the lamellar spacing and obtain an effective single phase problem with "effective single phase" parameters that are somewhat different than those of the simplified model of Mollard and Flemings. However, whereas the theory of Cline [32] leads to an effective liquidus slope that can be inserted in eq. (15.71) to give reasonable agreement with experiment, the work of Hurle and Jakeman [33] and Hunt, Hurle, Jackson and Jakeman [34] has shown that the theory of Cline is in error. In particular, perturbations in the relative spacing of the lamellae were not properly accounted for in Cline's theory; a proper treatment, in fact, predicts that the interface will be stable with respect to perturbations whose wavelengths are large compared to the lamellar spacing. For perturbations whose wavelengths are comparable to the lamellar spacing, no completely satisfactory analysis exists. Nevertheless, Hurle and Jakeman [33] have considered the application of a single-phase analysis to a single lamellae with some success. For a brief critique of the detailed assumptions made by various workers, see references [35] and [36].

15.5. Experimental tests of the theory

In the first four sections, attention was focussed primarily on theoretical aspects of the subjects of morphological stability. Occasional experimental observations were noted but a rigorous comparison of theory and experiment was not presented. We shall therefore devote this section to a discussion of recent experiments that provide some test of the validity of stability theory.

(i) *Tests of dynamical aspects*

The most significant tests of the perturbation analysis of morphological stability have resulted from the excellent work of S.R. Coriell and S.C. Hardy at the National Bureau of Standards [36–41]. This work is not concerned with a measurement of the conditions under which morphological instability will be encountered; instead, it is concerned with a measurement of the dynamical aspects of an unstable interface under conditions where instability is manifest.

Specifically, experiments were conducted on circular cylinders of *ice*, about 0.2 cm in diameter and 6 cm long. Each cylinder was a single crystal of ice, oriented such that the c axis of the ice corresponded to the cylinder axis. They were grown from either distilled water or from aqueous solutions with HCl, NaCl, NH_4OH and similar compounds as solutes. Instabilities developed and were directly observed and photographed.

Figs. 15.19a–d are photographs of an initially circular cylinder of ice growing into supercooled distilled water. They can be understood by reference to our previous equation

$$r = R + \delta_\Phi \cos(k\Phi) + \delta_z \cos(k\Phi)\cos(\omega z) \tag{15.56}$$

and to figs. 15.13–15.15. Fig. 15.19a depicts an unperturbed cylinder for which eq. (15.56) would simply become $r = R$. Fig. 15.19b shows a breakdown into Φ perturbations and should be compared with the schematic drawing in fig. 15.13. Both fig. 15.19b and fig. 15.13 correspond to eq. (15.56) with $\delta_z = 0$; however, fig. 15.19b corresponds to $k = 12$ while fig. 15.13 corresponds to $k = 6$. In general, k is found to be a multiple of 6

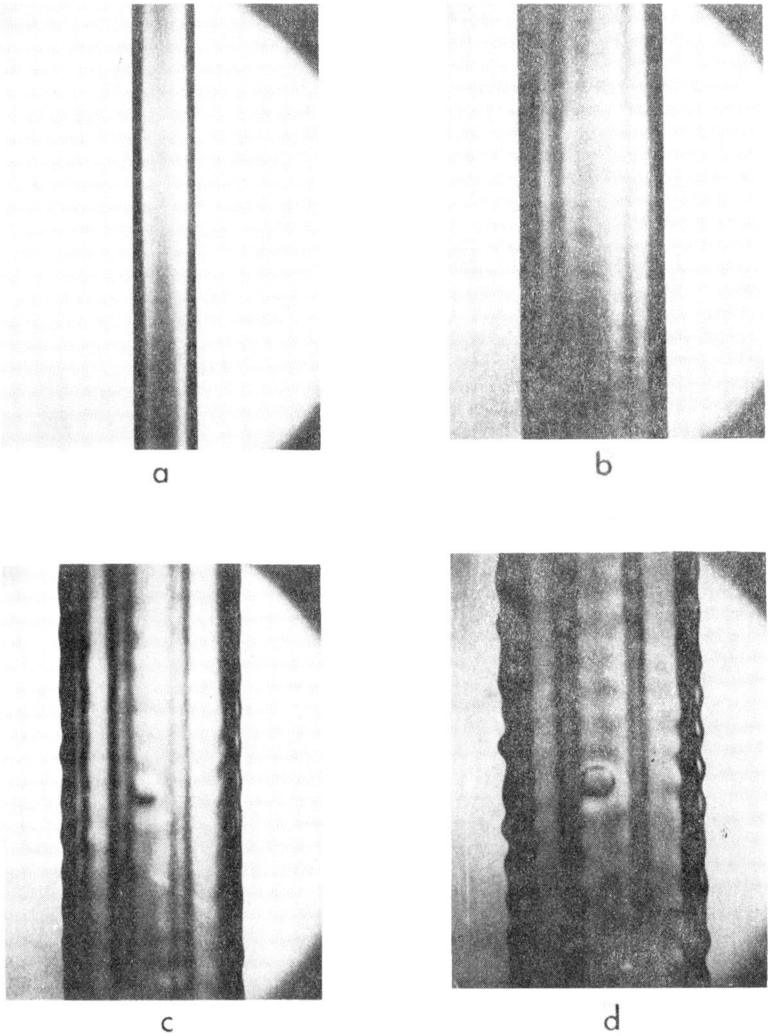

Fig. 15.19. Photographs of initially circular cylinders of ice growing into supercooled distilled water. (a) unperturbed cylinder, (b) cylinder with Φ perturbations only, (c) cylinder with Φ perturbations and small z perturbations, (d) cylinder with Φ perturbations and well developed z perturbations (Courtesy Dr. S.R. Coriell).

which is a consequence [42] of the hexagonal anisotropy of ice in the basal plane. In fact, a perturbed cylinder corresponding to fig. 15.14 is never observed experimentally for ice because this slight anisotropy of growth in the basal plane leads to preferred initial development of the Φ perturbations.

Figs. 15.19c and 15.19d correspond to the full eq. (15.56) and to fig. 15.15 with both Φ perturbations and z perturbations superposed. By the time that the z perturbations have grown to substantial size, as in fig. 15.19d, the Φ perturbations have become quite large and a special non-linear stability theory [43] is necessary for detailed analysis.

In the experiments of Hardy and Coriell, the following quantities can be measured and compared with theoretical predictions:

a) $(\dot{\delta}_\Phi/\delta_\Phi)/(\dot{R}/R)$, the relative growth rate of the Φ perturbations.

b) $(\dot{\delta}_z/\delta_z)/(\dot{R}/R)$, the relative growth rate of the z perturbations.

c) $\lambda = 2\pi/\omega$, the wavelength of the z perturbations.

Both "growth experiments" and "smoothing experiments" were performed; in the former, perturbations were caused to grow and in the latter, the undercooling was reduced to zero and well-developed perturbations were caused to decay. If the solid-liquid surface tension is taken as known, calculated values of the relative growth rates, as in a) and b) above agree with experiment to within 10% and 5% respectively; this is consistent with experimental accuracy. If, on the other hand, the relative growth rates are measured and combined with the predictions of morphological stability theory to calculate a value of the ice-water surface tension, a value $\gamma = 25$ erg cm^{-2} is obtained [43], in satisfactory agreement with values obtained in experiments on homogeneous nucleation. Somewhat different values of γ are obtained for aqueous solutions but these values are consistent for a given solute.

Theory predicts that the wavelength of the fastest-growing Fourier component of the z perturbations should vary inversely as the square root of the bath undercooling. Comparison with measured wavelengths of the z perturbations is in agreement with this predicted dependence on undercooling and with the theoretically-predicted constant of proportionality. The agreement is quite satisfactory considering the nature of the experiments.

The experiments of Hardy and Coriell are therefore in strong support of the dynamical aspects of morphological stability theory.

Other workers have also lent support to the validity of the dynamical aspects of the theory although the applicability of the theory to their experiments is somewhat less straightforward. Noteworthy were the measurements of Seidensticker and Hamilton [44] of the spacing of facets on germanium dendrites and the measurements of Townsend and Kirkaldy [45] on the spacing of Widmanstätten plates in hypo-eutectoid steels; these workers obtained rough agreement with theory with the aid of a few reasonable, but unproven, assumptions.

(ii) *Tests of the onset of instability*

Another experimental test of the theory of morphological stability consists in a determination, and comparison with the stability criterion (eq. (15.46)), of the growth conditions that mark the onset of instability. Many early experiments of this nature were, in fact, directed toward a test of the constitutional supercooling principle and are discussed in [10]. Suffice it to say that these early experiments confirmed the general trend of the constitutional supercooling principle but were insufficiently precise to enable us, at this date, to decide between the predictions of the constitutional supercooling principle and those of stability theory. We shall therefore confine our remarks below to a discussion of experiments conducted since the advent of stability theory.

a) Dilute alloys of Sb in Pb

Morris and Winegard have conducted experiments on the unidirectional solidification of dilute alloys of Sb in Pb in order to investigate the development of a cellular structure [46]. They determined the onset of instability by the accompanying solute segregation which they revealed by etching and also by autoradiography. Their results indicated that instability first occurs at defects, e.g. striation boundaries and grain boundaries, at a growth speed V_S^*. Then, at a somewhat higher growth speed, V_I^*, instability occurs at defect-free regions of the interface. Typically, $V_S^* \approx 0.8\ V_I^*$. This can be understood if one takes V_S^* to be associated with a criterion for the onset of solute-segregation in a steep walled groove as calculated by Tiller [47], and V_I^* to be associated with eq. (15.46). Tiller's [47] criterion resembles the constitutional supercooling criterion, eq. (15.35), except G_L is replaced by G_S. One therefore, has, approximately,

$$0.8 \approx \frac{V_S^*}{V_I^*} \approx \frac{G_S}{\frac{1}{2}(\mathcal{G}' + \mathcal{G})} \approx \frac{G_S}{\frac{2}{3}G_S + \frac{1}{3}G_L} \tag{15.72}$$

which is in agreement with measured values of the gradients.

b) Succinonitrile ($CNCH_2CH_2CN$)

Succinonitrile is a transparent organic material with a low entropy of fusion and a low melting point (54.5°C); it has been observed to freeze like a metal [1]. Schaefer and Glicksman [48] have observed the transition from planar to dendritic solidification in this substance by direct observation of the solid-liquid interface. No quantitative measurements were reported but instabilities were always found to develop first at grain boundaries and particulate matter on the solid-liquid interface. These direct observations tend to substantiate the aforementioned findings, based on solute segregation, of Morris and Winegard [46].

c) dilute alloys of In in Sn

Davis and Fryzuk [49] have employed autoradiography to monitor the transition from stability to instability in dilute alloys of In in Sn. Their results were compared with both the principle of constitutional supercooling and the results of morphological stability theory. Experiments at relatively high growth rates ($V = 5.3 \times 10^{-3}$ cm/sec) and low temperature gradients (G_L = (1.0°C/cm) were particularly sensitive to differences in the two theories. The results indicate that solute segregation, and hence interface instability, is present whenever eq. (15.46) predicts stability. However, if the effect of surface tension as a stabilizing force is neglected, $\mathcal{S} = 1$ in eq. (15.46) and one cannot then decide between eq. (15.46) and eq. (15.35). These results are, therefore, consistent with the hypothesis that defects play an important role in the initiation of morphological instability.

d) alloys of Bi in Sn

Hecht and Kerr [50] have investigated the transition from stability to instability in alloys of Bi in Sn and have found, contrary to the above findings, that the interface was more stable than predicted by either constitutional supercooling or stability theory. While it is possible that Hecht and Kerr's etching techniques failed to reveal solute segregation in some cases where it was actually present, an alternative possibility is that interface kinetics have led to some additional stabilization. Indeed, bismuth has an "intermediate" entropy of fusion [1] $(L/RT_M) = 2.4$, and solidifies with a faceted interface [1, 50]. Therefore some degree of interface control is obviously present, perhaps of sufficient magnitude to account for the enhanced stability [17–24, 51].

(iii) *Stability at a melting interface*

Up to now we have only discussed the stability at a freezing interface where there is some difference, but not a large experimentally realizable difference, in the predictions of the constitutional supercooling principle and stability theory. However, if stability at a melting interface is considered, the principle of "constitutional superheating" and the perturbation theory of instability lead to extremely different predictions [52, 53]. Under conditions where surface tension effects are not important, the results of the various theories can be summarized as follows (subscript L denotes liquid and subscript S denotes solid; all conditions are for stability);

a) perturbation theory, freezing

$$\frac{\mathcal{G}' + \mathcal{G}}{2} > \frac{(1-k)}{k} m_L c_\infty \left(\frac{V}{D_L}\right); \tag{15.73}$$

b) perturbation theory, melting

$$\frac{\mathcal{G}' + \mathcal{G}}{2} > (1-k) m_L c_\infty \left(\frac{V}{D_L}\right) ; \tag{15.74}$$

c) constitutional supercooling (freezing)

$$G_L > \frac{1-k}{k} m_L c_\infty \left(\frac{V}{D_L}\right) ; \tag{15.75}$$

d) constitutional superheating (melting)

$$G_S > (1-k) m_S c_\infty \left(\frac{V}{D_S}\right) = \frac{1}{k} \frac{D_L}{D_S} (1-k) m_L c_\infty \left(\frac{V}{D_L}\right) \tag{15.76}$$

Comparison of eqs. (15.74) and (15.76) shows that perturbation theory (melting) predicts a much more stable interface than constitutional superheating. If the difference between $\frac{1}{2}(\mathcal{G}' + \mathcal{G})$ and G_S is ignored, it is seen, in order to insure stability, that constitutional superheating requires a gradient that is a factor of D_L/kD_S larger than that required by perturbation theory. This great difference occurs because it is diffusion in the liquid, and not diffusion in the solid that determines the concentration of solute at the solid-liquid interface.

Experiments on the morphological stability at melting interfaces of mixtures of carbon tetrabromide (CBr_4) and hexachloroethane (C_2Cl_6) were conducted by Chen and Jackson [53] and found to agree with the perturbation analysis of stability. For those experiments, the factor $D_L/kD_S \approx 100$ so the results are quite decisive.

Acknowledgement

The author would like to acknowledge Dr. S.R. Coriell for many stimulating comments and the National Science Foundation (Grant GK–23871) for financial support during the preparation of this manuscript.

References

[1] K.A. Jackson and D.R. Uhlman, J. Crystal Growth 1 (1967) 1.
[2] R.F. Sekerka and M.E. Glicksman, in: Problems in materials science, ed. by H.S. Merchant (Gordon and Breach, to be published).
[3] H.S. Carslaw and J.C. Jaeger, in: Conduction of heat in solids (Clarendon Press, Oxford, 1959), ch. 11.
[4] J.W. Rutter and B. Chalmers, Can. J. Phys. 31 (1953) 51.
[5] W.A. Tiller, J.W. Rutter, K.A. Jackson and B. Chalmers, Acta Met. 1 (1953) 428.
[6] W.W. Mullins and R.F. Sekerka, J. Appl. Phys. 35 (1964) 444.
[7] R.F. Sekerka, in: Crystal Growth, ed. by H.S. Peiser (Pergamon Press, New York, 1967), p. 691.
[8] R.T. Delves, Phys. Stat. Sol. 16 (1966) 621 and 17 (1966) 119.
[9] R.F. Sekerka, J. Appl. Phys. 36 (1965) 264.
[10] R.F. Sekerka, J. Crystal Growth 3, 4 (1968) 71.
[11] R.F. Sekerka, J. Crystal Growth 10 (1971) 239.
[12] W.W. Mullins and R.F. Sekerka, J. Appl. Phys. 34 (1963) 323.
[13] S.R. Coriell and R.L. Parker, J. Appl. Phys. 36 (1965) 632.
[14] G.R. Kotler and W.A. Tiller, J. Crystal Growth 2 (1968) 287.
[15] V.V. Voronkov, Soviet Physics Solid State 6 (1965) 2378.
[16] R.F. Sekerka, J. Phys. Chem. Solids 28 (1967) 983.
[17] R.G. Seidensticker, J.Electrochem. Soc. 113 (1966) 152.
[18] R.G. Seidensticker, in: Crystal growth, ed. by H.S. Peiser (Pergamon Press, New York, 1967), p. 733.
[19] J.W. Cahn, in: Crystal growth, ed. by H.S. Peiser (Pergamon Press, New York, 1967), p. 681.
[20] S.R. Coriell and R.L. Parker, in: Crystal growth, ed. by H.S. Peiser (Pergamon Press, New York, 1967), p. 703.
[21] L.A. Tarshis and W.A. Tiller, in: Crystal growth, ed. by H.S. Peiser (Pergamon Press, New York, 1967), p. 709.
[22] G.R. Kotler and W.A. Tiller, in: Crystal growth, ed. by H.S. Peiser (Pergamon Press, New York, 1967), p. 721.
[23] P.G. Shewmon, Trans. AIME 233 (1965) 736.
[24] S.R. Coriell and R.L. Parker, J. Appl. Phys. 37 (1966) 1548.
[25] D.T.J. Hurle, Solid State Electronics 3 (1961) 37 and W. Bardsley, J.M. Callan, H.A. Chedzey and D.T.J. Hurle, Solid State Electronics 3 (1961) 142.
[26] D.T.J. Hurle, J. Crystal Growth 5 (1969) 162.
[27] R.T. Delves, J. Crystal Growth 3, 4 (1968) 562.
[28] R.T. Delves, J. Crystal Growth 8 (1971) 13.
[29] F.R. Mollard and M.C. Flemings, Trans. AIME 239 (1967) 1526.
[30] F.R. Mollard and M.C. Flemings, Trans. AIME 239 (1967) 1534.
[31] E.H. Myers and R.H. Hopkins, J. Crystal Growth 7 (1970) 231.
[32] H.E. Cline, Trans. AIME 242 (1968) 1613.
[33] D.T.J. Hurle and E. Jakeman, J. Crystal Growth 3, 4 (1968) 574.
[34] K.A. Jackson and J.D. Hunt, Trans. AIME 236 (1966) 1129.
[35] J.D. Hunt, D.T.J. Hurle, K.A. Jackson and E. Jakeman, Met. Trans. 1 (1970)318.
[36] R.F. Sekerka, Bull. Soc. Fr. Minér. Crist. 92 (1969) 540.

[37] S.C. Hardy and S.R. Coriell, J. Appl. Phys. 39 (1968) 3505.
[38] S.R. Coriell and S.C. Hardy, J. Appl. Phys. 40 (1969) 1652.
[39] S.C. Hardy and S.R. Coriell, J. Crystal Growth 3, 4 (1968) 569.
[40] S.C. Hardy and S.R. Coriell, J. Crystal Growth 5 (1969) 329.
[41] S.C. Hardy and S.R. Coriell, J. Crystal Growth 7 (1970) 147.
[42] S.R. Coriell and S.C. Hardy, J. Res. Nat. Bur. Stand. 73A (1969) 65.
[43] S.R. Coriell, S.C. Hardy and R.F. Sekerka, J. Crystal Growth 11 (1971) 53.
[44] R.G. Seidensticker and D.R. Hamilton, J. Appl. Phys. 34 (1963) 3113.
[45] R.D. Townsend and J.S. Kirkaldy, Trans. ASM 61 (1968) 605.
[46] L.R. Morris and W.C. Winegard, J. Crystal Growth 5 (1969) 361.
[47] W.A. Tiller, J. Appl. Phys. 33 (1962) 3106.
[48] R.J. Schaefer and M.E. Glicksman, Met. Trans. 1 (1970) 1973.
[49] K.G. Davis and P. Fryzuk, J. Crystal Growth 8 (1971) 57.
[50] M.V. Hecht and H.W. Kerr, J. Crystal Growth 7 (1970) 136.
[51] J.C. Brice, J. Crystal Growth 6 (1969) 9.
[52] D.P. Woodruff, Phil. Mag. 15 (1967) 283.
[53] H.S. Chen and K.A. Jackson, J. Crystal Growth 8 (1971) 184.

P. Hartman, ed., Crystal growth: an introduction. © (1973) North-Holland Publ. Co.

THE PROPERTIES AND OBSERVATION OF DISLOCATIONS

A.R. LANG

The presence of dislocations is the most common reason for the behaviour of real crystals departing from that expected of ideally perfect crystals. Discussion of crystal perfection therefore commences with a summary, in Section 1, of the more important properties of dislocations. Experimental methods for detecting dislocations are reviewed in Section 2. Dislocation observation techniques based upon electron or X-ray diffraction contrast can be more efficiently applied when at least some of the underlying diffraction theory is familiar to the experimenter. Section 3 sets forth a little basic diffraction theory which is common to both electrons and X-rays. It then describes certain diffraction phenomena which manifest themselves in X-ray studies of perfect crystals. Discussion of X-ray diffraction phenomena continues in Section 4 which deals first with the X-ray diffraction contrast images of individual dislocations and concludes with X-ray topographic illustrations of dislocation configurations in natural and synthetic crystals.

16.1. Crystal dislocations

16.1.1. The study of dislocations

Since 1945 dislocation theory has grown into an impressive body of knowledge, intricate in the detail of some of its ramifications. Many developments have been directed towards understanding mechanical problems of long-standing concern such as work-hardening, fracture and creep. Other have been prompted by direct observations of individual dislocations. For example, the discovery of beautifully regular dislocation helices, which were first revealed by "dislocation decoration" techniques, led to much theoretical analysis of interactions between dislocations and point defects such as vacancies. The thin-film transmission electron-microscope method has for many years provided a stream of novel observations at a rate which has always outstripped the supply of explanations devised by theoreticians, impressive both in number and in thoroughness though their efforts have been. The high resolving power of the electron microscope permits dense and complex dislocation configurations to be observed, but always these lie in close proximity to a specimen surface. Hence theoretical analyses of interactions between dislocations and free surfaces or other types of interface, and of dislocation reactions in both simple and complex crystal structures, have been consequences of stimulus from electron microscope observations. Studies currently expanding apace concern equilibrium dislocation configurations in elastically anisotropic crystals. This is again a development of theory in attempt to explain the electron microscope evidence. Complexities such as those just mentioned the introductory account in the following paragraphs must perforce ignore. Only basic properties of dislocations are described; and the selection of topics for discussion has been made with the interests of the crystal-grower and of the assessor of crystal perfection particularly in mind.

There are many good texts on dislocations. Here is an indication of the scope of some of the English language books on the subject. Good introductions to the study of dislocations, available in pocket-sized editions, are those by Hull [1] and by Weertman and Weertman [2]. The first three chapters of a well-established book by Cottrell [3] also offer an excellent introduction to the subject. Another well-established book, by Read [4], is designed for selfteaching. A longer book, by Friedel [5], covers a wide range of topics in an easily readable fashion. More comprehensive still is the text by Hirth and Lothe [6], in which the reader is led from simple classical elasticity theory right through to dislocation problems of considerable complexity. For later reading, Cottrell [7] provides a compact and precise summary of basic theory, and the authoritative treatise by Nabarro [8] is recommended for its combination of clarity with completeness and rigour. Valuable is the recent comprehensive review by Christian and Vítek [9].

The dislocation topics discussed below fall into two divisions. Sections 1.2 to 1.5 describe the geometrical properties of dislocations. Sections 1.6 to 1.12 deal with the elastic stresses in dislocated crystal lattices and the forces on dislocations that arise therefrom.

16.1.2. The need for dislocations

It was the remarkable properties of *"glide"* or *"slip"* exhibited during the *plastic deformation* of large, fairly pure, single crystals of simple-structured metals that led during the 1930's to the models of crystal dislocations nowadays familiar. The properties of this "glide" are as follows. (1), during glide one part of a crystal slides bodily relative to a neighbouring part: ideally the parts undergo relative translation only, with absence of relative rotation, (2), glide motion occurs most readily along lattice rows with small (or smallest) translation vectors (e.g. in face-centred cubic (f.c.c.) metals glide directions are ⟨110⟩, parallel to the primitive translation vectors), (3), the surface of slip is often a low-index crystallographic plane: in close-packed structures it will be the plane within which there is closest packing, e.g. (0001) in hexagonal close-packed (h.c.p.), and belonging to the form {111} in f.c.c. metals. (4), microscopic examination of the surface of the deformed crystal usually shows that deformation has been concentrated in localised slip-bands. Increasing resolution (optical microscope → scanning electron microscope → transmission electron microscopy of surface replicas) may reveal finer and finer slip steps within slip bands, but it remains generally true that, within a slip-band width of but a few inter-atomic distances, magnitudes of slip amounting to tens or hundreds of inter-atomic distances has occurred, (5), for pure crystals with close-packed structures, slip in "easy-glide" directions can commence when the "resolved shear stress" on the slip-plane is as low as $10^{-5}\,\mu$ (where μ is the shear modulus). Such a very low "yield stress" is incompatible with the mechanism of *rigid bodily* relative motion of the atom layers on either side of the slip-plane because such would give rise to theoretical shear strengths of not less than about $\mu/30$. This anomaly is removed by a slip mechanism requiring only narrowly localised strips of mis-fitting atoms, i.e. dislocation lines, to sweep one after the other over the slip-plane, each dislocation translating one half of the crystal relative to the other half by a unit lattice translation vector. Intuitively it is seen that this model of the glide process will give rise to low shear strengths: corresponding to each atom that is being raised in potential energy by moving out of an equilibrium position, another atom not far away is losing an equivalent amount of potential energy by falling back, with opposite movement, towards an equilibrium position.

16.1.3. Edge and screw dislocations

Fig. 16.1. shows G.I. Taylor's [10] sketches of slip in a simple cubic crystal resulting from the passage of a pure *edge* dislocation over the slip-plane in either the sequence a → b → c, or the sequence d → e → f. The extra halfplanes of atoms that are characteristic of edge dislocations have been emphasized by solid circles, and the conventional symbols for edge dislocations have been added; the negative dislocation is merely the mirror image in the slip-plane of the positive dislocation. A better picture of atomic displacements surrounding an edge dislocation is shown in fig. 16.2., taken from Nabarro [11]. Note especially (1), above the slip-plane the atoms surrounding the termination of the extra half-plane are in a state of compression, whereas below the slip-plane they are in a state of extension. This provides a simple mechanical driving force for the capture of impurity atoms having radii differing from that of the host lattice atoms. Such capture tends to lock the

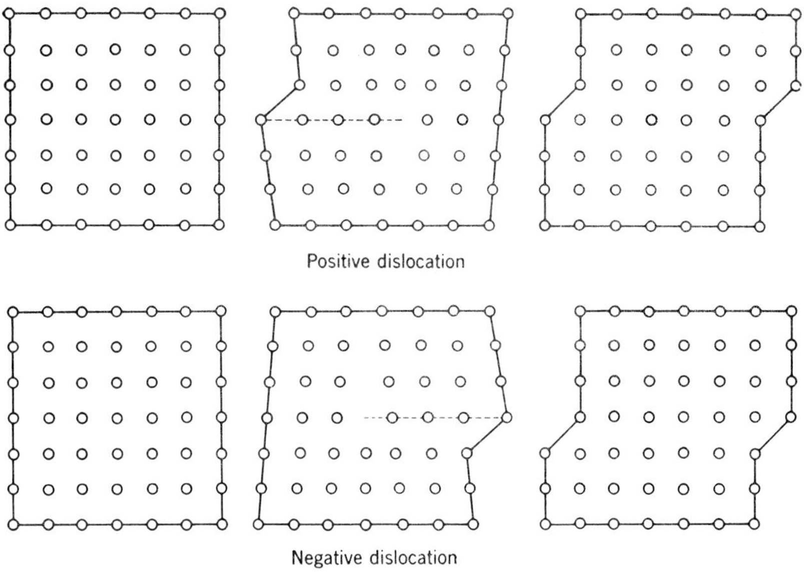

Fig. 16.1. G.I. Taylor's (1934) drawing of positions of atoms during passage of positive and negative edge dislocations over a slip plane (from ref. [10]).

dislocation in position (*"Cottrell locking"*). Note also, (2), that the slip-plane, and also the planes perpendicular both to the slip-plane and to the diagram, possess a characteristic type of curvature. This curvature is of importance in producing "diffraction contrast" with X-rays or electrons.

Fig. 16.3 is a plan view of a slip-plane containing a dislocation loop which encloses an area which has slipped by a unit slip vector (*the Burgers vector,* defined below), the slip occurring as a consequence of externally applied shear stress. The right-hand half of the loop is drawn to represent dislocation line segments of either *pure edge* or *pure screw* character, whereas the left-hand side is drawn as a dislocation line of *mixed character.* Drawings or mechanical models show the greater simplicity of the displacement field around a pure screw dislocation compared with that around an edge, or mixed, dislocation: the displacement field around a pure screw dislocation exhibits cylindrical symmetry about the dislocation line, but strictly so only when the line is *remote* from external crystal surfaces and when the crystal is *elastically isotropic.* Note that the applied shear stress will cause the loop in fig. 16.3 to expand; and if screw segments move *faster* than edge segments then the loop will become predominantly edge in character, and vice-versa. In

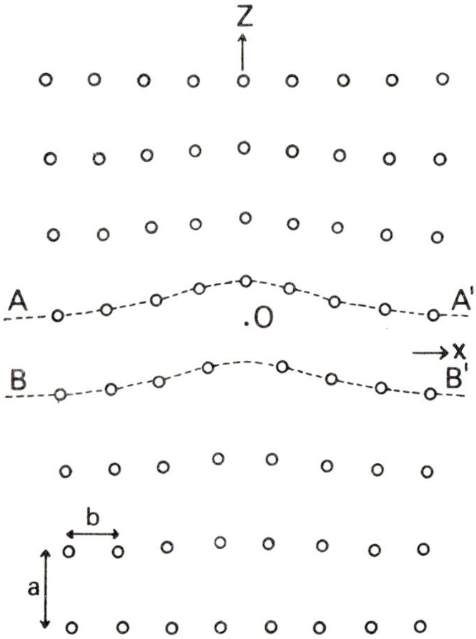

Fig. 16.2. The structure of an edge dislocation in a rectangular lattice, according to Peierl's model. The material above AA′ and below BB′ is treated as an elastic continuum. The force between the rows AA′ and BB′ is a periodic function of the displacement (after Nabarro [11]).

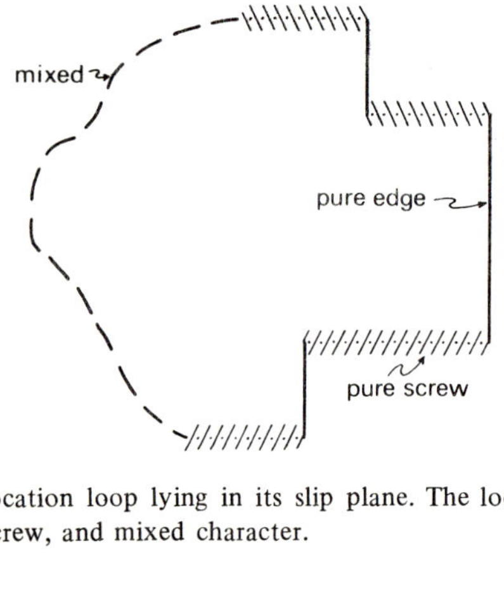

Fig. 16.3. A dislocation loop lying in its slip plane. The loop consists of segments of pure edge, pure screw, and mixed character.

the case of a stationary dislocation line in or near equilibrium configuration, the ratio of lengths of mainly screw to mainly edge segments, and the dimensional scale of such segments, are characteristics determined largely by the crystal structure, but they are dependent also on temperature and impurity concentration.

16.1.4. Definition of Burgers vector of crystal dislocations

The preferred method of definition (see, for example, Bilby [12], Frank [13] and Cottrell [7]) is as follows. Describe a *closed Burgers circuit* around the dislocation line in the *real* crystal by means of a given sequence of lattice vectors, keeping to a path sufficiently far from the dislocation line so that the chosen sequence of lattice vectors may be unambiguously repeated in a perfect, undislocated *reference* crystal. In the reference crystal the Burgers circuit will *not* close: the closure vector gives the magnitude of the Burgers vector. For *perfect* (or *complete*) dislocations the Burgers vector so defined must be precisely a lattice vector. It will not necessarily be a *primitive translation* vector, but is likely to be so. It remains to define the *sense* of the Burgers vector. Screw dislocations clearly are distinguishable as to sense of hand, so let us use a screw dislocation to illustrate the conventional definition of sense. Fig. 16.4a represents a closed circular path lying in a lattice plane (*hkl*). The circle is centred on a point O, and its diameter is large compared with the lattice translation vectors. Now introduce a screw dislocation line perpendicular to the plane, piercing it at O, so as to transform the plane into an infinite helicoidal surface. Suppose the dislocation be of such hand as to transform the circle of fig. 16.4a into a right-handed helix one turn of which is shown in fig. 16.4b. Now perform the Burgers circuit in the dislocated crystal as follows. First choose arbitrarily a positive direction for the unit vector **l** which is tangent to the dislocation line e.g. into the paper in fig. 16.4c. Then make the circuit in a *clockwise* sense when looking in the direction to which **l** points. Suppose the circuit starts and finishes at P. Next make the same sequence of lattice vector steps in the reference crystal, starting at P′ (corresponding to P) but now finishing at Q′ (fig. 16.4d). The closure vector Q′ → P′ defines the Burgers vector **b** in both sense and magnitude. For a *right-handed* screw dislocation **l** and **b** are similarly directed, and **l**·**b** is *positive*. This makes a reasonable convention for the sense of Burgers vector. Corresponding illustrations for sense definition in the pure edge case are given by Bilby [12] and Cottrell [7]. See also Bilby *et al.* [14] regarding sense definition.

Note, (1), the vector **b** is an invariant characteristic of a dislocation line, provided the line does not branch, or dissociate into *imperfect* (or *partial*)

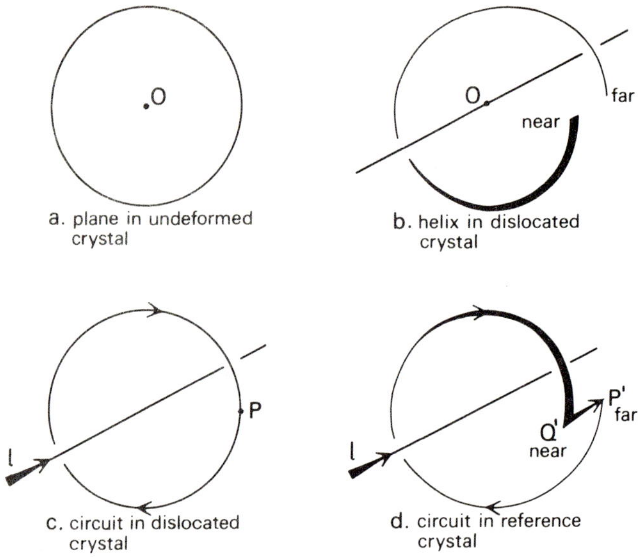

Fig. 16.4. Convention for defining Burgers vector according to Frank. a) a closed path lying in a perfect atomic plane in the undislocated crystal. b) when a dislocation line pierces the atomic plane at O and has a component of Burgers vector normal to the atomic plane the path corresponding to (a) becomes a turn of a helix. c) the Burgers circuit is performed in the dislocated crystal starting and finishing at P. d) the circuit corresponding step-by-step with circuit (c) is performed in the perfect, reference crystal. The Burgers vector is the additional step Q′ to P′ required to close the circuit.

dislocations. Also, (2), a dislocation line cannot terminate within the crystal medium: it must either form a closed loop, or run from one crystal surface to another. (These properties may be recognised intuitively but they are discussed more rigorously by Nabarro [8]). Next, (3), consider a dislocation that does branch; if one takes the sense of the unit vectors $\mathbf{l}_i$ as positive when directed towards the branch point, or *node*, it follows that $\Sigma \mathbf{b}_i = 0$, (Frank, [13]) according to the Burgers vector definition procedure given above.

16.1.5. Dislocation sources

Application of a suitably directed external shear stress to crystals which contain a dislocation of the type drawn in figs. 16.1 or 16.3 will cause the dislocation to sweep out of the crystal and bring the process of glide to a stop. But, as pointed out in § 1.2 (note (4)), there is experimental evidence for very many dislocations having swept a given lattice plane (or closely

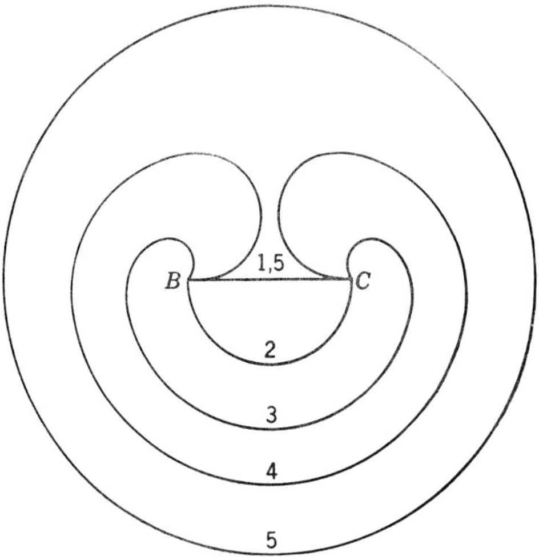

Fig. 16.5. The symmetrical Frank-Read dislocation source. Positions of the dislocation line between B and C follow the sequence 1 to 5 (from Nabarro [11]).

adjacent set of planes). Frank and Read [15] devised a model for an *internal source* of dislocations, capable in principle of operation ad infinitum. Fig. 16.5 (taken from Nabarro [11]) illustrates how the symmetrical Frank-Read source operates. The *pinning points* B and C could alternatively be dislocation nodes or impurity particles. Several variants of this source model have been proposed. For an account of these, see, e.g. Nabarro [8].

16.1.6. Dislocation dynamics

In order to proceed further in understanding the behaviour of dislocations it is necessary to know the stress fields that dislocations produce within an elastic medium. From these fields the elastic self-energy of the dislocation may be calculated, together with the interaction energy between dislocations, or between dislocations and crystal surfaces. When these energies are known, the forces acting on dislocations in particular configurations can be calculated, and this leads in turn to an understanding of why certain configurations, such as, for example, the low-angle tilt boundary, are often observed in real crystals. However, two characteristics of dislocation lines which are basic in determining dislocation configurations may be established by simple arguments. These characteristics are the force on a dislocation line due to an applied shear stress, and the tendency of dislocation lines to shorten themselves.

16.1.7. Force on a dislocation

The force F, acting on unit length of a dislocation whose Burgers vector is b when a shear stress σ is applied parallel to b is

$$F = \sigma b \tag{16.1}$$

The explanation is as follows. Suppose the total slip-plane area is A and suppose a segment ds of dislocation loop moves normal to itself by a distance dl so as to increase the slipped area by the rectangular element of area dsdl. Then, by simple proportion, the macroscopic slip displacement of the crystal as a whole is $b\mathrm{d}l\mathrm{d}s/A$. Since the externally applied force is σA the work done is $(\sigma A)(b\mathrm{d}l\mathrm{d}s/A)$, and this must equal $F\mathrm{d}s\mathrm{d}l$, the work done by F in moving the segment ds a length dl normal to ds. Hence follows equation (16.1).

16.1.8. Line tension of dislocations

Take rectangular coordinates Ox, Oy in the slip-plane. Suppose a dislocation initially lying straight and parallel to Ox is displaced in positive and negative y directions to form a zig-zag whose period is X measured along the x-axis and whose peak-to-peak y-coordinate range is $2Y$, the ratio Y/X being fairly small. At distances R from the dislocation which are large compared with X the strain *energy density* will be unaffected by the change from straight to zig-zag shape, but at distances R very close to the line ($R \ll X$) the *energy density* will be the same as that at the same small distance R from the line in its previous configuration. Thus the *total strain energy* of this zig-zag dislocation contains a term proportional to the total line length L (where $L^2 = X^2 + Y^2$), this term being the contribution to the strain energy arising in the volume within a radius $R \approx X$, and which contribution is obviously a function of X. However, this function is of the form $\ln(X/r_0)$ and so varies only by a small factor for a wide range of possible zig-zag periods X. (The radius r_0 is of order atomic dimensions). Thus the *increase*, ΔE, in energy resulting from the introduction of the zig-zag configuration is proportional to the *increase*, ΔL, in line length, for a given X. This is equivalent to a *constant line tension* $T = \Delta E/\Delta L$. It will be realised that expressions for T must bear a close relation to expressions for the total elastic energy of long, straight dislocations, which will be considered in ch. 16 § 1.11. At this point there will just be quoted an estimate of T, as may be seen derived by Cottrell [7] or Friedel [5]:

$$T \approx \frac{\mu b^2}{2\pi} \ln(X/r_0) . \tag{16.2}$$

Analogy with surface tension, or simple direct calculation (see Cottrell [3] or Cottrell [7]), show that the shear stress required to bend to radius r a dislocation stretched between two pinning points is

$$\sigma_c = T/br \,. \qquad (16.3)$$

It follows therefore that a certain minimum shear stress is required to operate a Frank-Read source: with the model of fig. 16.5 it is seen that the appropriate minimum value of r is about $\frac{1}{2}$BC and occurs when the dislocation line is in the position numbered 2.

16.1.9. Displacement and stresses due to a screw dislocation

Here, and in the following discussion of edge dislocations, only straight dislocations will be considered. Expressions will be given for the displacements and stresses in a hollow circular cylinder centred on the dislocation line. The inner radius, r_0, the *core* radius, may be identified with the distance from the location line at which stresses exceed the limit of Hookean behaviour or, as is virtually equivalent, the radius within which specification of the strain requires description of the actual atomic positions in a highly imperfect crystal region. The outer radius, R, is of order of the dimensions of the crystal, or of the average distance to adjacent dislocations, whichever is the smaller. Between r_0 and R the medium is taken to be an isotropically elastic continuum. The assumption of isotropic elastic properties applies well with crystals such as aluminium and tungsten, but cases are encountered when elastic anisotropy must be used to explain not only the mechanical behaviour of dislocations but also the X-ray and electron diffraction contrast effects they produce.

Take the straight dislocation as lying along the z axis. Whether Cartesian coordinates (x,y,z) or cylindrical-polar coordinates (r,θ,z) are to be preferred depends upon the problem concerned. With straight dislocations, parallel to the z-axis, the displacements do not vary with z. Calling u, v and w the displacements in the x,y and z directions, respectively, the helicoidal displacement field around a pure screw dislocation with **l** pointing in the positive **z** direction and **b** in the negative z direction is given by

$$\begin{aligned} u &= 0 \\ v &= 0 \\ w &= \frac{b}{2\pi}\,\mathrm{arc\,tan}\,(y/x) = \frac{b\theta}{2\pi}\,. \end{aligned} \qquad (16.4)$$

The strain components, and hence the stresses, are found by differentiating u, v and w. Only shear stresses are produced:

$$\sigma_{xz} = \frac{-\mu b}{2\pi}\left(\frac{y}{x^2+y^2}\right)$$

$$\sigma_{yz} = \frac{\mu b}{2\pi}\left(\frac{x}{x^2+y^2}\right) \tag{16.5}$$

or

$$\sigma_{\theta z} = \frac{\mu b}{2\pi r}\,.$$

The last equation displays the cylindrical symmetry and shows clearly the slow rate of decrease of stress with increasing r.

16.1.10. Displacement and stresses due to an edge dislocation

The complications attending the case of the edge dislocation are outlined by Cottrell [3], Read [4], and Weertman and Weertman [2]. The situation is one of *plane strain* ($w = 0$). Consider the positive dislocation, as defined by Taylor (fig. 16.1b), with $\mathbf{l}$ again in the positive z direction and directed towards the reader. The slip plane is $y = 0$ and the Burgers vector has components $(b,0,0)$. The displacement field can be expressed as

$$u = \frac{b}{2\pi}\left[\arc\tan\left(\frac{y}{z}\right) + \frac{1}{2(1-\nu)}\frac{xy}{x^2+y^2}\right]$$

$$= \frac{b}{2\pi}\left[\theta + \frac{\sin 2\theta}{4(1-\nu)}\right],$$

$$v = \frac{-b}{8\pi(1-\nu)}\left[(1-2\nu)\ln\frac{x^2+y^2}{b^2} - \frac{2y^2}{x^2+y^2}\right] \tag{16.6}$$

$$= \frac{-b}{4\pi}\left[\frac{1-2\nu}{(1-\nu)}\ln(r/b) - \frac{\sin^2\theta}{(1-\nu)}\right],$$

in which ν is Poisson's ratio. Some variation in expressions for u and v to be found in current texts arises from different choice of integration constants: the displacement field has no symmetry elements that suggest a unique choice of such constants.

The stresses are

$$
\begin{aligned}
\sigma_{xx} &= \frac{-\mu b}{2\pi(1-\nu)} \frac{y(3x^2+y^2)}{(x^2+y^2)^2} \\
\sigma_{yy} &= \frac{\mu b}{2\pi(1-\nu)} \frac{y(x^2-y^2)}{(x^2+y^2)^2} \\
\sigma_{zz} &= \nu(\sigma_{xx}+\sigma_{yy}) = \frac{-\mu\nu b y}{\pi(1-\nu)(x^2+y^2)} \\
\sigma_{xy} &= \frac{\mu b}{2\pi(1-\nu)} \frac{x(x^2-y^2)}{(x^2+y^2)^2} \\
\sigma_{xz} &= \sigma_{yz} = 0
\end{aligned}
\tag{16.7a}
$$

Note the non-zero value of σ_{zz}, even though $w = 0$. In the cylindrical-polar coordinates the expressions take a simpler form

$$
\begin{aligned}
\sigma_{rr} &= \sigma_{\theta\theta} = \frac{-\mu b}{2\pi(1-\nu)} \frac{\sin\theta}{r} \\
\sigma_{zz} &= \frac{-\mu\nu b}{\pi(1-\nu)} \frac{\sin\theta}{r}
\end{aligned}
\tag{16.7b}
$$

and

$$\sigma_{r\theta} = \frac{\mu b}{2\pi(1-\nu)} \frac{\cos\theta}{r}$$

A drawing that helps to visualise the stress field is given by Read [4] and is reproduced in fig. 16.6.

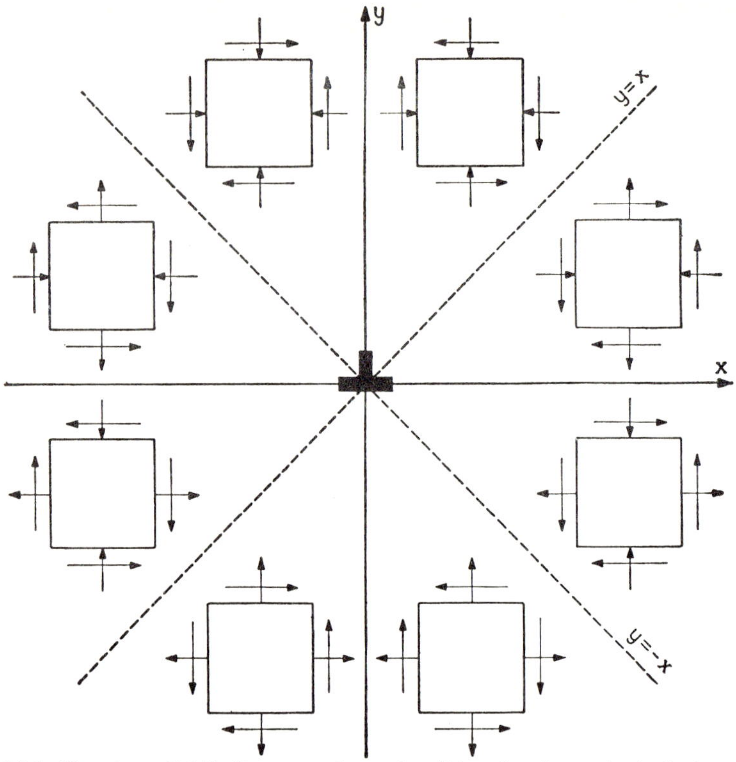

Fig. 16.6. The stress distribution around an edge dislocation in an elastically isotropic crystal, equations (16.7a) (from Read [4]).

16.1.11. Energy of dislocations

The elastic energy of a dislocation is the sum of, (1), the volume integral of the product ½ × stress × strain outside the core radius r_0, plus, (2), the energy of the highly distorted "bad" crystal within the radius r_0. The integral (1) can be calculated from the expressions for stresses, equations (16.5) and (16.7), using equations of type:

$$\text{elastic energy} = \tfrac{1}{2}\sum_i M_i e_i^2$$

where M_i is the appropriate elastic modulus for the strain of type e_i. However, the elastic energy of a pure screw dislocation may be quite simply found. Consider unit length of cylinder whose axis is the dislocation line. The strain energy in a cylinder shell between radii r and $r + \mathrm{d}r$ is $\frac{1}{2}\mu(b/2\pi r)^2\, 2\pi r\mathrm{d}r$ and hence the total strain energy is

$$E_\mathrm{s} = \frac{\mu b^2}{4\pi}\int_{r_0}^{R}\frac{\mathrm{d}r}{r} = \frac{\mu b^2}{4\pi}\ln\left(\frac{R}{r_0}\right). \tag{16.8}$$

A corresponding calculation may be made for a pure edge dislocation lying along the cylinder axis: in this case the cylindrical shells are cut along the slip-plane $y = 0$ and opposite sides of the cut made to slip the distance b relative to each other against the stress σ_{xy} (e.g. 16.7a) or $\sigma_{r\theta}$ (e.g. 16.7b). The total work done is

$$E_e = \frac{\mu b^2}{4\pi(1-\nu)} \ln\left(\frac{R}{r_0}\right), \tag{16.9}$$

giving $E_e \approx \frac{3}{2} E_s$ with $\nu \approx \frac{1}{3}$.

A mixed dislocation of Burgers vector **b** can be decomposed into screw and edge components, $\mathbf{b} = \mathbf{b}_e + \mathbf{b}_s$, where obviously the magnitudes b_s and b_e are given by $|\mathbf{b} \cdot \mathbf{l}|$ and $|\mathbf{b} \times \mathbf{l}|$ respectively. The energy E_m of a mixed dislocation is the sum $E_s + E_e$, the two energy components being evaluated for the appropriate Burgers vector components b_s and b_e.

Let the inner and outer "cut-off" radii, r_0 and R, be taken respectively as 10^{-7} cm and 0.1 cm, giving $\ln(R/r_0) \approx 12$. Then the contribution to the elastic strain energy of regions between $r = 1$ μm and $r = 1$ mm is as large as the contribution from the core outwards to a radius of one micron. This exemplifies the long-range nature of the stress fields of dislocations and indicates that their elastic energy may be significantly reduced by proximity within distances of μm order from stress-relieving agencies such as external surfaces, cavity surfaces, cracks or favourably directed stress fields of other dislocations.

Dislocation energies, E_s, in typical f.c.c. metals, expressed as electron volts per atom length, are, for example, about 3eV in aluminium and about 8eV in copper. These are in each case about five times the energy of a lattice vacancy, and are many times kT at room temperature ($\approx 1/40$ eV). Estimates of the contribution to the total energy arising from distorted material in the core ($r < r_0$) range from about ½ to 1eV per atom length of dislocation.

16.1.12. Forces between dislocations

The elastic self-energy of dislocations was derived above by taking a perfect crystal, making a cut in it and sliding the faces A of the cut relative to each other against the elastic stress generated by this displacement. The interaction energy between dislocations (say between two dislocations in an otherwise perfect crystal) may be found as follows. Introduce the first dislocation: this entails its self-energy E_1. Next introduce a second dislocation. The energy of the second dislocation can be divided into two parts, its self-energy E_2 which it would have if it were present alone in the crystal, and an interaction energy E_{12} which is the work done by the displacement $\mathbf{u}_2$ of A required in introducing the second dislocation when acting against the force T_1 produced on A by the stress field of the first dislocation. Thus the total energy is $E = E_1 + E_2 + E_{12}$

$$= E_1 + E_2 + \int_A (\mathbf{T}_1 \cdot \mathbf{u}_2)\,\mathrm{d}A\,. \tag{16.10}$$

In general E_{12} will vary with relative position of the dislocations, and there will then be a force acting between them given by

$$\mathbf{F} = -\operatorname{grad} E_{12}\,. \tag{16.11}$$

A case of zero interaction energy occurs with a pure screw dislocation lying parallel to a pure edge dislocation, for then $\mathbf{T}_1$ and $\mathbf{u}_2$ are orthogonal for all relative positions of the two dislocation lines.

Two simple situations of practical importance are the case of parallel screw dislocations and the case of parallel edge dislocations. In the case of parallel screw dislocations the radial symmetry of the stress field of each (for elasticity isotropic media) shows that the force between them is central. If they are of like sense, they repel each other, and vice versa. Orient the plane of the cut required to form the second dislocation so as to contain the first dislocation line. The force per unit length between the dislocation is then

$$F_r = -\frac{\mathrm{d}}{\mathrm{d}r}(\mathbf{u}_2 \cdot \mathbf{T}_1) = \frac{\mathrm{d}}{\mathrm{d}A}(b_2\,\sigma_{\theta z}\,A)$$

$$= \mu b_1 b_2 / 2\pi r\,, \tag{16.12}$$

paying regard to the sense in which the cut A is made.

The case of parallel edge dislocations on parallel slip planes is interesting. Let the two dislocations both have $\mathbf{l}$ parallel to Oz. Place dislocation (1) at the origin with slip-plane $y = 0$ and dislocation (2) at (xy), i.e. on a slip-plane distant y from the dislocation at the origin. First find the interaction energy by making the cut along the slip plane of dislocation (2), then $E_{12} = \int \sigma_{xy} b_2 \mathrm{d}A$. The energy per unit length parallel to $\mathbf{l}$ is then $\int_x^{x'} -\sigma_{xy} b_2\,\mathrm{d}x$, where $x' \approx R$. The negative of the derivative with respect to x gives the force on (2) acting along its slip-plane, i.e. in the x-direction,

$$F_x = \sigma_{xy} b_2 = \frac{\mu b_1 b_2}{2\pi(1-\nu)} \frac{x(x^2 - y^2)}{(x^2 + y^2)^2}\,. \tag{16.13a}$$

Alternatively, introduce dislocation (2) by making the cut parallel to the *yz* plane passing through (x,y), i.e. by pushing in the extra-half plane of atoms all the way from the external surface up to the point (x,y). Per unit length of dislocation line, the interaction energy is then $\int_y^{y'} -\sigma_{xx} b_2 \mathrm{d}y$, where $y' \approx R$; and the negative of the derivative with respect to y gives

$$F_y = \sigma_{xx} b_2 = \frac{\mu b_1 b_2}{2\pi(1-\nu)} \frac{y(3x^2+y^2)}{(x^2+y^2)^2} . \qquad (16.13b)$$

Equations (16.13a) and (16.13b) have important implications for dislocation configurations. Let us examine the cases when b_1 and b_2 are equal in magnitude ($= b$) but either of similar or of opposite sign.

(1) Suppose $y = 0$ so that dislocations (1) and (2) are on the same slip plane. If they are of like sign they repel each other with force inversely proportional to distance of separation. If they are of opposite sign the sign of the force is reversed and they may come together and mutually annihilate each other.

(2) For $y \neq 0$ (and $y \gg r_0$) the force F_x experienced by dislocation (2) (positive F_x directed to increasing x) is plotted in fig. 16.7, the continuous line for dislocations of like sign, the interrupted line for the opposite case. Note that like dislocations repel each other for $y/x < 1$ and attract for $y/x > 1$. F_x is zero both at $x = y$ and $x = 0$, but these are respectively unstable and stable equilibrium positions. The energy minimum at $x = 0$ leads to *polygonisation* and the formation of low-angle tilt boundaries, as discussed later. When the dislocations are of opposite sign, the direction of F_x is of

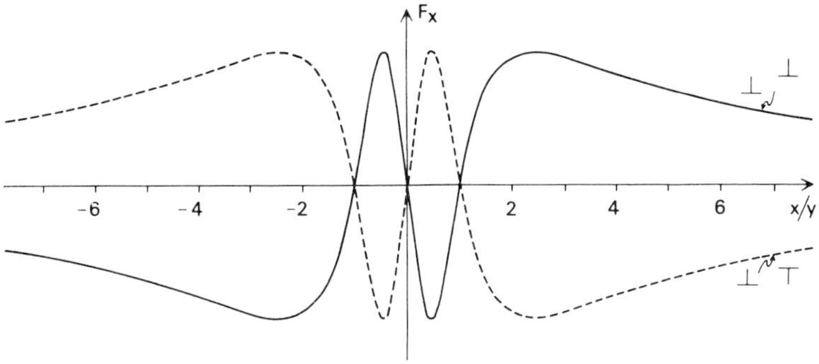

Fig. 16.7. Plot of the component parallel to the slip plane of the force between an edge dislocation at the origin and a parallel edge dislocation at (x,y). Burgers vectors of like sign, continuous line; unlike sign, interrupted line.

course reversed. The stable equilibrium position now lies along the line $y = x$. Pairs of long, straight, pure edge dislocations of opposite sign, thus stably oriented, are observed, and referred to as *dislocation dipoles.*

(3) Response of dislocation (2) to the force component F_y requires motion of (2) *out* of its slip-plane. Physically this involves the loss or addition of atoms at the edge of the extra half-plane of atoms. These processes are equivalent respectively to the absorption of vacancies from, or the emission of vacancies into the surrounding crystal. These are the processes of dislocation *climb*, and dislocation motion involving climb is termed *non-conservative motion.* Many factors influence the rate of climb of dislocations (see especially Friedel [5]) but, very roughly, it can be said that during movements towards lower-energy dislocation configurations *climb* motions of edge dislocations only become appreciable compared with their *glide* motions at temperatures above about $\frac{3}{4}T_m$, where T_m is the melting temperature on the absolute scale, in laboratory experiments of average duration.

16.2. Methods of observing dislocations

16.2.1. The physical bases underlying the detection of dislocations

The many methods nowadays established for detecting dislocations individually, either within crystals or at their outcrops on crystal surfaces, rely in different proportions upon the characteristic properties of a dislocated crystal lattice compared with the perfect lattice. In ch. 16 § 1 the geometrical properties of dislocated crystals, i.e. those properties essentially topological, were discussed first, and, subsequently, the states of stress and elastic energy within such crystals. One may attempt a crude division on similar lines regarding the causes of dislocation visibility. Firstly, there is the outstandingly important topological property that lattice "planes" intersected by dislocations possessing a screw component are in fact infinite helical surfaces. From this topological property derives the possibility of detecting dislocations by observing *growth spirals* on crystal faces (or analogous spirals revealed in the course of dissolution or evaporation). Again, it is the geometrical features of dislocated lattices that are important in generating *diffraction contrast* (electron or X-ray), for the distortion of the structure *within* individual unit cells (which must involve changes in bond angles and interatomic distances) is of small importance in creating diffraction contrast compared with the effects of bodily translation and rotation of volume elements of the dislocated crystal relative to those of the perfect, reference crystal. It is strain, on the other hand, or, more precisely, the strain energy density (which rises at a few atom distances from the dislocation line to an appreciable fraction of the lattice binding energy) that is responsible for the local reactivity from which results *etchpit* formation on surfaces, or the *decoration* of dislocations within the body of the crystal.

Several general texts on dislocations include sections dealing with methods of observing dislocations; examples from references given in ch. 16 § 1 are Friedel [5] and Nabarro [8]. The monograph by Amelinckx [16] and the review by Pashley [17] both have the direct observation of dislocations and other imperfections in crystals as their topic. Highly relevant, too, are many of the chapters in Amelinckx *et. al.* [18]; and since transmission electron microscopy is now the most important method of studying individual dislocations Hirsch *et. al.* [19] is also recommended as a general text concerning direct observation of dislocations.

To commence discussion of methods of detecting and observing dislocations, two ways in which the dislocation strain energy produces macroscopically evident effects will be mentioned: these are hollow dislocations and stress birefringence.

16.2.2. Hollow dislocations

Frank [20] showed that in the case of dislocations having large Burgers vectors the reduction of strain energy obtained by cutting out material so as to form a hollow tube surrounding the dislocation would outweigh the energy required to create the tube surface. Consider a pure screw dislocation, Burgers vector b. For unit length of dislocation the reduction in elastic energy by forming a hollow tube of radius r centred on the dislocation is $(\mu b^2/4\pi) \ln(r/r_0)$. The hole surface energy is $2\pi\gamma r$, where γ is the surface energy per unit area. The equilibrium hole radius is thus

$$r_e = \frac{\mu b^2}{8\pi^2\gamma}. \tag{16.14}$$

Now although γ has been determined experimentally for relatively few crystals, simple theoretical models relating surface energy to the ideal fracture strengths of solids show that for most structures γ/μ should be about $\frac{1}{4} \times 10^{-8}$ cm. It follows that only when $b \geqslant 10$Å, roughly, should hollow tubes form around screw dislocations, under equilibrium conditions. Holes centred on outcrops of screw dislocations with $b = 15$ Å on SiC platelet surfaces were observed by Verma in 1951 using optical phase contrast microscopy. (See Verma [21], and also Nabarro [8], for a general discussion of hollow dislocations). Holes along the axes of CsCl whisker crystals, shown by their macroscopic twist (the *Eshelby twist* [22]) to have axial screw dislocations with Burgers vectors of several hundred Ångstroms magnitude, were described by Webb and Bertolone [23]. Vapour-grown blade-like crystals of ZnS whose long axis is parallel to the crystallographic *c*-axis commonly contain a hollow tube a few microns in diameter which takes generally a slightly zig-zag course keeping within a small angle to the *c*-axis. X-ray topography has shown that in long-period polytypic ZnS crystals the tube coincides with a giant dislocation whose Burgers vector is the *c*-axis repeat period of the polytype family present in the crystal (Mardix, Lang and Blech [24]).

16.2.3. Stress birefringence

Since the late 1940's it has been recognised that the stress birefringence bands coinciding with slip-planes in plastically deformed halide crystals should be explicable in terms of the long-range stresses generated by arrays of many dislocations lying on the same slip-plane, in particular by arrays of edge dislocations all of the same sign (Nye [25]). The stress birefringence pattern due to a single dislocation was first observed by Bond and Andrus [26], using the transmission of silicon in the near infra-red. They looked along the length of a pure edge dislocation line, taking advantage of the cumulative effect of a stress field constant along the direction of view throughout the specimen thickness. However, stress birefringence due to small bundles of dislocations, seen in plan, has been detected in polished plates of diamond, the X-ray diffraction contrast showing that the resultant Burgers vector totalled not more than 3 or 4 unit vectors (Lang [27]).

16.2.4. Observation of surface outcrops of dislocations

16.2.4.1. Dislocations and crystal growth. Great interest in the microtopographical features on crystal faces was generated by Frank's proposal of the screw dislocation mechanism of crystal growth. Frank's mechanism solved the riddle of how low-index crystal faces could grow at rates proportional to super saturation of solution or vapour with growth on such faces proceeding at easily measurable rates down to supersaturations below 1%, whereas calculations of the expected rate of nucleation of new atomic layers upon completed, perfect crystal faces indicated that supersaturations of about 25% to 50% would be required to produce growth rates measurable at all. The emergence of a dislocation with a screw component normal to the face provided a step on the crystal surface, terminated at one end at the dislocation outcrop and free along the remainder of its length to sweep perpetually over the crystal face, rotating about the dislocation outcrop as pivot. On this topic the important source references are Frank [28] and Burton, Cabrera and Frank [29]. There are good brief summaries of the historical development of theory (Frank [30]) and of early observations of growth steps and etch pits (Mitchell [31]). Fuller accounts of both theory and experiment on the role of dislocations in crystal growth are provided by the books by Verma [21], and Dekeyser and Amelinckx [32]. See also Chapter 10 of this volume.

16.2.4.2. Growth steps and dislocation outcrops. The sequence of positions occupied by a growth step connecting the outcrops of a pair of screw dislocations of opposite hand is geometrically similar to that of dislocations multiplying at a symmetrical Frank-Read source (fig. 16.5). Closed loops encircling the pair of outcrops are formed. A lone screw dislocation outcrop gives rise to a spiral step, a *growth spiral.* If the rate of addition of atoms at the step is independent of local orientation of the step then the shape of the spiral is to a good approximation an Archimedian spiral, described in polar coordinates as

$$r = c\theta \,, \qquad (16.15)$$

where the "constant" c is a function of supersaturation, c is smaller for higher supersaturations, (Burton, Cabrera and Frank [29]). (*Vide* Ch. 10). Optical phase contrast microscopy, multiple-beam interferometry and transmission electron microscopy of surface replicas can reveal steps down to about 3Å in height (Sunagawa [33]). Hence individual atomic steps can be resolved in many crystal structures, and growth by the screw dislocation mechanism has been convincingly demonstrated. Conversely, a count of growth spirals gives an indication of the density of outcrops of dislocations with screw components normal to the surface. However, this will be a minimum estimate. Impurity *"poisoning"* may bring growth at some dislocation outcrops to a halt, and the small step-height spirals surrounding some outcrops may be obscured by the bunched steps from more active spirals sweeping over the area concerned. It is now realised that some of the early optical microscopic observations of growth spirals were aided by light scattering from minute droplets of condensible vapours "decorating" the atomic steps. The technique of deliberate decoration of unit steps on crystal faces, and thereby revealing dislocation outcrops, was first performed by Bassett [34] and has been developed with great success by Bethge (see, e.g. Bethge [35]). However, a warning must be given concerning the interpretation of large step-height spirals as

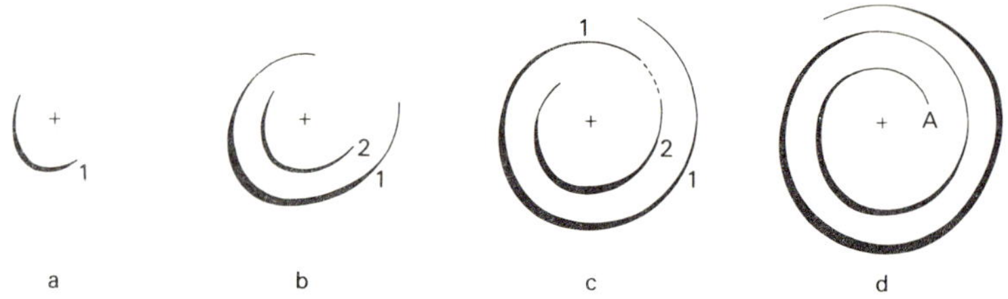

Fig. 16.8. Formation of a spiral by accidental merging of bunches 1 and 2 during stages (a) to (d) in the development of a set of bunches of steps (from Lang [37]).

unambiguous indicators of dislocation outcrops. Bunching of individual atomic layers into steps of multi-layer height is a common occurrence on crystal faces, both during growth and during dissolution. Frank [36] has analysed this phenomenon, applying the theory of kinematic waves which also describes road traffic jams and river flood waves. (This topic is discussed in ch. 10 §4). Given a tendency of individual steps to bunch, the bunches happening to develop first in one sector of a family of concentric steps, then it is easily visualised how spirals of large step-height may be formed by an accidental closure error as the developing bunches encircle the centre from which unit layers are expanding (in growth) or being removed (in etch-pit formation). The development is sketched in the sequence a → b → c → d in fig. 16.8 (Lang [37]). The critical stage is c when the "outer" bunch numbered 1 merges with the "inner" bunch numbered 2, just by accident, because of lack of perfect circular symmetry.

16.2.4.3. Etch pits and dislocation outcrops. Frank's mechanism outlined in §2.4.2 is reversible. Consider a crystal face which is almost perfectly flat, such growth spirals as exist having a very wide spacing between steps. This would be the likely condition if the final stages of growth took place under conditions of low supersaturation. Upon placing the crystal in undersaturated solvent, say, the motion of steps will be the reverse of that occurring during growth. Hollows will form surrounding dislocation outcrops and will be more pronounced the greater the undersaturation. Also to be taken into account is the lower chemical stability in the presence of suitable reagents that the higher energy regions at dislocation cores possess. It is thus readily seen qualitatively why etch pits can form at dislocation outcrops. A good "dislocation etch" reagent often contains (perhaps by accident) a "poison" that is adsorbed at the steps remote from the dislocation, thereby slowing down their rate of retreat from the dislocation outcrop. Then the slopes of the dislocation etch pit will steepen as individual steps are bunched together, and the walls of the pit may terminate sharply along lines where they meet the almost perfect low-index crystallographic plane of the surrounding crystal surface. Such conditions greatly enhance the optical visibility of pits, especially when dark-ground illumination or phase-contrast methods are employed. An introduction to the theory and phenomenology of dislocation etch-pit formation will be found in Nabarro [8].

The basic problem in studying dislocations by the etch-pit method is to establish with certainty a one-to-one correspondence between dislocation outcrops and etch-pits. Sometimes other observational techniques can offer confirmation, for example by decoration of the dislocations within the crystal interior (discussed in §2.5. below) or by X-ray topography (see §2.6. below). "Verification" by matching etch-pit distributions on the two faces produced by cleavage of a crystal is not infallible; the cleavage may cut through readily etchable impurity particles, leaving fragments imbedded in both faces. A good but tedious way of establishing correspondence is by alternate application of the "dislocation etchant" and then mechanical and/or chemical polishing to remove the old pit. Thus, step-by-step, the course of the dislocation line may be traced within the crystal, establishing thereby that a true line defect is being followed and that it has the geometrical properties expected of a dislocation line. Useful lists of dislocation etches for various crystals are given by Amelinckx [16] and of both dislocation etches and chemical polishes by W.G. Johnston [38]. The art of dislocation etching is now in a highly developed state. As an example of its possibilities it has been shown by Livingston [39] that positive and negative edge dislocations in copper can be distinguished by a difference in slope of the sides of their etch pits on crystal surfaces prepared close in orientations to (111).

A curiosity are the long dislocation etch tunnels or tubes that are found occurring naturally or can be produced by etching certain crystals. Alpha quartz etched for long periods in HF provides a notable example. Such etch tubes must not be confused with the much smaller-scaled etch tubes which may be produced along fission-fragment tracks. The latter may be observed alike in crystals (e.g. micas) and in glasses and plastics (see, e.g. Fleischer, Price and Walker [40]).

16.2.5. Decoration of dislocations within crystals

This method, the first to be developed for making dislocations within the interior of crystals visible, has played an important role in confirming the existence of theoretically predicted dislocation configurations, such as those in low-angle boundaries or Frank-Read sources. Many natural crystals have their dislocations decorated: this can be revealed by X-ray topography or transmission electron microscopy in opaque crystals, or by optical microscopy and ultramicroscopy in transparent crystals. However, even if the precipitates strung along the dislocation (which constitute the "*decoration*") are optically invisible, they may still be detectable by X-ray topography on account of the lattice deformations they produce. It may be mentioned that X-ray topography has shown that the needle-like "rutilations" so often seen in quartz are none other than strongly decorated dislocations.

Three decoration techniques have been of especial importance in fundamental studies of dislocations. The first followed from the recognition by Hedges and Mitchell [41] that particles of photolytic silver "decorated" networks of dislocations in thin silver chloride or silver bromide single crystals after exposure to light. This technique provided revealing demonstrations of dislocation movements during plastic deformation, an example being that caused by small glass spheres imbedded in the crystal. The spheres punched out edge dislocations in the form of a series of coaxial loops, all of pure edge dislocation and of the same sign. These loops glided away in a procession from the sphere by *prismatic glide*. Such loops repel each other, but with a fairly short-range force. When the distance d between loops is large compared with their radius, the force varies as d^{-4}. Large AgCl crystals may be decorated by sweeping photo-electrons into them by a pulsed electric field (Childs and Slifkin [42, 43]).

The second decoration technique of importance was developed by Amelinckx, Dekeyser and co-workers (see Amelinckx [16]) and is applicable to halide crystals stable on exposure to light, such as NaCl and KCl. The decorating impurity particles, silver or gold, and of colloidal size, are visible in the ultra microscope. They are precipitated by a high-temperature annealing of the crystal. This technique has been particularly revealing in the study of climb phenomena, especially in the case of the transformation of pure screw dislocations into *helical dislocations* with a large edge component.

Of importance not only in dislocation studies but also in the assessment of techniques for growing more perfect silicon crystals has been the copper precipitation method of Dash [44] which makes use of the near infra-red transmitting properties of silicon. Infra-red sensitive plates, or electronic infra-red to optical image converters must be used. Confirming the correspondence of dislocation outcrops with etch pits and the operation of Frank-Read dislocation sources have been among the applications of this technique.

When Dash's technique was first used, only dislocations with an edge component were found to be decorated by copper precipitates. Later the technique was developed so as to decorate dislocations of any character. The preferential nucleation of precipitates upon dislocations having some edge character is just what would be expected. The factors which favour the nucleation of precipitates of diffused impurity, and the subsequent growth of the precipitates, at sites located on dislocations include the following:

1. the dislocation stress field, where there is an edge component, will attract foreign atoms from either above or below the slip plane depending upon the size of the foreign atom relative to the atoms of the host lattice (See § 1.3 and § 1.10).

2. Along dislocations there will be in general an enhanced diffusion mobility (*pipe diffusion*). This helps to feed impurity atoms to nucleate and enlarge a precipitate.

3. By climbing, a dislocation can act as a source of vacancies of host lattice sites, thereby making room for growth of the precipitate.

16.2.6. Observation of dislocations by electron and X-ray diffraction contrast

In order to be trustworthily executed, the etchpit and decoration techniques require just the right "recipe" for chemical and/or thermal treatment of the crystal concerned. Some of the conspicuous successes of these techniques have been noted above. However, with many crystals much effort has been expended to try and develop dislocation etching or decoration techniques, but with only partial success. The advantage of diffraction techniques for dislocation observation is their generality of application: the apparatus required is the same for each crystal, and the appearance of the diffraction contrast image of a dislocation in one crystal does not differ fundamentally from that of a dislocation in another species of crystal (and the differences that can occur are now generally well understood and may be structurally informative). The diffraction methods and, say, the decoration method have a further dissimilarity. The decorations exhibit the loci of dislocation lines at the time when precipitation occurred; but the dislocation configuration present at other epochs in the crystal's history cannot be shown. However, both dislocation lines and the precipitates strung along them can be individually revealed in the diffraction methods, by virtue of the characteristic strain fields that each type of imperfection creates. The restrictions on applicability of the electron microscope method are basically only two: (1), the specimen must be sufficiently thin to transmit the electrons, at least locally in areas not less than about one micron square. For 100 kV electrons, specimen thicknesses should not exceed a few thousand Ångstroms for clear dislocation visibility, whereas for the 1 MeV microscope this limit is raised to about 5 microns.

(2), the specimen must not evaporate or decompose rapidly in the electron beam: this prohibits application to chemically labile crystals and to many organic crystals. Regarding the applicability of the X-ray methods, the restriction (2) above operates with much less force: both labile and volatile crystals have been successfully examined by X-ray topography. In place of restriction (1) above there exists the positive advantage that specimens sufficiently thick to be truly representative of bulk material may be used in experiments. But X-ray methods have their own severe restriction: the low maximum density of lattice defects at which they can be individually resolved. This, for example, precludes study of individual dislocation movements in plastic deformation in all but the earliest stages in nearly perfect crystals.

Studies of individual dislocations by transmission electron microscopy date from 1955 and by X-ray topography from 1957. Hirsch *et al.* [19] deal comprehensively with the electron microscope technique. The X-ray methods exist in a variety of experimental configurations which have been categorised and illustrated by Webb [45], Bonse, Hart and Newkirk [46] and Lang [47–49]. The outstanding difference between the electron microscope and the X-ray methods for observing dislocations is the much higher resolution of the former method. Take one important dislocation configuration, the *low-angle boundary,* as being the object of study with which to illustrate this difference. Consider the pure tilt boundary such as is commonly present in melt-grown crystals, or in crystals annealed after a plastic deformation. The dislocation model of such a boundary is explained in fig. 16.9. (Read [4]). The simple relation between boundary tilt angle, θ, dislocation spacing, D, and Burgers vector magnitude, b, is

$$\theta = b/D\,. \tag{16.16}$$

Now both in transmission electron microscopy and in X-ray topography, the apparent width of the diffraction contrast images of dislocations is determined by quantities fundamental in the diffraction process of the radiations concerned (assuming adequate instrumental design and technical competence in the performance of experiments). With common metals of medium atomic number, such as iron or copper, values of b are about 2½ Å, and the minimum resolvable values of D in images of low-angle boundaries are about 150 Å in the electron microscope case and about 1.5 microns in the X-ray case, a difference of two orders of magnitude. The tilt angles corresponding to these values of D are from equation (16.16) respectively about 1° and about 0.01°. For resolution of isolated defects such as small stacking faults the electron microscope is even more powerful than the X-ray method, the ratio of resolution limits being about three orders of magnitude.

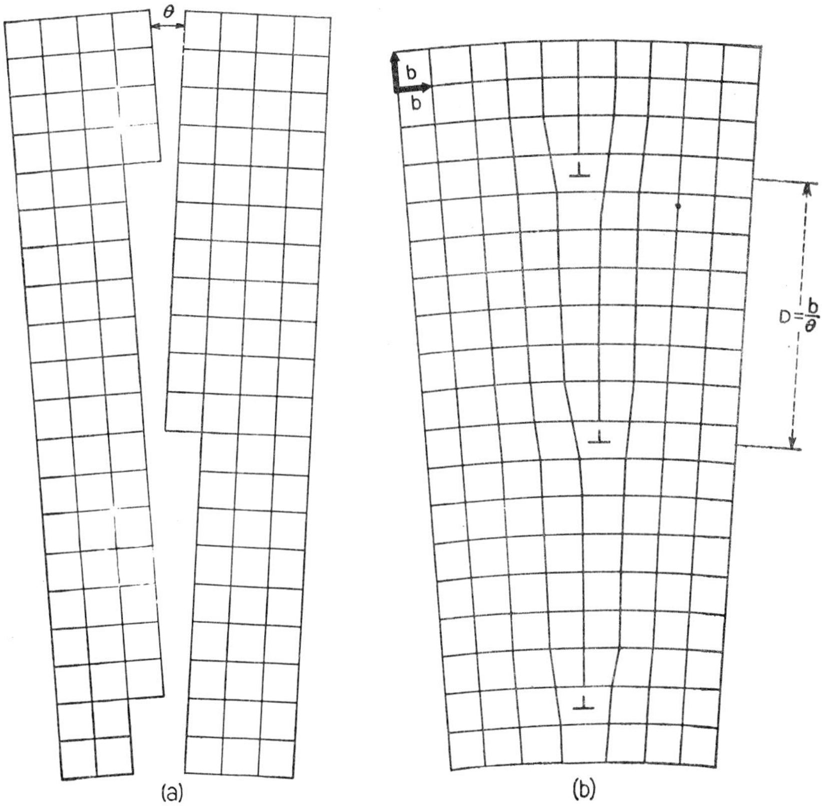

Fig. 16.9. Edge dislocation model of a pure tilt low-angle grain boundary (from Read [4]).

16.2.7. Direct resolution of crystal lattices

Two techniques of microscopy possess sufficiently high resolving power to allow the individual lattice planes in crystals to be resolved. These have enabled a variety of lattice defects to be seen directly, including dislocations. With standard 100 kV electron microscopes it is now possible to resolve in more than one direction the interplanar spacings in thin layers of crystals possessing interplanar spacings down to about 2 Å. The first direct resolution of lattice planes was achieved at the edges of thin flakes of platinum phthalocyanine crystals ($C_{32}H_{16}N_8Pt$), the electron beam being transmitted parallel to the $(20\bar{1})$ planes of spacing 12 Å. Edge dislocations were revealed by their extra halfplanes, and the actual width of the dislocation could be measured (about 12 interplanar spacings) (Menter [50]).

The second technique capable of resolving features down to about 2 Å in size is the field-ion microscope (For a recent full description of this technique see Müller [51]). The diameter of the field of view is limited to a few hundred atom diameters, and is located at an ultra-sharp tip of a metallic needle, the tip radius being about 1000 Å. Again a direct picture of the atomic position surrounding the edge of the extra half-plane of an outcropping edge dislocation can be seen (Müller [52], Brandon and Wald [53]).

16.2.8. Lattice moiré patterns

If two optical gratings differing either in orientation or spacing are superimposed, moiré fringes will be produced. The period and orientation of the moiré fringe system are neatly specified in terms of reciprocal vectors drawn in the plane of the gratings and normal to the grating rulings, thus:

$$G = g_1 - g_2 , \tag{16.17}$$

in which $|g_1|$ and $|g_2|$ are reciprocals of the ruling periods d_1 and d_2 of the two gratings. The difference vector **G** is normal to the moiré fringes which have spacing D equal to the reciprocal of **G**. The particular interest of moiré fringes is the way they make small differences in spacing and orientation of gratings (or crystals) visible. This follows from the reciprocal nature of the relationship

$$D = d/\theta \tag{16.18a}$$

for a *pure rotation* moiré, applicable to small rotations θ between gratings of like spacing, $d = d_1 = d_2$, and the relationship

$$\frac{1}{D} = \frac{1}{d_1} - \frac{1}{d_2} \tag{16.18b}$$

for the *pure compression* moiré in which gratings (1) and (2) are similarly oriented, but $d_1 \neq d_2$. The behaviour of moiré fringes can be demonstrated with optical models, such as are provided in the book by Bollmann [54], and optical analogues are useful for demonstrating how in electron microscopy or X-ray topography crystal lattice moiré patterns produce a magnified image of a dislocation, fig. 16.10.

The study of dislocations in electron microscope moiré patterns formed by

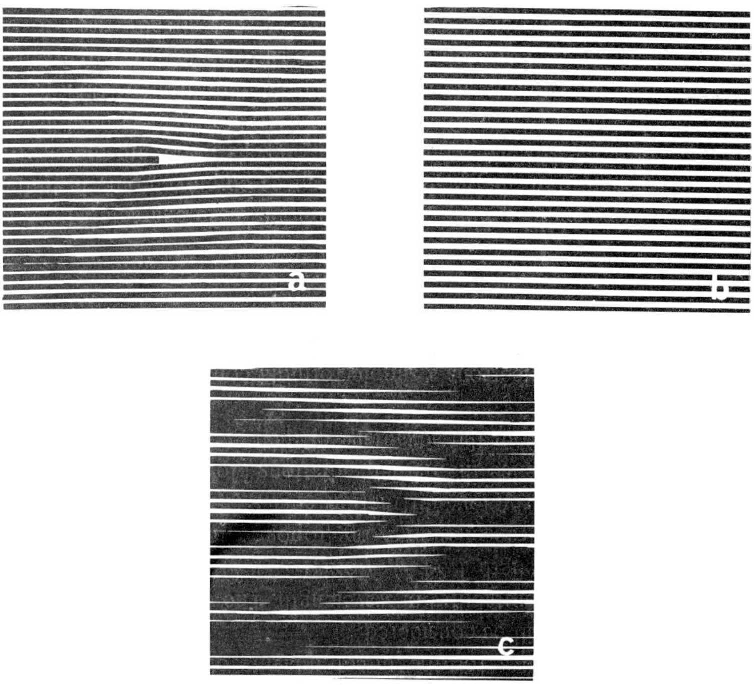

Fig. 16.10. Optical analogue demonstrating the production of an extra fringe in the moiré pattern (c) when a perfect grating (b) is applied to a grating of slightly different spacing (a) which also contains a single edge dislocation (from [56]; courtesy Dr. D.W. Pashley).

superimposed thin crystals commenced with the work of Hashimoto and Uyeda [55] and Pashley, Menter and Bassett (see Pashley, Menter and Bassett [56] and Bassett, Menter and Pashley [57]. In the X-ray case dislocations were first observed in lattice moiré patterns by means of the X-ray interferometer of Bonse and Hart [58], work with which device has been reviewed by Hart [59]. The X-ray analogue of the electron microscope experiment can be performed by the superposition of simultaneously X-ray diffracting crystal plates, mutually aligned by the method of Brádler and Lang [60]. The moiré technique permits a direct determination of both *magnitude* and *sense* of Burgers vectors of dislocations. Moreover, in its X-ray topographic embodiment it provides a direct picture of relative parameter differences and relative misorientations as small as 10^{-7}.

Fig. 16.11 is part of an X-ray moiré topograph formed by superposing one

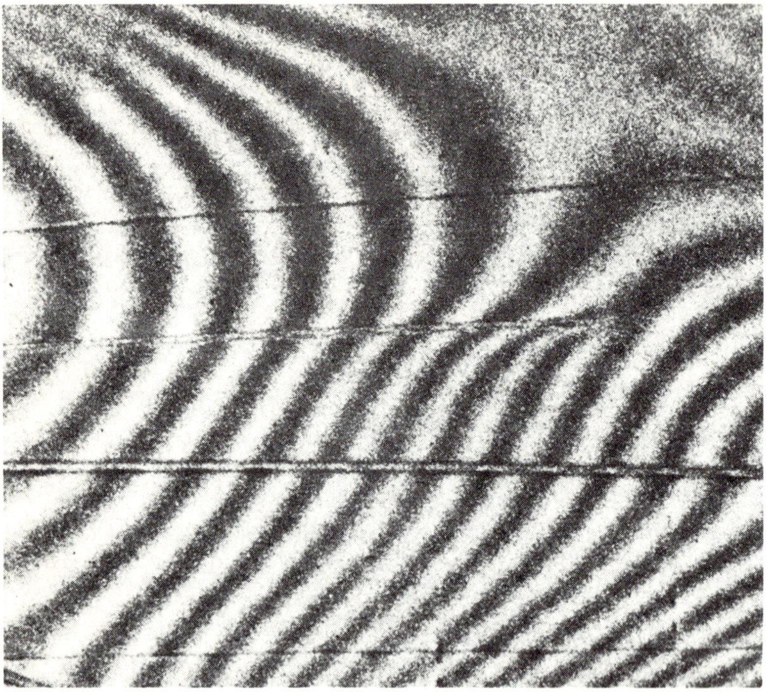

Fig. 16.11. Part of X-ray moiré topograph formed by superimposition of two slices of natural quartz. Width of field shown is 1 mm. Diffraction vector, $10\bar{1}1$, is approximately horizontal. Radiation AgKα.

crystal slice upon another, according to the technique of Brádler and Lang [60]. It was taken during a moiré topographic study of defects in rather perfect natural quartz [61]. The crystal slices used were each about 1½ mm thick and were cut normal to a major rhombohedral face of the crystal, (1011), so that the reflection $10\bar{1}1$ could be obtained in symmetrical transmission. The rather few grown-in dislocations ran through the specimen in directions not far from perpendicularity with the natural face $(10\bar{1}1)$. Their images therefore cross the field approximately horizontally in fig. 16.11. When dislocations outcrop from one or other slice at the slice surfaces which are opposed to each other, extra half-planes appear in the moiré pattern, just as in the optical analogue, fig. 16.10. A pair of dislocations which run parallel to each other, about 10 μm apart, make such an outcrop in fig. 16.11, and the 2 extra moiré fringes generated at their outcrop can be seen. Conventional X-ray topography established that each dislocation had the same basal-plane lattice translation vector, **a**, as Burgers vector, and thus with the $10\bar{1}1$ reflection $\mathbf{g}\cdot\mathbf{b} = \pm 1$ for each. The number of extra moiré fringes generated at a dislocation outcrop is also equal to $\mathbf{g}\cdot\mathbf{b}$ for a single dislocation, or to $\Sigma\mathbf{g}\cdot\mathbf{b}$ at a multiple outcrop. For the pair of dislocations under discussion, $\Sigma\mathbf{g}\cdot\mathbf{b} = 2$, showing that both had the same sign of Burgers vector.

The fringe pattern in the upper right-hand part of fig. 16.11 is similar to that given by superimposition of a parallel grating upon a radial grating [62].

16.3. Diffraction by perfect and almost perfect crystals

16.3.1. Introduction

In ch.16 §2.4 it was pointed out that methods of observing lattice imperfections based upon diffraction by crystal lattices were important because of their generality of application and their usually non-destructive nature. These methods are capable of presenting information in a direct and pictorial fashion: they produce images of an area of a crystal face, or of a crystal volume, in which local regions of lattice imperfection become conspicuous through *diffraction contrast.* That is, they diffract with a noticeably different intensity compared with that from the matrix of relatively perfect crystal in which they are imbedded. In order to interpret diffraction contrast patterns adequately it is necessary to be familiar with at least the essential features of the diffracting behaviour of perfect crystals. This involves the *dynamical theory of diffraction,* which has the multiple scattering of waves (electromagnetic, or material) as its central theme. An analysis along such lines is in sharp contrast to the more familiar *kinematical theory of diffraction* which is founded upon the assumption of single scattering only. The kinematic theory of diffraction is simply an extension to three dimensions of the optical theory of one-dimensional diffraction gratings, but over the years it has been developed elaborately (and sometimes quite elegantly) to cover many types of departure from perfect three-dimensional periodicity in a crystal. Many of such developments, together with accounts of the basic geometry of crystal diffraction as it applies in X-ray experiments, are described in English-language texts such as James [63], Warren [64], Wilson [65] and Wilson [66].

Turning to general texts on the dynamical theory of diffraction, one encounters a division between treatments specifically for X-rays and specifically for electrons. Recommended among the former are Batterman and Cole [67], and James [63], and among the latter Heidenreich [68] and Hirsch *et al.*

[19]. There are some stimulating articles and books which deal with the general principles of the interaction of waves with periodic media, among such are Brillouin [69], Ewald [70] and Slater [71]. Illustrations of both the common ground and also the significant differences between electron and X-ray diffraction theories, as exhibited in their particular developments which have accompanied the rise of diffraction techniques for direct observation of crystal imperfections, will be found by comparing their respective treatments in the relevant chapters of Amelinckx *et al.* [18].

This Section continues with a discussion of the choice of radiation to be used in a diffraction study of crystal imperfections. Sections 3.3 to 3.5 outline in a mainly non-mathematical way some of the features of dynamical diffraction theory, with special reference to the situations arising in the diffraction of X-rays by perfect or almost perfect crystals.

16.3.2. Choice of radiation

What waves (material or electromagnetic) should be chosen for a diffraction study of lattice imperfections? The choice could fall upon X-rays, electrons, neutrons, or protons, but only X-rays and electrons have been widely used so far in techniques for the direct imaging of lattice imperfections. Hence these two only will be considered.

The first question is suitability of wavelength. The well-known characteristic X-rays CuKα (wavelength 1.54 Å), MoKα (0.71 Å) and AgKα (0.56 Å) are eminently suitable. In transmission X-ray topographic techniques there are advantages in using fairly low diffraction angles (Bragg angles, θ_B, in the range 5° to 10°, say) hence the harder radiations MoKα and AgKα are to be preferred (quite apart from the advantage of their greater penetrating power, discussed below).

For non-relativistic electrons a simple formula connects accelerating voltage (V) in volts and wavelength (λ) in Ångstroms:

$$\lambda = (150/V)^{\frac{1}{2}} . \qquad (16.19)$$

Electrons of wavelength similar to the commonly used characteristic X-rays interact so strongly with matter that they can be employed only in studying reflections from the surface atom layers of crystals: these are the electrons used in Low Energy Electron Diffraction (LEED). Most widely used in transmission electron microscopy are electrons accelerated through 100 kV. Their wavelength is 0.037 Å. For such electrons the simple equation (16.19) needs a relativistic correction: at 100 kV the ratio v/c has already risen to 0.55 (v = velocity of electrons, c = velocity of light). The relativistically-corrected formula is

$$\lambda = (150/V)^{\frac{1}{2}} (1 + 0.98 \times 10^{-6} V)^{-\frac{1}{2}} . \qquad (16.20)$$

The 1 MeV high-voltage electron microscopes which are becoming increasingly common produce electrons of wavelength 0.0087 Å which are quite relativistic ($v/c = 0.94$). Bragg's Law shows that with such small wavelengths the angular separation of diffraction spectra is small. For this and other reasons it is virtually impossible to excite one Bragg relection only in transmission electron microscopy, a situation quite in contrast to the X-ray case.

The next question concerns the strength of scattering of the radiation by atoms. Notable is the relatively much stronger scattering of electrons compared with X-rays. For X-rays whose wavelength is small compared with any of the characteristic absorption edge wavelengths of the atom, the ratio of the amplitude of the spherical wave scattered by the atom to that of the wave incident upon it, measured at unit distance in the forward direction from the atom, is

$$\phi_X(0) = -Z(e^2/mc^2). \qquad (16.21)$$

In this formula Z is the atomic number, e the electronic charge in electrostatic units and m the electron mass. The quantity (e^2/mc^2) is equal to the classical radius of the electron, r_0, and is small, 2.8×10^{-13} cm. The absolute X-ray atomic scattering factor is essentially negative. Its magnitude is a decreasing function of $(\sin\theta/\lambda)$, where θ is the Bragg angle of the reflection concerned. Omitting the polarisation factor,

$$\phi_X(\sin\theta/\lambda) = -f_X r_0 , \qquad (16.22)$$

where f_X is the familiar X-ray atomic scattering factor expressed in "electron units", and is by definition positive. The form of its decrease with increasing $(\sin\theta/\lambda)$ for an atom of medium Z, say Fe, $Z = 26$, is sketched in fig. 16.12. (For information on X-ray atomic scattering factors see James [63] or the International Tables of Crystallography (1962) Vol. III.)

In the case of electrons, the relation corresponding to equation (16.22) is

$$\phi_e(\sin\theta/\lambda) = f_e = \frac{1}{8\pi^2 R_0}(\sin\theta/\lambda)^{-2}(Z - f_X) , \qquad (16.23)$$

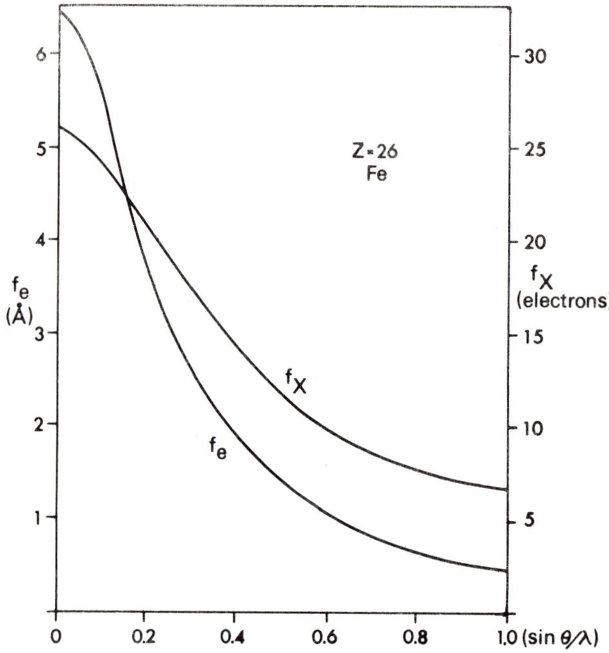

Fig. 16.12. Atomic scattering factors for X-rays, f_X, in electron units, and for electrons, f_e, in Å, as a function of $(\sin \theta/\lambda)$ in $Å^{-1}$.

in which $R_0 = h^2/4\pi^2 m e^2 = 0.53$ Å, the Bohr radius of hydrogen and f_X is the X-ray atomic scattering factor as defined above. For scattering by, say, iron atoms at $(\sin \theta/\lambda) = 0.25\ Å^{-1}$, one finds

$$\phi_X = 18 \times 2.8 \times 10^{-13} = 5 \times 10^{-12}\ \text{cm}\,,$$

whereas, writing equation (16.23) as

$$f_e = 2.38 \times 10^{-10} (\sin \theta/\lambda)^{-2} (Z - f_X)\,,$$

in which λ is now in Å, one obtains

$$f_e = 3.05 \times 10^{-8}\ \text{cm} = 3.05\ Å.$$

This shows the ratio (f_e/ϕ_X) to be about 10^4. The electron atomic scattering factors fall off more rapidly with increasing $(\sin \theta/\lambda)$ than do the X-ray atomic scattering factors, as can be seen in fig. 16.12. (Data on the atomic scattering factors for electrons are given in the International Tables of Crystallography (1962) Vol.III, and in Hirsch *et al.* [19].)

Suppose an incident wave exactly satisfies the Bragg condition for diffraction by a set of lattice planes. How many lattice planes must be traversed (or through what thickness of crystal must the incident beam pass) before the amplitude of the Bragg reflected beam builds up to the same value as the amplitude of the incident ray? This is a question of basic importance. A simple calculation using Fresnel zones can be made assuming that the crystal is perfect so that diffracted wavelets add coherently, but neglecting multiple scattering and the reduction of amplitude of the incident beam through diversion of its energy into the diffracted beam. This calculation of number of planes traversed involves the reticular density of scattering atoms on the plane, their individual scattering power, and the wavelength (James [63], Hirsch *et al.* [19]). A crystal thickness just double that obtained with such simple calculation is adopted as the standard parameter to denote diffracting strength. It is the *extinction distance,* ξ_g. For both X-rays and electrons

$$\xi_g^{-1} = \frac{\lambda |F|}{\pi V \cos \theta_B} \tag{16.24}$$

V is the unit cell volume, F is the structure factor in absolute units. Thus for X-rays F is the familiar structure amplitude expressed in electron units multiplied by (e^2/mc^2). Also, in the X-ray case, the expression should be multiplied by a polarisation factor C ($C = 1$ for the σ polarisation state and $C = |\cos 2\theta|$ for the π polarisation state). Extinction distances for 100 kV electrons are typically a few hundred Ångstroms for low-order reflections, but increase rapidly with increasing order because of the rapid way f_e falls off with increasing $\sin \theta/\lambda$. Table 16.1 gives a few representative values for 100 kV electrons, the figures are taken from the fuller listings in Hirsch *et al.* [19].

Table 16.1
Representative values of extinction distance for 100 kV electrons

Crystal	Reflection	ξ_g, Å
Diamond	111	476
Diamond	400	1215
α-iron	110	270
α-iron	400	1032

X-ray extinction distances are typically some tens of microns, but can rise to values of millimetre order when hard radiations such as AgKα and weak reflections (such as those given by organic structures with large cells) are employed. Table 16.2 shows some values, taken from Lang [49].

Table 16.2
Representative values of extinction distance for X-rays

Crystal	Reflection	Radiation	ξ_g, microns
Diamond	220	MoKα	48
Diamond	220	CuKα	27
Silicon	220	MoKα	38
Silicon	440	MoKα	60
LiF	111	MoKα	50
LiF	200	MoKα	33
α-Quartz	$10\bar{1}1$	MoKα	45
α-Quartz	0003	MoKα	205
α-iron	110	AgKα	13
α-iron	110	CoKα	5.7

Finally, in considering the choice of type of radiation, the attenuation of the incident beam arising from causes other than coherent scattering must be taken into account. With X-rays, photo-electric absorption is dominant. The quantity relevant in the experiment is the *linear absorption coefficient* μ. The quantities usually tabulated are the *mass absorption coefficients* μ/ρ, where ρ is the density, and the *atomic absorption coefficients* $\mu_a = \mu A/\rho N$, where A is the atomic weight and N is Avogadros number. Within wavelength ranges not containing an absorption edge, μ_a is proportional to $Z^4\lambda^3$, to a good approximation. Values of crystal thickness t such that $\mu t = 1$ are of order 1 cm for crystals of the lightest elements, using MoKα or AgKα radiations, whereas even for AgKα the value $\mu t = 1$ is reached for t not more than a few microns with the heavy elements. These values of t give an idea of the maximum thicknesses of crystals that can be used in transmission X-ray topographic experiments without heavy loss of beam intensity through absorption. In the electron case, there is a less steep dependence of attenuation upon Z. To take account of absorption empirically in diffraction contrast calculations, the extinction distance may be made a complex quantity, with its imaginary part about one tenth of its real part. The practical consequence of incoherent scattering processes is to make transmission electron micrograph images of lattice imperfections become weak and unclear for specimen

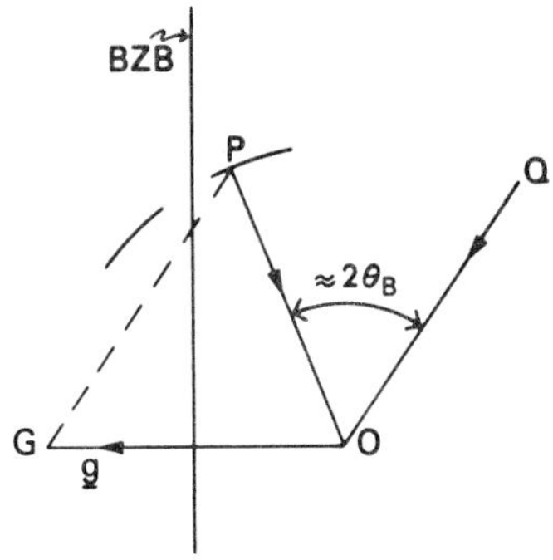

Fig. 16.13. Incident and diffracted wave vectors nearly satisfying the Bragg condition for the reciprocal lattice vector g, OG. The Brillouin Zone Boundary (BZB) bisects OG.

thicknesses greater than a few thousand Ångstroms when 100 kV electrons are transmitted. On the other hand, in 1 MeV electron microscopes dislocation images showing good definition can be obtained with specimens 5 to 10 microns thick.

16.3.3. Waves in infinite, perfect, non-absorbing crystals

Suppose a single plane wave of unique frequency is travelling through a crystal. Its wave vector, **k**, has a length determined by the frequency of the wave and the mean refractive index of the crystal. In reciprocal space its wave vector can be represented by a radius PO of a circle (radius $k = 1/\lambda$) centred on the origin O of the reciprocal lattice. Suppose the direction of OP brings P close to the Brillouin zone boundary bisecting the reciprocal lattice vector **g**(**g** is OG in fig. 16.13). Then a second plane wave is generated by Bragg reflection: its wave vector $\mathbf{k}_g$ may be represented by QO where the angle POQ must be close to $2\theta_B$, θ_B being the Bragg angle. In fact QO may be transferred to PG, with which vector it is identical. Why must the two wave vectors form a closed triangle with OG, so as to satisfy PO + OG = PG? Now the two plane waves must have the same frequency. Then they must interfere and form a standing wave pattern. Thus there is spatially-periodic variation of time-average of energy density of electromagnetic radiation (or of probability density in the case of de Broglie waves). Since the crystal has been postulated to be infinite and of perfect periodicity, all unit cells are equivalent and must have identical spatial relationships with the standing wave pattern. Stating this demand in more general terms, the standing wave pattern must have the same translation symmetry elements as the crystal lattice itself. In the present case the spatial periodicity of the standing wave pattern has as its reciprocal vector the vector difference between PO and QO; i.e. between $\mathbf{k}_0$ and $\mathbf{k}_g$. So this difference must be *exactly* **g**. More generally, if any number of plane waves of constant frequency is generated through interaction with the crystal lattice, the wave vectors of any pair of waves are related by $\mathbf{k}_m + \mathbf{g}_{mn} = \mathbf{k}_n$, where $\mathbf{g}_{mn}$ is a reciprocal lattice vector. This is a simple statement of *Bloch's theorem.*

Elementary arguments, based on the symmetry properties of a perfect crystal, can lead one further:

One would expect the symmetry planes of the standing wave pattern generated by the Bragg reflection g to coincide with those of the $\mathbf{g}^{\text{th}}$ Fourier component of the distribution of scattering matter in the crystal. This certainly should apply in the X-ray case with no anomalous dispersion and only one reciprocal lattice vector **g** active in Bragg reflection. The situation is most easily visualised in the simplest structures, and in the "full" reflections such as 220 and 400 in diamond structures. This symmetry condition still allows two possible standing wave patterns: the "Bragg plane" of atoms may coincide either with the maxima of disturbance, or with the minima, in the standing wave pattern. Writing the "$\mathbf{k}_0$" and "$\mathbf{k}_g$" waves as

$$C_0 \exp(2\pi i \mathbf{k}_0 \cdot \mathbf{r}) \text{ and } C_g \exp(2\pi i(\mathbf{k}_0 + \mathbf{g}) \cdot \mathbf{r}),$$

the position vector **r** being drawn from the chosen origin on the Bragg plane then $\alpha = C_g/C_0$ is a *real* number either positive or negative.

The final symmetry argument concerns the special case when P lies on the Brillouin zone boundary. With exact satisfaction of the Bragg condition the amplitudes of the "O" and "G" waves must be equal in magnitude. Therefore $\alpha = \pm 1$ for the two standing wave patterns. The "standing wave ratio" is unity and the electromagnetic energy density (or the probability density) distributions are proportional to $\cos^2(\pi \mathbf{g} \cdot \mathbf{r})$ for $\alpha = 1$ and $\sin^2(\pi \mathbf{g} \cdot \mathbf{r})$ for $\alpha = -1$, as shown in fig. 16.14.

16.3.4. The dispersion surface

The situation shown in fig. 16.14 has remarkable implications, which are most strikingly visualised if one makes the simplification that the atoms are of point size and are located exactly on the Bragg plane. The Bloch wave pair with $\alpha = -1$ represents photons (or particles) travelling parallel to the Bragg planes, and channelled midway between the atomic planes so that they have zero interaction with the atoms. The Bloch wave pair with $\alpha = +1$ again represents a flux parallel to the Bragg planes, but now peaked on the line of atoms with a density along this line twice the mean spatial density of the flux carried by these Bloch waves.

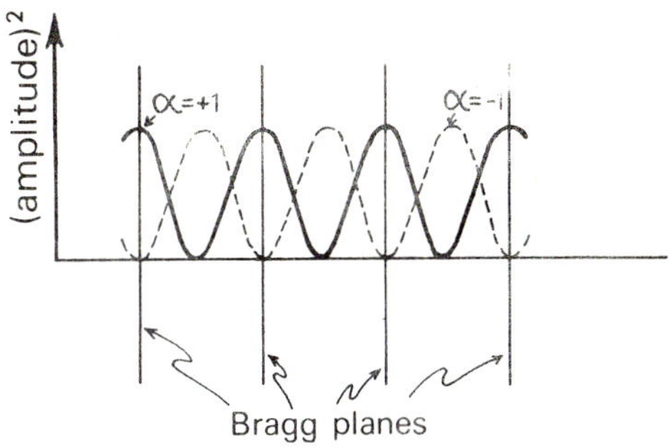

Fig. 16.14. Standing waves and Bragg planes of atoms when the Bragg condition is exactly satisfied. Branch (1) waves, dotted line; branch (2) waves, continuous line.

The above-described difference between the *symmetric* mode of propagation ($\alpha = +1$) and the *antisymmetric* mode ($\alpha = -1$) introduces a difference in their respective propagation velocities. In the case of a single wave far from satisfying the Bragg condition the refractive index $n = \lambda\,(\text{vacuum})/\lambda(\text{crystal}) = k/K$ where k is the wave vector magnitude in the crystal and K is the wave vector magnitude in vacuum. For both X-rays and high energy electrons n differs little from unity. For electrons

$$n = 1 + \frac{\lambda^2 F_0}{2\pi V}, \tag{16.25a}$$

F_0 being the structure factor for forward scattering of electrons (a quantity closely related to the diamagnetic susceptibility of the crystal (Ibers [72]). For X-rays

$$n = 1 - \frac{e^2}{mc^2}\,\frac{\lambda^2 F_0}{2\pi V}, \tag{16.25b}$$

in which F_0 is the zero-order structure factor, i.e. F_0/V is just the total number of electrons per unit volume. In both formulae λ can be taken to be the wavelength in vacuum.

Next consider what happens when Bragg reflection occurs. Fig. 16.15 represents an enlarged view of the vicinity of P in fig. 16.13. The interrupted lines AA′ and BB′ represent intersections of the plane of the diagram with spheres of radius k centred on O and G respectively, and the full lines CC′ and

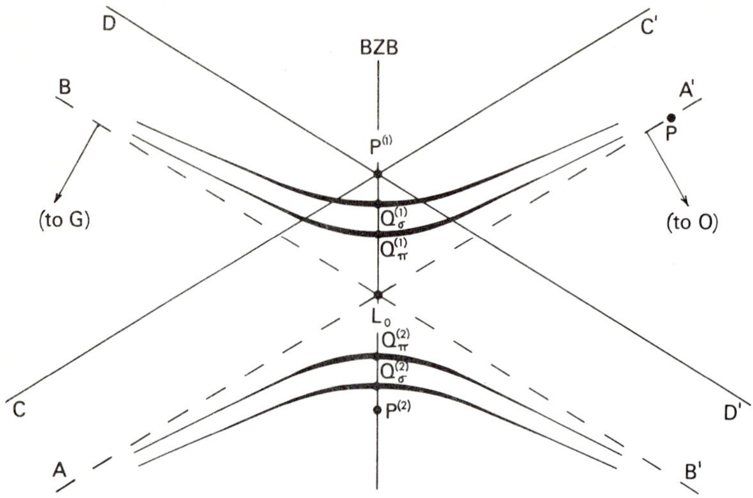

Fig. 16.15. Region of dispersion surface in reciprocal space close to the Brillouin Zone Boundary, explained in text.

DD′ represent corresponding intersections with spheres of radius K (The drawing is for the X-ray case, with K longer than k.)

Suppose P is sufficiently far from the Brillouin zone boundary that Bragg reflection is so weakly excited that the "standing wave pattern" is effectively non-existent. There is then no reason for P to lie elsewhere than on AA′. Now suppose the Bragg reflection condition is exactly satisfied for waves in the crystal, retaining for the moment the simplification that the atoms are point scatterers all on the Bragg plane so that all $F_g = F_0$. Then for propagation in the antisymmetric mode the crystal appears to the photons (or electrons, in the electron diffraction case) no different from a vacuum. Hence this Bloch wave pair has wave vectors $\mathrm{k}_0^{(1)}$ (= $\mathrm{P}^{(1)}\mathrm{O}$) and $\mathrm{k}_g^{(1)}$ (= $\mathrm{P}^{(1)}\mathrm{G}$) both equal in magnitude to the vacuum value K. On the other hand the symmetric mode, in which the electromagnetic energy density at the scattering centres is double the value when no Bragg reflection occurs, produces a doubling of the amplitude of scattered waves and hence a doubling of the departure of the refractive index from unity. The symmetric Bloch wave pair exactly satisfying the Bragg condition thus has wave vectors $\mathrm{k}_0^{(2)}$ (= $\mathrm{P}^{(2)}\mathrm{O}$) and $\mathrm{k}_g^{(2)}$ (= $\mathrm{P}^{(2)}\mathrm{G}$) where $\mathrm{P}^{(1)}\,\mathrm{P}^{(2)} = 2\,\mathrm{P}^{(1)}\,\mathrm{L}_0$.

Now

$$K - k \quad = K(1-n)$$

and so

$$P^{(1)} L_0 = K(1-n)/\cos\theta = \left(\frac{e^2}{mc^2}\right)\frac{F_0\lambda}{2\pi V\cos\theta}$$

for the X-ray case, giving

$$P^{(1)} P^{(2)} = \left(\frac{e^2}{mc^2}\right)\frac{F_0\lambda}{\pi V\cos\theta}\,.$$

By analogy with eq. (16.24) this may be written $P^{(1)} P^{(2)} = \xi_0^{-1}$. Removing the assumed simplification of point atoms just replaces F_0 by $|F_g|$ in the formula for $P^{(1)} P^{(2)}$, and $|F_g|$ is always less than F_0. For X-rays there are then four possible "wave points" on the Brillouin zone boundary: $Q_\sigma^{(1)}$ and $Q_\sigma^{(2)}$, for the σ polarisation state, with

$$Q_\sigma^{(1)} Q_\sigma^{(2)} = \left(\frac{e^2}{mc^2}\right)\frac{|F_g|\lambda}{\pi V\cos\theta}, \tag{16.26a}$$

and $Q_\pi^{(1)}$ and $Q_\pi^{(2)}$ for the π polarisation state, with

$$Q_\pi^{(1)} Q_\pi^{(2)} = \left(\frac{e^2}{mc^2}\right)\frac{|F_g\cos 2\theta|\lambda}{\pi V\cos\theta}. \tag{16.26b}$$

Thus at the Brillouin zone boundary the constant frequency wave surface is split into separate branches whose minimum separation is directly proportional to the structure amplitude of the reflection concerned. The wave surface, or *dispersion surface,* upon which points such as Q lie, merges asymptotically with the spheres, radius k, centred either on O or G, when the distance from the Brillouin zone boundary becomes large compared with ξ_g^{-1}. To calculate the shape of the dispersion surface in a fairly simple way relevant to the present discussion one may (a) consider scalar waves (such as electrons), (b) assume all F's to be zero except F_0 and F_g which are taken as real and positive quantities, and (c) assume $\lambda \ll \xi_g$: this last condition is in practice always satisfied (see tables 16.1 and 16.2). Take the Bloch wave

$$\psi = C_0 \exp(2\pi i\,\mathbf{k}_0\cdot\mathbf{r}) + C_g \exp(2\pi i\,\mathbf{k}_g\cdot\mathbf{r})$$

as a solution to the Schrodinger equation. The wave vectors k_0 and k_g will differ in magnitude from the mean wave vector k in the crystal, but always $\mathrm{k}_g = \mathrm{k}_0 + \mathrm{g}$. There are obtained the equations

$$\begin{aligned} (k_0^2 - k^2)\,C_0 - k^2\,Q_g\,C_g &= 0 \\ k^2 Q_g C_0 - (k_g^2 - k^2)\,C_g &= 0\,. \end{aligned} \tag{16.27}$$

The dimensionless quantity Q_g is given by

$$Q_g = \frac{\lambda^2 F_g}{\pi V}$$

for electrons when F_g is measured absolutely in units of length, and in the X-ray case by

$$Q_g = -\left(\frac{e^2}{mc^2}\right)\frac{\lambda^2 F_g C}{\pi V}, \text{ where}$$

F_g is expressed in the usual fashion in electron units. Note that Q_g is positive for electrons and negative for X-rays. The quantity Q_g is closely related to the extinction distance since

$$|Q_g| = \lambda \cos\theta\, \xi_g^{-1}\,.$$

Putting

$$k_0^2 - k^2 = 2kp_0$$

and

$$k_g^2 - k^2 = 2kp_g\,,$$

and setting the determinant of eq. (16.27) equal to zero gives

$$p_0 p_g = \tfrac{1}{4} k^2 Q_g^2 \tag{16.28}$$

as the equation of the dispersion surface. The parameters p_0 and p_g measure the perpendiculars dropped from a point on the surface to the "O" sphere of radius k and the "G" sphere of radius k, respectively, i.e. to the lines AA′ and BB′ in fig. 16.15. It follows from eq. (16.28) that the sections of the dispersion surface in fig. 16.15 are a pair of hyperbolae having AA′ and BB′ as asymptotes. The branch of the pair for which at the Brillouin zone boundary k_0 and k_g have magnitudes lying between k and K is termed "branch (1)", and it corresponds to the asymmetric standing wave mode. The other branch is called "branch (2)". Hence in fig. 16.15, which represents the X-ray case (refractive index less than unity), branch (1) is the upper branch.

Equation (16.27) also gives the ratio $\alpha = C_g/C_0$. This ratio is uniquely related to the position on the dispersion surface of the relevant wave point which defines the Bloch wave since

$$C_g/C_0 = 2p_0/k\,Q_g = k\,Q_g/2p_g. \tag{16.29}$$

It follows that each Bloch wave has a unique ratio α, and this ratio can be a useful parameter to plot along the dispersion surface, as is done in fig. 16.16. Again take note that in eq. (16.29) both Q_g and the p's change sign together when changing from the electron case to the X-ray case for a given mode, e.g. the antisymmetric, branch (1).

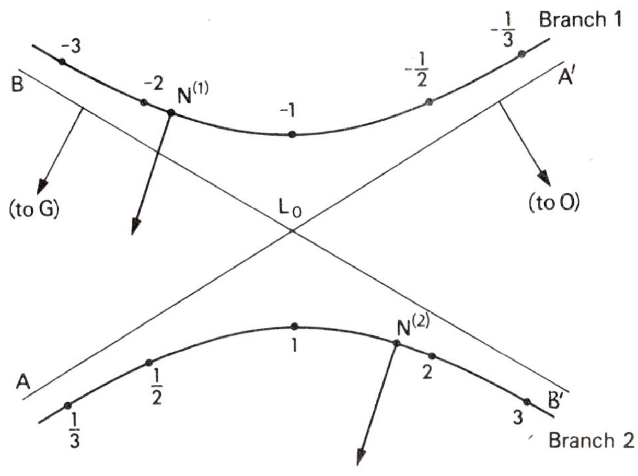

Fig. 16.16. Amplitude ratios α plotted as parameters along the dispersion surface. Bloch waves corresponding to the conjugate points $N^{(1)}$ and $N^{(2)}$ have parallel energy-flow vectors, normal to the dispersion surface at $N^{(1)}$ and $N^{(2)}$.

16.3.5. Energy flow

Consider one-dimensional waves in which the frequency ν is related to the wave number k by a dispersion relation, i.e. $\nu \equiv \nu(k)$. The group velocity u is $\mathrm{d}[\nu(k)]/\mathrm{d}k$. In three dimensions the group velocity vector **u** is

$$\mathbf{u} = \mathrm{grad}_k\, \nu(\mathbf{k})\,. \tag{16.30}$$

For waves of constant frequency, with which the present discussion of diffraction in crystals is concerned, the expression $\nu(k) = \text{const.}$ is an implicit description of a two-dimensional constant frequency surface (i.e. the dispersion surface) in reciprocal space, and the operation grad_k, which represents differentiation with respect to **k**, produces a vector whose magnitude may be found by algebraic methods or geometrical construction but which is *necessarily directed along the normal to the dispersion surface.* Thus the particle flux (in the case of material waves), or the Poynting vector (in the case of electromagnetic waves) is directed normal to the dispersion surface. This fundamental statement is discussed in both its general and particular aspects by Kato [73], Ewald [74] and Wagner [75].

Figs. 16.15 and 16.16 show that if all the Bloch waves in the vicinity of the Brillouin zone boundary are excited inside the crystal, as they would be in response to an externally incident beam of X-rays or electrons covering a sufficient angular range of incidence, then energy flow takes place inside the crystal in all directions within the angle $2\theta_\mathrm{B}$ enclosed between the directions $\mathrm{L_0\,O}$ and $\mathrm{L_0\,G}$ (fig. 16.15). In real space the situation is as shown in fig. 16.17. Suppose a spatially narrow beam of radiation passes into a crystal through a slit at O, and let the beam have sufficient angular divergence to cover the whole angular range in which Bragg reflection from the planes sketched is appreciably excited. Then there exists at least some energy flux carried by Bloch waves within every volume element of the "energy-flow triangle" OAB. This spatial spread of multiple-scattered waves is on a much greater dimensional scale in the X-ray experimental situation than in electron microscopy, and is of much greater practical significance with the former than with the latter. Consider the electron case. The specimen foil thickness may be about 1,000 Å and the angle $2\theta_\mathrm{B}$ for low-order reflections will be about 0.01 so that AB $\approx$ 10 Å only. This distance is comparable with the resolution limit of the instrument. It is only about one tenth the typical width of the diffraction contrast image of a dislocation. Hence for the purpose of performing diffraction contrast calculations the crystal foil can be divided into columns normal to its surfaces and of order 10 Å in diameter. In each of these columns the energy flow is effectively one-dimensional along the axis of the column.

In a typical X-ray topographic experiment the specimen may be 1 mm thick, $2\theta_B$ may be about 0.2 and so AB is about 200 microns. Compare this distance with the approximate resolution limit of the technique, say 1 to 2 microns, and with the diffraction contrast image width of a dislocation which may range from 2 to 20 microns. It is clear that in the X-ray case an extra dimension is added which makes all diffraction contrast calculations of lattice imperfections more difficult and laborious than in the electron case. Only certain simple types of lattice distortion allow calculation of the intensity distribution over AB to be made with a simplification equivalent to the "column approximation" valid in transmission electron microscope theory. Generally it is necessary to determine how lattice distortion modifies the propagation of Bloch waves on a two-dimensional mesh of points in the whole triangle OAB. One may specifically observe that if an imperfection occurs locally at X in an otherwise perfect crystal then it will modify the intensity distribution of emerging X-rays over a distance YZ on the exit face where the angle ZXY is $2\theta_B$. Alternatively, if a considerable width of entrance surface of the crystal is illuminated, either by a broad X-ray beam, or by scanning the crystal with a narrow beam as is done in the "projection topograph" technique, then in order to determine the intensity emerging at O′ (in either the incident beam direction or the diffracted beam direction) account must be taken of the way lattice imperfections affect the diffracting conditions at all points in the triangle A′O′B′. This diffraction problem has been investigated for rather general conditions of lattice distortion by Takagi [76, 77].

One of the early observations of transmission X-ray topography was to find how much structure of interest occurs in the "spatial intensity profiles" which are recorded by placing a photographic plate to receive either the multiply-scattered diffracted beam CD or the multiply-scattered incident beam EF. (In practice it is the diffracted beam CD that is recorded in the great majority of experiments). In the case of specimens in which the phenomenon of anomalous absorption (the Borrmann effect) is not prominent, but which are in other respects behaving as perfect crystals, the greater density of energy-flow vectors close to the directions L_0O and L_0G compared with that parallel to the Bragg planes, which follows as a consequence of the asymptotic shape of the dispersion surface hyperbolae, leads to an accumulation of energy-flow trajectories taking paths close to OB and to OA in fig. 16.17. This accumulation causes maxima of diffracted intensity to occur in the regions close to C and D in the diffracted beam intensity profile.

Suppose, however, that the crystal absorbs X-rays fairly strongly. If the structure is simple so that the atom centres are all located on the Bragg

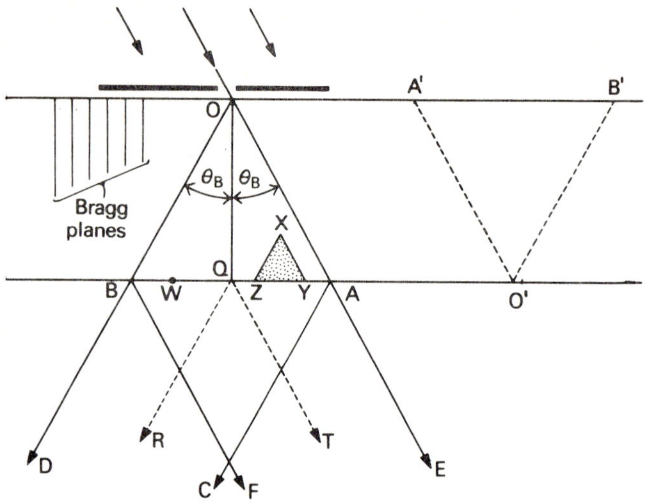

Fig. 16.17. Energy-flow triangle OAB in a parallel-sided crystal plate with Bragg planes parallel to OQ. Details explained in text.

planes, and if K-shell electrons are concerned in the absorption process, the assumption of "point" atoms is a fair first approximation. Then it will be apparent from fig. 16.14 that the branch (1) mode can pass through the crystal with less than average absorption. This difference between the two modes is greatest when the "standing wave ratio" is unity i.e. when $\alpha = \pm 1$, as it is for exact satisfaction of the Bragg condition. Then branch (2) waves are stopped but branch (1) waves travel with little attenuation parallel to the Bragg plane. This is the *"Borrmann effect"*. In fig. 16.17, therefore, as the Borrmann effect comes increasingly into play with increasing absorption, the energy flow is more and more restricted to the path OQ, and intensity maxima appear along QR and QT in the diffracted and transmitted beams emerging from the crystal exit surface. Figure 16.18 (from Kato [78]) illustrates the transition of the diffracted-beam spatial intensity profiles from the zero absorption to the high absorption cases.

There is an additional phenomenon of importance in diffraction by perfect crystals which would be experimentally observed on profiles such as CD, (but which is not shown in fig. 16.18). The angular range over which perfect crystals reflect X-rays is typically only a few seconds of arc. A coherent spherical wave coming from an X-ray emitting atom in the X-ray tube target, which may be distant say 50 cm from the crystal, will cover this angular range within a distance range centred on O in fig. 16.17 of only a few microns

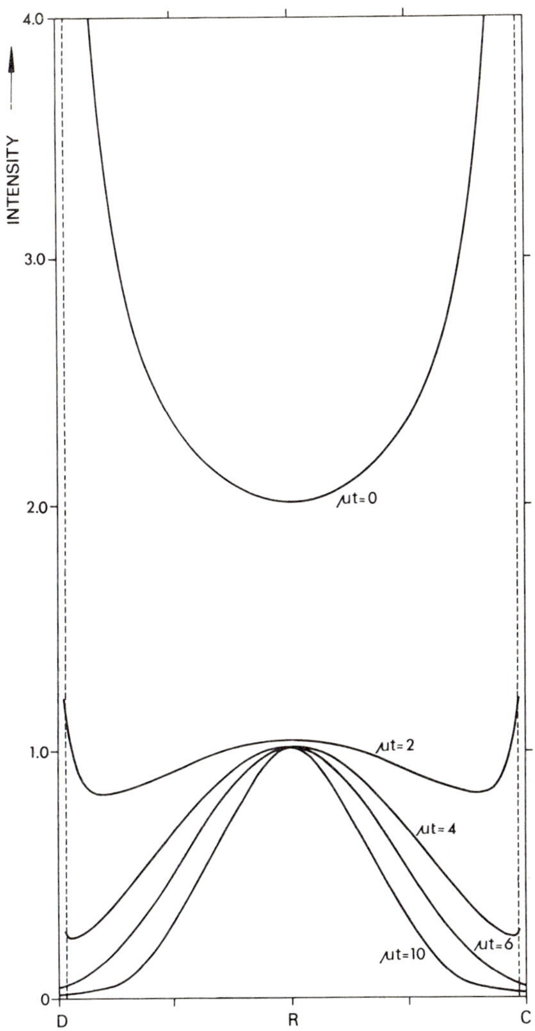

Fig. 16.18. Calculated spatial intensity profile across the diffracted beam CD in fig. 16.17. Intensity scale arbitrary. X-ray absorption centres all lie precisely on the Bragg planes. Pendellösung interference phenomena neglected. (From Kato [78]).

width. Thus at any point within the triangle AOB more than about 10 microns, say, from the X-ray entrance surface, one must regard all the Bloch waves flowing within OAB as produced by a single *coherent* source. Hence waves associated with different wave points on the dispersion surface are capable of interfering with each other. Consider a point W on the X-ray exit face of the crystal. It receives energy from O carried by the Bloch waves associated with wave points $N^{(1)}$ and $N^{(2)}$ on branch (1) and branch (2) (see fig. 16.16), for which points the normals to the dispersion surface are parallel to OW. The symmetry of fig. 16.16 shows that both Bloch waves have the same value of $|C_g/C_0|$, but their component wave vectors have different magnitudes. Consequently they interfere with a relative phase which is a function of position within the triangle AOB. This is known as *Pendellösung interference.* In the special case of exact satisfaction of the Bragg condition, energy flows parallel to OQ and the interference pattern along OQ has period ξ_g for both "O" waves and "G" waves. In X-ray work ξ_g is known as the *Pendellösung period.* It can be measured accurately in perfect undistorted crystals and it gives a precise, absolute value of F_g, via eq. (16.24). X-ray Pendellösung fringes were first observed and used to measure absolute values of F_g by Kato and Lang [79]. It was the particular form the pattern took on spatial intensity profiles such as CD that led to the development of the coherent spherical wave theory of X-ray diffraction in perfect crystals by Kato [80]. This theory is necessary to explain Pendellösung interference phenomena in the X-ray case, but in the electron case a plane-wave theory is adequate: the important differences between the X-ray case and the electron case are discussed in a review by Kato [81]. For proper calculation of X-ray diffraction contrast arising from imperfections such as dislocations and stacking faults the spherical wave theory should again be used; but as a first approximation many features on X-ray topographs can be accounted for by plane-wave theory with the deviation parameter, w, set equal to zero. (The deviation parameter is defined in §4.2.1., eq. (16.32)).

Returning to consider spatial intensity profiles, the profile of the multiply-scattered incident beam across EF is of as much importance as that across CD, when it comes to interpretation of diffraction contrast caused by lattice imperfections. Experimental recording of the profile across EF is rendered difficult because of scattering from the intense direct beam, containing all wavelengths emitted by the X-ray tube, that travels along OA in fig. 16.17. But satisfactory recording can be effected by using as incident beam a slightly diverging pencil of monochromatic X-rays produced by a curved crystal monochromator [79, 82]. Kato's [78] calculated profile across EF is shown in fig. 16.19. Like fig. 16.18, it averages over Pendellösung intensity oscilla-

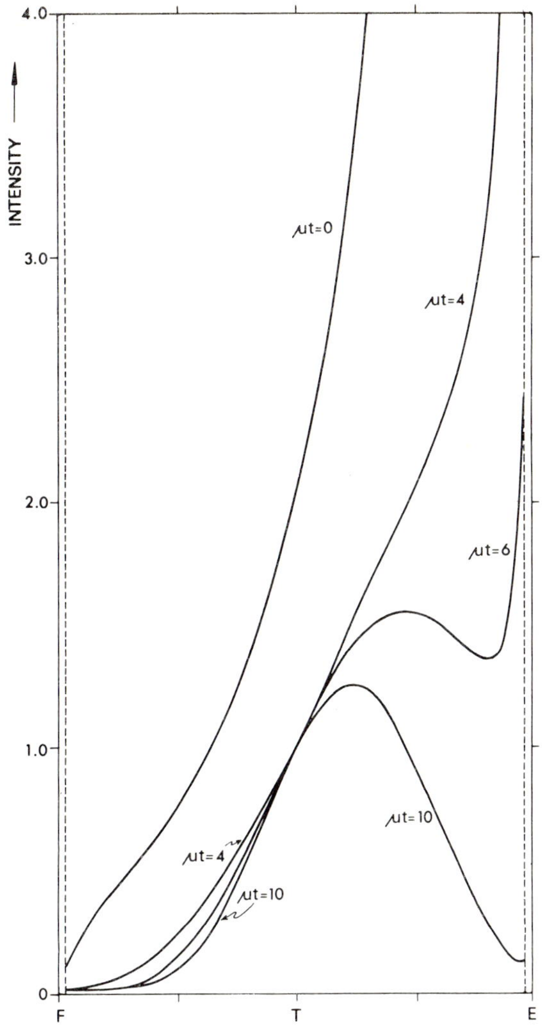

Fig. 16.19. Spatial intensity profile across the multiply-scattered primary beam EF in fig. 16.17, calculated with similar assumptions as curves in fig. 16.18. (From [78]).

tions. Also, both figs. 16.18 and 16.19 assume that the X-ray absorbing centres are point atoms all coinciding with the Bragg planes. Departures from this ideal situation have the consequence that to produce a given profile in figs. 16.18 or 16.19 a larger value of μt is required than that actually written alongside it in the figure. The reason is that the rapidity with which the profiles transform from those characteristic of low-absorption to those of the highly absorbing case, as μt increases, depends upon the degree to which branch (1) and branch (2) waves undergo respectively less and more absorption than that corresponding to the tabulated absorption coefficient.

16.4. Assessment of crystal perfection by X-ray topography

16.4.1. X-ray topographic techniques

Since several reviews of X-ray topographic techniques have been published [46–49] which describe the variety of X-ray diffraction geometries that may be employed, little need be said here except to point out a few fundamental ways in which the experimental techniques may differ from each other. The most fundamental is whether a surface reflection technique (such as the Berg-Barrett technique) or a transmission technique is employed. Circumstances such as irreducibly large specimen thickness and absorption may compel use of surface reflections only, but if at all possible the specimen should be thinned and sufficiently penetrating radiation used so that transmission topographs may be taken. It is with the transmission arrangement that the full potentialities of the X-ray method may be realised for the non-destructive assessment of lattice imperfection within crystal interiors which are otherwise inaccessible to examination. In the transmission arrangement it is advantageous to work under conditions of low X-ray absorption, that is with values of X-ray linear absorption coefficient, μ, and specimen thickness, t, such that the product μt does not much exceed unity. The interpretation of the diffraction contrast patterns generally becomes simpler as μt is reduced. On the other hand, when the thickness of the specimen is less than about one third the extinction distance (i.e. one third of the quantity ξ_g given by equation (16.24)) diffraction contrast of the kind often called *extinction contrast*, such as is produced by individual dislocations, disappears. This type of diffraction contrast depends essentially on the different diffracting power of a distorted crystal region (such as that surrounding a dislocation) compared with the matrix of perfect crystal in which it is embedded; and it clearly cannot be produced when the whole crystal is so thin that the conditions of kinematical diffraction (i.e. negligible multiple scattering) apply to it, however perfect it may be. The minimum-thickness criterion suggested above applied to the extinction distances listed in table 16.2 shows that in certain cases the minimum thickness for X-ray extinction contrast is less than the maximum thickness for fair dislocation visibility in the 1 MeV microscope. X-ray diffraction contrast that may be termed *orientation contrast* arises from gross misorientations (as at low-angle boundaries, for example) that are comparable with or greater than the angular divergence of the incident beam which is typically 1 to 2 minutes of arc. Orientation contrast can still occur with specimen thicknesses in the "kinematical diffraction" range, and would in fact be observable with specimen crystals even so thin that they manifested observable "diffraction broadening" due to their small crystal thickness (say $t < 1{,}000$ Å).

The most widely used transmission techniques are those of the "*section topograph*" described in 1957 [83] and the "*projection topograph*" described in 1959 [84]. The topographs shown in this chapter are all projection topographs. Several factors which conspire together to limit X-ray topographic resolution to not better than 1 to 2 microns under practical conditions have been discussed in the recent review cited [49]. This resolution is indeed poor compared with the electron microscope, but when it is borne in mind that there is in principle no upper limit to topograph image area, and that topograph images 80 mm in diameter are routinely recorded from the silicon slices used in the semiconductor industry, it will be recognised that the quantity of information carried by an X-ray topograph compares favourably with that in an electron micrograph.

16.4.2. Diffraction contrast by dislocations

16.4.2.1. A simple calculation of dislocation image widths. Figure 16.20 is part of an X-ray topograph of a silicon crystal containing a single-ended Frank-Read spiral dislocation source. This dislocation source, or dislocation "mill" as it is sometimes called, has operated for about 10 turns. (In silicon, higher numbers of turns than about 10 are rarely observed.) The three-dimensional geometry of dislocations in this specimen has been described [85]. Let the pattern serve here just to illustrate some points relating to diffraction contrast. The left-hand region of the specimen is wedge-shaped, the wedge edge coinciding with the left-hand margin. The Pendellösung fringes mark depth contours at intervals of the Pendellösung period which is 53 microns for AgKα radiation and a 111-type reflection as used in this topograph. The dislocation contrast diminishes as the specimen gets thicker, and the dislocation images become diffuse as the distance from dislocation line to X-ray exit surface increases. (This diffusion arises from multiple scattering within the energy-flow triangle OAB of fig. 16.17.) The plane of the Frank-Read source lies about 350 microns from the X-ray exit surface in fig. 16.20. The intensity profile across dislocation images is generally asymmetric (it is frequently bimodal, as explained in §4.2.2. below). A reasonable measure of

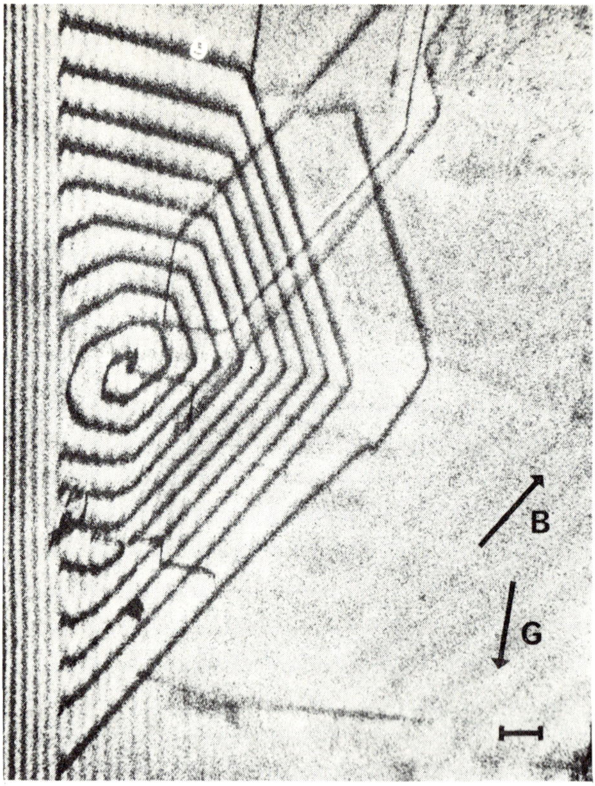

Fig. 16.20. X-ray topograph showing single-ended Frank-Read source in silicon crystal. Slip plane is $(1\bar{1}\bar{1})$. Arrow B, parallel to [110] shows direction of Burgers vector; arrow G shows direction of diffraction vector $\bar{1}1\bar{1}$. Scale mark 100 microns.

the image width is specified as the width between points where the photographic density has risen to double that of the local average perfect-crystal background intensity on the topograph. Estimates of this width range from about 8 microns to about 20 microns on various segments of the spiral shown.

Now consider a naive analysis which can account for image widths of such magnitudes. In fig. 16.21, ABCD represents an area of Bragg-reflecting plane which is pierced normally at O by a pure screw dislocation with Burgers vector **b**. Thus **b** is parallel to the reciprocal lattice vector **g** of *undistorted* regions. Let EF be a trace of the plane of incidence of the X-rays. Consider

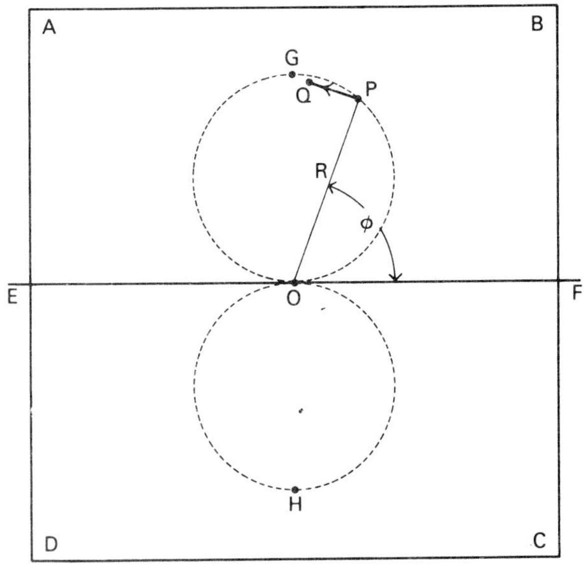

Fig. 16.21. Contours of equal effective misorientation surrounding a screw dislocation as used in simple calculation of dislocation image width. The screw dislocation pierces the Bragg plane at the origin of coordinates O. Other features described in text.

the point P, coordinates (R,ϕ), in the helicoidal sheet on which atom centres in the dislocated crystal lie. Its local reciprocal lattice vector, $\mathbf{g}'$, is rotated relative to $\mathbf{g}$, the rotation axis being OP and the rotation angle $b/2\pi R$. The difference vector $\mathbf{g}'-\mathbf{g}$ is represented by PQ on the drawing. The effective misorientation of P with respect to Bragg reflection of X-rays incident in planes parallel to EF is proportional to the projection of PQ on EF, and is $b\sin\phi/2\pi R$. On the Bragg plane ABCD contours of equal misorientation (positive or negative) can be plotted: they form figure-of-eight-shaped curves such as that shown by the interrupted line.

In the standard topographic technique the crystal is illuminated by a beam whose divergence is generally 10 or more times greater than the angular range of reflection of the perfect crystal. The integrated intensity of reflection of the perfect crystal is proportional to its angular reflecting range [63, 67], in the low X-ray absorption case. In the approximation for zero absorption, the reflection curve in the case of symmetrical transmission through a parallel-plate specimen is given by [19, 67]

$$I = I_{\max}/(1+w^2) \qquad (16.31)$$

where w, the *deviation parameter*, is related to the deviation, $\Delta\theta$, from the exact Bragg angle, the interplanar spacing, d, and the extinction distance, ξ_g, by

$$\Delta\theta = \theta - \theta_B = wd/\xi_g \,. \tag{16.32}$$

The full width at half maximum intensity is the range of $\Delta\theta$ corresponding to the range $-1 < w < +1$. In the case under consideration, $d = 3.13$ Å, and the range is $2 \times 3.13 \times 10^{-8}/53 \times 10^{-4}$ radians, or 2.4 seconds of arc.

Let it now be assumed that crystal regions close to the dislocation whose effective misorientation $b\sin\theta/2\pi R$ is equal to or greater than the full width at half-maximum intensity of the perfect matrix are able to diffract X-rays incoherently with respect to the matrix (since they can "see" incident X-rays outside the angular range of those reflected by the matrix) and that the strongly distorted regions behave more as imperfect or mosaic crystals than as perfect crystals with respect to their diffracting power. Then the diffracted intensity on the topograph should rise steeply above its background perfect-crystal value at linear distances OG or OH on either side of the geometrical image of the dislocation O. The linear width, V_S, of this pure screw dislocation image will then be given by

$$b/\pi V_S \approx 2d/\xi_g$$

or

$$V_S \approx (2\pi d)^{-1} b\,\xi_g \,. \tag{16.33a}$$

For a pure edge dislocation, with b parallel to g, the corresponding estimate of image width, V_E, is about 1.75 V_S, again derived by a simple "maximum tilt" calculation using the displacement field given by equation (16.6), for elastically isotropic media and Poisson's ratio, ν, equal to 1/3. In the screw dislocation case, if the Burgers vector **b** makes an angle ψ with **g**, then the expected image width V_S is reduced to

$$V_S \approx (2\pi d)^{-1} b\,\xi_g \cos\psi$$

or

$$V_S \approx (2\pi)^{-1} \xi_g\, \mathbf{b}\cdot\mathbf{g} \,. \tag{16.33b}$$

Consider the pure screw segment in fig. 16.20. The angle ψ is that between [111] and [110], and is 35¼°. The Burgers vector magnitude $b = 3.84$ Å. The estimate of V_S derived from equation (16.33b) is 8½ μm, and is in good agreement with the observed value.

16.4.2.2. Double images of dislocations. In certain circumstances X-ray topograph images of dislocations appear as double dark lines. Before interpreting such an observation as an indication of a dislocation dipole, or a dislocation dissociated into a pair of partials, it should be checked whether there is not a straightforward explanation in terms of the simple dislocation image model outlined in §4.2.1 above, for such straightforward explanation will in all probability account for the observation. On the crude assumption that the blackness at each point of the topograph image is proportional to the thickness of the crystal region traversed in which the misorientation exceeds $2d/\xi_g$, it will be seen from the figure-of-eight plots in fig. 16.21 that this thickness has a narrow minimum at O. A central dip in blackness can often be resolved in the X-ray topographic images of dislocations when geometrical conditions such as dislocation line orientation and depth below X-ray exit surface are favourable, and a good topographic technique is practised. Since the whole bi-lobed contour plot scales up as $\mathbf{g}{\cdot}\mathbf{b}$ and/or ξ_g increase (*vide* equations (16.33a) and (16.33b)) the bimodal image intensity profile becomes quite obvious in high-order reflections and weak reflections. Some published examples of double images of dislocations which arise as a direct consequence of the dislocation strain-field, and whose scale can be satisfactorily accounted for in terms of the bi-lobed contours of fig. 16.21, relate to dislocations in triglycine sulphate [86], CdS [87], silicon [88], muscovite [89], quartz [90], potassium alum [91] and ZnS [24]. Authier [92, 93] has discussed double images in some detail. They should not be a cause of misinterpretation of X-ray topograph images of dislocation configurations.

16.4.2.3. Factors contributing to diffraction contrast. It is quite surprising how well the simple formulae, equations (16.33a) and (16.33b), account for the observed widths of dislocation images over a wide range of values of the variables ξ_g and $\mathbf{g}{\cdot}\mathbf{b}$. This must not lead to the impression that the X-ray diffraction contrast theory of dislocation images really can be despatched with the naive treatment given in §4.2.1; for the diffraction phenomena involved are in reality quite complex and difficult to treat analytically and completely. Only the briefest indication of some of the diffraction processes involved can be attempted here, just for the purpose of providing guidance in the interpretation of topographic images of dislocation structures.

Much of the difficulty in calculating the expected diffraction contrast of dislocations arises from the type of long-range strain field they possess, in particular the way the local lattice plane tilt and curvature change continuously in the neighbourhood of the dislocation. Consider a Bloch wave such as that denoted by the wave point $N^{(1)}$ or $N^{(2)}$ in fig. 16.16 as it takes its trajectory through an almost perfect crystal along the path OW within the energy-flow triangle OAB of fig. 16.17. As the trajectory approaches a dislocation it enters regions where the lattice planes are curved, the curvature increasing with increasing proximity to the dislocation core. How the Bloch waves respond to gentle curvature was first established by Penning and Polder [94, 95] and later analysed more fully by Kato [96–98]. Their theory applies to the case when the spatial rate of variation of the displacement vector whose components are given by equations (16.4) or (16.6) is sufficiently small for the orientation of a *local* reciprocal lattice vector $\mathbf{g}'$ to be specified within an angular range less than a few tenths of the perfect-crystal reflection-curve width, equation (16.32), throughout a volume which is defined by a cross-sectional area not so small that a pencil of X-rays passing through an aperture of the same area would be diffraction-broadened through the angular range of $\mathbf{g}'$ across the aperture, and a length, parallel to the pencil of rays, not less than a few tenths of ξ_g. Under such conditions, the Bloch wave behaves like a light ray in a medium of continuously varying refractive index. Curvature of Bragg planes causes wave-points such as $N^{(1)}$ and $N^{(2)}$ to migrate along the dispersion surface hyperbolae. If the sense of lattice plane curvature is that which makes the radius drawn towards the centre of curvature point in the same direction as **g**, then branch (1) Bloch waves migrate so that the energy-flow vector normal to the dispersion surface swings from the OA direction towards the OB direction, whereas energy-flow vectors on branch (2) swing oppositely. For zero absorption, there is no net change in integrated reflection intensity with very weak curvatures, but stronger curvatures of either sense cause an increase in diffracted intensity [98]. When the Borrmann effect is appreciable the dominance of branch (1) waves ensures that when the Bragg-plane radius of curvature, R_{plane}, drawn towards the centre of curvature has the same sense as **g** there is an *increase* in diffracted intensity, and *vice versa.* Application of this effect to determining *senses* of Burgers vectors will be discussed below (§4.3.2). Bloch waves whose wave points are closest to the Brillouin zone boundary are 'refracted' most strongly by lattice-plane curvature. At the point on their trajectory through the triangle OAB where such waves exactly satisfy the Bragg condition the radius of curvature of the ray pencil, R_{ray}, is related to R_{plane} by

$$R_{\text{ray}} = \tfrac{1}{2}(d/\xi_g)R_{\text{plane}} \,. \tag{16.34}$$

For a typical value of (d/ξ_g), such as that quoted following equation (16.32), then

$$R_{\text{ray}} \approx 10^{-5} R_{\text{plane}} ,$$

a clear demonstration of the extreme sensitivity to lattice curvature of energy-flow trajectories within the energy-flow triangle OAB.

As the X-rays enter the regions of stronger curvature a new process takes place, called "*interbranch scattering*". In this process some of the energy associated with Bloch waves on branch (1) is diverted into branch (2) waves, and *vice versa.* This process begins to be appreciable when the lattice curvature becomes so high that, very roughly [95],

$$\frac{1}{R_{\text{plane}}} \geqslant d\,\xi_g^{-2} . \tag{16.35}$$

This condition is attained when R_{plane} is of metre order in the case of strong X-ray reflections. Referring again to fig. 16.16, a diversion of energy in the Bloch wave associated with $N^{(1)}$ to a Bloch wave on branch (2) below it, or from $N^{(2)}$ to the Bloch wave on branch (1) above it, will switch energy flow from the direction OW in OAB (fig. 16.17) to a direction making an equal angle with OQ but on its other side, i.e. in the triangle OQA. Now, looking at the important spatial energy distribution curves figs. 16.18 and 16.19, it is evident that in the zero-absorption case there is very strong energy flow in the triangle OQA, increasing rapidly as OA is approached. Interbranch scattering will switch some of this energy into a path in the triangle OQB, and hence put more energy into the diffracted beam direction. In X-ray topographs taken in the low-absorption case, the strong excess intensity in images of stacking faults and analogous defects arises entirely from interbranch scattering, and a great part of the excess intensity in dislocation images arises also from this mechanism.

16.4.2.4. The effects of X-ray absorption. The pronounced effect that increasing absorption has on the character of the images of dislocations and other lattice imperfections is a consequence of the Borrmann effect. It is natural to use strong X-ray reflections for taking X-ray topographs, and when F_g is relatively large the g-th Fourier component of the distribution of X-ray absorbing centres will in general also be relatively large. The latter condition ensures a significant difference between the effective absorption coefficient of branch (1) and branch (2) waves when the standing wave ratio is close to unity. The following discussion will assume this condition applies. Two ranges of μt may be considered: that when μt is in the range of, say, 2 to 6; and that of higher values, say $\mu t \geqslant 10$. Reference again to fig. 16.19 shows that when μt is more than a few units the energy flow in directions close to OA is much reduced. As a consequence, possible additions to diffracted beam intensity through interbranch scattering are also diminished. Positive contrast (i.e. excess diffracted intensity) at stacking faults, and that part of dislocation image contrast that arises from interbranch scattering, become much weaker. In certain circumstances dislocation contrast may become very feeble in this range of values of μt. However, the contrast that arises in the outer parts of dislocation images, where "wave-point migration" but not interbranch scattering is occurring, can become very pronounced in this absorption range, and can be positive or negative according to the sense of the lattice plane curvature with respect to the direction of **g**.

In the high absorption range, all defects except those very close to the X-ray exit surface will become visible with negative contrast, and only those fairly close to the exit surface will show well-defined images and strong contrast. When μt is large, energy which is transferred from branch (1) waves to branch (2) waves by interbranch scattering is soon lost by absorption, and defects which cause such scattering give rise to negative contrast in both diffracted-beam topographs and in forward-beam topographs, i.e. in recordings either of the beams CD or EF in fig. 16.17. When μt is quite high, negative contrast is also produced in both CD and EF beams by defects producing lattice plane curvatures too weak to cause interbranch scattering. This is because wave-point migration prevents the wave-point of any branch (1) ray from remaining very close to the Brillouin Zone Boundary for more than a fraction of the ray's path through the crystal. Hence only during a fraction of its trajectory can it enjoy minimum absorption.

16.4.3. Analysis of dislocation configurations

16.4.3.1. Determination of Burgers vector direction. The general rule for *invisibility* of a dislocation (or any other defect) in a given Bragg reflection is that its displacement field has no component normal to the Bragg plane used. The art of determining Burgers vector directions by X-ray topography consists of choosing X-ray diffraction conditions to obtain maximum significant differences in diffraction contrast of an image of a given dislocation in various reflections, and to obtain data of sufficient quality and quantity to permit unambiguous identification of Burgers vector directions. The equations (16.4) and (16.6) for the displacement fields surrounding screw and edge dislocations respectively are those for an elastically isotropic medium: discussion will be confined to this case. (In practice, elastic anisotropy has not prevented unambiguous Burgers vector identification by X-ray topography). The displacement fields of equations (16.4) and (16.6) are of course modified by proximity to the specimen surface. This modification is of less consequence in X-ray topography than in thin-film transmission electron microscopy. Nevertheless, dislocation images do exhibit well recognisable special diffraction contrast effects at the dislocation outcropping points, but such effects interfere little with Burgers vector identification using transmission X-ray topographs.

Let a dislocation of Burgers vector **b** and line vector **l** cause a displacement $\mathbf{u}(x,y,z)$, $(\mathbf{u} \equiv u\mathbf{i} + v\mathbf{j} + w\mathbf{k})$; **b** has a screw component $\mathbf{l}\cdot\mathbf{b}$ and edge component of magnitude $|\mathbf{l} \times \mathbf{b}|$, and it gives rise to displacement components parallel to **b**, to $\mathbf{l} \times \mathbf{b}$ and to $\mathbf{l} \times \mathbf{b} \times \mathbf{l}$, as can be seen by combining equations (16.4) and (16.6). There are two special cases when $\mathbf{u}\cdot\mathbf{g}$ is zero: these can be recognised in the diagram fig. 16.22. Firstly, when **b** and **l** are collinear, $\mathbf{l} \times \mathbf{b}$ and $\mathbf{l} \times \mathbf{b} \times \mathbf{l}$ are zero and so there is the single invisibility condition $\mathbf{b}\cdot\mathbf{g} = 0$. This is the case of the screw dislocation for which there is a zone of complete invisibility, i.e. the zone of all Bragg planes which contain **b**. Secondly, if **b** and **l** are orthogonal, $\mathbf{l} \times \mathbf{b} \times \mathbf{l}$ is collinear with **b**. Then $\mathbf{u}\cdot\mathbf{g}$ is zero when $\mathbf{b}\cdot\mathbf{g} = 0$ and $\mathbf{l} \times \mathbf{b}\cdot\mathbf{g} = 0$, that is for the single Bragg plane normal to the dislocation line. Mixed dislocations never give $\mathbf{u}\cdot\mathbf{g} = 0$. However, under low-absorption conditions, it is the displacement component parallel to **b** that dominates

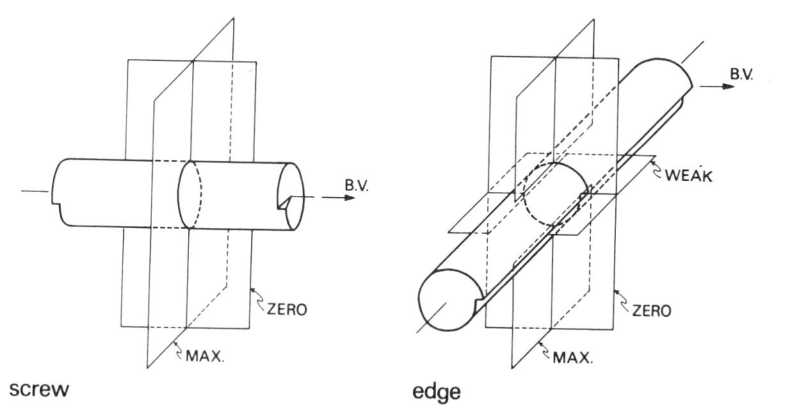

Fig. 16.22. Diagram showing visibility of pure screw and pure edge dislocations in reflections from various Bragg planes.

in producing diffraction contrast, and consequently dislocations are either invisible or at most weakly visible if $\mathbf{b}\cdot\mathbf{g} = 0$. When $\mathbf{b}\cdot\mathbf{g}$ is not zero, then to a very rough approximation the integrated excess density in a dislocation image is proportional to $(\mathbf{b}\cdot\mathbf{g})^2$, i.e. proportional to the area within the lobes of the critical misorientation contours of fig. 16.21. In experiments to identify Burgers vectors it is helpful to compare observed dislocation visibilities with values of $\mathbf{b}\cdot\mathbf{g}$ tabulated for the Bragg reflections used and the Burgers vectors deemed likely to be present.

16.4.3.2. Determination of Burgers vector sense. Some of the very earliest X-ray topographic observations showed that a given dislocation, or, very often, an array of dislocations of like sign on a slip plane, displayed contrast of opposite sense in the outer regions of their images when hkl and $\bar{h}\bar{k}\bar{l}$ topographs were compared. This phenomenon became noticeable when $\mu t \geqslant 1$. It was evident that the sense of contrast was in some way related to the sense of lattice plane curvature in the dislocation strain field. Quantitative analysis of the effect became possible following the development of Penning and Polder's theory [94]. The theoretical predictions were checked by introducing dislocations of known sign by plastic deformation of germanium crystals [99]. The determination of *sense* of Burgers vectors through comparison of their images in hkl and $\bar{h}\bar{k}\bar{l}$ reflections has now become a routine and useful adjunct to X-ray topographic studies of dislocations.

Consider a dislocation which is pure edge, or has a substantial edge component. The "wedging" effect of inserting an extra half plane of atoms causes a long-range curvature of the dislocation slip-plane (this is the displacement v of equation (16.6)). The effect of this curvature on the X-ray image will be most marked if the slip-plane is used as Bragg-plane. Then, as described in §4.2.3, when the Bragg-plane radius of curvature, $\mathbf{R}_{\text{plane}}$, drawn towards the centre of curvature, is parallel to $\mathbf{g}$, branch (1) waves are swung into the diffracted beam direction and branch (2) waves away from it. When μt is equal to a few units there results strong positive contrast when $\mathbf{R}_{\text{plane}}$ is parallel to $\mathbf{g}$, and strong negative contrast when $\mathbf{R}_{\text{plane}}$ and $\mathbf{g}$ are antiparallel.

Under favourable conditions the sense of pure screw dislocations can be determined, either from an asymmetry in strengths of the two components of their bimodal image intensity profile [88] or from the local lattice plane curvature arising from stress relief at their surface outcrops.

16.4.3.3. Determination of Burgers vector magnitude. Comparison of the visibility of a given dislocation in a set of Bragg reflections involves essentially a comparison of relative values of $\mathbf{b}\cdot\mathbf{g}$. If at the same time the dislocation image widths are measured, then a set of estimates of *magnitudes* of **b**·**g** is obtained, for, as already discussed in §4.2.1, the agreement between observation and the naive formulae (16.33a) and (16.33b) is sufficiently good to make such estimates realistic. Certainly, on a given topograph, there will usually be little doubt in distinguishing between images for which **g**·**b** = 1 and **g**·**b** = 2: this discrimination can be regarded as an essential part of the process of Burgers vector identification. In complex structures information on the magnitudes of **b** may be much desired, and it may be obtainable through X-ray topography. The special case of giant dislocations, such as are found in ZnS, is certainly amenable to quantitative study by X-ray topography, and measurements of Burgers vector magnitudes which are in the range 50 to 100 Å can be made with an uncertainty not greater than 30% [24].

16.4.4. Defects in natural and synthetic crystals

To conclude this review of perfection of crystals a few illustrations of defects in natural and man-made crystals will be discussed.

Among natural crystals *diamond* is of outstanding interest. Diamond is the stable phase of carbon from a depth of about 120 km down to 1,200 km or more below the Earth's surface, and there is no reason to suppose that diamond was crystallised under conditions other than those in which it was the thermodynamically stable phase. Whether diamond is at the present time being crystallised far below the Earth's surface, or whether all diamonds brought up to the Earth's surface by eruptive processes were drawn from a reservoir at great depth [100] in which they may have resided since an early stage in the Earth's history there is no means of ascertaining. However, there is no doubt that diamonds are products of a crystallisation process that took place under temperature and pressure conditions greatly different from those operative in most crystal-growing laboratories. Yet these products can range in their content of dislocations and other imperfections all the way from the closest similarity with some of the best man-made crystals, to, at the other extreme, examples of "growth pathology" quite bizarre in character and baffling in origin. Some accounts of the variety of imperfections in diamond that various optical and X-ray topographic techniques reveal have been given in reviews [101, 102]; the diamond chosen for description here would certainly be classed as an example of normal and indeed rather perfect growth, but nevertheless it contains imperfections well meriting detailed study.

Fig. 16.23 is an X-ray topograph of a diamond of plate-like habit in which a pair of opposite faces of the form {111} have developed much larger than the others. The topographic image gives a view nearly perpendicular to the large faces. Natural diamonds have acquired surface damage such as ring-cracks resulting from percussion, and scratches. Damage along edges and at corners is particularly severe. These damaged regions show as flecks of blackening on the topograph. They are easily distinguished from internal imperfections by examination of stereo-pairs of topographs. Conspicuous in fig. 16.23 are the bundles of dislocations which radiate from a point near the centre of

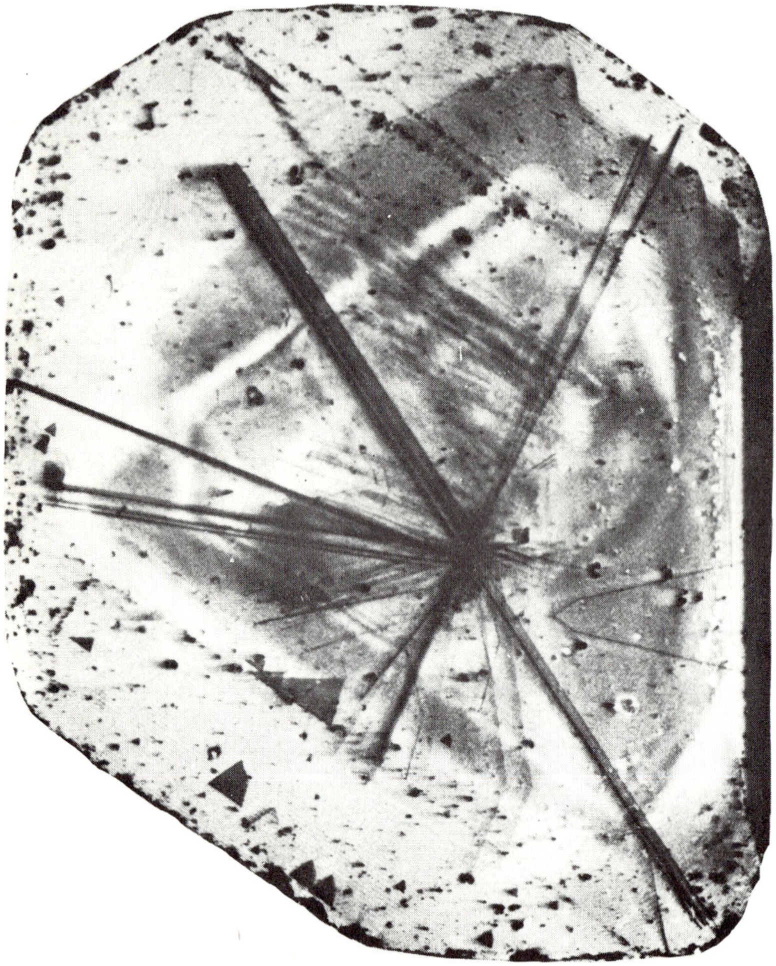

Fig. 16.23. X-ray topograph of dislocations in a natural diamond. Reflection 111, diffraction vector horizontal, image width 3.2 mm, radiation MoKα.

the crystal and terminate at outcrops on the crystal faces. These are clearly grown-in dislocations, generated in the earliest stages of crystallisation, probably by lattice closure mistakes during over-growth on a nucleus either of diamond or of another crystal species. Notable is the straightness of many dislocations. Broad diffuse light and dark bands covering a large part of the image area, and triangular or hexagonal in outline, indicate strain due to gradients of lattice parameter normal to growth horizons. The outer layers of this stone contain stacking faults; their images appear in fig. 16.23 as small triangular patches of extra diffracted intensity, and they are most noticeable in the lower left sector of the topograph. Such faults are not common in diamonds. When they do appear they give rise to interesting diffraction effects [103].

A configuration of generally straight dislocations radiating outward from a central nucleus is again to be seen in fig. 16.24. The crystal is *hexamethylenetetramine* ("hexamine") grown by slow evaporation of a solution in ethanol at constant laboratory temperature. This very soft crystal is easily damaged in handling and in the necessary operation of attaching it to a support which will hold it very steady during the duration of recording of the topograph. The patches of intense blackening on the topograph arise from such accidental damage. In this crystallisation process there can be little doubt that the crystal grew by heterogeneous nucleation upon some small particle floating in the solution. Lattice closure mistakes when growth enveloped the foreign nucleus can account for the bundles of grown-in dislocations radiating from the centre of the crystal. In parts of both topographs figs. 16.23 and 16.24 in which wedge-shaped regions of the crystal are being imaged, Pendellösung fringes can be seen running as equal thickness contours. Near dislocations and severe surface damage the Pendellösung fringes are distorted because strain gradients perturb the Pendellösung interference period [81]. However, good visibility of these fringes demonstrates that the diffraction behaviour of the crystals is representative of that of perfect or almost perfect lattices. In undistorted regions of the hexamethylenetetramine crystals, the measurement of the Pendellösung fringe spacing in images of wedge-shaped regions enclosed between adjacent faces of the rhombic dodecahedral form has provided an absolute measure of the structure factor of the 110 reflection [104], the measurement technique being that of Kato and Lang [79] previously discussed in ch. 16 §3.5.

A crystal much less perfect than either the diamond or the hexamethylenetetramine is shown in fig. 16.25. This basal-plane section of a "finger" or "candle" of natural *quartz* shows quite marked growth banding in some growth sectors, and at certain stages of its growth. There is a narrow band of

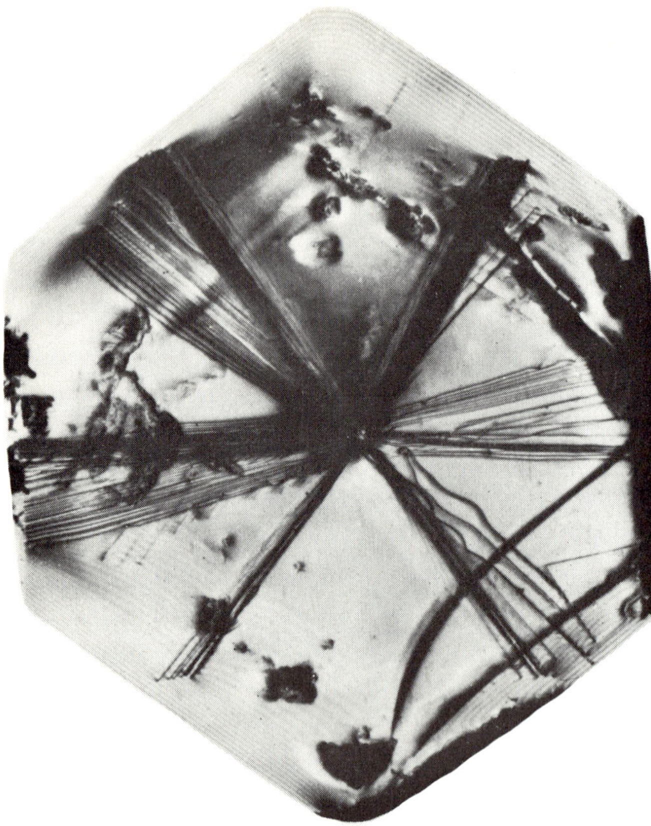

Fig. 16.24. X-ray topograph of dislocations in a crystal of hexamethylenetetramine grown from solution. Reflection 110, diffraction vector horizontal, crystal image width 4 mm, radiation MoKα.

intense diffraction contrast arising from a fault surface which encloses the central region of the section, irregularly hexagonal in shape. In this inner region the dislocation images appear as fairly short lines because they are inclined to the basal plane at angles not less than about 40°, and run through the crystal slice from X-ray entrance surface to X-ray exit surface accordingly. These dislocations generally lie in orientations within about 10° to 15° to the normal of the rhombohedral face upon which they outcropped during the relevant stage of crystal growth. However, in the outer parts of the crystal many of the dislocation images are much longer: some of these dislocations

Fig. 16.25. X-ray topograph of dislocations, Brazil twin boundaries and other imperfections in a plate of natural quartz about 1 mm thick cut parallel to the basal plane. Reflection $10\bar{1}0$, diffraction vector horizontal, crystal image width 2.3 cm. Radiation $MoK\alpha$.

make angles of only about 10° with the basal plane. Also to be seen in the outer parts of the crystal is a profusion of images of Brazil twin boundaries. These display "fault-fringes" like stacking faults, and with the conditions of low X-ray absorption under which this topograph was taken they produce a net increase in diffracted intensity through interbranch scattering, as mentioned in §4.2.3. Some of the Brazil twins are wedge-shaped, others are lath-like, and some are lamellae but a few microns thick.

Very different from fig. 16.25 is the dislocation pattern shown in fig. 16.26. The crystal is again quartz, and the specimen is again a slice cut parallel to the basal plane. However, it is hydrothermally grown synthetic quartz. The seed plate was parallel to (0001). Most of the dislocations were generated at the interface between the seed plate and the subsequent growth. That external surface of the crystal which lies parallel to the seed plate, i.e. parallel to (0001), has the hummocky relief given the descriptive name "cobble" texture. Each cobble represents an impurity cell. By taking topographs of crystal slices at various distances between the seed plate and the external surface of mean (0001) orientation it is possible to trace the gradual

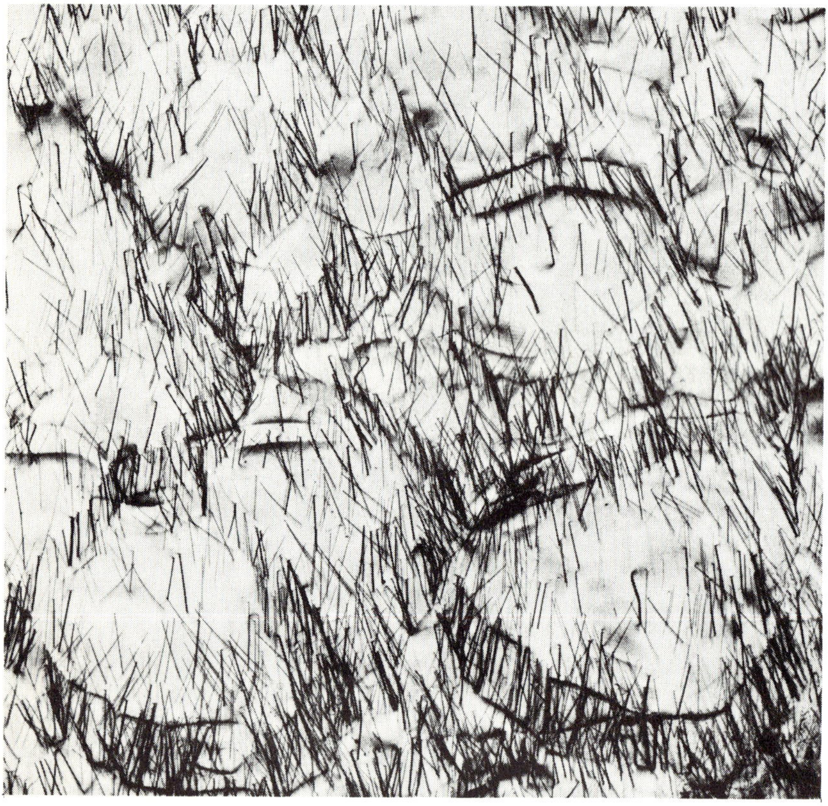

Fig. 16.26. X-ray topograph showing dislocations and contrast due to impurity cell walls in synthetic quartz. Crystal slice thickness about 1 mm, cut parallel to basal plane. Reflection $10\bar{1}1$, diffraction vector vertical, field size 1 mm^2, radiation AgKα.

development of these impurity cells [90]. Slices near the external surface (fig. 16.26 is an example of such a slice) show the cells in a well-developed state. The abnormal lattice spacing of impurity-containing material concentrated in the impurity cell walls gives rise to fault-surface fringe contrast, and to locally intense diffraction at the outcrops of cell walls with the surfaces of the specimen slice. The latter contrast is generated by lattice plane curvature resulting from relief of stresses where the impurity cell wall meets the free surface. In fig. 16.26, which is a reflection from the major rhombohedron plane an oblique view of the cell walls is obtained since they, like the dislocations, lie generally within small angles with [0001]. As the cell walls develop, they tend to capture a fair proportion of the dislocation population. It was definitely established that the outcrops on the external crystal surface of the fault surfaces seen in fig. 16.26 coincide with the grooves between cobble protuberances.

References

[1] D. Hull, Introduction to dislocations (Pergamon Press, London, 1965).
[2] J. Weertman and J.R. Weertman, Elementary dislocation theory (Macmillan Co., New York, 1964).
[3] A.H. Cottrell, Dislocations and plastic flow in crystals (Oxford University Press, Oxford, 1953).
[4] W.T. Read Jr., Dislocations in crystals (McGraw-Hill, New York, 1953).
[5] J. Friedel, Dislocations (Pergamon Press, London, 1964).
[6] J.P. Hirth and J. Lothe, Theory of dislocations (McGraw-Hill, New York, 1968).
[7] A.H. Cottrell, Theory of crystal dislocations (Blackie and Sons, London, 1964).
[8] F.R.N. Nabarro, Theory of crystal dislocations (Oxford University Press, Oxford, 1967).
[9] J.W. Christian and V. Vítek, Rep. Prog. Phys. 33, part 1, (1970) 307.
[10] G.I. Taylor, Proc. Roy. Soc. 145A (1934) 363.
[11] F.R.N. Nabarro, Advances in Physics 1 (1952) 269.
[12] B.A. Bilby, Research 4 (1951) 387.
[13] F.C. Frank, Phil. Mag. 42 (1951) 809.
[14] B.A. Bilby, R. Bullough and E. Smith, Proc. Roy. Soc. A231 (1955) 263.
[15] F.C. Frank and W.T. Read Jr., Phys. Rev. 79 (1950) 722.
[16] S. Amelinckx, The direct observation of dislocations, Solid State Physics, Suppl. 6 (1964).
[17] D.W. Pashley, Rep. Progr. Phys. 28 (1965) 291.
[18] S. Amelinckx, R. Gevers, G. Remaut and J. Van Landuyt, Modern diffraction and imaging techniques in material science (North-Holland Publ. Co., Amsterdam, 1970).
[19] P.B. Hirsch, A. Howie, R.B. Nicholson, D.W. Pashley and M.J. Whelan, Electron microscopy of thin crystals (Butterworths, London, 1965).

[20] F.C. Frank, Acta Cryst. 4 (1951) 497.
[21] A.R. Verma, Crystal growth and dislocations (Butterworths, London, 1953).
[22] J.D. Eshelby, J. appl. Phys. 24 (1953) 176.
[23] W.W. Webb and N.P. Bertolone, J. appl. Phys. 31 (1960) 207.
[24] S. Mardix, A.R. Lang and I. Blech, Phil. Mag. 24 (1971) 683.
[25] J.F. Nye, Proc. Roy. Soc. A198 (1949) 190, A200 (1949) 47.
[26] W.L. Bond and J. Andrus, Phys. Rev. 101 (1956) 1211.
[27] A.R. Lang, Nature, 213 (1967) 248.
[28] F.C. Frank, Disc. Faraday Soc. 5 (1949) 48.
[29] W.K. Burton, N. Cabrera and F.C. Frank, Phil. Trans. Roy. Soc. A243 (1951) 299.
[30] F.C. Frank, in: Growth and perfection of crystals, ed. by R.H. Doremus et al. (Wiley, New York, 1958) p. 3.
[31] J.W. Mitchell, in: Direct observation of imperfections in crystals, ed. by J.B. Newkirk and J.H. Wernick (Interscience, New York, 1962) p.3.
[32] W. Dekeyser and S. Amelinckx, Les dislocations et la croissance des cristaux (Masson et Cie., Paris, 1955).
[33] I. Sunagawa, Amer. Miner. 46 (1961) 1216.
[34] G.A. Bassett, Phil. Mag. 3 (1958) 1042.
[35] H. Bethge, Phys. Stat. Sol. 2 (1962) 3, 775.
[36] F.C. Frank, in: Growth and perfection of crystals, ed. by R.H. Doremus et al. (Wiley, New York, 1958) p. 411.
[37] A.R. Lang, J. appl. Phys. 28 (1957) 497.
[38] W.G. Johnston, Progr. Ceram. Sci. 2 (1962) 1.
[39] J.D. Livingston, in: Direct observation of imperfections in crystals, ed. by J.B. Newkirk and J.H. Wernick (Interscience, New York, 1962) p. 115.
[40] R.L. Fleischer, P.B. Price and R.M. Walker, Science 149 (1965) 383.
[41] J.M. Hedges and J.W. Mitchell, Phil. Mag. 44 (1953) 223.
[42] C.B. Childs and L.M. Slifkin, Phys. Rev. Lett. 5 (1960) 502.
[43] C.B. Childs and L.M. Slifkin, Brit. J. appl. Phys. 16 (1965) 771.
[44] W.C. Dash, J. appl. Phys. 27 (1956) 1193.
[45] W.W. Webb, in: Direct observation of imperfections in crystals, ed. by J.B. Newkirk and J.H. Wernick (Interscience, New York, 1962) p. 29.
[46] U. Bonse, M. Hart and J. Newkirk, in: Encyclopaedic dictionary of physics, ed. by J. Thewlis (Pergamon, Oxford, 1966) Suppl. Vol. I, p. 391.
[47] A.R. Lang, in: Encyclopaedia of X-rays and gamma rays, ed. by G.L. Clark (Reinhold, New York, 1963) p. 1053, 1058.
[48] A.R. Lang, Acta Geol. Geogr. Univ. Comenianae 14 (1968) 75.
[49] A.R. Lang, in: Modern diffraction and imaging techniques in material science, ed. by S. Amelinckx et al. (North-Holland Publ. Co., Amsterdam, 1970) p. 407.
[50] J.W. Menter, Proc. Roy Soc. A236 (1956) 119.
[51] E.W. Müller, in: Modern diffraction and imaging techniques in material science, ed. by S. Amelinckx et al. (North-Holland Publ. Co., Amsterdam, 1970) p. 683, 701, 717.
[52] E.W. Müller, Acta Met. 6 (1958) 620.
[53] D.G. Brandon and M. Wald, Phil. Mag. 6 (1961) 1035.
[54] W. Bollmann, Crystal defects and crystalline interfaces (Springer, Berlin, 1970).
[55] H. Hashimoto and R. Uyeda, Acta Cryst. 10 (1957) 143.

[56] D.W. Pashley, J.W. Menter and G.A. Bassett, Nature 179 (1957) 752.
[57] G.A. Bassett, J.W. Menter and D.W. Pashley, Proc. Roy Soc. A246 (1958) 345.
[58] U. Bonse and M. Hart, Appl. Phys. Lett. 6 (1965) 155.
[59] M. Hart, Sci. Progr. 56 (1968) 429.
[60] J. Brádler and A.R. Lang, Acta Cryst. A24 (1968) 246.
[61] A.R. Lang, Nature 220 (1968) 652.
[62] G. Oster, M. Wasserman and C. Zwerling, J. Opt. Soc. Amer. 54 (1964) 169.
[63] R.W. James, The optical principles of the diffraction of X-rays (Bell and Sons, London, 1965).
[64] B.E. Warren, X-ray diffraction (Addison-Wesley Publ. Co., Reading, Mass., 1969).
[65] A.J.C. Wilson, X-ray optics (Methuen, London, 1962).
[66] A.J.C. Wilson, Elements of X-ray crystallography (Addison-Wesley Publ. Co., Reading, Mass., 1970).
[67] B.W. Batterman and H. Cole, Rev. Mod. Phys. 36 (1964), 681.
[68] R.D. Heidenreich, Fundamentals of transmission electron microscopy (Wiley-Interscience, New York, 1964).
[69] L. Brillouin, Wave propagation in periodic structures (McGraw-Hill, New York, 1946).
[70] P.P. Ewald, Rev. Mod. Phys. 37 (1965) 46.
[71] J.C. Slater, Rev. Mod. Phys. 30 (1958) 197.
[72] J.A. Ibers, Acta Cryst. 11 (1958) 178.
[73] N. Kato, Acta Cryst. 11 (1958) 885.
[74] P.P. Ewald, Acta Cryst. 11 (1958) 888.
[75] E.H. Wagner, Acta Cryst. 12 (1959) 345.
[76] S. Takagi, Acta Cryst. 15 (1962) 1311.
[77] S. Takagi, J. phys. Soc. Jap. 26 (1969) 1239.
[78] N. Kato, Acta Cryst. 13 (1960) 349.
[79] N. Kato and A.R. Lang, Acta Cryst. 12 (1959) 789.
[80] N. Kato, Acta Cryst. 14 (1961) 526, 627.
[81] N. Kato, in: Crystallography and crystal perfection, ed. by G.N. Ramachandran (Academic Press, London, 1963) 153.
[82] N. Kato, Acta Met. 6 (1958) 647.
[83] A.R. Lang, Acta Met. 5 (1957) 358.
[84] A.R. Lang, Acta Cryst. 12 (1959) 249.
[85] A. Authier and A.R. Lang, J. appl. Phys. 35 (1964) 1956.
[86] A. Authier and J.F. Petroff, C.R. Acad. Sci. Paris 258 (1964) 4238.
[87] J.-I. Chikawa, J. appl. Phys. 36 (1965) 3496.
[88] A.R. Lang, Z. Naturforschung 20a (1965) 636.
[89] C. Williame and A. Authier, Bull. Soc. franç. Minér. Crist. 89 (1966) 269.
[90] A.R. Lang and V.F. Miuscov, J. appl. Phys. 38 (1967) 2477.
[91] S.H. Emara, B.R. Lawn and A.R. Lang, Phil. Mag. 19 (1969) 7.
[92] A. Authier, in: Advances in X-ray analysis (Plenum Press, New York, 1967) Vol. 10, p. 9.
[93] A. Authier, in: Modern diffraction and imaging techniques in material science, ed. by S. Amelinckx et al. (North-Holland Publ. Co., Amsterdam, 1970) 481.
[94] P. Penning and D. Polder, Philips Res. Rep. 16 (1961) 419.
[95] P. Penning, Philips Res. Rep. Suppl. No. 5 (1966) 1.
[96] N. Kato, J. phys. Soc. Jap. 18 (1963) 1785.

[97] N. Kato, J. phys. Soc. Jap. 19 (1964) 67.
[98] N. Kato, J. phys. Soc. Jap. 19 (1964) 971.
[99] M. Hart. Ph.D. Thesis, Bristol, 1963.
[100] A.R. Lang, Nature 226 (1970) 345.
[101] F.C. Frank and A.R. Lang, in: Physical properties of diamond, ed. by R. Berman (Clarendon Press, Oxford, 1965) ch. 3.
[102] F.C. Frank, in: Proc. Intern. Industrial Diamond Conf. (Industrial Diamond Information Bureau, London, 1967) p. 119.
[103] B. Lawn, Y. Kamiya and A.R. Lang, Phil. Mag. 12 (1965) 177.
[104] R.A. Duckett, M.Sc. Thesis, Bristol, 1966.

LIST OF SYMBOLS

For each symbol the corresponding chapter in which it is used, is indicated. When no chapter number is mentioned, the symbol is used in several chapters.

symbol	chapter	definition
a	1	activity of ions in solution
a	2	q'/kT
a	8	thermal diffusivity
a	10	distance between growth units
a_0	2	lattice parameter of substrate
a_0	13	activity of particles in solution
$\boldsymbol{a}_i$	11	unit normal vector of i'th crystallographic face
$a_{\mathrm{L}}, a_{\mathrm{S}}$	15	thermal diffusivity of liquid (solid)
a'	2	parameter in Van der Waals equation
$a^{\mathrm{c,l}}$	13	activity of particles in saturated solution of infinite dimensions
A	2	area of substrate on which n' atoms are adsorbed
A	8	constant
A	16	atomic weight
A_0	11	surface area of atom
A_i	1	surface area of cluster of size i
$A_{\mathrm{i}}, A_{\mathrm{s}}$	2	eqs. (2.63) and (2.49′)
A_i	11	surface area of i'th crystallographic face
$A_{\mathrm{L}}, A_{\mathrm{S}}$	15	constants
b	1	geometrical coefficient
b	2	area occupied per adsorbed atom
b	10	eq. (10.97)
$\boldsymbol{b}$		Burgers vector
$\boldsymbol{b}$	11	indefinite vector
B	8	factor
B	15	constant
c	1	concentration of clusters
c	10	component of growth velocity
c_0	10	eq. (10.33)
c_0	15	equilibrium concentration of solute

symbol	chapter	definition
c_∞	15	bulk concentration of solute
c_j	13	geometrical factor
c_R	15	c_0 corrected for curvature
c_S	15	concentration of solute freezing out in solid
$c(1)$	2	concentration of ad-atoms per unit area
$c(i,t)$	2	concentration of clusters of i atoms at time t
C	10	parameter in BCF theory
C	15	concentration of solute
C	16	polarisation factor
C_0	8	solute concentration in liquid zone at $t = 0$
C_∞	15	solute concentration far from interface
C_E	8	eutectic composition
C_f	8	solute concentration in solidified material
C_L, C_S		solute concentration in liquid (solid)
C_m	8	solute concentration in charge to be melted
C_n	10	fraction of blocks in n'th layer that are solid
C_p		molar heat capacity at constant pressure
C_v	15	specific heat per unit volume
C^i	10	actual concentration of i-th growth unit in solution
C_0^i	10	equilibrium concentration of i-th growth unit in solution
d		interplanar spacing
d	8	depth of Bénard cell
d	10	thickness of growth layer
d	11	distance between nearest neighbours
d	16	ruling period of grating
D		diffusion coefficient
D	16	spacing of moiré fringes
D_i	2	diffusion coefficient of cluster of i atoms
D_s		surface diffusion coefficient
D_v	10	volume diffusion coefficient
D_i^+, D_i^-	1	frequency of addition (separation) of a single molecule to (from) a cluster of i molecules
$\mathcal{D}$	15	effective diffusivity
e		electronic charge
e	10	average distance between screw dislocations of the same sign
e	16	strain
$\mathbf{e}$	12	unit vector along intersection line of two faces
$\mathbf{e}^*$	12	reciprocal unit vector perpendicular to face

symbol	chapter	definition
$\dot{E}$	1	overvoltage
E	2	electric potential
E	3	elastic modulus
E	12	number of broken bonds per mesh area of (100)
E	16	energy
E_{att}	14	attachment energy
E_{cr}	14	half-crystal energy
E_e	16	total strain energy of edge dislocation
E_i	2	binding energy necessary to dissociate nucleus of i atoms
E_{jk}	14	interaction energy of ionic chains j and k
E_s	16	total strain energy of screw dislocation
E_{sl}	14	slice energy
E_{ss}	10	total crystallization energy
E_{ss}^{att}	10	vertical bond energy between adjacent solid blocks
E_{ss}^{slice}	10	horizontal bond energy between adjacent solid blocks
$\dot{E}$	2	desorption rate
ΔE^*	6	energy of activation
f_0	10	surface occupied by one growth unit in the crystal
$f_1(\theta), f_2(\theta)$		geometrical factor of spherical cap
f_A	12	fractional projected area of face A
f_e	16	atomic scattering factor for electrons
f_X	16	X-ray atomic scattering factor
$f(r, t)$	2	number of particles of size r at time t per unit area
F		Helmholtz free energy
F	12	number of broken bonds per mesh area of (010)
F	16	force on a dislocation line
F	16	structure factor
F_{fm}, F_{sm}	10	Helmholtz free energy of fluid (solid) phase at melting point
F_{st}	10	flux of steps
g		acceleration due to gravity
g	8	fraction solidified at time t
g	15	departure of temperature from thermodynamic equilibrium
g	16	reciprocal of ruling period of grating
g	16	reciprocal lattice vector
g_i', g_s	2	eqs. (2.64′) and (2.50′)
$g(t)$	2	size of nucleus at time t
G		Gibbs free energy (or free enthalpy)
G	12	number of broken bonds per mesh area of (001)

symbol	chapter	definition
G	16	reciprocal of moiré fringe period
G_c	15	concentration gradient at interface
G_L, G_S		temperature gradient in liquid (solid)
G_+	15	function defining interface shape (eq. 15.50)
ΔG	10	activation free energy
ΔG_{des}		desorption free energy of ad-atoms from substrate
ΔG_{diff}		diffusion free energy of ad-atoms on substrate
ΔG_i^*		free energy of formation for critical nucleus containing i atoms
ΔG_v	3	change in Gibbs free energy per unit volume of guest material
$\mathcal{G}, \mathcal{G}'$	15	conductivity weighted temperature gradient in liquid (solid)
h		Planck's constant
h_i		length of normal to i'th crystallographic face
h_j	13	central distance
H_m	1	molar enthalpy of dissolution
$H_0^{(1)}$	14	Hankel function
ΔH_{des}	2	heat of desorption of single molecule
i		number of atoms in a cluster
i^*		number of atoms in the critical cluster or nucleus
I	16	intensity of diffracted X-rays
Im	8	imaginary part of complex function
j_s	10	flux of growth units towards a step
j_v	10	flux of growth units from the bulk to the surface
J		rate of nucleation
J	10	flux of steps past a point
J_i	2	nucleation rate of clusters of i atoms
J_i	2	incident vapour flux
J_r	2	reflected (desorbed) flux
J_s	2	number of atoms attaching themselves to the nucleus per second by surface diffusion
J_t	2	number of atoms attaching themselves to the nucleus per second by interface transfer
J_v	2	impingement flux
k		Boltzmann constant
k		segregation coefficient (or distribution constant)
k	9	proportionality constant
$\boldsymbol{k}$	16	wave vector
k_1, k_2	2	constants

symbol	chapter	definition
k_{hkl}	6	velocity constant of face (*hkl*)
$k\uparrow, k\downarrow$	11	pre-exponential factors
k_{eff}		effective segregation coefficient
k^*	8	interface segregation coefficient
K, K_p, K_i		chemical equilibrium constants
K	15	parabolic rate constant (eq. 15.11)
K	16	wave vector magnitude in vacuum
K_1	1	pre-exponential constant
K_L, K_S		thermal conductivity of liquid (solid)
$\bar{K}$	15	average thermal conductivity in solid-liquid system
l	2	screening distance
l	8	length of molten zone
l	10	mean free path of a growth unit in solution
l	11	length of normal to edge in two-dimensional crystal
l	16	unit vector
l_0	8	length of molten zone at time $t = 0$
L		latent heat of fusion
L	8	length of ingot
L	16	length of dislocation line
L_{gr}	10	length of grain boundary
L_{st}	10	distance traveled by a step from its source to its annihilation
$L^{(2)}L^{(3)}$	11	linear size of quadratic (cubic) crystal
m		molecular mass
m		slope of liquidus line
m	1	number of nuclei as a random quantity
m	2	density of film material
m	10	number of growth units in a block
m	10	mass of growth unit
m	14	number of slice
m	16	mass of electron
m_0	2	volume of ad-atom
M	16	elastic modulus
M_{hkl}		mesh area of face (*hkl*)
ΔM	10	change in free energy of a growth unit for the formation of a solid block
MI	10	morphological importance
n	1	number of moles
n	10	number of layer

symbol	chapter	definition
n	11	number of atoms in a row of a cube
n	14	number of ionic chains within a slice
n	16	refractive index
n_0	3	concentration of monomers
n_0	13	number of particles in a supersaturated solution
n_1	1	concentration of single atoms
n_1	2	ad-layer coverage
n_∞	13	number of particles in saturated solution of infinite dimensions
n_E	13	number of particles in crystal at equilibrium in finite volume
n_{eq}	2	ad-atom concentration corresponding to equilibrium vapour pressure
n_i	2	number of nuclei of i atoms
n_r	2	concentration of nuclei of radius r
n_s	10	actual concentration of growth units on a face
n_{s0}	10	equilibrium concentration of growth units on a face
n_{step}	10	number of adsorbed molecules per unit surface adjacent to a step
$\hat{\boldsymbol{n}}$	15	unit normal to interface
$n(i)$	1	equilibrium distribution function of clusters of i atoms
$n(t)$	2	total number of discrete nuclei at time t
$n(R)$	2	local concentration of ad-atoms
n'	2	local concentration of ad-atoms at boundary of nucleus
n'	2	number of atoms adsorbed on area A
n'	11	number of atoms in a row of a square
$\overline{n}$	2	average ad-atom population
n^c, n^l	13	number of particles in crystal (liquid)
n^*	13	number of particles in critical nucleus
N		Avogadro's number
N	1,9	total number of nuclei
N	10	total number of blocks per unit surface
N	14	number of point charges on a row
N_0	2	number of adsorption sites per cm^2
N_0	10	concentration of growth units per cm^3
N_A	1	Avogadro's number
N_m	10	number of growth units per mesh area
N_{nf}, N_{ns}	10	number of fluid (solid) blocks in n'th layer per unit surface

symbol	chapter	definition
N^*	2	number of critical nuclei per cm^2
N^i_f, N^i_s	10	number of growth units of type i in fluid (solid) phase
p		vapour pressure
p	10	slope of growth spiral in x-direction
p	14	period of row of ions
p	15	$1 - k$
p_0		equilibrium pressure
p_g, p_0	16	parameters of dispersion surface
$\mathfrak{p}$	8	Prandtl number
q	10	slope of growth spiral in y-direction
q'	2	rate of impingement from vapour
Q_{ad}	2	adsorption energy of ad-atom
Q_g	16	dimensionless quantity (eq. 16.27)
Q_m	1	molar heat of fusion
r		polar coordinate
r	1,2	cluster radius
r	8,15	rod radius or cylinder radius
r	13	radius of drop of solution
r	14	distance to row of ions
$\mathbf{r}$	16	position vector
r_0	16	core radius of dislocation
r_0	16	classical radius of electron
r^*		radius of critical nucleus
R		gas constant
R	2	distance from centre of nucleus
R	8	Rayleigh number (dimensionless temperature gradient)
R	10	growth rate of crystal face
R	15	radius
R	16	distance from dislocation line
R_0	16	Bohr radius of hydrogen
R_1, R_2	15	radius of curvature
R_{plane}	16	radius of curvature of plane
R_{ray}	16	radius of curvature of X-ray pencil
Re	8	real part of complex function
s	1	summation index
s	2	supersaturation of ad-atoms on substrate
s	6	solubility
s	13	surface area

symbol	chapter	definition
s, s'	10	number of co-operating spirals
s_t	10	sticking coefficient
S		entropy
S	8	supercooling
ΔS	6	supersaturation
ΔS	15	entropy of fusion
S^*	2	critical supersaturation
$\mathcal{S}(A,k)$	15	function given in fig. 15.11
t		time
t	16	thickness of crystal
$\bar{t}$	1	mean time for the appearance of the first nucleus
t_i, t_s	2	eqs. (2.71) and (2.56)
T		temperature
T	8	Taylor number
T	10	lifetime of a step
T	16	line tension of dislocation
T_0, T_m, T_M		melting point temperature
T_a	2	effective temperature of material leaving the surface
T_E	8	eutectic temperature
T_i	15	temperature of interface
T_L, T_S	15	temperature of liquid (solid)
T_s	2	surface temperature
$T_s(\tau_s)$	2	eq. (2.52)
T_v	2	temperature of impinging vapour
ΔT		undercooling
T^*	15	thermodynamic freezing temperature
TAC	2	thermal accomodation coefficient
u	10	distance of real step from its unperturbed position
u	16	displacement in x-direction
u	16	group velocity
$\mathbf{u}$	16	displacement
ΔU	1	activation energy of self-diffusion
v	1	molar volume
v	8	velocity of solidification
v	10	step velocity
v	16	displacement in y-direction
v_0	11	volume of molecule

symbol	chapter	definition
v_∞	10	velocity of advance of a straight step
v_f, v_m	8	velocity of freezing (melting) interface
v_m	2	volume of ad-atom
v_p	15	phase velocity
$\mathbf{v}$	15	local growth velocity
$\widetilde{v}_n$	11	mean vibrational volume
V		volume
V		unit cell volume
V	2	molar volume of condensed phase
V	10	velocity of shock wave
V	11	volume of a crystal
V	14	electrostatic potential
V	15	constrained velocity of solidification
V	16	voltage
V_E, V_S	16	linear width of edge (screw) dislocation
V^c	13	molar volume of crystal
$\overline{V}^1$	13	partial molar volume of solute
w	2	collision factor
w	10	number of configurations
w	10	function defined by eq. (10.97)
w	16	displacement in z-direction
w	16	deviation parameter
W	12	work required to create a surface
W_m	1	probability of formation of m nuclei
$W_{\geqslant 1}$	1	probability of formation of a minimum of one nucleus
$W\uparrow, W\downarrow$	11	probability of evaporation (condensation)
x		coordinate
x_0	10	mean distance between kinks in a step
y		coordinate
y	10	coordinate in surface perpendicular to step
y_i, y_s	2	eqs. (2.68) and (2.54)
Y_{lm}	15	spherical harmonic function
z		coordinate
z		valence of ion
z	8,10,15	coordinate perpendicular to interface
z_n	10	eq. (10.16a)
Z		Zeldovitch factor
Z	16	atomic number
α	6	dimensional conversion constant

symbol	chapter	definition
α	8	expansion coefficient of fluid
α	10, 12	$4\epsilon/kT$
α	16	standing wave ratio
α_0	2	typical atomic dimension
α_2	2	addition rate to cluster of two atoms
α_i	1	frequency of separation of single molecules per unit area
α_J	10	Jackson factor
α'	10	supersaturation ratio in the bulk
α'_s	10	supersaturation ratio on the surface near step
α_{cond}	2	differential condensation coefficient
β	1	impingement frequency of single molecules per unit area
β, β'	2, 3	pre-exponential factors
β	8	adverse temperature gradient
β	10	$\Delta\mu/kT$
β	11	specific adhesion energy
β_0	2	probability constant
β_0	13	supersaturation in solution
β_i	2	rate of decay of cluster of *i* atoms
β_i, β_s	2	constants
β_r	3	relative β
β'	1	sticking probability
γ		specific surface free energy or interfacial tension
γ	2	kinetic factor
γ_0	13	activity coefficient of n_0 particles in supersaturated solution
$\gamma^{c,1}$	13	activity coefficient of *n* particles in saturated solution of infinite dimensions
γ'	10	edge free energy of a growth unit in a step
Γ		Lothe-Pound factor
Γ	15	γ/L
δ	1	difference in radius between equimolecular surface and surface of tension
δ	10	thickness of unstirred layer
δ	15	amplitude of perturbation
δ_{ij}		Kronecker delta
δ_m	8	thickness of diffusion layer
δ_t	8	thickness of thermal boundary layer

symbol	chapter	definition
Δ	15	thickness of solute boundary layer
ϵ	2	hard sphere interaction potential
ϵ	10	energy gain upon formation of a ϕ_{sf} bond
ζ	10	retardation factor $(\sigma - \sigma_{step})/\sigma$
η		viscosity
θ		polar coordinate
θ	2	fractional coverage
θ	2, 3	contact angle
θ	12	angle between stepped face and F face
θ	16	Bragg angle
θ	16	angle of low-angle grain boundary
κ	11	specific edge energy
λ		wavelength of radiation or of perturbation or of eutectic structure
λ	10	distance between steps
λ	11	indefinite scalar
λ_0	10	distance between parallel equidistant steps
λ_i, λ_s	2	eqs. (2.63) and (2.49′)
λ_s	10	mean displacement of growth unit on the surface
λ_+, λ_-	10	distance to step in front (behind)
λ'	10	distance between steps of single spiral
μ		chemical potential
μ	15	kinetic coefficient
μ	16	shear modulus
μ	16	linear absorption coefficient
μ_a	16	atomic absorption coefficient
ν		vibration frequency
ν		Poisson's ratio
ν	8	kinematic viscosity
ν	15	exponent
ν_i	10	number of growth units of type i in block
ξ	10, 15	Jackson anisotropy factor or crystallographic factor
ξ	15	$(z^2/4a_s t)^{\frac{1}{2}}$
ξ_g	16	extinction distance
ρ		density
ρ	2	dimensionless variable (eq. 2.48)
ρ	10	local density of steps
ρ_{st}	10	linear density of steps
ρ'	2	distance between two adsorbed atoms

symbol	chapter	definition
ρ_0'	2	minimum distance between two adsorbed atoms
σ		specific surface free energy
σ	10	relative supersaturation
σ	16	shear stress
$\sigma_{c-v}, \sigma_{c-s}$ σ_{s-v}	2	interfacial energy per unit area (c = cluster, s = substrate, v = vapour)
σ_{hg}	3	interfacial free energy per unit area between guest and host crystal
σ_s	10	relative supersaturation on the surface
σ_v	10	relative supersaturation at the top of unstirred layer
τ	1	induction time
τ	2	dummy time variable
τ	10	relaxation time
τ_1, τ_s	2	dimensionless pseudo-time variables, eqs. (2.61) and (2.48)
τ_s	2	lifetime of adsorbed atom on a substrate
τ_0	2	inverse of ad-atom vibrational frequency
τ_k'	10	relaxation time for entering the kink directly
$\tilde{\varphi}$	2	two-dimensional gas pressure
φ	2	total amount of film deposited
φ	3	misorientation angle
φ	15	azimuthal angle
$\overline{\varphi}$	11	mean work of separation
$\varphi_{\frac{1}{2}}$	11	work needed to separate an atom from the half-crystal position
ϕ	1	surface energy
ϕ	11	nearest neighbour interaction
ϕ	12	polar coordinate
ϕ_e, ϕ_X	16	scattering strength of atoms for electrons (X-rays)
$\phi_{ff}, \phi_{sf}, \phi_{ss}$	10	bond energy between blocks (subscripts: s = solid; f = fluid)
Φ	11	total surface energy of a crystal
Φ	15	eq. (15.37)
ψ	10	potential function $(\sigma - \sigma_s(y))$
ψ	16	Bloch wave
ψ	16	angle
$\psi_i(0), \psi_s(0)$	2	eqs. (2.66) and (2.52)
Ψ	14	function $d \ln x/dx$

symbol	chapter	definition
ω	2	probability of addition of a single atom to a critical nucleus
ω	8	angular velocity of rotating crystal
ω	15	spatial frequency of perturbation
Ω		atomic volume or volume of growth unit
Ω	8	angular velocity of fluid

AUTHOR INDEX

SUBJECT INDEX